Fundamentals of Perovskite Oxides

Fundamentals of Perovskite Oxides

Synthesis, Structure, Properties and Applications

Gibin George
Sivasankara Rao Ede
Zhiping Luo

CRC Press
Taylor & Francis Group
Boca Raton London New York

CRC Press is an imprint of the
Taylor & Francis Group, an informa business

First edition published 2021
by CRC Press
6000 Broken Sound Parkway NW, Suite 300, Boca Raton, FL 33487-2742

and by CRC Press
4 Park Square, Milton Park, Abingdon, Oxon OX14 4RN

Visit the Taylor & Francis Web site at
http://www.taylorandfrancis.com

and the CRC Press Web site at
http://www.crcpress.com

ISBN: 978-0-367-35448-0 (hbk)
ISBN: 978-0-367-55865-9 (pbk)
ISBN: 978-0-429-35141-9 (ebk)

Typeset in Times
by codeMantra

Dedicated to all our family members

Contents

List of Figures

List of Tables

List of Symbols

Symbol	Explanation
3DOM	Three-dimensionally ordered macroporous
A	Constant of proportionality
AA	Ascorbic acid
AACVD	Aerosol-assisted chemical vapor deposition
AFEs	Antiferroelectrics
AFM	Antiferromagnetic
AIE	A-site ionic electronegativity
BSA	Bovine serum albumin
BSCF	$Ba_{0.5} Sr_{0.5} Co_{0.8}Fe_{0.2}O_{3-\delta}$
c	Cubic packed layer
C	Curie constant
C	Specific heat capacity
CB	Conduction band
CCD	Charge-coupled device
CCT	Colloid crystal template
CL	Chemiluminescence
C_m	Molar Curie constant
CMR	Colossal magnetoresistance
CNT	Carbon nanotubes
CPE/$SrPdO_3$	Carbon paste and $SrPdO_3$
CVD	Chemical vapor deposition
C_p	Heat capacity at constant pressure
d	Thickness of dielectric layer
D	Diffusivity
D_0	Pre-exponential factor
D_i	Diffusion coefficient
DA	Dopamine
d-block	Transition elements
DJ	Dion-Jacobson
DLICVD	Direct liquid injection chemical vapor deposition
DMFC	Direct methanol fuel cells
DTPA	Diethylenetriaminepentaacetic acid
dT/dx	Temperature gradient
E	Electric field
e	Electronic charge
e^-	Electron
E_0	Activation energy

E_a	Activation energy for conduction
EDLCS	Electric double-layer capacitors
EDTA	Ethylenediaminetetraacetic acid
EL	Electroluminescence
e_g	Upper pair of *d*-orbitals
F	Faraday
f-block	Lanthanides and actinides
FEs	Ferroelectrics
FiM	Ferrimagnetic
FM	Ferromagnetic
g	Gyromagnetic factor
GKA	Goodenough–Kanamori–Ander
GMR	Giant magnetoresistance
h	Hexagonal packed layer
H	Applied magnetic field
h^+ or $h^\bullet$	Hole
HA	Hydroxyapatite
HER	Hydrogen evolution reaction
HRS	High resistive state
HRTEM	High resolution transmission electron microscope
HS	High spin
IS	Intermediate spin
ITO	Indium tin oxide
J	Particle current
j_i	Diffusion flux
J_q	Isothermal energy current
J_{O_2}	Oxygen flux
JT	Jahn–Teller
k	Thermal conductivity
k_B	Boltzmann's constant
KBNNO	$[KNbO_3]_{1-x}[BaNi_{1/2}Nb_{1/2}O_{3-\delta}]_x$
K_{ox}	Equilibrium constant
LB	Langmuir-Blodgett
LFMR	Low-field magneto resistance
Ln	Rare-earth elements
LRS	Low resistive state
LS	Low spin
LSC	$La_{1-x}Sr_xCoO_{3-\delta}$
LSM	$La_{1-x}Sr_xMnO_{3-\delta}$
LTA	$LaTiO_3$-$Ag_{0.2}$
LY	Light yield
m.p.	Melting point

MAMs	Microwave absorption materials
MBE	Molecular-beam epitaxy
MCE	Magnetocaloric effect
ML	Mechanoluminescence
MOCVD	Metal-organic chemical vapor deposition
MOR	Methanol oxidation reaction
MR	Magnetoresistance
MSS	Molten-salt synthesis
n	Number of slabs of BO_6 octahedron
n	Refractive index
N_A	Avogadro's number
NN	Nearest-neighbor
O/W	Oil-in-water
OER	Oxygen evolution reaction
oh	Octahedron
ORR	Oxygen reduction reaction
O_O^x	Oxygen in the lattice
P	Polarization density
P_s	Saturation polarization
PBSCF	$PrBa_{0.5}Sr_{0.5}Co_{2-x}Fe_xO_{5+\delta}$
PCEs	Power conversion efficiencies
PCMO	$Pr_{0.7}Ca_{0.3}MnO_3$
PEs	Paraelectrics
PL	Photoluminescence
PLD	Pulsed laser deposition
PMMA	Polymethyl methacrylate
P_s	Electric polarization
PVC	Poly(vinyl alcohol)
PVD	Physical vapor deposition
PVP	Poly(vinylpyrrolidone)
P_{O_2}	Partial pressure for oxygen
q	Charge of the specific cation A, A′, A″, B, B′ or B″
Q_x	Heat flow along a certain direction x
R	Gas constant
R	Surface reflectivity
r	Electro-optic coefficient
R_0	Magnetic resistance in zero magnetic field
R_H	Magnetic resistance in applied magnetic field
r_O	Radius of O-ion
r_A	Radius of A-ion
r_B	Radius of B-ion
RAM	Random-access memory

RCP	Relative cooling power
RHEED	High-energy electron diffraction
RP	Ruddlesden–Popper
r-PBM	Reduced $PrBaMn_2O_{6-\delta}$
RRAM	Resistive Random-Access Memory
RT	Room temperature
S	Spin state
S	Seebeck coefficient
SC	$SrCoO_3$
SEM	Scanning electron microscopy
SOC	Spin orbital coupling
SOFC	Solid oxide fuel cells
SPH	Small-polaron hopping
SSD	Solid-state storage disk
T	Temperature
t	Thickness of dielectric structure
t	Tolerance factor
t_k	Transport number
TCO	Transparent conductive oxides
TEM	Transmission electron microscope
TGA	Thermogravimetric analysis
TL	Thermoluminescence
TMSS	Topochemical molten-salt synthesis
TOF	Turnover frequency
tp	Trigonal prism
t_{2g}	Lower *d*-orbitals
T_c	Curie temperature
T_f	Frozen temperature
T_g	Glass forming temperature
T_m	Maximum permittivity
T_{M-I}	Metal-insulator transition temperature T_{M-I}
T_N	Néel temperature
T_{SC}	Superconducting transition temperature
T_{SG}	Spin-glass temperature
U	Hubbard or Mott–Hubbard gap
UA	Uric acid
UVO	Ultraviolet/Ozone
v	Valence
V	Voltage
VB	Valence band
$V_O^{\bullet\bullet}$	Oxygen vacancy
V_{OC}	Open-circuit voltage

YSZ	Yttria-stabilized zirconia
zT	Figure of merit
$\lvert\Delta S_M\rvert$	Isothermal magnetic entropy change
α	Thermal diffusivity
α_H^V	Magneto-electric coefficient
α_L	Linear thermal expansion
α_P	Lattice parameter
β	Thermal expansion
χ	Paramagnetic susceptibility
χ_e	Dielectric susceptibility
ΔH_D	Activation enthalpy for diffusion
ΔT_{ad}	Adiabatic temperature change
ε_0	Dielectric permittivity in vacuum
ε_r	Relative dielectric permittivity
μ_B	Bohr magneton
μ	Electronic chemical potential
μ_{eff}	Effective paramagnetic moment
θ	Curie-Weiss constant
ρ	Resistivity
ρ	Density
ρ_e	Electron concentration
σ_0	Pre-exponential ionic conductivity factor
σ_e	Electronic conductivity
σ_{ion}	Ionic conductivity
σ_{total}	Mixed conductivity
Ω	Number of polar states accessible in the system
ψ	Bond angle

Preface

Perovskites are increasingly attracting the interest of researchers around the world. The research on perovskite materials is currently extended to many science and engineering fields, and it is rapidly expanding since 1950s. The term perovskite is used for representing materials with a crystal structure that is similar to calcium titanate ($CaTiO_3$) mineral, known as perovskite. A simple perovskite structure adopts a general formula of ABX_3, where A and B represent cations, and X is the anion. The BX_6 octahedra determine the major structural characteristics of the perovskites, as the deviation of perovskite structure from the ideal cubic structure is due to the different arrangement of BX_6 octahedra. However, the cationic and anionic defects, the stoichiometry of cations and anions, doping, synthesis route, etc. all make influences on the functional aspects of perovskites.

Perovskites are unique by virtue of their customizable chemical composition. Besides the fact that the majority of the elements (alkali, alkaline, transition, rare earth, semiconductors, etc., ~90%) in the periodic table can occupy either A or B sites of the perovskite lattice, they can be partially substituted with several combinations of other elements without disturbing their crystal structure significantly. The presence of several such ions in the same lattice can bring distinct functionalities that cannot be obtained by a single oxide material or their mixtures. Therefore, they offer a facile way of correlating materials chemistry and physics with electronic, magnetic, transport, optical, and catalytic properties. More importantly, their structural variants such as double perovskites, Ruddlesden-Popper phase, Dion–Jacobson phase, Aurivillius phase, hexagonal, and $A_nB_nO_{3n+2}$ are exceptional candidates for applications like separators for energy scavenging and storage, fuel cells, superconductors, etc.

Perovskites and their derivatives are a crucial class of materials for a wide range of unique applications. For instance, $LaGaO_3$ is the most sought material as the solid electrolyte in solid oxide fuel cells due to its high-temperature stability and high ionic conductivity, which can be tuned easily by doping. Additionally, the layered superconducting cuprates with perovskite structure is a breakthrough in the history of high-T_c superconductors, and some layered perovskites exhibit negative colossal magnetoresistance. Many perovskites are identified as suitable candidates for ferroelectrics, piezoelectrics, superconductivity, catalysis, magnetic ordering, fuel cells, and optical related application.

At Fayetteville State University, we are currently conducting research on the physical and chemical properties of perovskite materials. In spring of 2019, our group planned to write a book on perovskite oxides, as a comprehensive reference book. In this book, we exclusively tried to incorporate a complete framework of the perovskite oxides starting from the synthesis, fundamentals to the applications. Akin to the other compounds, perovskites can be easily synthesized, with the immediately available assistance of oxygen in air during calcination. A brief overview of the several techniques available for the synthesis of perovskite oxide materials and the techniques for size control are provided in the respective sections. The subsequent section is on the structures of perovskites and their structural variants. Their unique structural characteristics play an important role in their chemical, dielectric, magnetic, electronic, thermal, ion transport, and optical properties, and, ultimately, in the potential applications.

This book was written as a group effort, and the current stay-at-home order in our local area in the coronavirus season facilitated our writing process. The book is designed as a reference book on perovskite oxides, targeting a broad range of audiences in science and engineering. The perovskite oxide is a vast area, and we sincerely hope that this book provides handy useful information on perovskite oxides.

Gibin George, Sivasankara Rao Ede, and Zhiping Luo
Fayetteville, North Carolina
April 2020

Authors

Dr Gibin George is a postdoctoral fellow at Fayetteville State University, North Carolina. He was an assistant professor in the Department of Mechanical Engineering at Jyothi Engineering College, Thrissur, India, before joining Fayetteville State University. Dr George received his Ph.D. in 2015 from the National Institute of Technology, Karnataka, India. His research interest is in nanomaterials synthesis and characterization, and oxide nanomaterials for energy storage and production, photoluminescence, sensors, and scintillators.

Dr Sivasankara Rao Ede is a postdoctoral fellow at Fayetteville State University, North Carolina. Dr Ede received his Ph.D. in 2018 from Central Electrochemical Research Institute, Karaikudi, India. His research mainly focuses on the perovskite oxide nanomaterials for energy conversion and energy storage.

Dr Zhiping Luo is a professor in material science at Fayetteville State University, North Carolina. He received a Ph.D. from the Chinese Aeronautical Establishment in 1994, followed by postdoc research at Okayama University of Science, Japan. From 1998 to 2001, he worked at Materials Science Division, Argonne National Laboratory, on structure of perovskite materials as a visiting scholar, with a promotion to an assistant scientist. From 2001 to 2012, he worked at Texas A&M University as a research scientist. Dr Luo joined the current position in 2012 and established his research on the nanomaterials for energy-related applications. Dr Luo has coauthored over 300 articles in peer-reviewed journals.

1 Introduction to Perovskites

Perovskites have been an important class of materials, exhibiting unusual promising functionalities in various transport and physical properties over other traditional ceramic or composite materials. They have received extensive attention more recently in the fields of materials science, physics, chemistry, geology, and engineering. In this chapter, we give a brief introduction to perovskites regarding their history, formation, and classification.

1.1 HISTORY OF PEROVSKITES

The term perovskite originated from the mineral perovskite ($CaTiO_3$) named after the Russian mineralogist Mr. Lev Alekseyevich von Perovski (1792–1856) (Liu et al. 2017). A photo of the $CaTiO_3$ mineral is shown in Figure 1.1. Nowadays, perovskites represent a wide class of materials with similar or derived crystal structure as that of $CaTiO_3$ with a general formula ABX_3, as shown in Figure 1.2. In the structure of perovskites, A and B are cations and X is the anion, and A cations are larger than B cations (Figure 1.2a). Six X anions form an octahedron covering the smaller cation B (Figure 1.2b). Therefore, this perovskite lattice is formed by such apex-connected octahedra with A cations between them (Figure 1.2c).

FIGURE 1.1 A photo of $CaTiO_3$ mineral. (From https://en.wikipedia.org/wiki/Perovskite.)

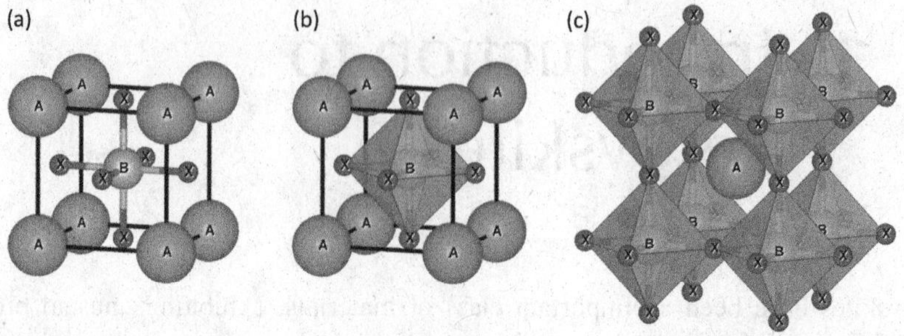

FIGURE 1.2 Perovskite ABX_3 lattice. (a) Large A and small B cations, with X anions; (b) octahedron by X anions; (c) octahedra lattice.

Often A-site ion provides the structural integrity to the perovskite structure, and B-site ion determines the properties. $CaTiO_3$ crystallizes into an orthorhombic structure; however, the ideal perovskite structure belongs to the cubic space group $Pm\bar{3}m$ (No. 221). Apart from $CaTiO_3$, many naturally existing minerals adopt a perovskite structure, for instance, ilmenite ($FeTiO_3$), $MgSiO_3$, etc. The silicate perovskites containing Mg, Ca, and Fe are assumed as the most abundant solid phase in the earth's lower mantle at 670–2,900 km from the surface. Moreover, $MgSiO_3$ is stable above 23 GPa (Zhang et al. 2014), and the same is believed to constitute the big planets other than Mars (Umemoto et al. 2006). Many perovskite materials exhibit exceptional properties such as high absorption coefficient, long-range ambipolar charge transport, low exciton-binding energy, high dielectric constant, and ferroelectric properties.

$BaTiO_3$ is the first manmade perovskite synthesized during World War II in 1941; however, it's naturally existing counterpart benitoite is an extremely rare mineral. $BaTiO_3$ was used as a ferroelectric and piezoelectric material for a wide range of applications (Eshita et al. 2014). Its application as a piezoelectric material is later replaced by lead zirconate titanate Pb(Zr, Ti)O_3, another important perovskite. Similarly, $SrTiO_3$ is the first insulator, and the first oxide reported as superconductive below 0.35 K. Even though $SrTiO_3$ was synthesized in the 1950s, its natural counterpart was discovered in 1982. $SrTiO_3$'s first-ever use was as a simulant for diamond due to its resemblance to a diamond, but the softness and high cost resulted in their replacement with other simulants such as yttrium aluminum garnet (YAG), gadolinium gallium garnet (GGG), and cubic zirconia.

1.2 FORMATION OF PEROVSKITES

Since the discovery of $BaTiO_3$, a large number of perovskites with general formula ABO_3 have been synthesized and studied for their unique properties pertaining to the presence of two different cations in their lattice. In general, the new perovskite structured materials possess a wide range of properties such as optical, magnetic, thermoelectric, piezoelectric, photochemical, thermochromic, electrochromic, and electrochemical, which are not observed in their ancestors. Besides the general physical and chemical properties, these materials are extensively studied for their applicability as electrode materials for energy storage and scavenging, e.g., $LaMnO_3$,

$Ba_{0.5}Sr_{0.5}Co_{0.8}Fe_{0.2}O_3$, etc. Despite the low-cost and easy synthesis, the primary advantage of perovskite materials is the flexibility of using different cations in A and B sites and partial substitution of the A- and B-site ions to enhance the properties originated by the defects and the distortion in the lattice structure.

The perovskite structures can accommodate most of the metallic elements in the periodic table and a significant number of anions in their structure. Most of the perovskite structured materials are oxides and halides; however, some hydrides, oxyfluorides, oxynitrides, and oxyhalides also adopt perovskite structures. The periodic tables shown in Figures 1.3–1.5 present the possible A-site, B-site, and X-site ions, respectively, that can form the perovskite structures. Oxide perovskites are the most widely reported perovskite structured materials, and most elements in the periodic table in Figures 1.3 and 1.4 can be either an A-site or B-site element in the perovskite oxide structure. Based on the theoretical prediction, there are 49 A-site and 68 B-site elements available on the periodic table that can form an oxide perovskite structure (Zhang et al. 2007; Emery et al. 2016; Bartel et al. 2019), and the possible

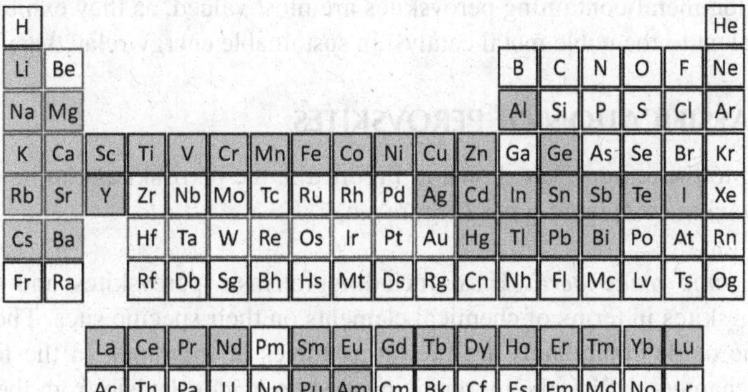

FIGURE 1.3 Periodic table representing the possible A-site cations in the perovskite structured materials.

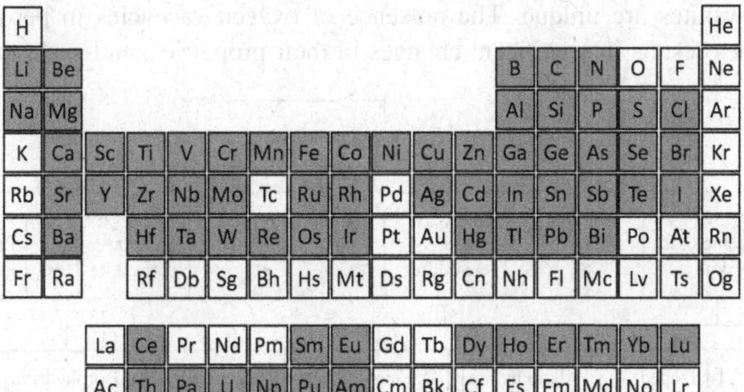

FIGURE 1.4 Periodic table representing the possible B-site cations in the perovskite structured materials.

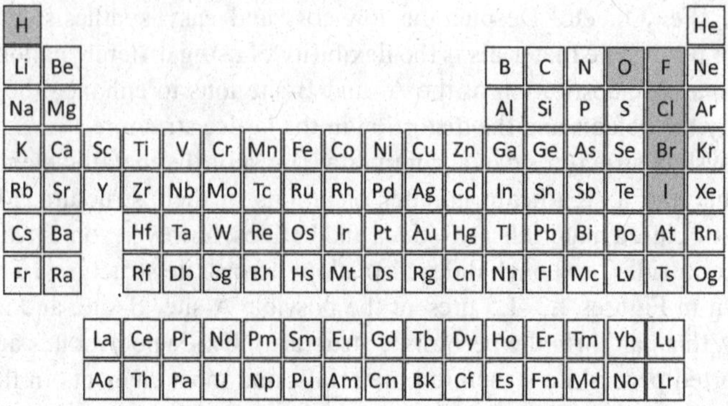

FIGURE 1.5 Periodic table representing the possible X-site anions in the perovskite structured materials.

combinations of ABO_3 perovskite oxides are 3,332. Among the oxide perovskites, the transition metal-containing perovskites are most valued, as they exhibit properties equivalent to the noble metal catalyst in sustainable energy-related applications.

1.3 CLASSIFICATION OF PEROVSKITES

According to the anion X, as shown in Figure 1.5, the perovskites can be classified as the following compound types (Figure 1.6):

1. *Inorganic oxide perovskites*, including intrinsic perovskites and doped perovskites in terms of chemical elements on their specific sites. The inorganic oxide compounds are the target topics of this book in the following chapters. Often, the inorganic perovskite oxides deviate from the ideal cubic structure depending on the size of A- and B-site cations, which we will see in Chapter 3. Additionally, such structural changes are sensitive to temperature and sometimes pressure, resulting in a subsequent change in the properties with temperature or pressure. The oxygen vacancies of oxide perovskites are unique. The presence of oxygen vacancies in perovskite oxides results in significant changes in their properties, and such changes

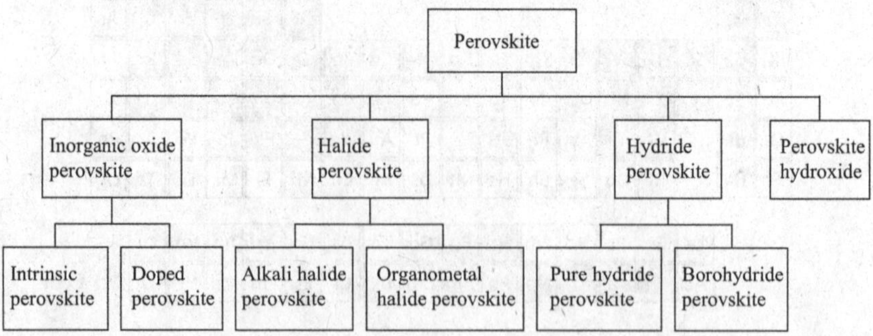

FIGURE 1.6 Classification of perovskites, according to anion X.

are rarely observed in oxides with other structures. The oxygen vacancies in perovskite oxides are easily manipulated by doping. The discovery of some oxide perovskites is a breakthrough in the history of ferroelectricity, piezoelectricity, solid oxide fuel cells, superconductivity, etc.

2. *Halide perovskites*, including alkali halide perovskites and organometal halide perovskites. The attempts are also made to synthesize perovskite structured fluorides with the general formula AMF_3 (where A = Na, K, Cs, Rb, etc., M = Mn, Fe, Co, Ni, Zn, Mg, Cu, etc.) (Scatturin et al. 1961). Generally, these perovskite fluorides exhibit antiferromagnetic properties, and some of them are identified as an excellent host for luminescent ions. Perovskite structured halides with the general formula ABX_3 (where A = Cs, B = Pb, Sn, X = Cl, Br, and Cl) gained remarkable attention because of their superior optical properties as compared to many organic small molecules (Shpatz Dayan et al. 2018). The tractability on the perovskite lattice sites allows the occupancy of inorganic cations such as CH_3NH_3 in the A site resulting in the formation of organic-inorganic hybrid halides such as $(CH_3NH_3)PbX_3$ (where X = Cl, Br, I or a combination of these anions). The perovskite structured $APbX_3$ (A = Cs, CH_3NH_3, etc.; X = Br, I, etc.) exhibit excellent properties suitable for optoelectronic applications. Due to the toxicity of Pb and instability in the open air, lead-based halide perovskites have a limited commercial application. Lead-free environmentally friendly double perovskite halides are the recent breakthrough among the materials for optoelectronics with exceptional stability and tunable optoelectronic properties.
3. *Hydride perovskites*, including ABX_3-type pure hydride perovskites, such as $NaMgH_3$ (Pottmaier et al. 2011), $MgXH_3$ (X = Fe, Co) (Candan and Kurban 2018), and borohydride perovskites $AB(BH_4)_3$ (Schouwink et al. 2014). Hydride perovskites are formed when the stoichiometric ratio of the hydrides of A- and B-site cations are heated (~673 K) under a high-pressure H_2 atmosphere. This process is reversible, and the hydrogen is released at ~673. Therefore, these materials are extensively studied as hydrogen storage materials (Komiya et al. 2008).
4. *Perovskite hydroxide* adopts the formula $AB(OH)_6$ with double perovskite structure. These materials exhibit catalytic properties due to their unique optical and electronic properties. Some of the examples are, $ASn(OH)_6$ (A=Mg, Sr, Ba, Zn, Cu, Co, Fe, Mn, etc.)

In terms of perovskite structures, perovskites can be classified as

1. *Single perovskites*. As shown in Figure 1.7a, the octahedra are identical or randomly distributed, showing no ordering in their structure. The single perovskites adopt a low symmetrical triclinic to high symmetric cubic phases. Single perovskites are the most studied perovskites, and their properties can be easily modified by doping. Single perovskite oxide structures with alkaline earth metal or rare earth metals at the A-site and transition metal at the B-site are the most studied among single perovskites. A list of important single oxide perovskites with different structures and applications are

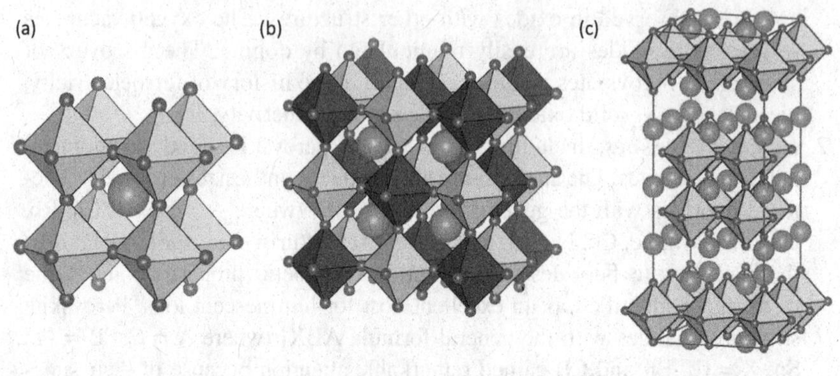

FIGURE 1.7 (a) Single perovskite; (b) double perovskite; (c) layered perovskite.

presented in Table 1.1. Unlike complex perovskites, most single perovskites can be synthesized easily at low temperatures using conventional techniques.

2. *Double perovskites.* As shown in Figure 1.7b, two different types of octahedra produce doubled lattice spacing. Double perovskites with the formula $ABB'X_6$ are also successfully synthesized following the advantages of ABX_3 perovskite structured materials. The presence of two property determining B-site cations is likely to exhibit superior properties than ABX_3 perovskite materials, especially among the halide perovskites for opto-electronic applications. Some double perovskite oxides outperform single perovskites in electrocatalysis water splitting and thermoelectric properties par to the commercial noble metal and chalcogen-based materials.
3. *Layered perovskites.* As shown in Figure 1.7c, the octahedra can form a layered structure. The layered perovskites are further classified as the Ruddlesden-Popper phase, the Dion–Jacobson phase, the Aurivillius phase, and $A_nB_nO_{3n+2}$ layered phase. Their structures are described in Chapter 3. The layered perovskites exhibit exceptional characteristics, such as superconductivity, that are not observed in single or double perovskite counterparts, due to the oxygen-rich separating layers between the perovskite slabs. In general, the layered perovskites exhibit anisotropy in their properties along the *ab*-plane and *c*-axis.
4. *Anion deficient phase*, such as the well-known brownmillerite, which is composed of alternating BO_6 octahedra and BO_4 tetrahedra layers. This structure is described in Chapter 3.
5. *Hexagonal perovskites*, which possess hexagonal close packing of AX_3 layers, instead of cubic close packing of AX_3 layers (Fop et al. 2019). More details can be found in Chapter 3.

The group of materials that adopt a perovskite structure or a related phase is huge, and a large number of publications are added to the literature every year from the 1950s. A significant increase in the number of publications is observed in recent years. A comparison of the number of publications that appeared in the literature for the last 50 years is shown in Figure 1.8. Considering an average number of

TABLE 1.1
List of Important Single Oxide Perovskites and Their Properties/Applications

Perovskite Material	Structure at Room Temperature	Properties/Applications	Reference
$BaBiO_3$	Orthorhombic	Diamagnetic, superconductivity	Sleight (2015)
$BaCeO_3$	Orthorhombic	Proton and oxygen-ion conductivity suitable for fuel cell electrolyte	Medvedev et al. (2014)
$BaCoO_3$	Hexagonal	Cathode for solid-oxide fuel cells	Felser et al. (1999)
$BaFeO_3$	Hexagonal	Magnetic, suitable for non-volatile memory storage	Hayashi et al. (2011)
$BaGeO_3$	Orthorhombic	Hosts for luminescent rare-earth ions	Noor et al. (2018)
$BaMnO_3$	Hexagonal	Antiferromagnetic, catalysis	Chamberland et al. (1970a)
$BaNiO_3$	Rhombohedral	Diamagnetic	Lee et al. (2016)
$BaSnO_3$	Cubic	Thermoelectric and electro-photocatalysis	Avinash et al. (2019)
$BaTiO_3$	Tetragonal	Piezoelectric and dielectric	Acosta et al. (2017)
$BaThO_3$	Orthorhombic	Optoelectronic applications	Murtaza et al. (2011)
$BaZrO_3$	Cubic	Proton conducting, suitable for high-temperature fuel cells	Thananatthanachon (2016)
$BaSiO_3$	Orthorhombic	Luminescent and thermoelectric	Xu et al. (2017)
$BiFeO_3$	Rhombohedral	Photovoltaics, photocatalysis	Jia et al. (2009)
$BiMnO_3$	Triclinic	Magnetic and ferroelectric	Hanif et al. (2017)
$BaVO_3$	Cubic	Metallic conductivity	Nishimura et al. (2014)
$BiCoO_3$	Tetragonal	Magnetoelectric	Pan et al. (2019)
$BiNiO_3$	Triclinic	Multiferroics	Liu et al. (2016)
$BiCrO_3$	Orthorhombic	Multiferroic, ferro-electromagnet	Wang et al. (2011)
$BiAlO_3$	Trigonal	Non-volatile memory storage	Belik et al. (2006a)
$BiGaO_3$	Orthorhombic	Magnetic	Belik et al. (2012)
$BiInO_3$	Orthorhombic	Nonlinear optoelectronics	Belik et al. (2006b)
$BiScO_3$	Monoclinic	Piezoelectric	Dai et al. (2018)
$CaCrO_3$	Orthorhombic	Antiferromagnet	Goodenough et al. (1968)
$CaFeO_3$	Orthorhombic	Ferromagnet	Takeda et al. (1978)
$CaMnO_3$	Orthorhombic	Electrocatalyst for water splitting, thermoelectric	Zhou and Kennedy (2006)
$CaSiO_3$	Triclinic	Optical applications, additive in paints, inert to chemicals	Swamy and Dubrovinsky (1997)
$CaTiO_3$	Orthorhombic	Diamagnetic	Moreira et al. (2009)
$CaZrO_3$	Orthorhombic	Oxygen sensing	Stoch et al. (2012)
$CaGeO_3$	Orthorhombic	Garnet, optical applications	Liu et al. (1991)
$CeFeO_3$	Orthorhombic	Photocatalyst	Petschnig et al. (2016)

(Continued)

TABLE 1.1 (*Continued*)
List of Important Single Oxide Perovskites and Their Properties/Applications

Perovskite Material	Structure at Room Temperature	Properties/Applications	Reference
$CdTiO_3$	Orthorhombic	Dye sensitized solar cell, antibacterial, photocatalysis	Kennedy et al. (2011)
$CaRuO_3$	Orthorhombic	Ferromagnetic, catalysis	Nanda et al. (2007)
$DyCrO_3$	Orthorhombic	Photocatalytic water splitting, water treatment	Ahsan et al. (2018)
DyFeO	Orthorhombic	Ferromagnetic, multiferroic	Zhao et al. (2014)
$DyCoO_3$	Orthorhombic	Photocatalyst, UV sensor, thermoelectric	Michel et al. (2019)
$GdCrO_3$	Orthorhombic	Multiferroic	Yoshii (2001)
$GdMnO_3$	Orthorhombic	Paraelectric and paramagnetic	Wagh et al. (2015)
$GdCoO_3$	Orthorhombic	Gas sensor	Gildo-Ortiz et al. (2019)
$GdFeO_3$	Orthorhombic	Magnetic, dielectric, photo and electrocatalysis	Wu et al. (2014)
$HoMnO_3$	Hexagonal	Multiferroic, memory, magnetic field sensors	Fiebig et al. (2002)
$HoCrO_3$	Orthorhombic	Multiferroic, sensitivity to humidity and gases	Yin et al. (2017)
$HoCoO_3$	Orthorhombic	Thermoelectric	Muñoz et al. (2012)
$KTaO_3$	Cubic	Piezoelectric, superconductivity	Ueno et al. (2011)
$KNbO_3$	Orthorhombic	Ferroelectric, nonlinear optics	Skjærvø et al. (2018)
$LaRuO_3$	Orthorhombic	Catalyst for oxidation in fuel cells	Labhsetwar et al. (2003)
$LaCrO_3$	Orthorhombic	High oxygen ion conductivity, magnetic	Zhang et al. (2015)
$LaMnO_3$	Cubic	Super capacitor, electrode material, magnetic, biomedical applications	Rivero et al. (2016)
$LaGaO_3$	Orthorhombic	Catalysis, oxygen separating membranes	Ishihara et al. (1994)
$LaFeO_3$	Orthorhombic	Antiferromagnetic insulator	Rao et al. (2019)
$LaCoO_3$	Rhombohedral	Magnetic, ionic conductor	Dragan et al. (2019)
$LaNiO_3$	Hexagonal	Reforming of methane, magnetic	Golalikhani et al. (2018)
$LaVO_3$	Orthorhombic	Photovoltaic absorber	Jellite et al. (2018)
$LaScO_3$	Orthorhombic	Host for luminescent ions, proton conducting	Lybye et al. (2000)
$LaTiO_3$	Orthorhombic	Dielectric	Cwik et al. (2003)
$LaHoO_3$	Orthorhombic	Luminescence	Siaï et al. (2016)
$LaAlO_3$	Rhombohedral	Thermoluminescence, substrate for the epitaxial growth of single crystal perovskites, ionic conductivity	Rizwan et al. (2019)

(*Continued*)

TABLE 1.1 (*Continued*)
List of Important Single Oxide Perovskites and Their Properties/Applications

Perovskite Material	Structure at Room Temperature	Properties/Applications	Reference
$LiNbO_3$	Trigonal	Optical waveguides, mobile phones, piezoelectric sensors, optical modulators	Nitta (1968)
$LiVO_3$	Monoclinic	Electrode for Li-ion batteries	Jian et al. (2013)
$LiTaO_3$	Rhombohedral	Ferroelectric	Yang et al. (2014)
$MgTiO_3$	Rhombohedral	Photocatalyst	Zhang et al. (2012)
$MgSnO_3$	Trigonal	Thermoelectric	Rashad and El-Shall (2008)
$MgGeO_3$	Rhombohedral	Luminescent	Merkel et al. (2006)
$MgMnO_3$	Cubic	Antiferromagnetic	Chamberland et al. (1970b)
$NdFeO_3$	Orthorhombic	Magnetic, non-volatile memories and sensors	Shanker et al. (2018)
$NdCrO_3$	Orthorhombic	Multiferroic, cathode for intermediate temperature fuel cell	Liu et al. (2008)
$NdNiO_3$	Orthorhombic	Metal–insulator transition, sensor	Li et al. (2019)
$NdMnO_3$	Orthorhombic	Multiferroic	Muñoz et al. (2000)
$NaNbO_3$	Orthorhombic	Antiferroelectric	Ji et al. (2014)
$NaTaO_3$	Orthorhombic	Photocatalyst, antiferroelectric	Li et al. (2016)
$PrMnO_3$	Orthorhombic	Multiferroic	Mansouri et al. (2017)
$PrCoO_3$	Orthorhombic	Catalyst	Knížek et al. (2009)
$PbTiO_3$	Tetragonal	Piezoelectric, ferroelectric	Chaudhari and Bichile (2013)
$PbZrO_3$	Tetragonal	Antiferroelectric	Tagantsev et al. (2013)
$PbTaO_3$	Cubic	Paramagnetic	Khandy and Gupta (2016)
$SrMnO_3$	Hexagonal	Supercapacitor electrode, magnetic	Bai et al. (2017)
$SrRuO_3$	Orthorhombic	Electrocatalyst	Koster et al. (2012)
$SrTiO_3$	Cubic	Thermistors, displays	Ahadi et al. (2019)
$SrZrO_3$	Cubic	Photocatalysts, dielectric	Weston et al. (2015)
$SrFeO_3$	Orthorhombic	Catalyst for methane combustion	Falcón et al. (2002)
$SrSnO_3$	Orthorhombic	Transparent conductor	Lee et al. (2012)
$SrVO_3$	Cubic	Transparent conducting oxide	Macías et al. (2016)
$SmNiO_3$	Orthorhombic	Metal-to-insulator transition with temperature	Chen et al. (2019)
$SmCoO_3$	Orthorhombic	Catalyst methane reforming	He et al. (2018)
$SmCrO_3$	Orthorhombic	Oxygen transport membranes in fuel cells	Panwar et al. (2019)
$SmFeO_3$	Orthorhombic	Electrode for solid oxide fuel cells	Liu et al. (2019)

(*Continued*)

TABLE 1.1 (*Continued*)
List of Important Single Oxide Perovskites and Their Properties/Applications

Perovskite Material	Structure at Room Temperature	Properties/Applications	Reference
$TbMnO_3$	Orthorhombic	Ferroelectric	Aoyama et al. (2014)
$TbFeO_3$	Orthorhombic	Antiferromagnetic	Gupta et al. (2020)
$TbCoO_3$	Orthorhombic	Magnetic	Knížek et al. (2014)
$YFeO_3$	Hexagonal	Visible light photocatalysis	Shang et al. (2013)
$YMnO_3$	Hexagonal	Ferroelectric and antiferromagnetic	Van Aken et al. (2004)
$YCrO_3$	Monoclinic	Multiferroic, non-volatile memory applications	Gervacio-Arciniega et al. (2018)
$YAlO_3$	Orthorhombic	Phosphors, scintillators	Diehl and Brandt (1975)
$YCoO_3$	Orthorhombic	Sensors	Buassi-Monroy et al. (2004)
$YNiO_3$	Orthorhombic	Antiferromagnetic	Xu et al. (2004)
$YCuO_3$	Cubic	Superconductivity	Yanase et al. (1987)
YVO_3	Orthorhombic	Thermoelectric	Tsvetkov et al. (2004)
$YTiO_3$	Orthorhombic	Mott-insulator	Garrett et al. (1981)
$ZnMnO_3$	Cubic	Antiferromagnetic	Chen et al. (2016)

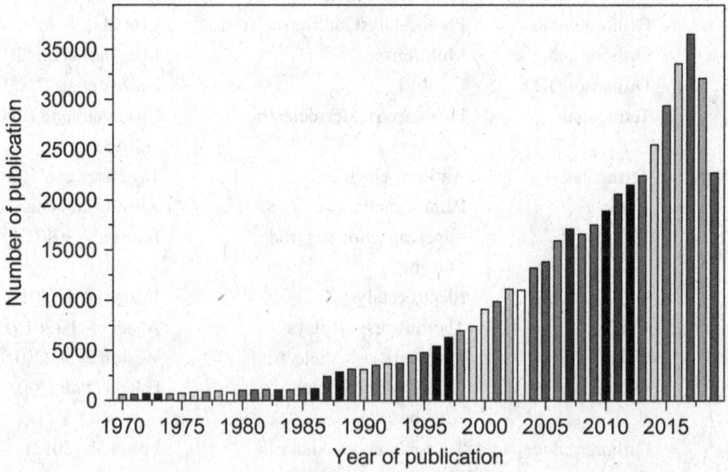

FIGURE 1.8 The histogram on the number of publications in the past 50 years on perovskite from institute for scientific information (ISI) web of science. (Search keyword: perovskite; search date: 01/06/2020.)

10 publications on each perovskite structured materials, one can imagine the huge variety of perovskite materials and their derivatives potentially available for different applications.

REFERENCES

Acosta, M., N. Novak, V. Rojas, et al. 2017. $BaTiO_3$-Based Piezoelectrics: Fundamentals, Current Status, and Perspectives. *Appl. Phys. Rev.* 4:041305.

Ahadi, K., L. Galletti, Y. Li, et al. 2019. Enhancing Superconductivity in $SrTiO_3$ Films with Strain. *Sci. Adv.* 5:eaaw0120.

Ahsan, R., A. Mitra, S. Omar, et al. 2018. Sol–Gel Synthesis of $DyCrO_3$ and 10% Fe-Doped $DyCrO_3$ Nanoparticles with Enhanced Photocatalytic Hydrogen Production Abilities. *RSC Adv.* 8:14258–14267.

Aoyama, T., K. Yamauchi, A. Iyama, et al. 2014. Giant Spin-Driven Ferroelectric Polarization in $TbMnO_3$ under High Pressure. *Nat. Commun.* 5:1–7.

Avinash, M., M. Muralidharan, and K. Sivaji. 2019. Structural, Optical and Magnetic Behaviour of Cr Doped $BaSnO_3$ Perovskite Nanostructures. *Phys. B Condens. Matter* 570:157–165.

Bai, J., J. Yang, W. Dong, et al. 2017. Structural and Magnetic Properties of Perovskite $SrMnO_3$ Thin Films Grown by Molecular Beam Epitaxy. *Thin Solid Films* 644:57–64.

Bartel, C. J., C. Sutton, B. R. Goldsmith, et al. 2019. New Tolerance Factor to Predict the Stability of Perovskite Oxides and Halides. *Sci. Adv.* 5:eaav0693.

Belik, A. A., D. A. Rusakov, T. Furubayashi, et al. 2012. $BiGaO_3$-Based Perovskites: A Large Family of Polar Materials. *Chem. Mater.* 24:3056–3064.

Belik, A. A., S. Y. Stefanovich, B. I. Lazoryak, et al. 2006b. $BiInO_3$: A Polar Oxide with $GdFeO_3$-Type Perovskite Structure. *Chem. Mater.* 18:1964–1968.

Belik, A. A., T. Wuernisha, T. Kamiyama, et al. 2006a. High-Pressure Synthesis, Crystal Structures, and Properties of Perovskite-like $BiAlO_3$ and Pyroxene-like $BiGaO_3$. *Chem. Mater.* 18:133–139.

Buassi-Monroy, O. S., C. C. Luhrs, A. Chávez-Chávez, et al. 2004. Synthesis of Crystalline $YCoO_3$ Perovskite via Sol–Gel Method. *Mater. Lett.* 58:716–718.

Candan, A., and M. Kurban. 2018. Electronic Structure, Elastic and Phonon Properties of Perovskite-Type Hydrides $MgXH_3$ (X = Fe, Co) for Hydrogen Storage. *Solid State Commun.* 281:38–43.

Chamberland, B. L., A. W. Sleight, and J. F. Weiher. 1970a. Preparation and Characterization of $BaMnO_3$ and $SrMnO_3$ Polytypes. *J. Solid State Chem.* 1:506–511.

Chamberland, B. L., A. W. Sleight, and J. F. Weiher. 1970b. Preparation and Characterization of $MgMnO_3$ and $ZnMnO_3$. *J. Solid State Chem.* 1:512–514.

Chaudhari, V. A., and G. K. Bichile. 2013. Synthesis, Structural, and Electrical Properties of Pure $PbTiO_3$ Ferroelectric Ceramics. *Smart Mater. Res.* 2013:147524

Chen, H., L.-X. Ding, K. Xiao, et al. 2016. Highly Ordered $ZnMnO_3$ Nanotube Arrays from a "Self-Sacrificial" ZnO Template as High-Performance Electrodes for Lithium Ion Batteries. *J. Mater. Chem. A* 4:16318–16323.

Chen, J., W. Mao, B. Ge, et al. 2019. Revealing the Role of Lattice Distortions in the Hydrogen-Induced Metal-Insulator Transition of $SmNiO_3$. *Nat. Commun.* 10:1–8.

Cwik, M., T. Lorenz, J. Baier, et al. 2003. Crystal and Magnetic Structure of $LaTiO_3$: Evidence for Nondegenerate t_{2g} Orbitals. *Phys. Rev. B* 68:060401.

Dai, Z., W. Liu, D. Lin, et al. 2018. Electrical Properties of Zirconium-Modified $BiScO_3$-$PbTiO_3$ Piezoelectric Ceramics at Re-Designed Phase Boundary. *Mater. Lett.* 215:46–49.

Diehl, R., and G. Brandt. 1975. Crystal Structure Refinement of $YAlO_3$, a Promising Laser Material. *Mater. Res. Bull.* 10:85–90.

Dragan, M., S. Enache, M. Varlam, et al. 2019. Perovskite-Type Lanthanum Cobaltite $LaCoO_3$: Aspects of Processing Route toward Practical Applications. In *Cobalt Compounds and Applications*, ed. Y. Yıldız, and A. Manzak, Chapter 5. London, UK: IntechOpen.

Emery, A. A., J. E. Saal, S. Kirklin, et al. 2016. High-Throughput Computational Screening of Perovskites for Thermochemical Water Splitting Applications. *Chem. Mater.* 28:5621–5634.

Eshita, T., T. Tamura, and Y. Arimoto. 2014. Ferroelectric Random Access Memory (FRAM) Devices. In *Advances in Non-volatile Memory and Storage Technology*, ed. Y. Nishi, 434–454. Waltham, MA: Woodhead Publishing.

Falcón, H., J. A. Barbero, J. A. Alonso, et al. 2002. $SrFeO_{3-\delta}$ Perovskite Oxides: Chemical Features and Performance for Methane Combustion. *Chem. Mater.* 14:2325–2333.

Felser, C., K. Yamaura, and R. J. Cava. 1999. The Electronic Structure of Hexagonal $BaCoO_3$. *J. Solid State Chem.* 146:411–417.

Fiebig, M., C. Degenhardt, and R. V. Pisarev. 2002. Magnetic Phase Diagram of $HoMnO_3$. *J. Appl. Phys.* 91:8867–8869.

Fop, S., K. S. McCombie, E. J. Wildman, et al. 2019. Hexagonal Perovskite Derivatives: A New Direction in the Design of Oxide Ion Conducting Materials. *Chem. Commun.* 55:2127–2137.

Garrett, J. D., J. E. Greedan, and D. A. MacLean. 1981. Crystal Growth and Magnetic Anisotropy of $YTiO_3$. *Mater. Res. Bull.* 16:145–148.

Gervacio-Arciniega, J. J., E. Murillo-Bracamontes, O. Contreras, et al. 2018. Multiferroic $YCrO_3$ Thin Films: Structural, Ferroelectric and Magnetic Properties. *Appl. Surf. Sci.* 427:635–639.

Gildo-Ortiz, L., V. M. Rodríguez-Betancourtt, O. Blanco-Alonso, et al. 2019. A Simple Route for the Preparation of Nanostructured $GdCoO_3$ via the Solution Method, as Well as Its Characterization and Its Response to Certain Gases. *Results Phys.* 12:475–483.

Golalikhani, M., Q. Lei, R. U. Chandrasena, et al. 2018. Nature of the Metal-Insulator Transition in Few-Unit-Cell-Thick $LaNiO_3$ Films. *Nat. Commun.* 9:1–8.

Goodenough, J. B., J. M. Longo, and J. A. Kafalas. 1968. Band Antiferromagnetism and the New Perovskite $CaCrO_3$. *Mater. Res. Bull.* 3:471–481.

Gupta, P., P. K. Mahapatra, and R. N. P. Choudhary. 2020. $TbFeO_3$ Ceramic: An Exciting Colossal Dielectric with Ferroelectric Properties. *Phys. Status Solidi B* 257:1900236.

Hanif, S., M. Hassan, S. Riaz, et al. 2017. Structural, Magnetic, Dielectric and Bonding Properties of $BiMnO_3$ Grown by Co-Precipitation Technique. *Results Phys.* 7:3190–3195.

Hayashi, N., T. Yamamoto, H. Kageyama, et al. 2011. $BaFeO_3$: A Ferromagnetic Iron Oxide. *Angew. Chem. Int. Ed.* 50:12547–12550.

He, J., W. Zhou, J. Sunarso, et al. 2018. 3D Ordered Macroporous $SmCoO_3$ Perovskite for Highly Active and Selective Hydrogen Peroxide Detection. *Electrochim. Acta* 260:372–383.

Ishihara, T., H. Matsuda, and Y. Takita. 1994. Doped $LaGaO_3$ Perovskite Type Oxide as a New Oxide Ionic Conductor. *J. Am. Chem. Soc.* 116:3801–3803.

Jellite, M., J.-L. Rehspringer, M. A. Fazio, et al. 2018. Investigation of $LaVO_3$ Based Compounds as a Photovoltaic Absorber. *Sol. Energy* 162:1–7.

Ji, S., H. Liu, Y. Sang, et al. 2014. Synthesis, Structure, and Piezoelectric Properties of Ferroelectric and Antiferroelectric $NaNbO_3$ Nanostructures. *CrystEngComm* 16:7598–7604.

Jia, D.-C., J.-H. Xu, H. Ke, et al. 2009. Structure and Multiferroic Properties of $BiFeO_3$ Powders. *J. Eur. Ceram. Soc.* 29:3099–3103.

Jian, X. M., J. P. Tu, Y. Q. Qiao, et al. 2013. Synthesis and Electrochemical Performance of $LiVO_3$ Cathode Materials for Lithium Ion Batteries. *J. Power Sources* 236:33–38.

Kennedy, B. J., Q. Zhou, and M. Avdeev. 2011. The Ferroelectric Phase of $CdTiO_3$: A Powder Neutron Diffraction Study. *J. Solid State Chem.* 184:2987–2993.

Khandy, S. A., and D. C. Gupta. 2016. Structural, Elastic and Thermo-Electronic Properties of Paramagnetic Perovskite $PbTaO_3$. *RSC Adv.* 6:48009–48015.

Knížek, K., Z. Jirák, P. Novák, et al. 2014. Non-Collinear Magnetic Structures of $TbCoO_3$ and $DyCoO_3$. *Solid State Sci.* 28:26–30.

Knížek, K., J. Hejtmánek, Z. Jirák, et al. 2009. Neutron Diffraction and Heat Capacity Studies of $PrCoO_3$ and $NdCoO_3$. *Phys. Rev. B* 79:134103.

Komiya, K., N. Morisaku, R. Rong, et al. 2008. Synthesis and Decomposition of Perovskite-Type Hydrides, $MMgH_3$ (M = Na, K, Rb). *J. Alloys Compd.* 453:157–160.

Koster, G., L. Klein, W. Siemons, et al. 2012. Structure, Physical Properties, and Applications of $SrRuO_3$ Thin Films. *Rev. Mod. Phys.* 84:253–298.

Labhsetwar, N. K., A. Watanabe, and T. Mitsuhashi. 2003. New Improved Syntheses of $LaRuO_3$ Perovskites and Their Applications in Environmental Catalysis. *Appl. Catal. B Environ.* 40:21–30.

Lee, C. W., D. W. Kim, I. S. Cho, et al. 2012. Simple Synthesis and Characterization of $SrSnO_3$ Nanoparticles with Enhanced Photocatalytic Activity. *Int. J. Hydrog. Energy* 37:10557–10563.

Lee, J. G., H. J. Hwang, O. Kwon, et al. 2016. Synthesis and Application of Hexagonal Perovskite $BaNiO_3$ with Quadrivalent Nickel under Atmospheric and Low-Temperature Conditions. *Chem. Commun.* 52:10731–10734.

Li, J., J. Pelliciari, C. Mazzoli, et al. 2019. Scale-Invariant Magnetic Textures in the Strongly Correlated Oxide $NdNiO_3$. *Nat. Commun.* 10:1–7.

Li, S., H. Qiu, C. Wang, et al. 2016. Highly Efficient $NaTaO_3$ for Visible Light Photocatalysis Predicted from First Principles. *Sol. Energy Mater. Sol. Cells* 149:97–102.

Liu, M., Y. Shen, Y. Ji, et al. 2008. Structures and Properties of Sr-Doped $NdCrO_3$ Solid Solutions. *J. Alloys Compd.* 461:628–632.

Liu, Q., X.-X. Wang, C. Song, et al. 2019. Magnetic Properties of La Doped $SmFeO_3$. *J. Magn. Magn. Mater.* 469:76–80.

Liu, X., Y. Wang, R. C. Liebermann, et al. 1991. Phase Transition in $CaGeO_3$ Perovskite: Evidence from *X*-Ray Powder Diffraction, Thermal Expansion and Heat Capacity. *Phys. Chem. Miner.* 18:224–230.

Liu, Y., Z. Wang, D. Chang, et al. 2016. Charge Transfer Induced Negative Thermal Expansion in Perovskite $BiNiO_3$. *Comput. Mater. Sci.* 113:198–202.

Liu, Y., Z. Yang, and S. Liu. 2017. Recent Progress in Single-Crystalline Perovskite Research Including Crystal Preparation, Property Evaluation, and Applications. *Adv. Sci.* 5:1700471.

Lybye, D., F. W. Poulsen, and M. Mogensen. 2000. Conductivity of A- and B-Site Doped $LaAlO_3$, $LaGaO_3$, $LaScO_3$ and $LaInO_3$ Perovskites. *Solid State Ion.* 128:91–103.

Macías, J., A. A. Yaremchenko, and J. R. Frade. 2016. Enhanced Stability of Perovskite-like SrVO3-Based Anode Materials by Donor-Type Substitutions. *J. Mater. Chem. A* 4:10186–10194.

Mansouri, S., S. Jandl, A. Mukhin, et al. 2017. A Comparative Raman Study between $PrMnO_3$, $NdMnO_3$, $TbMnO_3$ and $DyMnO_3$. *Sci. Rep.* 7:13796.

Medvedev, D., A. Murashkina, E. Pikalova, et al. 2014. $BaCeO_3$: Materials Development, Properties and Application. *Prog. Mater. Sci.* 60:72–129.

Merkel, S., A. Kubo, L. Miyagi, et al. 2006. Plastic Deformation of $MgGeO_3$ Post-Perovskite at Lower Mantle Pressures. *Science* 311:644–646.

Michel, C. R., M. A. Lopez-Alvarez, A. H. Martínez-Preciado, et al. 2019. Novel UV Sensing and Photocatalytic Properties of $DyCoO_3$. Research Article. *J. Sens.* 2019:5682645.

Moreira, M. L., E. C. Paris, G. S. do Nascimento, et al. 2009. Structural and Optical Properties of $CaTiO_3$ Perovskite-Based Materials Obtained by Microwave-Assisted Hydrothermal Synthesis: An Experimental and Theoretical Insight. *Acta Mater.* 57:5174–5185.

Muñoz, A., J. A. Alonso, M. J. Martínez-Lope, et al. 2000. Magnetic Structure Evolution of $NdMnO_3$ Derived from Neutron Diffraction Data. *J. Phys. Condens. Matter* 12:1361–1376.

Muñoz, A., M. J. Martínez-Lope, J. A. Alonso, et al. 2012. Magnetic Structures of $HoCoO_3$ and $TbCoO_3$. *Eur. J. Inorg. Chem.* 2012:5825–5830.

Murtaza, G., I. Ahmad, B. Amin, et al. 2011. Investigation of Structural and Optoelectronic Properties of $BaThO_3$. *Opt. Mater.* 33:553–557.

Nanda, B. R. K., S. Satpathy, and M. S. Springborg. 2007. Electron Leakage and Double-Exchange Ferromagnetism at the Interface between a Metal and an Antiferromagnetic Insulator: $CaRuO_3/CaMnO_3$. *Phys. Rev. Lett.* 98:216804.

Nishimura, K., I. Yamada, K. Oka, et al. 2014. High-Pressure Synthesis of $BaVO_3$: A New Cubic Perovskite. *J. Phys. Chem. Solids* 75:710–712.

Nitta, T. 1968. Properties of Sodium-Lithium Niobate Solid Solution Ceramics with Small Lithium Concentrations. *J. Am. Ceram. Soc.* 51:623–630.

Noor, N. A., Q. Mahmood, M. Hassan, et al. 2018. Physical Properties of Cubic $BaGeO_3$ Perovskite at Various Pressure Using First-Principle Calculations for Energy Renewable Devices. *J. Mol. Graph. Model.* 84:152–159.

Pan, Z., X. Jiang, T. Nishikubo, et al. 2019. Pronounced Negative Thermal Expansion in Lead-Free $BiCoO_3$-Based Ferroelectrics Triggered by the Stabilized Perovskite Structure. *Chem. Mater.* 31:6187–6192.

Panwar, N., I. Coondoo, S. Kumar, et al. 2019. Structural, Electrical, Optical and Magnetic Properties of $SmCrO_3$ Chromites: Influence of Gd and Mn Co-Doping. *J. Alloys Compd.* 792:1122–1131.

Petschnig, L. L., G. Fuhrmann, D. Schildhammer, et al. 2016. Solution Combustion Synthesis of $CeFeO_3$ under Ambient Atmosphere. *Ceram. Int.* 42:4262–4267.

Pottmaier, D., E. R. Pinatel, J. G. Vitillo, et al. 2011. Structure and Thermodynamic Properties of the $NaMgH_3$ Perovskite: A Comprehensive Study. *Chem. Mater.* 23:2317–2326.

Rao, M. P., S. Musthafa, J. J. Wu, et al. 2019. Facile Synthesis of Perovskite $LaFeO_3$ Ferroelectric Nanostructures for Heavy Metal Ion Removal Applications. *Mater. Chem. Phys.* 232:200–204.

Rashad, M. M., and H. El-Shall. 2008. Effect of Synthesis Conditions on the Preparation of $MgSnO_3$ Powder via Co-Precipitation Method. *Powder Technol.* 183:161–168.

Rivero, P., V. Meunier, and W. Shelton. 2016. Electronic, Structural, and Magnetic Properties of $LaMnO_3$ Phase Transition at High Temperature. *Phys. Rev. B* 93:024111.

Rizwan, M., S. Gul, T. Iqbal, et al. 2019. A Review on Perovskite Lanthanum Aluminate ($LaAlO_3$), Its Properties and Applications. *Mater. Res. Express* 6:112001.

Scatturin, V., L. Corliss, N. Elliott, et al. 1961. Magnetic Structures of 3*d* Transition Metal Double Fluorides, $KMeF_3$. *Acta Crystallogr.* 14:19–26.

Schouwink, P., M. B. Ley, A. Tissot, et al. 2014. Structure and Properties of Complex Hydride Perovskite Materials. *Nat. Commun.* 5:1–10.

Shang, M., C. Zhang, T. Zhang, et al. 2013. The Multiferroic Perovskite $YFeO_3$. *Appl. Phys. Lett.* 102:062903.

Shanker, J., G. Narsinga Rao, K. Venkataramana, et al. 2018. Investigation of Structural and Electrical Properties of $NdFeO_3$ Perovskite Nanocrystalline. *Phys. Lett. A* 382:2974–2977.

Shpatz Dayan, A., B.-E. Cohen, S. Aharon, et al. 2018. Enhancing Stability and Photostability of $CsPbI_3$ by Reducing Its Dimensionality. *Chem. Mater.* 30:8017–8024.

Siaï, A., K. Horchani-Naifer, P. Haro-González, et al. 2016. Effects of the Preparation Processes on Structural, Electronic, and Optical Properties of $LaHoO_3$. *Mater. Res. Bull.* 76:179–186.

Skjærvø, S. L., K. Høydalsvik, A. B. Blichfeld, et al. 2018. Thermal Evolution of the Crystal Structure and Phase Transitions of $KNbO_3$. *R. Soc. Open Sci.* 5:180368.

Sleight, A. W. 2015. Bismuthates: $BaBiO_3$ and Related Superconducting Phases. *Phys. C Supercond. Its Appl.* 514:152–165.

Stoch, P., J. Szczerba, J. Lis, et al. 2012. Crystal Structure and Ab Initio Calculations of $CaZrO_3$. *J. Eur. Ceram. Soc.* 32:665–670.

Swamy, V., and L. S. Dubrovinsky. 1997. Thermodynamic Data for the Phases in the $CaSiO_3$ System. *Geochim. Cosmochim. Acta* 61:1181–1191.

Tagantsev, A. K., K. Vaideeswaran, S. B. Vakhrushev, et al. 2013. The Origin of Antiferroelectricity in $PbZrO_3$. *Nat. Commun.* 4:1–8.

Takeda, Y., S. Naka, M. Takano, et al. 1978. Preparation and Characterization of Stoichiometric $CaFeO_3$. *Mater. Res. Bull.* 13:61–66.

Thananatthanachon, T. 2016. Synthesis and Characterization of a Perovskite Barium Zirconate ($BaZrO_3$): An Experiment for an Advanced Inorganic Chemistry Laboratory. *J. Chem. Educ.* 93:1120–1123.

Tsvetkov, A. A., F. P. Mena, P. H. M. van Loosdrecht, et al. 2004. Structural, Electronic, and Magneto-Optical Properties of YVO_3. *Phys. Rev. B* 69:075110.

Ueno, K., S. Nakamura, H. Shimotani, et al. 2011. Discovery of Superconductivity in $KTaO_3$ by Electrostatic Carrier Doping. *Nat. Nanotechnol.* 6:408–412.

Umemoto, K., R. M. Wentzcovitch, and P. B. Allen. 2006. Dissociation of $MgSiO_3$ in the Cores of Gas Giants and Terrestrial Exoplanets. *Science* 311:983–986.

Van Aken, B. B., T. T. M. Palstra, A. Filippetti, et al. 2004. The Origin of Ferroelectricity in Magnetoelectric $YMnO_3$. *Nat. Mater.* 3:164–170.

Wagh, A. A., K. G. Suresh, P. S. A. Kumar, et al. 2015. Low Temperature Giant Magnetocaloric Effect in Multiferroic $GdMnO_3$ single Crystals. *J. Phys. Appl. Phys.* 48:135001.

Wang, H., X. Chen, X. Chen, et al. 2011. Stable Antiferromagnetism of Orthorhombic $BiCrO_3$ under Pressure: A Theoretical Study. *Adv. Mater. Res.* 298:243–248.

Weston, L., A. Janotti, X. Y. Cui, et al. 2015. Structural and Electronic Properties of $SrZrO_3$ and Sr(Ti, Zr)O_3 Alloys. *Phys. Rev. B* 92:085201.

Wu, A., Z. Wang, B. Wang, et al. 2014. Crystal Growth and Magnetic Properties of $GdFeO_3$ Crystals by Floating Zone Method. *Solid State Commun.* 185:14–17.

Xu, J., Y. Zhao, J. Chen, et al. 2017. Insights into the Discrepant Luminescence for $BaSiO_3:Eu^{2+}$ Phosphors Prepared by Solid-State Reaction and Precipitation Reaction Methods. *Lumin. J. Biol. Chem. Lumin.* 32:957–963.

Xu, X., X. Meng, C. Wang, et al. 2004. Charge Disproportionation in $YNiO_3$ Perovskite: An Ab Initio Calculation. *J. Phys. Chem. B* 108:1165–1167.

Yanase, A., T. Yamasaki, M. Onozaki, et al. 1987. Electronic Structure of Superconducting Oxide-Effect of Non-Metallic Elements. Phys. BC 148:385–387.

Yang, T., Y. Liu, L. Zhang, et al. 2014. Powder Synthesis and Properties of $LiTaO_3$ Ceramics. *Adv. Powder Technol.* 25:933–936.

Yin, S., M. S. Seehra, C. J. Guild, et al. 2017. Magnetic and Magnetocaloric Properties of $HoCrO_3$ Tuned by Selective Rare-Earth Doping. *Phys. Rev. B* 95:184421.

Yoshii, K. 2001. Magnetic Properties of Perovskite $GdCrO_3$. *J. Solid State Chem.* 159:204–208.

Zhang, H., N. Li, K. Li, et al. 2007. Structural Stability and Formability of ABO_3-Type Perovskite Compounds. *Acta Crystallogr. B* 63:812–818.

Zhang, K. H. L., Y. Du, A. Papadogianni, et al. 2015. Perovskite Sr-Doped $LaCrO_3$ as a New p-Type Transparent Conducting Oxide. *Adv. Mater.* 27:5191–5195.

Zhang, L., Y. Meng, W. Yang, et al. 2014. Disproportionation of (Mg, Fe)SiO_3 Perovskite in Earth's Deep Lower Mantle. *Science* 344:877–882.

Zhang, M., L. Li, W. Xia, et al. 2012. Structure and Properties Analysis for $MgTiO_3$ and $(Mg_{0.97}M_{0.03})TiO_3$ (M = Ni, Zn, Co and Mn) Microwave Dielectric Materials. *J. Alloys Compd.* 537:76–79.

Zhao, Z. Y., X. Zhao, H. D. Zhou, et al. 2014. Ground State and Magnetic Phase Transitions of Orthoferrite $DyFeO_3$. *Phys. Rev. B* 89:224405.

Zhou, Q., and B. J. Kennedy. 2006. Thermal Expansion and Structure of Orthorhombic $CaMnO_3$. *J. Phys. Chem. Solids* 67:1595–1598.

2 Synthesis of Perovskite Oxides

To experimentally study a material's properties for an application, the first step is to synthesize the material. A variety of techniques have been used for the synthesis of inorganic perovskite oxide materials, with diverse choices of their structure, size, and shape. Based on the nature of precursors, the synthesis can be classified as a solid-state, liquid-phase, and gas-phase reactions. The selection of the right synthesis method is important to match the functionalities of the perovskite materials since the structure and morphology of the materials can significantly affect their performance. For example, to synthesize high-performance perovskite catalysts, one has to select a synthesis technique suitable for the fabrication of nanostructured perovskite oxide catalysts with controllable structures.

2.1 SOLID-STATE REACTIONS

2.1.1 High-Temperature Synthesis

The traditional method to synthesize perovskite oxides is the solid-state reaction. Perovskite materials are successfully synthesized by heating the salts of the individual elements correspond to the A- and B-sites of the perovskite structure at a high temperature. The precursor materials for solid-state synthesis can be nitrates, carbonates, oxides, and acetates. The term solid-state reaction is used to represent the reactions in which the starting materials and the final products are in solid-state. In a typical solid-state reaction, the solid raw materials at a stoichiometric ratio are thoroughly mixed by making them as fine powders to increase the surface area of the reactants and thereby maximizing the contact between reactants. Pelletizing is also performed in several instances to affirm the contact between reagent particles. The reagents are then heated to an intermediate temperature, where all the volatile part or the reagents are eliminated. At the same time, all the salts are converted to the respective oxides. If the precursors are salts, such as carbonates or nitrates, defects are introduced during their decomposition process prior to the actual reaction. The presence of such defects can increase the rate of reaction in the further processing. Since the reagent crystals have different orientations and each plane of orientations have different reactivity or diffusion rates, the mixing of reagents in every possible way is needed to increase the reaction rate to obtain the final phase.

In solid-state reactions, the final products are formed by the interdiffusion of cations; therefore, a high-temperature treatment (above 1,000°C) for 12–24 h is normally required to form the perovskite oxides. The high temperature for calcination can be determined by Tamman's Rule. The rule suggests that a solid-state reaction occurs in a reasonable amount of time, only when the temperature of the reaction is about two-thirds of the melting point of the lower melting reactant. At this temperature, the nucleation of the final product phase is intended to take place, and the crystal continues to convert the reagents to the final product. Therefore, the reaction can be completed in a short period of time by increasing the rate of nucleation and diffusion of cations. The diffusion of cations can be faster in the presence of defects or at a high temperature.

Besides inducing the defects or using high temperatures for the reactions, the rate of nucleation can be maximized by using the reagents of similar crystal structures to that of the final product. Such reactions are known as topotactic transition and epitaxy transition. In topotactic transition, the atoms of the reagents are displaced by accommodating or removing the material, but the crystal structure of the reagents and products are identical. In epitaxy transition, the product crystal growth takes place on a certain reagent crystal plane so that both the reagents and products have the same crystal structure.

In the typical synthesis of perovskite oxides, as schematically shown in Figure 2.1, the precursors with the appropriate stoichiometric ratio are dry or wet-milled, with the help of a ball mill. Mechanochemical synthesis routes using ball mills significantly increase the diffusion rates through increasing the interaction between the grains of the solid-state precursors. During the ball milling process, energy in the range of 0.1–100 MJ kg^{-1} is applied to the precursors by controlling the milling parameters. During the grinding process, the kinetic energy of the mill is applied to the ground material through collisions, strike/hit, compression, and friction between the grinding media and the ground material. This elastic/plastic energy transfer results in the formation of cracks in the precursor bulk, and an ultimate reduction of the grain size and integration of the precursor particles.

The thoroughly mixed precursors are then pelletized, dried, and calcined in air to a certain intermediate temperature between 600°C and 900°C, for up to 8 h. The sample container for high-temperature treatment should be inert. Ceramic crucibles such as Al_2O_3 (melting point (m.p.) 1,950°C), ZrO_2/Y_2O_3 (m.p. 2,000°C) and SiO_2 (m.p. 1,710°C) and noble metals such as Pt (m.p. 1,770°C) and Au (m.p. 1,063°C) are the choices of sample containers for such reactions. The heating and cooling are performed at a slow rate of 2°C min^{-1}. The calcined samples are then crushed well and heated again at a higher temperature (>1,000°C) at a slow heating rate. Sometimes, the product after intermediate temperature treatment is pelletized to enhance the diffusion rate of ions. The temperature at this step can be as high as 1,800°C. An

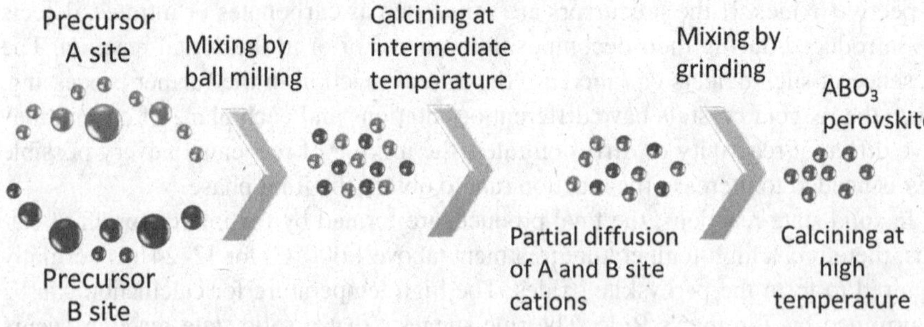

FIGURE 2.1 Steps in solid-state processing of ABO_3 perovskites.

oxidizing heating atmosphere is also critical for perovskite oxides, and the atmosphere can be air or O_2. The reactions are also performed in an inert or reducing atmosphere depending on the nature of the final product.

However, the solid-state reaction method has the following limitations. It is challenging to use cations that are volatile at high temperatures; for instance, Mn^{4+}, Cr^{3+}, etc. The metastable (kinetically stabilized) products cannot be formed since the high oxidation states of the cations are unstable at high temperatures; for instance, MnO_2 (Mn^{4+}) is converted to Mn_2O_3 (Mn^{3+}) and Mn_3O_4 at 700°C (Lan et al. 2015).

The solid-state processing has been successfully employed in the fabrication of simple to layered and complex perovskite structured oxides. By the selection of appropriate calcination/annealing temperature, most of the perovskite oxides can be fabricated by the solid-state method.

2.1.2 Solution-Assisted Synthesis

The solid-state methods for the fabrication of perovskite oxides involve heating at a high temperature to initiate the diffusion of cations. In order to reduce the solid-state reaction temperature, the precursors mixed at a molecular level can be used. For this purpose, precursors are dissolved in a common solvent and mixed at a molecular level and then dried to obtain a well-mixed starting material for solid-state synthesis. The solution-based methods can be divided into two stages: in the first stage, the precursor materials (acetate, citrate, hydroxide, oxalate, alkoxide, etc.) are prepared in a stoichiometric ratio, and in the second stage these precursors are heated at a high temperature. The temperature used in solution methods is often close to the intermediate temperatures used in the solid-state reactions. The ultimate aim of solution-based treatment before heating is to improve the mixing of the cationic reagents so as to reduce the paths of cation diffusion, thereby reducing the required reaction temperatures. Such techniques are also helpful in controlling the morphology of the final products.

The simplest solution-based processing is very similar to solid-state processing, in which the precursor salts are dissolved in a suitable solvent and then dried to form the starting material of well-mixed cationic salts. Due to the thorough mixing of the cations at a molecular level, the diffusion distance of the cations is considerably decreased, and the efficiency of the heating process is tremendously improved. As a result, the final products can be obtained at lower reaction temperatures.

Besides the low reaction temperature, the products in the metastable state can be obtained through this method. Moreover, the impurity phases are eliminated tremendously, and often small crystallites are formed with a high surface area. However, the solubility of the precursor salts and hydrolysis of certain salts in the solvents limit its application to certain materials.

2.1.3 Combustion Synthesis

In combustion synthesis, fuel is added to the precursor material to enhance the reaction rate by increasing the local temperature at the reaction zone. The flow chart of combustion synthesis is shown in Figure 2.2, and a representative scheme is shown in Figure 2.3. The combustion synthesis combines the advantages of exothermic,

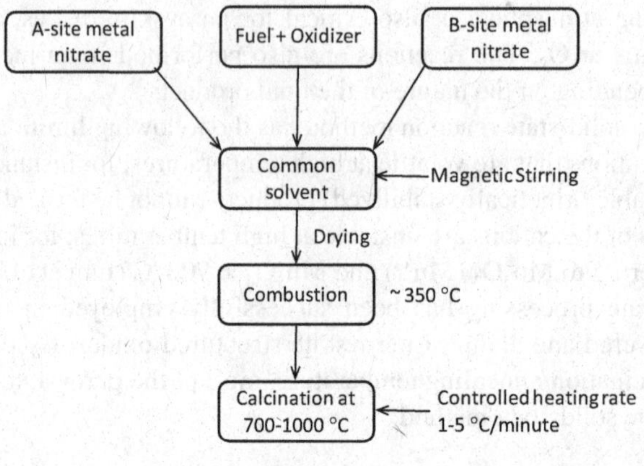

FIGURE 2.2 Flow chart of combustion synthesis of perovskite oxides.

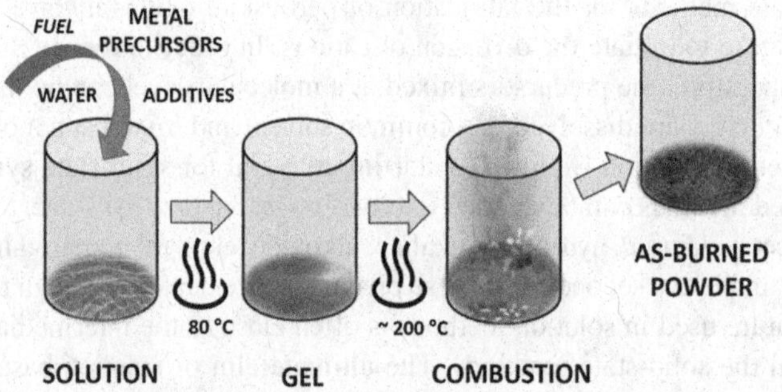

FIGURE 2.3 Schematic of steps in a combustion process, reproduced with permission of the Elsevier (Deganello 2017).

fast, and self-sustaining chemical reactions between metal salts and suitable organic fuels. As a result, most of the heat required for the production of oxide is supplied by the synergic reaction; and the mixture of the reactants only needs heating up to a temperature that is significantly lower than the actual phase formation. The highly exothermic and self-sustaining reaction in a combustion synthesis can be achieved using metal nitrates as precursors (since the nitrate groups act as oxidizing agents), and an organic fuel produces CO_2, H_2O, and N_2 during the combustion reaction (Civera et al. 2003). The organic fuel molecules such as glycine, citric acid, urea, etc. play double roles: they react with the precursors, especially metal nitrates, and form complexes with metal cations to improve the homogeneity, and at the same time avoid the precipitation of new species (Specchia et al. 2004).

The precursors that are susceptible to exothermic reactions can lead to a self-propagating combustion reaction, and local temperature in the reaction zone can be considerably higher than the actual temperature used in the process. As a result, the reaction can occur rapidly at a lower operating temperature. Ammonium dichromate is one such precursor that undergoes exothermic decomposition to Cr_2O_3, N_2, and H_2O. Ammonium dichromate, along with a small quantity of fuel, glycine, is successfully used for the synthesis of $LnCrO_3$ (Ln = rare earth elements) perovskites at

250°C (Kingsley and Pederson 1993). Metallic materials such as gallium, magnesium, and aluminum are also capable of undergoing self-propagating combustion reactions because of their ability to undergo exothermic oxidation under high temperatures. In such cases, an oxidizing agent, such as sodium perchlorate, is added to promote the oxidation process, e.g., $La_{0.7}Sr_{0.3}Ga_{0.7}Mg_{0.1}Fe_{0.2}O_3$ (Hsieh et al. 2009).

Besides heating, the combustion reaction can be triggered using electrical means. $LaFeO_3$ is prepared using benzoic acid as a fuel from the dried precursor pellets containing $La(NO_3)_3$, $Fe(NO_3)_3$, and citric acid. The pellets are fired in a stainless steel calorimetric bomb filled with oxygen. The reaction was then triggered by means of an electric circuit, and $LaFeO_3$ is rapidly synthesized (Milt et al. 1995). There are several organic molecules that can be a potential fuel for the combustion synthesis of perovskite, and some of them are listed in Table 2.1.

TABLE 2.1
Representative Perovskite Oxides Synthesized by Combustion

Perovskite	Fuel	Oxidizer	Application	References
$LaMO_3$ (M = Fe, Co, Mn)	Glycine	Metal nitrate	Electrocatalyst for metal-air battery	Zhu et al. (2013)
$LaMnO_3$	Glycine, alanine, or glycerol	Metal nitrate	Catalyst for methane oxidation	Specchia et al. (2004)
$LnCrO_3$ (Ln = La, Pr, Nd, Sm, Dy, Gd, and Y)	Tetraformal trisazine	Metal nitrate	Catalyst	Manoharan and Patil (1993)
$LaMO_3$ (M = Mn, Cr, and Al), $La_{0.67}Sr_{0.33}MnO_3$, and $La_{0.67}Ca_{0.33}MnO_3$	Hexamethylenetetramine	Metal nitrate	-	Prakash et al. (2002)
$MTiO_3$, $MZrO_3$ (where M = Ca, Sr, Ba, and Pb),	Tetraformyl triazine	Metal nitrate	Dielectric	Sekar and Patil (1992)
$Sr(Fe_{0.5}Ti_{0.5})O_3$	Urea	Metal nitrate	Electrochemical applications	Fumo et al. (1997)
$La_{1-x}Sr_xM_{1-y}Co_yO_3$	Alanine	Metal nitrate	Cathode of intermediate temperature solid oxide fuel cells	Dutta et al. (2009)
$LaFeO_3$	Glycine	Metal nitrate	-	Wang et al. (2006)
	Citrate, urea, sucrose, egg whites, gelatin, or chitosan			Gabal et al. (2019)
	Urea	Metal nitrate	Diesel engine exhaust gas purification	Dhal et al. (2017)
	Cyano complex precursor	-	Oxygen separation	Sánchez-Rodríguez et al. (2017)

(Continued)

TABLE 2.1 (*Continued*)
Representative Perovskite Oxides Synthesized by Combustion

Perovskite	Fuel	Oxidizer	Application	References
$CaMnO_3$ and $CaTiO_3$	Citric acid	Metal nitrate	-	Vieten et al. (2018)
Sm^{3+}- and Eu^{3+}-doped $BaZrO_3$	Citric acid	Ammonium nitrate	Luminescent material	Gupta et al. (2016)
$CaTiO_3$	Tetraformyl trisazine	Metal nitrate	Nuclear waste immobilization	Muthuraman et al. (1994)
$A_{0.5}Sr_{0.5}MnO_{3-\delta}$ (A = La, Nd, Sm, Gd, Tb, Pr, Dy, and Y)	Glycine	Metal nitrate	Thermochemical splitting of CO_2	Takalkar et al. (2019)
$SrFe_{0.8}Ti_{0.2}O_{3-\delta}$	Glycine	Metal nitrate	Cathode for solid oxide fuel cells	Baharuddin et al. (2019)
Tb^{3+}-doped $CaZrO_3$	Urea	Metal nitrate	Luminescent material	Singh et al. (2012)

2.2 LIQUID-PHASE REACTIONS

2.2.1 Coprecipitation

Precipitation involves the formation of intended materials by the nucleation and subsequent growth in a solvent medium. Therefore, solvent or liquid medium is chosen according to the solubility of precursor salts, and the obtained products must be insoluble in the medium. The number of nucleation sites determines the growth of uniform particles, and the use of a suitable capping agent controls the growth. Else, the growth continues to form large particles by means of Ostwald ripening processes, which may ultimately result in aggregates with non-uniform particles in terms of size, shape, morphology, and properties. Interestingly, the shape of the structures obtained using coprecipitation is determined by the crystal structure of the perovskite materials. For instance, perovskite structure with a cubic symmetry adopts a cubic shape, hexagonal or trigonal symmetry as rods, and orthorhombic structure as spheres. Figure 2.4 shows the representative examples of coprecipitated perovskite particles belonging to different crystal structures.

A typical coprecipitation reaction process is schematically shown in Figure 2.5. In a coprecipitation method, the metal salts are dissolved in a suitable solvent, such that the metal salt is supersaturated in the solvent. The solution is then mixed with precipitating agents like cyanide, oxalate, carbonate, citrate, hydroxides, etc. to precipitate a homogenous single-phase material. The obtained precipitate is then washed thoroughly and heated at a high temperature to obtain the perovskite oxide materials with a high surface area. The temperature and pH are adjusted as required upon mixing the precipitating agents to control the morphology and particle size. The precursors can be chloride, nitrate, or acetate, and the solvent can be distilled water,

FIGURE 2.4 Coprecipitated perovskite particles with different crystal structures (a) cubic $SrTiO_3$, reproduced with permission of the American Chemical Society (Calderone et al. 2006), (b) hexagonal $LaAlO_3$, and (c) orthorhombic $LaFeO_3$ (Haron et al. 2018).

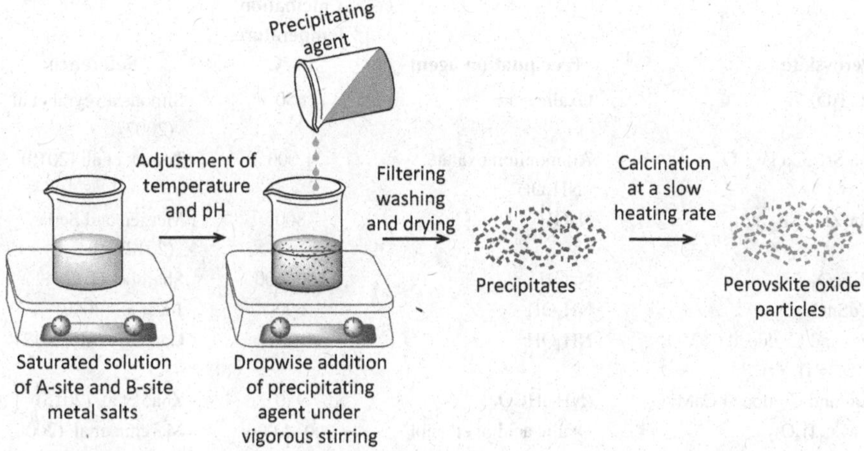

FIGURE 2.5 Steps in the co-precipitation process for the fabrication of perovskite oxides.

ethanol, cyclohexane, *N*,*N*-dimethyl formaldehyde, or a mixture of them which can completely dissolve the precursors.

Many perovskite oxides obtained through coprecipitation are nanoscale-microscale structures, depending on the precipitating agents. A strong precipitating agent, such as NaOH, result in larger particles than the ones prepared using a weak precipitating agent, like ammonium carbonate (Varandili et al. 2018). Lanthanum-based perovskite materials are the most widely studied perovskite materials using precipitation methods. For example, nanostructured $LaMO_3$ (M = Al, Co, Fe) can be fabricated using K_2CO_3 and NaOH as precipitating agents. In the representative synthesis process, the nitrate salts of La and the B-site elements are dissolved in distilled water, followed by the addition of K_2CO_3 (2 M) and a small amount of NaOH (1 M) solution. The obtained precipitate is then filtered, washed with distilled water, and dried, and then calcined at 900°C (Haron et al. 2018).

There are different precipitating agents and techniques for the fabrication of perovskite oxide materials, which can result in the formation of different sizes and shapes. For instance, $LaCoO_3$ synthesized by precipitating the nitrate salts with

NaOH at 60°C followed by calcination at 600°C–800°C, resulted in the formation of $LaCoO_3$ nanorods (Junwu et al. 2007). Occasionally, the precursor metal salt themselves act as a precipitating agent as in the case of $La_{1-x}Ni_xVO_3$ synthesis, where sodium metavanadate, the source of vanadium, serves as the precipitating agent (Rajmohan et al. 2016). Table 2.2 lists some important perovskite oxides fabricated through coprecipitation.

TABLE 2.2
Representative Perovskite Oxides Synthesized by Coprecipitation

Perovskite	Precipitation agent	Calcination Temperature in °C	Reference
$BaTiO_3$	Oxalic acid	650	Simon-Seveyrat et al. (2007)
$Ba_xSr_{1-x}Co_yFe_{1-y}O_3$	Ammonium oxalate, NH_4OH	1,000	Toprak et al. (2010)
$Ba(Zn_{1/3}Nb_{2/3})O_3$	NaOH	800	Mergen and Sert (2012)
$BiFeO_3$	NaOH	400–600	Shami et al. (2011)
$CdSnO_3$	NH_4OH	600	Jia et al. (2009)
Fe- and Cr-doped $CaMO_3$ (M = Ti, Zr)	NH_4OH	1,000	Gargori et al. (2012)
La- and Co-doped $CaMnO_3$	$(NH_4)HCO_3$	950	Zhao et al. (2015)
$CaCu_3Ti_4O_{12}$	Oxalic acid in ethanol	900–1,000	Marchin et al. (2008)
$LaAl_{1-x}Ni_xO_{3-\delta}$	NaOH	700	Djoudi and Omari (2015)
$LaCoO_3$	Ammonium oxalate	1,000	Nakayama et al. (2003)
$LaFeO_3$	NH_4OH, NaOH, and $(NH_4)_2CO_3$	800–1,000	Varandili et al. (2018)
Ag-doped $LaMO_3$ (M = Co, Mn, Ni)	Na_2CO_3	750	Choudhary et al. (1996)
$LaCr_{1-x}Co_xO_3$	NaOH	800	Madoui and Omari (2016)
$LaNiO_3$	Oxalic acid in alcohol	600	Takahashi et al. (1990)
$La_{1-x}Ca_xFeO_3$	$(NH_4)_2CO_3$	900	Abdel-Khalek and Mohamed (2013)
$LaMnO_3$	$(NH_4)_2CO_3$	1,350	Pelosato et al. (2013)
$LaRuO_3$	NH_4OH	800	Labhsetwar et al. (2003)
$La_{0.8}Sr_{0.2}Ga_{0.8}Mg_{0.2}O_{3-\delta}$	NH_4OH	600	Pelosato et al. (2010)
Ag-doped $LaFeO_3$ and $LaFe_{0.5}Co_{0.5}O_3$	Na_2CO_3	750	Choudhary et al. (1999)

(Continued)

TABLE 2.2 (*Continued*)
Representative Perovskite Oxides Synthesized by Coprecipitation

Perovskite	Precipitation agent	Calcination Temperature in °C	Reference
Sr- and Mg-doped $LaGaO_3$	$(NH_4)_2CO_3$ or $(NH_4)HCO_3$	1,100	Chae et al. (2008)
$La_{2(1-x)}NiSr_{2x}O_4$	NaOH	700	Kao and Jeng (2000)
$La(Cr,Fe,Mn)O_3$	NaOH	700	Fabian et al. (2015)
$LaBO_3$ (B = Mn, Fe, and Co)	Tetramethylammonium hydroxide	950	Einaga et al. (2016)
$MSnO_3$ (M = Zn, Cd, Ni)	NH_4OH	600	Wu et al. (2002)
$Pb(Fe_{0.5}Nb_{0.5})O_3$	NH_4OH	800	Tang et al. (2007)
$Pb(Zr_{0.52}Ti_{0.48})O_3$	NH_4OH	500–700	Xu et al. (2005)
$(Pb,La)(Zr,Sn,Ti)O_3$	NH_4OH	600	Yang et al. (2011)

2.2.2 Hydrothermal/Solvothermal Synthesis

The hydrothermal reaction represents any heterogeneous chemical reaction occurring in the presence of a solvent media above atmospheric pressure and temperature. The term "hydrothermal" is generally used if the solvent is water, and "solvothermal" if the reaction is carried out using any other solvent. Frequently, such reactions are performed in air-tight constant volume vessels called hydrothermal autoclave reactors. During hydrothermal reactions, the reactors are heated to a fixed temperature above the boiling temperature of the solvent, and the pressure builds up autogenously in the reactor by the vapor pressure of the solvent. Under such conditions, solid phases can crystallize from the reactants at the specific temperature and composition from the hydrothermal solution. Therefore, any hydrothermal process can be distinguished as a combination of two steps: the first step is the preparation of the precursor solution in a liquid phase with the accurate stoichiometric composition of elements, and the second step is the heat treatment that determines the morphology.

The hydrothermal synthesis of perovskite oxides can be classified into two methods. In the first method, reactants and a mineralization agent are dissolved in water. The above solution is then heated beyond the supercritical temperature of the water, >350°C. In the second method, autoclave vessels lined with Teflon are used; therefore, the temperature conditions for using such reactors are limited to 200°C with pressures less than 10 MPa. Under severe reaction conditions that involve highly concentrated acidic or basic solutions, metallic reactor materials are not recommended due to possible rapid corrosion. The hydrothermal method involves the dynamic interaction of processing parameters such as solvent type, temperature, and pressure that govern the chemical reactions by controlling ionic mobility. During the reaction, the particles are grown from liquids, and the size and shape of the particles depend on the number of nuclei formations and the subsequent particle growth in the controlled atmosphere. The nucleation rate and growth can be significantly accelerated by the supersaturation of the reactant solutions. The presence of surfactants, adjusting pH, and growth inhibitors can control the morphology of the particles obtained using the hydrothermal process. The simultaneous precipitation, reduction, oxidation, hydrolysis, etc. also determines the morphology and composition of the final products.

The hydrothermal method is the simplest processing method that does not require any specialized instruments and complicated processes; nonetheless, the method allows good control over particle size, chemical composition, phase, and morphology of the resultant products. Hydrothermal processes are often combined with microwave, reverse micelle, microemulsion, colloids, or sol–gel processes in order to obtain characteristic nanostructures. The hydrothermal process followed by high-temperature annealing is a prerequisite for the obtention of perovskite nanostructures.

The generalized synthesis of oxides using a hydrothermal process can be described as follows. The stoichiometric quantities of the precursor salts are dissolved in a suitable solvent, sometimes with the assistance of an acid, preferably at room temperature. Capping agents or surfactants are added to the above solution to control the growth or morphology. Capping agents also stabilize the nanostructured materials from agglomeration. Finally, an oxidizing or reducing agent is also added to the precursor solutions to accelerate precipitation. Then, the reaction is further carried out in hydrothermal reaction vessels to obtain the precursor for the perovskite oxides. The obtained product is then calcined at an elevated temperature attained at a controlled heating rate for several hours to obtain the final perovskite material. Many single and double perovskite nanomaterials are successfully synthesized using hydrothermal methods, and the reactants and reaction conditions are unique for each perovskite oxide. Hydrothermal processing of perovskites using templates is also successfully employed for the fabrication of predetermined structures. For instance, during the hydrothermal processing of titanates, TiO_2 templates are used for the formation of $ATiO_3$ (Ba, Sr, Ca, etc.) structured nanowire, nanotubes, etc. (Mao et al. 2003). Some important hydrothermally synthesized perovskite nanomaterials are listed in Table 2.3.

2.2.3 Sol–Gel Processing

Sol–gel technique is a "bottom-up" process, in which the particle or porous structures are assembled from molecular or elemental components, like the construction of a wall using bricks and mortar. By definition, "sol" is a suspension of colloidal particles or molecules in a liquid, and "gel" is a three-dimensional network formed when the sol is mixed with a liquid, which can help to form the network. Advanced ceramic materials such as perovskites with high purity, homogeneity, and accurate composition can be obtained using sol–gel processing. In many bottom-up approaches, for the fabrication of nanomaterials, building blocks are molecules, molecular complexes, atoms, or aggregates, whose sizes vary from tenths of a nanometer to tens of nanometers, and the mortar is their assembling technique. Sol–gel process is one such technique to assemble the molecular building blocks. The sol–gel process needs the inclusion of polymers or in-situ polymerization of organic or inorganic entities, which essentially contribute to the final composition of the materials (Niederberger and Pinna 2009).

TABLE 2.3
Hydrothermally Processed Perovskite Oxides

Perovskite	Control	Hydrothermal Reaction Condition	Calcination Temperature	Morphology	Reference
$BaTiO_3$	KOH	200°C for 6h	-	Nanostructured dendrites	Wang et al. (2009)
	NaOH	40°C–60°C for 4–72h	-	Films	Slamovich and Aksay (1996)
	TiO_2 nanotubes, NaOH	150°C, 48h	-	Nanocrystals	Maxim et al. (2008)
$(Ba,Sr)TiO_3$	EDTA/KOH	90°C 72h	-	Submicron particles	Gersten et al. (2004)
$BaSnO_3$	KOH	250°C for 6h	1,650°C for 30min	Nanoparticles	Lu and Schmidt (2008)
$BaZrO_3$	NaOH	450°C, 31h	-	Nanoparticles	Aimable et al. (2008)
$BiFeO_3$	KOH and polyethylene glycol	200°C, 3 days	-	Micropills, rods	Fei et al. (2011)
	KOH	200°C for 6h	900°C in nitrogen ambient	Nanoparticles	Chen et al. (2006)
	NH_4OH as precipitant and NaOH as mineralizer	180°C for 12h	-	Flakes	Miao et al. (2008)
	KOH	120°C for 8h	-	Microspheres	Sun et al. (2013)
$CaTiO_3$	Urea	180°C for 4h	450°C for 4h	Rectangular prism	Lozano-Sánchez et al. (2013)
$CaSnO_3$	PVP	140°C, 10h	500°C for 5h	Micro-/nanocubes	Lu et al. (2004)

(Continued)

TABLE 2.3 (*Continued*)
Hydrothermally Processed Perovskite Oxides

Perovskite	Control	Hydrothermal Reaction Condition	Calcination Temperature	Morphology	Reference
$KTaO_3$	Water–ethanol and water–hexane systems/ KOH	160°C, 24 h	-	Nanocubes	He et al. (2004)
$LaFeO_3$	Citric acid	180°C, 24 h	800°C for 2 h	Microsphere composed of nanoparticles	Thirumalairajan et al. (2012)
$LaCrO_3$	KOH	260°C, 7 days	-	Microparticles	Zheng et al. (1999)
$La_{0.5}Ba_{0.5}MnO_3$	KOH	270°C for 25 h	-	Nanowires	Zhu et al. (2002)
Sr-doped $LaCrO_3$	KOH	260°C, 4 days	-	Cubes	Wang et al. (2013b)
$LiNbO_3$	Benzyl alcohol	220°C, 4 days	-	Nanoparticles	Modeshia et al. (2009)
$La_{1-x}Sr_xMnO_3$	NaOH	220°C, 24 h	-	Nanowires, dendrites	Makovec et al. (2013)
$NaNbO_3$	NaOH	240°C, 3 h		Microparticles	Modeshia et al. (2009)
$AtaO_3$ and $AnbO_3$ (A = Na and K)	NaOH or KOH	240°C, 24 h	-	Cubes	Liu et al. (2007)
$(Na_{0.8}K_{0.2})_{0.5}Bi_{0.5}TiO_3$	NaOH	160°C, 48 h	600°C	Nanowires	Hou et al. (2007)
$REFe_{0.5}Cr_{0.5}O_3$ (RE = La, Tb, Ho, Er, Yb, Lu and Y)	KOH	240°C, 5 days	-	Cubes, plates, spheres	Yuan et al. (2014)
$PbTiO_3$	KOH	200°C, 1 h	-	Film	Cho and Yoshimura (1997)
$Pb(Zr_{0.5},Ti_{0.5})O_3$	Hydroxypropyl cellulose	220°C, 30 h	-	Spherical powders	Choi et al. (1998)
$Pr_{1-x}Ca_xMnO_3$	KOH	240°C, 3 days	-	Microcubes	Chen et al. (2007c)
$RECrO_3$ (RE = La, Pr, Nd, Sm)	KOH	240°C–260°C, 3–5 days	800°C	Micocubes	Wang et al. (2015)

(*Continued*)

TABLE 2.3 (*Continued*)
Hydrothermally Processed Perovskite Oxides

Perovskite	Control	Hydrothermal Reaction Condition	Calcination Temperature	Morphology	Reference
$RCrO_3$ (R = La, Pr, Sm, Gd, Dy, Ho, Yb, and Lu) and $YCrO_3$	KOH	300°C, 24 h,	-	Submicrometer particle	Sardar et al. (2011)
$RMnO_3$ (R = Sm–Ho)	KOH	240°C, 3 days	-	Microparticles	Chen et al. (2007b)
$SrTiO_3$	NaOH	130°C, 24 h	-	Nanocubes	Huang et al. (2014)
	PVA/KOH	200°C, 12h		Nanoparticles	Wei et al. (2008)
$SmFeO_3$	KOH	240°C, 3 days	700°C, 24 h	Microcubes	Zhang et al. (2016)
$SrTiO_3$	NH_4OH	170°C, 3 days	-	Nanowires	Joshi and Lee (2005)
$YAlO_3$	NH_4OH	600°C, 120 h	-	Crystals	Basavalingu et al. (2008)
$ZnSnO_3$	NaOH	200°C, 24 h		Microparticles	Xu et al. (2006)
Ba_2SbMO_6	KOH	240°C–260°C, 5–7 days	-	Microcubes	Zheng et al. (1998)
Ba_2YSbO_6	H_2O_2 and KOH	240°C–260°C, 7 days		Nanoparticles	Wu et al. (2006)
Y_2NiMnO_6	KOH	260°C, 3 days	500–1,000°C	Microcrystals	Zhang et al. (2014)

In general, sol–gel processing has four important steps: (1) mixing of molecules, during which the precursor chemicals for the intended ceramic material, a gel-forming medium (or a polymer), a solvent, and a catalyst are mixed together; (2) chemical treatment, which will accelerate the polymerization of the mixture to form a gel; (3) shaping of the polymeric gel from the previous step to the final morphology; and (4) high-temperature treatment to remove the volatile and organic phases from the shaped amorphous gel or xerogel to crystalline ceramic material of required shape. Spinning, freeze-drying, etc. are the shaping process to transform the polymeric sol to a gel before it is calcined at a high temperature. If a dense material is desired, drying followed by hot or cold isostatic pressing, and then sintering is required at a higher temperature. The precursor for a sol–gel process is a metal salt or a ligand. Metal salts, such as nitrates, acetates, formates, citrates, and tartarates, which are soluble in a wide range of solvents, are commonly used, and the more thermally stable salts such as chlorides and sulfates are avoided. Metal salts provide a viable source of oxides in sol–gel processing because they are readily converted to the oxide by thermal or oxidative decomposition. The sol–gel processing requires a temperature lower than that for a solid-state process since the mixing of reagents takes place at an atomic/molecular level, and the interdiffusion of ions from one phase into the other is much easier.

In a typical sol–gel process, the metal salts and a complexing medium are subsequently polymerized to form a hybrid material. Based on the processing of the polymeric gels obtained after hydrolysis, different morphologies can be obtained, as shown in Figure 2.6. A scanning electron microscope (SEM) image of $Sr_2Co_2O_5$ double perovskite obtained through the sol–gel process without any shaping of gel is shown in Figure 2.7.

The homogeneity of the solutions is important in a sol–gel process. The metal salts should be uniformly dispersed, and subsequent gelation should then freeze all elements in the gel network (Livage et al. 1988). In the sol–gel synthesis of perovskite oxides, Pechini's process is widely used, and the conventional Pechini's

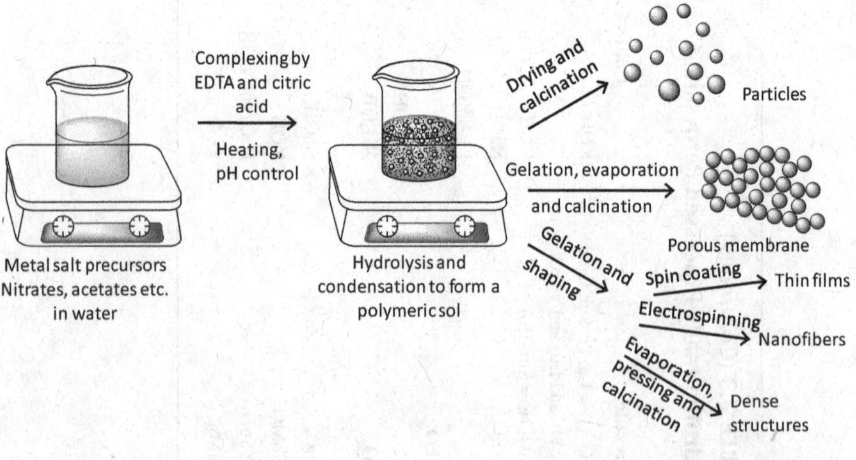

FIGURE 2.6 Steps in sol–gel process to achieve different final morphologies.

FIGURE 2.7 The SEM image of $Sr_2Co_2O_5$ obtained using sol–gel processing.

process is shown in Figure 2.8. In the typical fabrication of perovskite materials, a modified Pechini's process is used, where the metallic ions are complexed using ethylenediaminetetraacetic (EDTA) and citric acid, and ethylene glycol is added as a stabilizing agent. The pH of the solution is maintained between 8 and 11

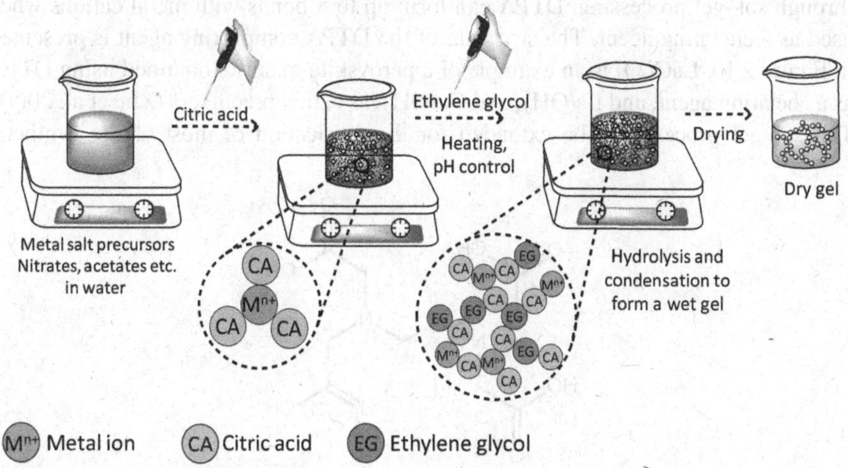

FIGURE 2.8 Schematic representation of *Pechini's* process.

using ammonium hydroxide. The electron-donating carboxylic and aliphatic amine groups of EDTA are linked to metal cation by six bonds and the citric acid links with the metal cation by three bonds. The structure of the EDTA is shown in Figure 2.9. Chelating agents like EDTA, citric acid, glucose, etc. arrest the movement of the metallic ions in the sol and prevent the partial segregation of cations in the final compound, possible by the interaction between the metallic ions. The addition of ethylene glycol or mannitol (polyhydroxy alcohols), under heating, will lead to poly-esterification of the chelates, forming a cross-linked network of metal atoms and organic radicals. All the above processes take place in a basic medium, and thus a $pH>7$ must be maintained during the reaction.

Since the number of chelating molecules used is more than the number of metal ions, all the metallic ions undergo complexation; as a result, a stable chelate complex system is formed, which is important to avoid the precipitation of any new phases. The addition of ethylene glycol promotes polymerization, which facilitates homogeneity. The major role of citric acid is to form a polymeric network during the condensation reactions. As like EDTA, glycine can be used as the chelating agent, since glycine contains carboxylate and aliphatic amine groups. Both carboxylate and aliphatic amine groups can participate in the complexation of metal ions; therefore, glycine has a stronger chelating ability than citric acid, which contains only carboxylate groups. Therefore, the amount of glycine required for the fabrication of perovskites is less than that of citric acid (Liu et al. 2002). Additionally, glycine and citric acid undergo exothermic pyrolysis, and thereby, increasing the temperature in the reaction zone than the surrounding temperature during the calcination. The Pechini method is a popular method adopted by many researchers working on perovskite oxides because of its simplicity and the great control on the composition of the perovskite structures, in addition to the lower processing temperature requirements.

Diethylenetriaminepentaacetic acid (DTPA) is another chelating agent with a similar structure of EDTA, which is also used in the fabrication of perovskite oxides through sol–gel processing. DTPA can form up to 8 bonds with metal cations when used as a chelating agent. The structure of the DTPA complexing agent is presented in Figure 2.10. $LaCoO_3$ is an example of a perovskite material obtained using DTPA as a chelating agent, and $La(OH)_3$ and $Co(OH)_2$ as cation precursors (Zhu et al. 2000). The sol–gel process can be extended for the fabrication of most of the synthetic

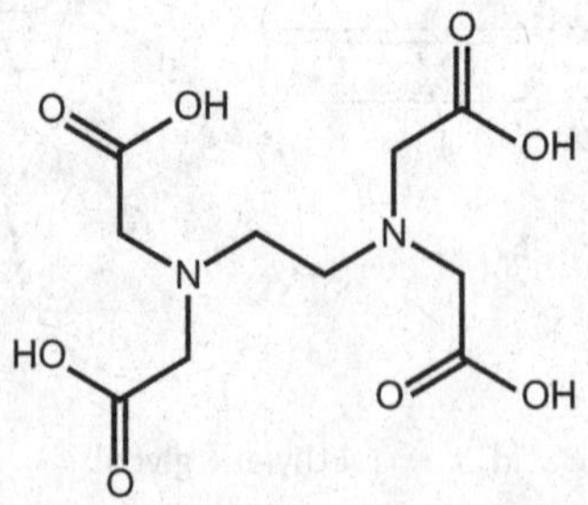

FIGURE 2.9 Structure of ethylenediaminetetraacetic (EDTA).

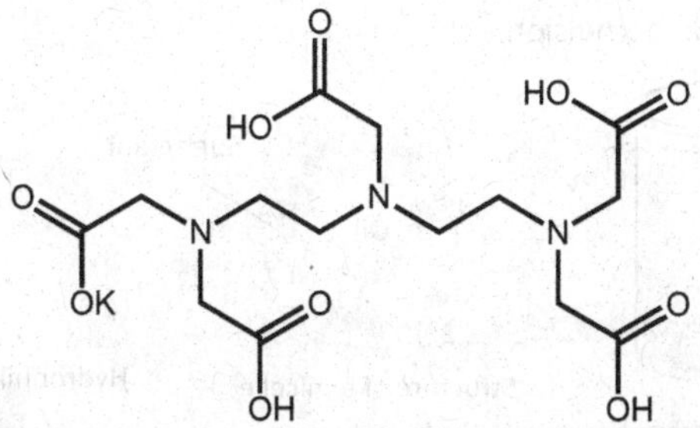

FIGURE 2.10 Structure of diethylenetriaminepentaacetic acid (DTPA).

perovskite oxides. Moreover, the polymeric sol is also used as the precursor for various shaping processes such as spin casting, freeze-drying, electrospinning, etc.

2.2.4 Microemulsion

Microemulsions are thermodynamically stable isotropic dispersion of two immiscible liquids, stabilized by an interfacial film of surfactant molecules. Unlike emulsion with macroscopic domains, domains in microemulsion are microscopic, with size 100 times smaller than that of emulsions, typically in the range of 1–100 nm. Such nanosized dispersions are called a micelle, and the monolayer of surfactant molecules provide a consistent shape and dimensions. The microemulsions are composed of at least three components: a dispersed non-polar solvent, which is in a continuous polar solvent (water) in the liquid state (or *vice versa* for reverse micelle), and a surfactant. The dispersed phases always carry the soluble precursors of the end product. In reverse microemulsions, in addition to the surfactant with long-chain hydrocarbons typically with 8–18 carbon atoms, co-surfactants (short-chain aliphatic alcohols) are also added.

The schematic representations of the micelles and reverse micelles are shown in Figure 2.11. Surfactants are molecules with amphiphilic (polar) functional groups on one end, and a hydrophobic tail. The hydrophobic hydrocarbon chains weakly interact with the polar solvent, while the polar functional groups interact strongly with the polar solvent. Apparently, the surfactant molecules are oriented with the hydrophobic tail towards the non-polar phase (oil), and the hydrophilic part oriented towards the polar phase, forming a clearly distinguishable boundary between the above phases. As a result, a microscopic microheterogeneous domains of oil-in-water (O/W) or water-in-oil droplets (W/O) forms. The microscopic emulsion O/W or W/O acts as tiny reactors which allow the nucleation and growth of the nanostructured products in a confined space.

The conventional formation of micelle or reverse micelle microemulsions is associated with the hydrophobic and hydrophilic interaction between dispersed and continuous phases. A possible schematic of nanostructured perovskite synthesis through the microemulsion route is shown in Figure 2.12. The formation of micelles is a function of surfactant concentration, which is determined by the critical micelle concentration. The microemulsion method is recognized for the synthesis of inorganic

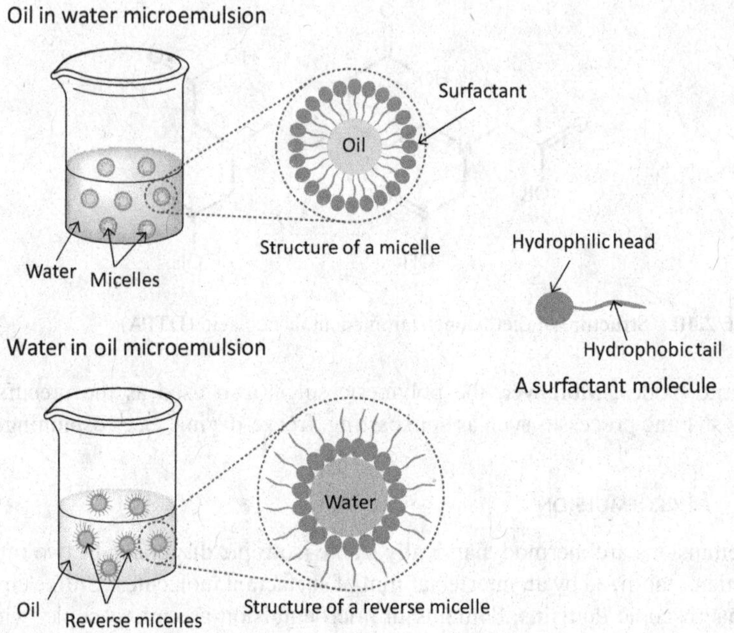

FIGURE 2.11 Structure of microemulsions.

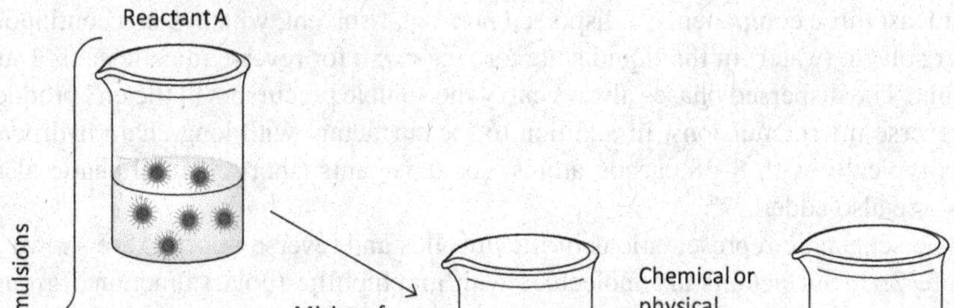

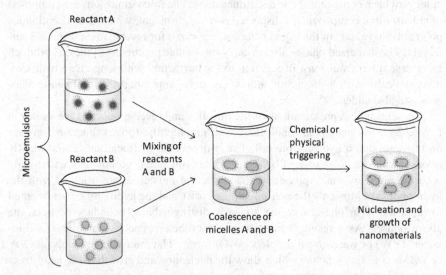

FIGURE 2.12 Microemulsion synthesis of perovskite oxide nanostructures.

nanoparticles such as metals, metal oxides, and other inorganic materials. The major advantage of microemulsions synthesis is the homogeneity and monodispersion of nanoparticles. In the microemulsion-mediated synthesis of nanostructured perovskites, two identical solutions containing the continuous solvent, a surfactant, and a co-surfactant are prepared and mixed thoroughly. Then, the salts of A- and B-site cations are dissolved in suitable solvents separately. The precursor solutions are then added separately to the formerly prepared solvent solution at a controlled rate to form the respective microemulsions. Finally, both the microemulsions are mixed thoroughly, during which the micelles of A- and B-site precursors coalescent together to form a larger micelle. The nucleation and the subsequent growth to nanostructured materials in the nanosized micelles are either triggered by chemical means through the introduction of additional reagents or physical means *via* temperature, pressure, microwave, etc.

Many perovskite oxides with size in the nanoscale domain are widely synthesized by the reverse micelle technique. The reverse microemulsion technique is very sensitive to the amount of water present in the reaction mixture. If the amount of water is very high, the water droplets join together to form a continuous phase resulting in large particles. The change in particle size obtained using a reverse micelle, and a continuous phase is depicted in Figure 2.13. In some cases, the emulsions are used as a shaping process of the precursors, which are subsequently calcined at a high temperature to obtain the intended perovskite structures (Afsharikia et al. 2017). Table 2.4 lists a few examples of perovskite nanostructures obtained using microemulsion synthesis.

2.2.5 Template-Assisted Synthesis

In template-assisted synthesis, a hard or soft template is used to achieve nanosized perovskite structures. Porous anodic alumina, porous silica, carbon, etc. are few examples for hard templates, and amphiphilic molecules such as block copolymers and surfactants, for soft templates. When a hard template is used, the sol containing A- and B-site precursors is prepared first, which is then poured onto the porous structure. For example, $PbTiO_3$ and $BaTiO_3$ nanotubes can be synthesized

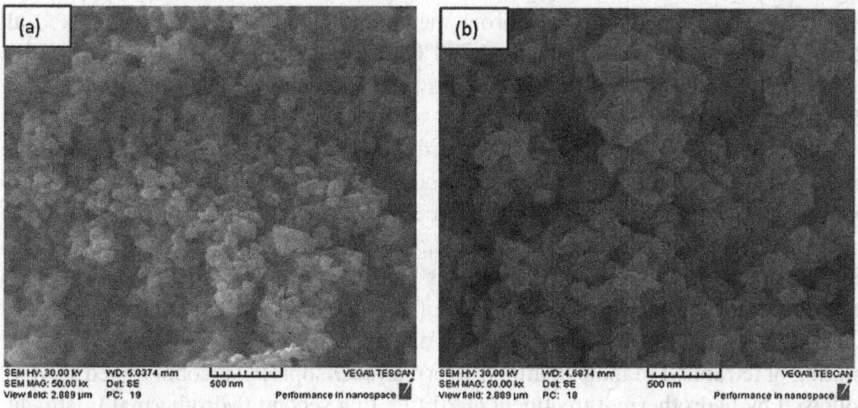

FIGURE 2.13 SEM images of the $LaMn_xV_{1-x}O_{3\pm\delta}$ perovskites prepared by the (a) reverse and (b) continuous microemulsions, reproduced with permission of the Elsevier (Afsharikia et al. 2017).

TABLE 2.4
Representative Perovskite Oxides Synthesized by Microemulsion

Perovskite	Surfactant	Nature of Surfactant	Oil Phase	Reference
Zr- and Mg-doped $BiFeO_3$	Cetlytrimethylammonium bromide	Cationic	-	Sharif et al. (2016)
$Ca_{0.5}Sr_{0.5}MnO_3$	Alkyl polyglycol ether	Non-ionic	-	López-Trosell and Schomäcker (2006)
$LaFeO_3$	Dioctyl sulfosuccinate	Anionic	Isooctane	Abazari and Sanati (2013)
$LaNiO_3$	Cetrimonium bromide/1-butanol	Cationic	Cyclohexane	Aman et al. (2011)
$La_{1-x}Sr_xMnO_3$	Cetrimonium bromide/1-hexanol	Cationic	-	Usković and Drofenik (2006)
$LaMnO_3$ and $LaFeO_3$	Cetrimonium bromide /1-butanol	Cationic	N-octane	Giannakas et al. (2004)
$Nd_{0.67}Sr_{0.33}CoO_{3-\delta}$	Cetrimonium bromide /1-butanol	Cationic	2,2,4-Trimethylpentane	Lim et al. (2018)
$SrCoO_{3\pm\delta}$	Alkyl polyglycolethers/ octoxinol-9	Non-ionic	Cyclohexane	Langfeld et al. (2011)
$SrTiO_3$	Oleic acid	Anionic	-	Hu et al. (2013)

using commercial anodic alumina membranes. During the synthesis, a sol containing metal (barium or lead) acetate and titanium (IV) tetrabutoxide is prepared first, which is then dropped onto the porous membrane taped on one end (Rørvik et al. 2009). The template is then heated at 700°C for converting the sol to the respective oxide nanotubes, and then the template is removed using NaOH (Hernandez et al. 2002).

Similarly, mesoporous silica-based hard templates (e.g., SBA-15 and KIT-6) are also used for the fabrication of mesoporous perovskite oxide structures without a well-defined morphology, but a high surface area, e.g., $LaCoO_3$ (Nguyen et al. 2002), $LaNiO_3$, $LaFeO_3$ (Zhao et al. 2013), $La_{0.8}Ca_{0.2}NiO_3$, $La_{0.8}Ca_{0.2}Ni_{0.6}Co_{0.4}O_3$ (Rivas et al. 2010), etc. The reactant themselves can be a template, for instance, $SrTiO_3$ nanocubes are fabricated on the surface of flower-like layered titanate hierarchical spheres as templates, which is the source of Ti. These spheres are fabricated by the mixing of tetrabutyl titanate in dimethylformamide/isopropyl alcohol mixed solvent, followed by hydrothermal treatment at 200°C. In a second hydrothermal treatment, the spheres are heated at 150°C–180°C for 10h in an aqueous solution containing NaOH, cetyltrimethylammonium bromide (CTAB), and $SrCl_2$ (Kuang and Yang 2013).

Eggshell membranes can be used as a porous template for the fabrication of perovskite oxides, e.g., $Sm_{0.5}Sr_{0.5}CoO_3$. The template membranes are separated from the eggshell by removing $CaCO_3$ by soaking them in a 1M HNO_3 solution for 5 min. The membranes are then washed, dried, and dipped into the precursor solution. The membranes saturated with metal ions are rinsed with deionized water and subsequently dried at 95°C. The membrane/metal salt composite is calcined at 1,000°C (Dong et al. 2011).

In a soft-template method, polymers and surfactants are used as the templates, which can be removed by calcination. The surfactants can be cationic, anionic, or non-ionic. Three-dimensionally ordered macroporous (3DOM) structured perovskite oxides are synthesized using monodisperse polymethyl methacrylate (PMMA) or polystyrene microsphere colloid crystal template (CCT) (Xu et al. 2014). In this technique, PMMA microspheres are synthesized using emulsifier-free emulsion polymerization, and the obtained microspheres are then centrifuged to form a CCT, as shown in Figure 2.14a and b. Then the precursor salt solution in ethylene glycol/methanol mixture is added to the CCT to fill the gaps between the spheres, which form a hard-inorganic precursor framework upon drying. When the CCT/precursor composites are calcined, the polymer degrades, and the precursor is converted to the respective oxide with 3DOM structure, as shown in Figure 2.14c. The perovskite structured catalysts $La_{1-x}Sr_xFeO_3$ (Sadakane et al. 2005), $LaCo_xFe_{1-x}O_3$ (Xu et al. 2010), and $LaFe_{0.7}Co_{0.3}O_3$ (Shen et al. 2016) are fabricated using CCT as a template. Carbon nanospheres obtained by hydrothermal treatment of glucose can be used as a template for the synthesis of hollow perovskite spheres (Sennu et al. 2017).

Templates with biological origin are also used for the synthesis of nanostructured perovskite materials. One such example is the use of genetically engineered M13 virus for the fabrication of $SrTiO_3$ and $BiFeO_3$ nanowires. To prepare the nanowires, at first, the precursor salts are dissolved in ethylene glycol, then mixed with the virus solution, to which NaOH is added and heated at 80°C for 4 h. The virus has an affinity to ethylene glycol attached to the precursors at a pH of 10. After incubating for a day, the virus attached to the precursors is separated and calcined at 600°C (Nuraje et al. 2012).

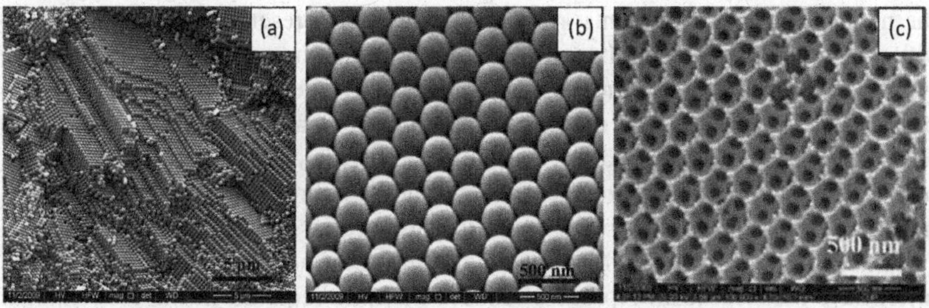

FIGURE 2.14 (a, b) Microsphere colloid crystal template and (c) 3DOM structured perovskite oxide, reproduced with permission of the Royal Society of Chemistry (Xu et al. 2014).

2.2.6 Microwave Synthesis

The microwave synthesis is a widely used technique for the bulk and rapid synthesis of organic and inorganic materials. During microwave synthesis, microwave radiation is applied to simulate the chemical reactions. Even though microwaves contain electric and magnetic fields, only electric fields contribute to the heating process. In such reactions, the microwaves employ high-frequency electric fields to heat the materials through dipolar polarization, dipole rotation, and ionic conduction. The dipole rotation takes place in the case of polar molecules during which the molecules try to align themselves with the alternating electric field of the microwave. Due to the rapid change in the electric field applied by the microwave, polar molecules attempt to align themselves to the field to create rotational motion. The rotational motion of the molecules generates heat by molecular interactions. In the case of materials containing free ions, the ions are rapidly oscillated by aligning themselves with the rapidly alternating electric field of the microwaves. During such rotation of the molecules, the energy is lost in the form of heat through dielectric loss.

Microwave heating of a material is based on its characteristic to absorb electromagnetic energy from the microwave and transform that into heat. Based on the interaction of the materials to the microwave, the materials can be classified into three groups. They are reflectors, transmitters, and absorbers of microwaves. A microwave reflector is a material that can reflect the microwave, and it is not effectively heated by microwaves. The best example of reflectors is metals. Microwave transmitters transmit the microwaves without any absorption, losses, or heat generation. Teflon, quartz, plastics, and some ceramics are few examples for transmitters, and the containers for chemical synthesis can be one among them. Microwave absorbers absorb energy from the microwaves and transform it into heat energy instantly; they are high-loss materials.

The conventional chemical reactions can be performed with the assistance of microwaves. Sol–gel, coprecipitation, microemulsion, hydrothermal, combustion, complexing, etc. are some examples of perovskite synthesis techniques that can be assisted by microwave heating. When used for the above chemical reactions, polar solvents such as dimethylformamide (DMF), acetonitrile, CH_2Cl_2, ethanol, and H_2O can be used. The non-polar solvents such as toluene, carbon tetrachloride, diethyl ether, and benzene are microwave-inactive (Sariah et al. 2012). In a microwave-assisted reaction, heating occurs by the direct interaction of microwave energy with the molecules, such as solvents, reagents, or catalysts present in the recipe, which ultimately result in the simultaneous heating of the entire reaction mixture. In the typical hotplates, the reaction progress by conductive heat transfer from the walls of the reaction vessel to the interior through the bombardment of molecules. The mechanism of microwave and the conventional heating of water molecules is compared in Figure 2.15. Therefore, the reaction time for microwave-assisted synthesis is more rapid than the conventional synthesis using heaters. However, in conventional synthesis, the reaction occurs irrespective of the polar nature of the reactants. The microwave-assisted heating process reduces the time required in the conventional process significantly. For instance, hydrothermal processing of perovskite requires 24–72 h of heating in a conventional heating method, whereas, when the microwave method is used, the reaction time is reduced to 10–30 min.

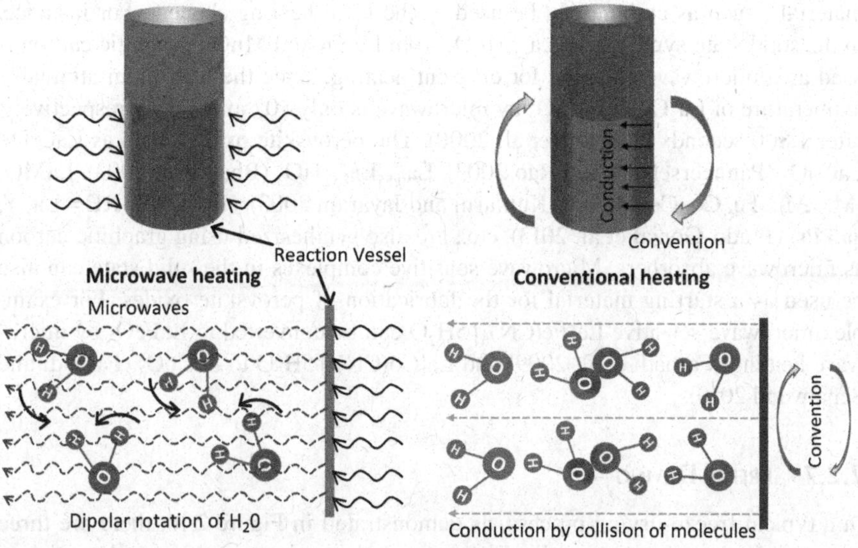

FIGURE 2.15 Microwave *vs.* conventional heating of water.

In sol–gel processing of perovskite oxide synthesis, microwave-assisted heating is often used. During the microwave-assisted sol–gel process, the microwaves are applied to convert the organic or carbon-rich gel precursors to the respective oxides. $LaMnO_3$ (Weifan et al. 2006), $LaCoO_3$ (Yi et al. 2005), $Sm_{0.5}Sr_{0.5}Co(Fe)O_{3-\delta}$ (Li et al. 2014b), $La_{0.8}Sr_{0.2}Co_{0.5}Fe_{0.5}O_3$ (Liu et al. 2007), $ZnTiO_3$ (Phani et al. 2007), $La_{0.7}Sr_{0.3}MnO_{3+\delta}$ (Ran et al. 2007), etc. are some perovskite materials synthesized by microwave-assisted sol–gel method.

In hydrothermal and coprecipitation processes, all the ingredients for the successful formation of perovskite materials (through hydrothermal or coprecipitation synthesis route) is added to the polar solvent, and the heating of the reagents is carried out in a closed Teflon vessel or a beaker with the assistance of microwaves to form the final precipitate of the perovskite oxides. $BaTiO_3$ (Zhu et al. 2008), $CaTiO_3$ (Moreira et al. 2009), $BiFeO_3$ (Biasotto et al. 2011), $BaZrO_3$ (Macario et al. 2010), $BaZrO_3$ (Moreira et al. 2009), $La_{1-x}Ag_xMnO_{3+\delta}$ (Ifrah et al. 2007), etc. are some perovskite oxides obtained by microwave-assisted hydrothermal process and $LaFeO_3$ (Tang et al. 2013), $LaCoO_3$ (Alvarez-Galvan et al. 2018), $La_{0.75}Sr_{0.25}Cr_{0.93}Ru_{0.07}O_{3-\delta}$ (Combemale et al. 2009), etc. for microwave-assisted coprecipitation method.

Besides the liquid phase synthesis, the solid-state synthetic process of perovskite oxides can also be assisted by microwave heating. The generation of heat is achieved by the strong microwave absorbance of the constituent reactants. During the solid-state synthesis of $Ba_2Cu_3O_7$ (Baghurst et al. 1988) and $La_{2-x}Sr_xCuO_4$ (Baghurst and Mingos 1988) from the respective oxides, the strong microwave absorbing properties of CuO is made use, as CuO reaches a temperature of ca. 700°C after only 30 seconds of microwave irradiation in a household microwave oven. However, if the constituent reactants do not absorb microwaves, strong microwave absorbing materials, such as carbon, can be used as the local heating element. For instance, in the solid-state synthesis of $La_{0.9}MnO_3$ from La_2O_3 and MnO_2, graphitic carbon is used as a microwave absorber for efficient heating, since the maximum attainable temperature of La_2O_3 and MnO_2 by microwave is only 107 and 321°C, respectively, after 1,800 seconds (Gibbons et al. 2000). The perovskite oxides, such as $LaCrO_3$, $LaCoO_3$ (Panneerselvam and Rao 2003), $La_{0.55}Li_{0.35}TiO_3$ (Bhat et al. 2003), $LaMO_3$ (M = Mn, Fe, Co, Cr, and Al) (Kulkarni and Jayaram 2003), (RE)CrO_3 (RE = La, Y, and Pr) (Prado-Gonjal et al. 2013), etc., are also synthesized using graphitic carbon as microwave absorbers. Microwave sensitive complexes in the solid state can also be used as a starting material for the fabrication of perovskite oxides. For example, microwave sensitive $La[Fe(CN)_6]\cdot 5H_2O$ can be converted to $LaFeO_3$ by microwave heating (Farhadi et al. 2009) and $La[Co(CN)_6]\cdot 5H_2O$ to $LaCoO_3$ (Farhadi and Sepahvand 2010).

2.2.7 Freeze Drying

In a typical freeze-drying process, as demonstrated in Figure 2.16, there are three steps involved: preparation of the precursor solution, primary drying, and secondary drying. The secondary drying involves freezing at temperatures and pressures below the triple point of the precursors to enable sublimation, i.e., without undergoing a liquid state. In the typical synthesis process of perovskite materials, at first, the precursors are dissolved in a suitable solvent, such as water. The solution containing the precursors are then frozen rapidly using liquid nitrogen to retain the chemical homogeneity. The obtained frozen precursor mixture is then freeze-dried in a freeze dyer under a high vacuum, during which the unstable solvent part of the precursor mixture is dehydrated or sublimed without undergoing a liquid phase. The obtained porous mixture is then calcined at a high temperature at a slow heating rate to decompose the precursor materials to obtain the intended highly porous perovskite material.

A sol–gel network type precursor solution can result in highly porous structures during the freeze-drying process. Therefore, EDTA-citric acid sol containing the nitrate metal salt is an ideal starting material for freeze-drying. A polymer can also result in a similar effect and is more advantageous than the EDTA-citric acid method since metal salts, other than nitrates, can be used as the metal source in polymer-based technique. Freeze spraying is also adopted for the fabrication of perovskite oxides. In freeze spraying, the metal salts are simply dissolved in a suitable solvent,

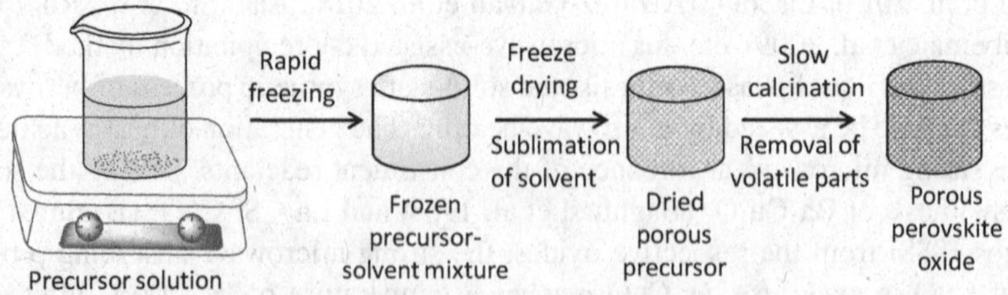

FIGURE 2.16 Steps in the freeze-drying process for the fabrication of perovskite oxides.

and then they are sprayed to liquid nitrogen to form precursor particles; they are then freeze-dried. These particles are then calcined at a high temperature to obtain the resulting perovskite oxide.

A large number of perovskites with a high surface area are synthesized using the freeze-drying technique. The freeze-drying can be applied to a simple solution of the metal salts in a solvent for particle synthesis, or sol–gel networks like structure for porous membranes or aerogels. Single and double perovskite structured materials are fabricated using freeze-drying and freeze spraying techniques. $LaMeO_3$ (Me = Fe, Co, Ni, and Cr), $La_{0.85}Me_{0.15}CoO_3$ (Me = Ca, Sr, Ba, and Ce) (Wachowski 1986), $Sm_{0.5}Sr_{0.5}CoO_{3-\delta}$ (Kim et al. 1998), $LaRuO_3$ (Labhsetwar et al. 2003), $La_{0.9}Sr_{0.1}Ga_{0.8}Mg_{0.2}O_{2.85}$ (Traina et al. 2007), $NdCoO_3$ (González et al. 1997), $La_{0.66}Sr_{0.34}Ni_{0.3}Co_{0.7}O_3$ (Klvana et al. 1999), $La_{1-x}K_xMnO_3$ (x = 0–0.15), $La_{1-x}Sr_xMnO_3$ (x = 0–0.3), $Nd_{1-x}Sr_xMnO_3$ (x = 0–0.4), $Nd_{1-x}K_xMnO_3$ (Lee et al. 2001), $Sr_3CoSb_2O_9$, and $Sr_2CoSbO_{6-\delta}$ (Primo-Martín and Jansen 2001) are some examples of perovskite oxides synthesized through freeze-drying technique. $La_{0.7}Ca_{0.3}MnO_3$ (Shlyakhtin et al. 2000), $La_{1-x}Sr_xCoO_{3-\delta}$ (Mizusaki et al. 1989), $La_{1-x}Me_xCoO_{3-\delta}$ (Me = Sr and Ca) (Zipprich et al. 1997), and $La_{0.8}Ca_{0.2}Fe_{0.8}Ni_{0.2}O_{3-\delta}$ (Ortiz-Vitoriano et al. 2010) are some examples of perovskite oxide particles obtained through freeze spraying process.

2.3 GAS-PHASE REACTIONS

2.3.1 Physical Vapor Deposition

2.3.1.1 Pulsed Laser Deposition

Physical vapor deposition (PVD) is a process used to produce material vapors that can be deposited on a substrate. Pulsed laser deposition (PLD) is a PVD process, in which a high-power pulsed laser beam is used to vaporize the target material to be deposited. A schematic of the PLD process is illustrated in Figure 2.17. Pulsed laser beams, typically in 10–30 ns, are used during this process, and the entire deposition is performed under a high vacuum for the deposition of metals; however, the chamber can be filled with the reactive gases depending on the nature of the thin films formed during the process. For instance, oxygen is used as the background gas for oxide deposition, nitrogen for nitrides, and acetylene for carbides. As the laser beam strikes the target material, the target is evaporated and forms a plume and ultimately deposits on a heated substrate facing the target.

The mechanism of film growth in PLD is complex, as several forms of energy conversion take place during the process. As the laser light is absorbed by the target, a thin layer on the target is superheated, forming an advancing luminous laser plume resulting from evaporation, ablation, and plasma. The plume forms at a high temperature, and it contains ionic species, atoms, molecules, electrons, particulates, etc. The deposition thickness varied from ~0.25 to1.0 nm s^{-1} under typical deposition conditions. The stoichiometric simultaneous ablation of multiple target or the target exchange systems allows the achievement of the desired chemical composition of the film. Heterostructures and superlattices can also be prepared by the precise control of the targets with individual growth rate regulation. The quality of the obtained

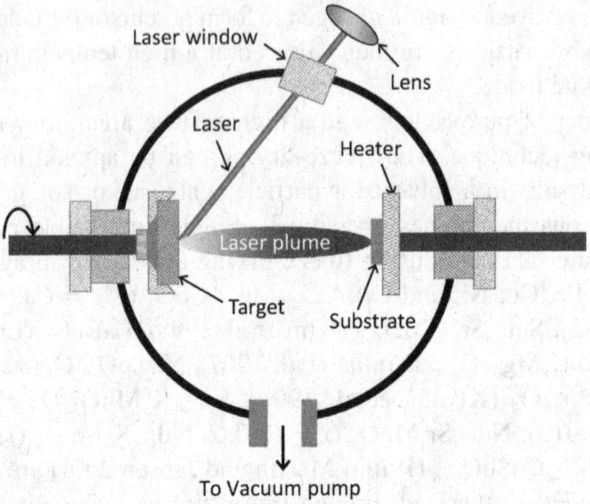

FIGURE 2.17 Schematic of a pulsed laser deposition process.

films is dependent on the quality of the laser source and the frequency of pulses. Even though the PLD is simple and versatile, the area of the deposition is limited up to 2 cm^2. Moreover, the high kinetic energies of the target particles impacting the substrate can result in a resputtering effect.

The laser ablation in PLD is analogous to laser machining. In both cases, material removal from the target surface is through the conversion of energy from photons to electronic excitations and then into thermal, chemical, and mechanical energy (Miotello and Kelly 1999). In the PLD process, the laser's wavelength/energy should be absorbed by a very shallow layer near the surface of the target to avoid subsurface boiling, which otherwise can lead to a large number of particulates at the film surface (Christen and Eres 2008). In perovskite synthesis using PLD, retaining the stoichiometry is a challenge, as the stoichiometry depends on the rate of material removal from the target. The vaporization temperature of the A- and B-site elements can be dissimilar, then the elements with high volatility must be evaporated in excess or keeping the substrate at an angle from the axial direction of the plume. In the case of $KNbO_3$, K ions are highly volatile; thus, the stoichiometric composition is achieved when a target with a K/Nb ratio of 2.85 is used (Martín et al. 1997). Similarly, by keeping the substrate at an angle (3°–15°) from the plume axis, the stoichiometric composition of $KNbO_3$ can be achieved (Yang et al. 2008).

The PLD deposited perovskite oxide films exhibit an epitaxial growth, meaning the deposited film is grown in the orientation of substrates. Pr-doped $Ca_{0.6}Sr_{0.4}TiO_3$ films deposited by PLD on $SrTiO_3$ (100) substrate are grown in (100) direction (Takashima et al. 2006). These thin films emit intense red emission under UV excitation. The thin films of ferroelectric perovskite oxides, such as $BaTiO_3$ (Watanabe et al. 1994) and $Pb_{0.6}Sr_{0.4}TiO_3$ (Chou et al. 1999), are fabricated using the PLD process, which is also suitable for the fabrication of antiperovskite oxides such as Ca_3SnO (Minohara et al. 2018). Table 2.5 lists the perovskite structured thin films fabricated using the PLD.

TABLE 2.5
List of Perovskite Oxide Thin Films Fabricated Using Pulsed Laser Deposition Process

Perovskite	Substrate	Characteristics/Applications	Reference
$BaTiO_3$	Si (100)	Metal-insulator-metal type capacitor	Lee et al. (1995)
$Ba_{0.25}Sr_{0.75}TiO_3$	Si (100)	Electrochemical pH sensing	Buniatyan et al. (2010)
Bi4Ti3O12	$SrTiO_3$ substrates of (100), (110), and (111)	Ferroelectric	Garg et al. (2004)
$La_{1-x}Sr_xCoO_{3-\delta}$	Pt electrode	Potentiometric hydrogen peroxide sensors	Anh et al. (2004)
	Si/SiO_2(100) and MgO(100)	Electrode material for capacitors	Cillessen et al. (1993)
$LaNiO_3$	Si(100)	Electrode in ferroelectric capacitors.	Sánchez et al. (2001)
$LaMnO_3$	(001) $SrTiO_3$ single crystal	Ferromagnetic insulator	Marton et al. (2010)
$La_{1-x}Ca_xCoO_{3-\delta}$	MgO (100), (110), (111) single crystals	Bifunctional electrocatalyst for metal-air batteries	Lippert et al. (2007)
$La_{0.5}Sr_{0.5}TiO_3$	$LaAlO_3$(001)	Dielectric	Wu et al. (2000)
$LaGaO_3$	Mixture of NiO, Fe_3O_4, and Sm-doped CeO_2	Solid oxide fuel cell electrolyte	Yan et al. (2006)
$LaTiO_{3-x}N_x$	(001)-oriented MgO and $LaAlO_3$	Strong visible light absorption at wavelengths below ~500 nm	Marozau et al. (2011)
$La_{0.6}Sr_{0.4}Co_{0.2}Fe_{0.8}O_{3-\delta}$	$LaAlO_3$	Electrical conductivity at high temperature ~600°C	Zomorrodian et al. (2010)
$La_{0.6}Ca_{0.4}CoO_3$	MgO (001)	Catalyst for metal-air batteries	Montenegro et al. (2002)
$PbTiO_3$	Platinum foil	Piezoelectric, dielectric	Chaoui et al. (1999)
Nb-doped $SrTiO_3$	Si substrates	Non-volatile resistance memory applications	Xiang et al. (2007)
$SrZrO_3$	$SrRuO_3$ on chemically treated $SrTiO_3$	Arrays of $SrZrO_3$ Nanowires	Karthäuser et al. (2004)
$SrBi_2Ta_2O_9$	$Pt/Ti/SiO_2/Si$	Dielectric and ferroelectric	Kim et al. (2002)
Sr_2FeMoO_6	(100) $SrTiO_3$	Negative magnetoresistance	Santiso et al. (2002)
Sr_2TiO_4	$SrTiO_3$ (100) single crystals	High substrate temperatures >900°C and low oxygen pressures <0.1 mbar yield the perfect stoichiometric films	Shibuya et al. (2008)

2.3.1.2 Reactive Sputtering

Sputtering is a thin film deposition process where the films are stoichiometric to the specific target. A schematic of the reactive sputtering system is shown in Figure 2.18. During the process, the atoms are ejected from the target by energized ions to form a plasma that is directed to the substrate under a high vacuum. Argon is the commonly used sputtering gas, and the sputtering is carried out with a DC power source, radio frequency alternating current, or ion-assisted deposition. In reactive sputtering, a reactive gas such as oxygen for oxides and nitrogen for nitrides is also passed to the reaction chamber along with argon. These reactive gases react with the target atoms in the plasma to form the desired composition.

While forming perovskite oxides films, multiple targets with different elements are simultaneously sputtered, which are reacted with oxygen and deposited as the desired film. Since the target elements and oxygen exhibit a large electronegativity difference, the formed ions can be negatively charged and accelerated towards the substrate due to the difference in the potential of negatively charged target and the grounded substrate. These ionic fluxes possibly act as sputtering ions to resputter the growing thin film on the substrate or modifying the composition of the films or etching the substrate. To avoid such problems, the following solutions are recommended. They are increasing the working gas pressures to reduce the kinetic energy of the ions striking the substrate, avoid placing the substrate directly facing the target, minimizing the discharge voltage, and designing the target composition to compensate for the resputtering effects (Habermeier 2007). The perovskite oxides synthesized by sputtering are presented in Table 2.6.

2.3.1.3 Reactive Molecular Beam Epitaxy

In molecular-beam epitaxy (MBE), the respective elements to be deposited are sublimed by heating in individual cells called effusion cells or furnaces. A typical MBE process setup is shown in Figure 2.19. The heating can also be carried out using a laser source or an electron-beam evaporator. The reaction chamber is always

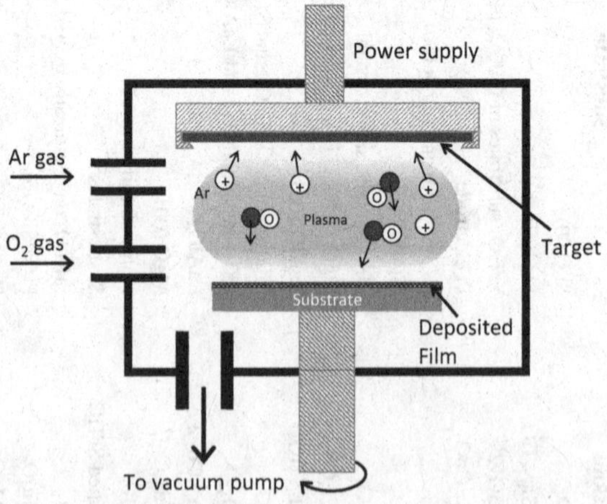

FIGURE 2.18 Reactive sputtering of perovskite oxide.

TABLE 2.6
Representative Perovskite Oxides Synthesized by Sputtering

Perovskite	Substrate	Mode of Sputtering	Application	Reference
$BaTiO_3$	Si (100)	Radio frequency (RF) magnetron sputtering	Ferroelectric	Qiao and Bi (2008)
$(Ba_xSr_{1-x})TiO_3$	Copper	RF magnetron sputtering	Dielectric and ferroelectric	Laughlin et al. (2005)
$BaPbO_3$	$Pb(Zr_{0.53}Ti_{0.47})O_3$	RF magnetron sputtering	Electrode for lead zirconate titanate ferroelectric	Luo and Wu (2001)
Y-doped $BaZrO_3$	Si wafers	RF magnetron sputtering	High temperature humidity sensors	Chen et al. (2009)
$BaSnO_{3-\delta}$	$SrTiO_3$ (001) and MgO (001)	High pressure oxygen RF sputter-deposition	Photovoltaics	Ganguly et al. (2015)
$Ba_{0.95}Sr_{0.05}Pb_{0.75}Bi_{0.25}O_3$	Silica, quartz, and sapphire	RF sputtering	Superconducting	Suzuki et al. (1981)
$BiFeO_3$	$Pt/Ti/SiO_2/Si$ (100)	RF sputtering	Photovoltaic	Chang et al. (2013)
Cd_3TeO_6 In-doped Cd_3TeO_6	Silica glass	RF magnetron sputtering	Photovoltaics	Tetsuka et al. (2005)
$(K,Na)NbO_3$	$SrRuO_3$ buffered $SrTiO_3$ (001)	RF magnetron sputtering	Piezoelectric	Li et al. (2014c)
$LaFeO_3$	MgO (001)	RF magnetron sputtering	-	Lee and Wu (2004)
	Silicon substrate		Selective NO_2 gas sensor	Thirumalairajan et al. (2014)
$LaNiO_3$	Si (100) substrate	RF magnetron sputtering	Ferroelectric capacitors	Yang et al. (2009)
$LaCrO_3$	Ferritic stainless steel 446 and Crofer 22 APU	DC magnetron sputtering	Coatings for metallic solid oxide fuel cell interconnects	Johnson et al. (2004); Johnson et al. (2009)
$La_{1-x}A_xMnO_{3-\delta}$ (A = Ca, Sr)	YSZ	RF magnetron sputtering	Cathode materials for solid oxide fuel cells	Takeda et al. (1994)
$LaTiO_2N$	(001) $SrTiO_3$	RF magnetron sputtering	Dielectric	Lu et al. (2013)

(Continued)

TABLE 2.6 (*Continued*)
Representative Perovskite Oxides Synthesized by Sputtering

Perovskite	Substrate	Mode of Sputtering	Application	Reference
$La_{0.67}Sr_{0.33}MnO_3$	Stainless steel Crofer22APU	Pulsed DC magnetron sputtering	Protective coatings for solid oxide fuel cell interconnect	Jan et al. (2008)
$La_{1-x}Sr_xCoO_{3-\delta}$	YSZ	RF magnetron sputtering	High electrode activity for oxygen reduction	Takeda et al. (1987)
Ag-$(La_{0.7}Sr_{0.3})CoO_3$ and Ag-$(La_{0.7}Sr_{0.3})MnO_3$	Y-stabilized Bi_2O_3	DC magnetron cosputtering	Solid oxide fuel cell air-electrode	Wang and Barnett (1995)
$La_{1-x}Sr_xMO_3$ (M = Cr, Mn, Fe, Co)	YSZ	RF sputter deposition	Oxygen electrodes for high temperature solid oxide fuel cells	Yamamoto et al. (1987)
$Ln_{0.4}Sr_{0.6}Co_{0.8}Fe_{0.2}O_{3-\delta}$ (Ln = La, Pr, Nd, Sm, Gd)	YSZ	RF sputtering	Electrode in solid oxide fuel cells	Tu et al. (1999)
$Pb(Zr,Ti)O_3$	Pt/Ti/Si	RF magnetron sputtering	Dielectric and piezoelectric	Thomas et al. (2002)
$SrTiO_3$	MgO (100) and $SrTiO_3$ (100) single crystals	RF magnetron sputtering	Ferroelectric	Fujimoto et al. (1989)
$Tl_2Ba_2Ca_1Cu_2O_8$	MgO (100) and $SrTiO_3$ (100)	DC diode sputtering	Superconducting	Hong et al. (1988)

Note: YSZ – Yttria stabilized zirconia.

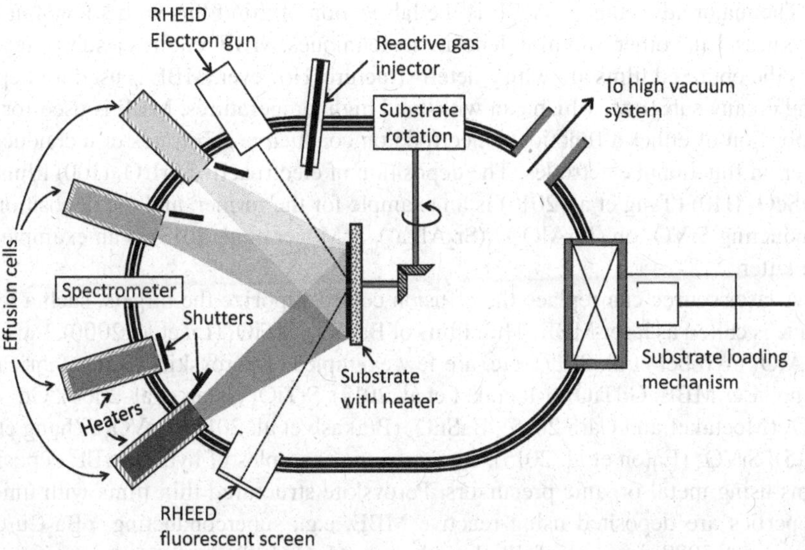

FIGURE 2.19 MBE reaction chamber.

maintained under a very high vacuum. The elemental vapors are then condensed on the substrate, which is continuously rotated and heated. Precise control on the temperature of the target and the substrate is necessary to control the rate of material deposition on the substrate. Molecular beam constitutes the evaporated atoms that may or may not interact with each other depending on the composition of the thin film. A typical MBE system is composed of four main parts: (1) an ultra-high vacuum epitaxial chamber with a background pressure of 1.3×10^{-8} Pa, containing a four-target holder and a substrate holder; (2) a device for heating the target; (3) a scanning device for the composition monitoring; and (4) a high-energy electron diffraction (RHEED) system and the charge-coupled device (CCD) camera.

RHEED accompanies the MBE set up for the continuous monitoring of the film growth. The shutters control the release of vaporized elements from the effusion cell. The oxide or nitride films are achieved by injecting the respective reactive gases into the chamber, as in the case of any other reactive deposition technique. In reactive MBE, the pressure is maintained sufficiently low to guarantee the reaction between the vaporized atoms and the reactive gases. A hybrid MBE technique is also used for the deposition of perovskite oxide films. In a hybrid MBE, A-site element is evaporated in the effusion cells, and a chemical beam produced by the thermal evaporation of the respective precursor is used as the source for the B-site element, which also acts as the source of the anion (Jalan et al. 2009). In reactive MBE, the pressure is reduced to 10^{-3} Pa; thus, the path of the atoms evaporated from the cells is much larger than the source–substrate distance. For the effective deposition of perovskite oxides, it is essential to retain the stoichiometric ratios of the molecular beams of the different constituent elements and the corresponding fluxes. To do the same spectroscopic techniques, such as atomic absorption spectroscopy, mass spectroscopy, quartz-crystal monitors, and electron microscopy, are often employed.

The major advantage of MBE is the fabrication of thin films with a few unit cell thickness. Like other thin film deposition techniques, MBE requires a substrate, and thus the obtained films are with a heterostructure. However, MBE is used for depositing on any substrate, which can withstand high temperatures. MBE is used for the deposition of either a functional electrode on conductive substrates or a conductive layer on functional electrodes. The deposition of electroactive $SrIrO_3$ (100) films on $DyScO_3$ (110) (Tang et al. 2016) is an example for the former, and the deposition of conducting $SrVO_3$ on $(LaAlO_3)_{0.3}(Sr_2AlTa_6)_{0.7}$ (Moyer et al. 2013) is an example for the latter.

A laser source can replace the effusion cell to vaporize the targets, such a technique is called as laser MBE. Thin films of $BaTiO_3$, $SrTiO_3$ (Lu et al. 2000), $LaNiO_3$/$LaAlO_3$ (Wrobel et al. 2017), etc. are few examples of perovskite oxides fabricated using laser MBE. $GdTiO_3$ (Moetakef et al. 2012), $SrTiO_3$ (Jalan et al. 2009), $Gd_{1-x}Sr_xTiO_3$ (Moetakef and Cain 2015), $BaSnO_3$ (Prakash et al. 2017), $LaVO_3$ (Zhang et al. 2015), $SrVO_3$ (Eaton et al. 2015), etc. are some examples of hybrid MBE deposited films using metal-organic precursors. Perovskite structured thin films with unique properties are deposited using reactive MBE, e.g., superconducting $YBa_2Cu_3O_{7-x}$ (Kwo et al. 1988) is successfully deposited on MgO (100) single crystals using oxygen as the reactive source.

2.3.1.4 Reactive Thermal Evaporation

In reactive thermal evaporation thin-film coating, the target material is physically evaporated by means of heat under a vacuum. The evaporated target material will condense directly to the substrate in the solid-state, similar to the condensation of water vapor on a lid. The thermal evaporation system consists of the sealed chamber connected to a vacuum pump with the provision to heat the target material resistively, and holders for the substrate. A schematic of the thermal evaporation system is shown in Figure 2.20. In typical thermal evaporation, the metallic target materials

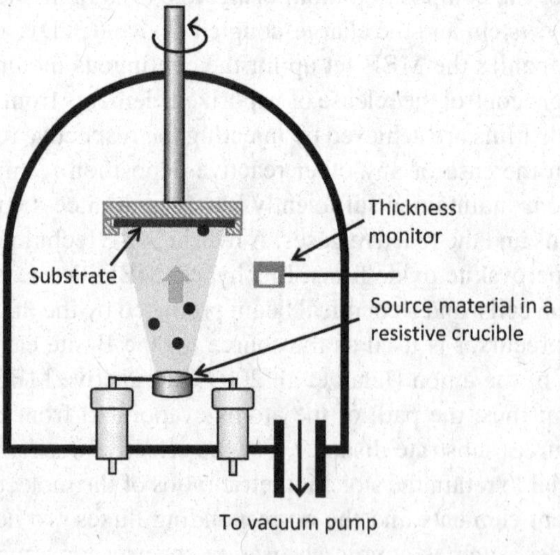

FIGURE 2.20 Reactive thermal evaporation process.

are fed into evaporators called "boats," due to their unique shape. The target materials are then heated above the melting temperature and subsequently evaporated to deposit on the substrates. Often, the boats themselves are highly resistive, and the adequate power supply allows the boats to be heated above the melting temperature of the targets. Alternatively, the target material in a crucible is heated radiatively by an electric filament, or it is fed continuously onto a heated element which allows the evaporation.

In the thermal evaporation process, a high vacuum is mandatory because the presence of gas molecules in the chamber can either redirect the travel of vaporized molecules towards the substrate or interact with the vapors to affect the purity of the films. However, in the reactive thermal deposition of oxide films, the evaporated metal atoms are allowed to interact with oxygen to form the respective oxide film, which obviously reduces the deposition rate and, ultimately, the control on the film thickness. The oxygen must be introduced only after attaining a high vacuum in the chamber, as the presence of water molecules in the moisture can quench the active functional properties of the films.

2.3.2 Chemical Vapor Deposition

In the chemical vapor deposition (CVD) process, thin films are deposited from the precursor chemicals in the vapor phase onto a substrate. The decomposition of the chemicals to the intended material occurs at the neighborhood of the heated substrate. The deposition process is often thermally driven; however, photo- and plasma-assisted methods can also be used. Since the deposition process is through chemical interaction, the availability of a wide choice of starting materials makes this technique more versatile than PVDs. CVD is a non-equilibrium process, and the control on the precursor vapor or solution determines the phase and morphology of the deposited films or nanostructures. CVD process is a bottom-up approach and allows the preferential growth of materials over a large area with a high degree of purity and reproducibility with an adjustable morphology, composition, tribology, and chemical properties. The significant experimental parameters of the CVD process are; nature of substrate, the temperature of the substrate and composition of the reaction vapor, the pressure of gas flows, the stability of the precursor during the transport, the efficiency of the precursor to be absorbed and decomposed on the substrate, etc. The accurate control of the above parameters in a CVD process can yield products in the form of thin films, powders, or single crystals. The CVD technique is promising for the fabrication of thin films with uniform thickness, even on substrates with an irregular surface profile. It also allows selective deposition on patterned substrates.

CVD process of oxides involves the chemical reaction of gaseous or vapor phase reactants with oxygen and the deposition on the heated substrate. Depending on the vaporization of the reactants and their transport to the substrate, CVD processes are classified. Aerosol-assisted CVD (AACVD) and direct liquid injection CVD (DLICVD) are two widely used techniques for the fabrication of perovskite oxide films. In AACVD, precursors are transported to the substrate by means of a liquid/gas aerosol. In DLICVD, the precursors are dissolved in a suitable solvent, and the precursor solutions are injected into the vaporization chamber through injectors, which is similar to the fuel injectors in automobiles. The construction of the DLICVD chamber is shown in Figure 2.21. The precursor vapors are then transported to the substrate and subsequently deposited as films of required composition. The interaction of precursor vapors with oxygen can be precisely controlled during the transport to the substrate; therefore, a subsequent annealing/calcination in the presence of oxygen is not necessary to form oxide layers or particles. If AACVD or DLICVD process uses metal-organic materials as the reactants, it is commonly called as metal-organic CVD (MOCVD). A representative flow chart of the MOCVD process for the fabrication of perovskites is shown in Figure 2.22. In MOCVD, the metal-organic precursor is either evaporated to a vapor phase, dissolved into a suitable solvent or atomized, before it is carried to the CVD chamber by means of an

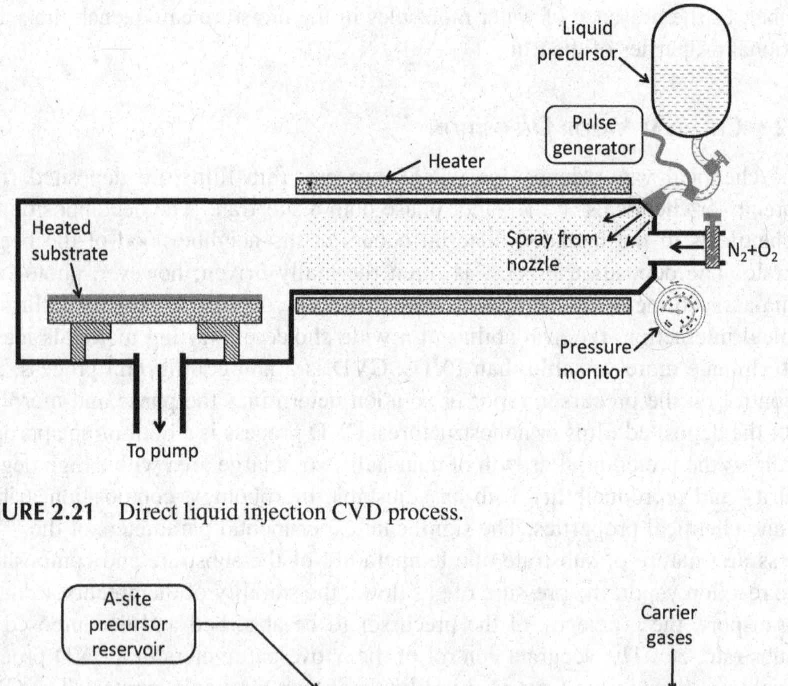

FIGURE 2.21 Direct liquid injection CVD process.

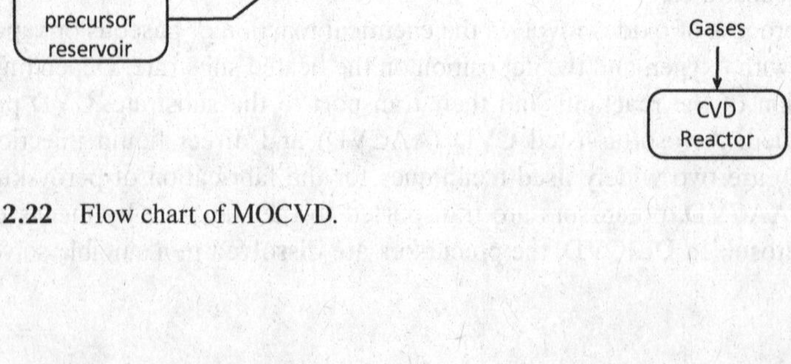

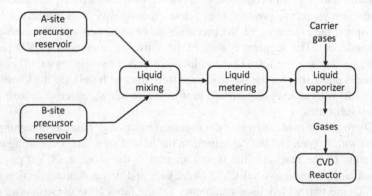

FIGURE 2.22 Flow chart of MOCVD.

inert/reactive carrier gas (Argon, N_2/O_2, N_2, etc.). Since the deposition of the films takes place by the interaction of decomposed reactants at a high temperature, both AACVD and DLICVD are known as thermal CVD processes.

The typical CVD process of perovskite oxide materials has the following steps: (1) dissolution of precursors in a stoichiometric ratio to a suitable solvent, and atomized or vaporization of target materials with the required composition; (2) transport of the vaporized reactants to the heated substrate; (3) reaction of the target atoms and the reactive gas (such as oxygen, nitrogen, or methane) during the transport; and (4) deposition of the vapor phase reactants on the substrate. An inert carrier gas is used to dilute the reactants in the vapor phase and to carry them to the substrate. The growth rate of oxides in a CVD process is much higher than a reactive PVD process. MOCVD is the most widely used CVD process for the fabrication of perovskite oxides, and a few examples are listed in Table 2.7.

TABLE 2.7
Representative Perovskite Oxides Synthesized by CVD Process

Perovskite	Precursors	Substrate	References
$BaTiO_3$	Barium β-diketonate and titanium isopropoxide	Fused quartz	Kwak et al. (1991)
Nb-doped $BaTiO_3$	Barium β-diketonate, niobium β-diketonates, and $Ti(i\text{-}OC_3H_7)_4$ β-diketonates	(012) $LaAlO_3$	Lemée et al. (2002)
$BiFeO_3$	$Bi(thd)_3$ and $Fe(thd)_3$	$SrRuO_3/SrTiO_3$ and $SrRuO_3/SrTiO_3/Si$	Yang et al. (2005)
$LaMO_3$ (M = Co, Fe, Cr, and Mn)	Acetonyl acetates of Co, Cr Fe, Mn and La	Silicon	Ngamou and Bahlawane (2009)
$La_{0.8}Sr_{0.2}MnO_3$	$La(NO_3)_3$, $Sr(NO_3)_2$, $Mn(NO_3)_2$, and glycine	YSZ wafers	Wang et al. (2000)
$LaNiO_3$	$La(thd)_3$ 1,2-dimethoxyethane and $Ni(thd)_2$	R-sapphire, (100) Si, and $LaAlO_3$	Kuprenaite et al. (2019)
$PbTiO_3$	$Pb(C_2H_5)_4$, $Ti(i\text{-}OC_3H_7)_4$	(100) MgO	Hirai et al. (1993)
$Pb(Zr_xTi_{1-x})O_3$	$Pb(thd)_2$, $Zr(thd)_4$, and titanium ethoxide	$Pt/Ti/SiO_2/Si$ and $RuO_x/SiO_2/Si$	Peng and Desu (1992)
$RNiO_3$(R = Pr, Nd, Sm, and Gd)	$Ni(thd)_2$ and $R(thd)_3$	MgO, ZrO_2-Y_2O_3, $SrTiO_3$, $LaAlO_3$	Novojilov et al. (2000)
$SrIrO_3$	$Sr(C_{11}H_{19}O_2)_2(C_8H_{23}N_5)_x$, $Ir(C_7H_9)(C_6H_8)$	(100) $SrTiO_3$	Kim et al. (2005)

Notes: acac – acetonyl acetate, thd = 2,2,6,6-tetramethylheptane-3,5-dionate, $Ti(i\text{-}OC_3H_7)_4$- ttanium isopropoxide.

2.4 MISCELLANEOUS TECHNIQUES

2.4.1 Molten-Salt Synthesis

If the solubility of the precursor salts is limited in solvents, then they cannot be used for the synthesis of perovskite oxides through solution methods at a low temperature. To overcome insolubility of the precursors in the solvents, molten salts can be used as alternate solvents, provided the salt must be soluble in water, and the final product must not. In spite of the solubility in water, salt must be stable at high temperatures. During the molten-salt synthesis (MSS) process, salts with a low melting point are added to the precursors and heated above the melting point of the salt, and the molten salts dissolve the precursors. Once the precursors are entirely dissolved in the molten salt, the salt can be cooled down slowly to room temperature. The slow cooling of the melt allows the crystallization of the final product. As the reactant materials are completely dissolved in the molten salt, during cooling, solid particles nucleate homogeneously in the liquid phase. Therefore, by controlling the nucleation sites, one can adjust the size of the crystals. In the MSS, the reaction rate is increased at a lower temperature than solid-state processing. By using molten salt as the solvent, the particle size, shape, and level of agglomeration can be controlled. The diffusion rate in solid-state reactions is very slow, but in the molten-salt method, mass transport is faster because, in the liquid phase, mass transport is through convection and diffusion. Many salts used for MSS are soluble in water; thus, MSS has the advantage of easy isolation of the product.

A single crystal with a large size can be obtained by reducing the number of nuclei formed during the cooling of molten salt. Then, the water-insoluble final product is separated from water-soluble salt flux using an excess of water. To prepare particles using MSS, a large number of nuclei must be present during the cooling process. The surfaces of the reactants are the heterogeneous nucleation sites. In such cases, reactants must be partially soluble in the molten salts, or in other words, complete solubility for all reactants is not desirable. When insoluble reactants are used as raw materials, the process is called topochemical MSS (TMSS). In TMSS, the morphology of the final product is identical to that of the insoluble raw materials; therefore, the starting materials themselves act as templates. During the process, the starting reactants are transformed into perovskite oxide materials through the exchange, deletion, or insertion of individual atoms. The rate of cooling determines the shape and size of the powder particles, especially during the single-crystal growth.

In the typical perovskite synthesis process, as shown in Figure 2.23, a stoichiometric mixture of the cationic precursors in the salt bath is heated above the melting temperature of the salt, at which the salt melts, and the product particles form. The size and shape of the product powders are influenced by the appropriate temperature and duration of the reaction. The reacted mixture is then allowed to cool down to atmospheric temperature and washed to remove the salt and separate the product. This process is scalable, like a conventional molten-salt metallurgical process. The nanosized particles can be obtained using a surfactant in the molten salt. The SEM images of the nanoparticles of $La_{2/3}Sr_{1/3}MnO_3$ obtained using KNO_3 as the molten salt (Gonell et al. 2019) is shown in Figure 2.24. In TMSS, the reactants of desired

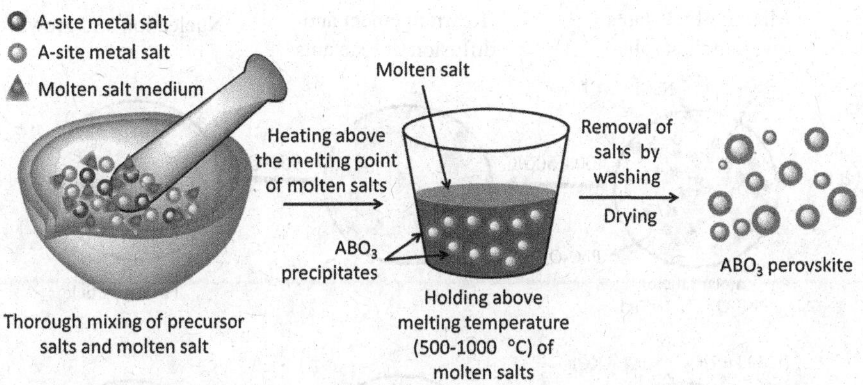

FIGURE 2.23 Steps in the molten-salt synthesis.

FIGURE 2.24 SEM images of $La_{2/3}Sr_{1/3}MnO_3$ nanoparticles obtained at (a) 600 and (b) 750°C using molten-salt method, reproduced with permission of the John Wiley and Sons (Gonell et al. 2019).

morphology are synthesized first and then added to the salt mixture, and the above process is followed. $PbTiO_3$ particles obtained using spherical TiO_2 particles are spherical in nature, whereas TiO_2 nanorods result in $PbTiO_3$ nanorods (Ji et al. 2019), as shown in Figure 2.25.

Formerly, a small quantity of molten salts was used as an additive in solid-state reactions to improve the reaction rate. When molten salt is used as a solvent, a large amount of salt (80–120 wt% of the reactants) is used to control the morphology and size of the final product. Chlorides and sulfates are frequently used as a salt bath. The eutectic mixtures of the salts are often used to lower the melting temperature of the salts. The salt mixture 0.5NaCl–0.5KCl (eutectic composition) with a melting point of 650°C and 0.635 Li_2SO_4–0.365 Na_2SO_4 with the melting point 594°C is the most commonly used salt mixtures for low melting temperature, and Na_2SO_4–K_2SO_4 mixture is used for high temperature, which has a melting point of 823°C.

Perovskite oxides with different compositions, morphologies, and dimensions can be synthesized using a molten-salt method. One, two, and three-dimensional structured materials are successfully fabricated using this method. The appropriate

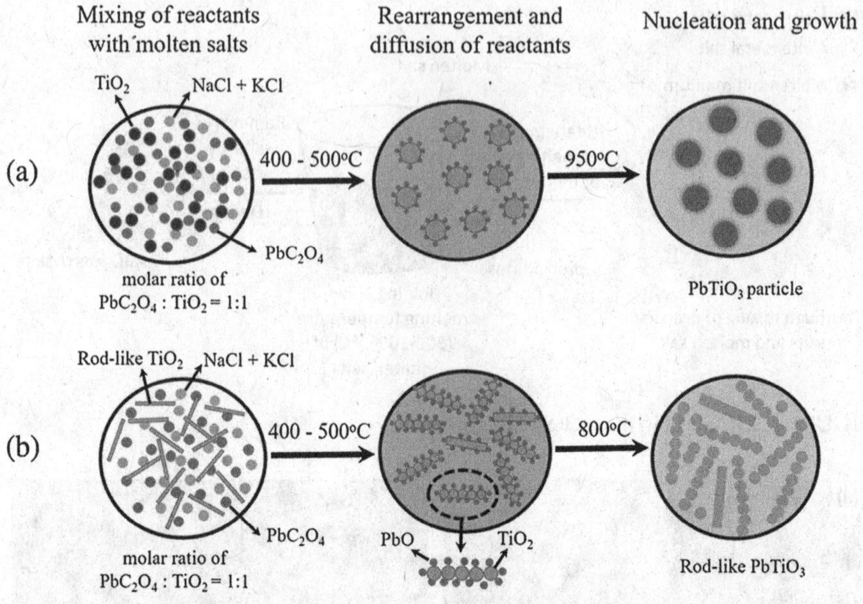

FIGURE 2.25 Schematic diagrams illustrating the formation of (a) PTO particles in the MSS process and (b) rod-like PTO powders in the template MSS process, reproduced with permission of the Springer Nature (Ji et al. 2019).

selection of the molten salt, starting materials, temperatures, heating rate, and reaction time is important to yield the anticipated composition of the final products. For instance, $LaMO_3$ (M = Al, Sc, Cr, Mn, Fe, Co, Ni, Ga, and In) microparticles are obtained using oxides and acetates of the starting materials such as α-Al_2O_3, Sc_2O_3, Fe_2O_3, Co_3O_4, Ga_2O_3, $Mn(CH_3COO)_3$, and $La_2(CO_3)_3$ by controlling the ratio of the molten salt bath constituting Li_2CO_3, Na_2CO_3, and K_2CO_3 at various temperatures starting from 500°C to 800°C (Kojima et al. 2006). More examples of perovskite oxides obtained by MSS are listed in Table 2.8.

2.4.2 Electrochemical Synthesis

Electrodeposition, also known as electroplating, is mainly used for depositing metals to a conducting surface from an ionic salt solution of the intended metal. Electrodeposition is a generally adopted technique for the fabrication of thin films on the metallic surfaces to induce corrosion protection, abrasion resistance, esthetics, and to reduce contact resistance. Electroplating can be extended for the deposition of multiple metallic elements as well. In the electroplating setup, the working electrode, electrolyte, power source or a potentiostat, and the counter electrode forms a closed circuit. A schematic of the electrodeposition process is shown in Figure 2.26. Sometimes, a reference electrode is also used as a reference potential for the potentiostat; however, the current flows only between the working and counter electrodes. The working electrode is the conducting material that is being plated, and the counter electrode is an inert material such as platinum or graphite. The flow of electrons

TABLE 2.8
Examples of Perovskite Materials Fabricated through MSS and TMSS

Perovskite	Reagents	Molten Salt	Temperature (°C)	Shape	Reference
		MSS			
$BaTiO_3$	$BaCO_3$ and TiO_2	NaCl–KCl	700–1,000	Submicron particles	Xue et al. (2017)
	$Ba(NO_3)_2$ and TiO_2	KCl	800	Nanorods	Li et al. (2014a)
$Ba(Zn_{1/3}Nb_{2/3})O_3$, $Ba(Mg_{1/3}Nb_{2/3})O_3$ and $Ba(Zn_{1/3}Ta_{2/3})O_3$	MgO, Nb_2O_5, ZnO, and $BaCO_3$	NaCl–KCl	900	Microparticles	Thirumal et al. (2001)
$BaFe_{0.5}Nb_{0.5}O_3$	Ba_2CO_3, Fe_2O_5, and Nb_2O_5	NaCl–KCl	700–900	Nanoparticles	Tawichai et al. (2012)
Ba_2BNbO_6 (B = Gd, Sm, Y)	$BaCO_3$ and B_2O_3 (B = Gd, Sm, and Y) Nb_2O_5	NaOH–KOH	350	Particles	Meng and Virkar (1999)
$Bi_{1-x}La_xFeO_3$	Bi_2O_3, La_2O_3, and Fe_2O_3	NaCl	820–900	Submicron single-crystals	Chen et al. (2008)
$BaZrO_3$	Barium oxalate and ZrO_2	NaOH–KOH	720	Submicrometer-sized particles	Zhou et al. (2007)
$BaTiO_3$–$SrTiO_3$	$BaCO_3$, $SrCO_3$, and TiO_2	NaOH–KOH	200–400	-	Gopalan et al. (1996)
$BiFeO_3$	Bi_2O_3 and Fe_2O_3	NaCl–Na_2SO_4	800	Nanoparticles	Chen et al. (2007a)
$CaCu_3Ti_4O_{12}$	$CaCO_3$, CuO and TiO_2	Na_2SO_4–K_2SO_4	1,000	Microparticles	Chen and Zhang (2010)
$CaZrO_3$	$CaCl_2$ and ZrO_2	Na_2CO_3	1,050	Submicrometer-sized particles	Li et al. (2007)
$LaFeO_3$	La_2O_3	Li_2CO_3 +Na_2CO_3	600	Film <5 µm thickness on stainless steel	Frangini et al. (2011)

(*Continued*)

TABLE 2.8 (*Continued*)
Examples of Perovskite Materials Fabricated through MSS and TMSS

Perovskite	Reagents	Molten Salt	Temperature (°C)	Shape	Reference
$La_2Ti_2O_7$	TiO_2 and La_2O_3	Na_2SO_4/K_2SO_4	1,100	Nanostructured particles	Arney et al. (2008)
$La_{0.6}Sr_{0.4}Co_{0.2}Fe_{0.8}O_{2.9}$	La_2O_3, $SrCO_3$, Co_3O_4, and Fe_3O_4	NaCl and KCl	850	Nanoparticles	Song et al. (2018)
$LaMO_3$ (M = Mn, Fe, Co, Ni)	$MnSO_4$, $Fe(NO_3)_3$, $Co(NO_3)_2$, $Ni(NO_3)_2$ and $La(NO_3)_3$	$NaNO_3$ and KNO_3	450–500	Nanoparticles	Matei et al. (2007); Yang et al. (2010)
La_2BMnO_6 (B = Co, Ni)	$La(NO_3)_3$, $Ni(NO_3)_2$ or $Co(NO_3)_2$, and $Mn(NO_3)_2$	$NaNO_3$ and KNO_3	700	Nanoparticles	Mao (2012)
$LaCoO_3$	$La(NO_3)_3$ and $Co(NO_3)_2$	NaCl–KCl	800	Cubic Microparticles	Vradman et al. (2017)
$LaMnO_3$	$La(NO_3)_3$ and $Mn(NO_3)_2$	NaCl–KCl	700	Submicron particles	Vradman et al. (2013)
Sr-doped and Mg-doped $LaAlO_3$	$Al(NO_3)_3$, $Mg(NO_3)_2$, $La(NO_3)_3$, and $Sr(NO_3)_2$	NaOH	350	Microparticles	Mendoza-Mendoza et al. (2012)
$Pb(Mg_{1/3}Nb_{2/3})O_3$ and $Pb(Fe_{1/2}Nb_{1/2})O_3$	PbO, MgO, Nb_2O_3	Li_2SO_4–Na_2SO_4	750	Microparticles	Yoon et al. (1998)
$Pb(Fe_{0.5}Nb_{0.5})O_3$	PbO, $FeZO_3$, and Nb_2O_5	NaCl and KCl	800	Microparticles	Chiu et al. (1991)
$(Pb_{0.95}La_{0.05})(Zr_{0.52}Ti_{0.48})O3$	PbC_2O_4, La_2O_3, $ZrO(NO_3)_2$ and TiO_2	NaCl–KCl	850	Submicron particles	Cai et al. (2008)

TMSS

Perovskite	Template	Salt	Temperature (°C)	Shape	Reference
$BaTiO_3$	TiO_2 rods	NaCl–KCl	700	Rod shaped	Huang et al. (2009)
	$BaBi_4Ti_4O_{15}$ platelets	$BaCl_2$–KCl	1,080	Plate like	Liu et al. (2007)
$KNbO_3$	Nb_2O_5 nanowires	KCl	850	Nanorods	Li et al. (2009)

(*Continued*)

TABLE 2.8 (*Continued*)
Examples of Perovskite Materials Fabricated through MSS and TMSS

Perovskite	Reagents	Molten Salt	Temperature (°C)	Shape	Reference
$LiNbO_3$	Nb_2O_5 nanowires	LiCl	800	Nanowires	Santulli et al. (2010)
$LaMnO_3$	Spherical Mn_2O_3	$NaNO_3$–KNO_3	550	Spherical particles	Wang et al. (2014)
$NaNbO_3$	$K_2Nb_8O_{21}$ nanowires	NaCl or CaCl	800	Nanorods	Xu et al. (2007)
	Nb_2O_5 nanowires	NaCl	850	Nanorods	Li et al. (2009)
$PbTiO_3$	Spherical TiO_2 and PbO	NaCl–KCl	900	Spherical	Cai et al. (2007)
	Rod-shaped TiO_2 and PbO			Rods	
	Rod shaped $K_2Ti_4O_9$ and PbO			Cubic	
	$PbBi_4Ti_4O_{15}$ plates	KCl	1,050	Plate-like	Poterala et al. (2010)
$SrTiO_3$	TiO_2 nanoparticles	NaCl–KCl	700	Nanoparticles	Li et al. (2010)

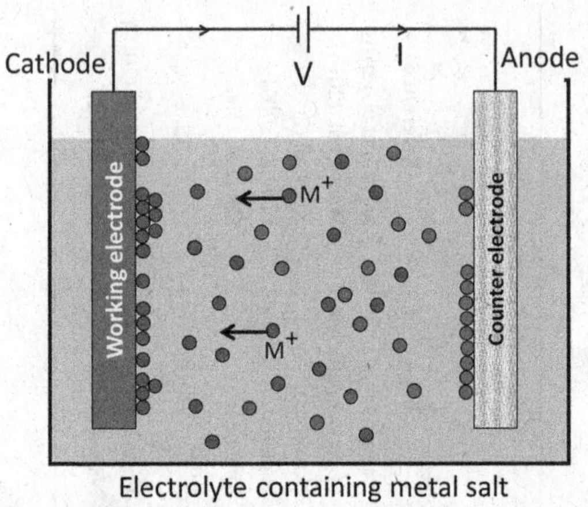

FIGURE 2.26 Schematic of an electrodeposition unit.

take place through the ionic electrolyte by the movement of metallic cations in the electrolyte. In other words, as the working electrode is connected to a negative potential, the positively charged metallic ions in the electrolyte move towards the working electrode or cathode to adhere on the surface evenly to form a thin film on the surface. The process continues until all the cations are deposited on the surface of the cathode, and the rate of deposition in majorly affected by the applied current and cell potential.

The electrochemical deposition can be typically classified as galvanostatic and potentiostatic synthesis. In galvanostatic experiments, the potential of the working electrode is selected with respect to a reference electrode such that the current density is constant, but the cell potential drifts as the reactant activity decreases and ultimately results in different products. In a potentiostatic synthesis, the voltage across the working electrode-counter electrode pair is adjusted to maintain a constant potential difference between the working and reference electrodes in a three-electrode assembly. In this case, the cell current usually decays rapidly as the reaction proceeds, due to low rates of diffusion of the reactant molecules from the bulk to the electrode surface, and therefore, a decrease in the activity of the reactant occurs. However, the reaction is likely to yield a pure single-phase product selected by the applied potential (Therese and Kamath 2000).

The preparation of perovskite oxides using the electrochemical process can be subdivided into anodic oxidation and electrogeneration of the base by cathodic reduction.

2.4.2.1 Anodic Oxidation

Anodic oxidation is used to deposit a metal ion with a lower oxidation state to a higher oxidation state by oxidation anodically. The oxidation process is controlled by pH, and the pH of the electrolyte is adjusted in such a way that the lower oxidation state is stable, and the higher oxidation state readily undergoes hydrolysis to yield the metal oxide or hydroxide under the applied potential. The overall reaction can be written as follows (Therese and Kamath 2000).

$$M^{n+} \rightarrow M^{(n+m)+} + me^- \quad (2.1)$$

$$M^{(n+m)+} + (n+m)OH^- \rightarrow M(OH)_{n+m} \rightarrow MO_{(n+m)/2} + \frac{(n+m)}{2}H_2O \quad (2.2)$$

During the synthesis of perovskite oxides using anodic oxidation, two identical electrodes are used as working and counter electrodes. The galvanostatic synthesis is followed under a constant current density. Perovskite structured $LaCoO_3$ and $LaMnO_3$ (Sasaki et al. 1991) are successfully synthesized using anodic oxidation. The final oxides are attained after heating to a temperature of 1,000°C, since the precursor for the perovskites is deposited during the process. The process can be depicted as follows in the case of $LaCoO_3$ (Matsumoto et al. 1992b):

$$2Co^{2+} + 6H_2O \rightarrow Co_2O_3 \cdot 3H_2O + 6H^+ + 2e^- \quad (2.3)$$

$$Co_2O_3 \cdot 3H_2O + 2La(NO_3)_3 \rightarrow Co_2O_3 \cdot 2La(NO_3)_3 \cdot 3H_2O \quad (2.4)$$

2.4.2.2 Electrogeneration of the Base by Cathodic Reduction

The simple example of cathodic reduction is metal deposition on the cathode as the electric current is passed through a metal salt solution.

$$M^{n+} + ne^- \rightarrow M \quad (2.5)$$

But in the case of electrolytes containing metal nitrates as the metal source, the anion reduction (nitrate reduction) reactions dominate over the other reactions in the cell because of the most positive $E°$ value of nitrate reduction reactions. The following nitrate reduction reactions are observed during such processes (Matsumoto et al. 1992b):

$$NO_3^- + H_2O + 2e^- \rightarrow NO_2^- + 2OH^- \quad (2.6)$$

$$NO_3^- + 7H_2O + 8e^- \rightarrow NH_4^+ + 10OH^- \quad (2.7)$$

As a result, the metal hydroxides are deposited on the cathode rather than pure metals. The nature of deposited films are amorphous; therefore, a high-temperature treatment is necessary to obtain the intended perovskite material. During the formation of perovskites in aqueous nitrate solution, two classes of reactions take place: nitrate reduction (Eqs. 2.6 and 2.7) and hydrogen evolution. The hydrogen evolution reactions can be represented as follows:

$$2H_2O + 2e^- \rightarrow 2OH^- + H_2 \quad (2.8)$$

The above reactions release OH^- ions and lead to an increase in pH in the electrode vicinity. As a result, the deposited metal ions on the cathode form a hydroxide or oxide coatings.

$$M^{n+} + OH^- \rightarrow M(OH)_n \tag{2.9}$$

Matsumoto et al. (1992a) electrodeposited the first perovskite structured oxide, $BaTiO_3$. The recommended cathodes for perovskite deposition are steel, Fe, Ni, and Ti. The deposition on Pt and Al electrodes are very difficult due to the weak interactions of the metal ions to the electrode surface. The electrodeposition process of perovskites can be carried out in a single cell or using a divided cell. A divided cell helps to separate the product of the cathodic reaction from that of the anodic reaction (Therese and Kamath 2000). The electrodeposition process can be briefed by considering $LaMnO_3$ as a representative perovskite. In the typical process, nitrate salts of La and Mn are dissolved in water to form the electrolyte solution, which is placed in the cathodic chamber, and the anodic chamber is filled with KNO_3 solution. The current density is kept constant galvanostatically as 0.5 mA cm^{-2}. The deposition continuous up to 1 h, and the deposits are $La(OH)_3$ and MnO(OH). The deposited layer is then calcined at 900°C to obtain perovskite structured $LaMnO_3$ (Therese and Kamath 1998).

Perovskite oxide films or particles can be prepared using electrochemical processing. Perovskite structured $SrTiO_3$ (Tao et al. 2006), $LnMO_3$ (Ln = La, Pr, Nd; M = Al, Mn, Fe), $LaMO_3$ (M = Co and Ni) (Therese et al. 2005), $KNbO_3$ (Papa et al. 2015), and $La_{1-x}Sr_xMnO_3$ (Sasaki et al. 1990) are successfully synthesized through the electrochemical route.

The electrochemical processes of perovskite fabrication through anodic oxidation is successfully combined with hydrothermal synthesis. The hydrothermal-electrochemical method possesses the advantage of high working temperatures. The hydrothermal-electrochemical method allows the deposition on substrates with different shapes, in-situ film growth, selective growth (only on active metallic substrates), and nanostructured polycrystalline films (Piticescu et al. 2006). In the typical process, the electrochemical reaction is performed inside a Teflon lined autoclave, and the reaction is carried out at 100°C–200°C. The obtained films are oxides, and thus no post-heating is required at a high temperature. Thin films of $BaTiO_3$ (Kajiyoshi et al. 1997), $SrTiO_3$ (Kajiyoshi et al. 1994), $NaTaO_3$ and $Na_2Ta_2O_6$ (Lee et al. 2005), and $Pb[Zr_xTi_{1-x}]O_3$ (*PZT*) (Piticescu et al. 2006) are successfully prepared using this method.

2.4.3 Electrospinning

Electrospinning works on the principle of electrostatic spraying. In electrostatic spraying, charged liquid droplets are deposited on a substrate that is grounded. In electrospinning, a charged polymeric solution containing the precursor material is drawn as fibers to a collector under a high potential difference. Electrospinning is a bottom-up approach, where the polymerized precursors dissolved in an appropriate solvent is spun as fibers. At present, electrospinning is a leading technique used for the fabrication of polymer as well as ceramic nanofibers when compared with the other techniques, such as vapor phase transport, laser ablation, arc discharge, template-assisted lithography, phase separation, and hydro/solvothermal methods. Electrospinning is unique for the fabrication of nanofibers because of its simple construction, large-scale productivity, versatility, high aspect ratio and uniformity of the nanofibers, the possibility of automation, large-scale production, low cost, and independence on the operator skills.

An electrospinning unit has three major components, which are (1) a syringe pump, which feeds the polymer solution or melts with the help of a syringe or any other means; (2) a high-voltage DC supply; and (3) a grounded collector plate, on which the nanofibers are collected. The geometry of the spinneret is not always analogous to a simple syringe, as shown in Figure 2.27, the purpose of spinneret is to deliver the solution at a controlled rate. The high voltage unit must be able to supply a DC voltage ranging from 1 to normally 30 kV or even higher. Similar to the spinneret, the geometry of the collector plate is also varied to meet the requirements of the product. Besides a static collector plate, a rotating drum collector is also commonly used to collect aligned nanofibers.

Electrospinning is successfully employed in the fabrication of nanoscale ceramic fibers, with controllable composition, morphology, and tunable properties. Electrospinning of ceramic nanofibers involves a combination of electrospinning and sol–gel processing, as illustrated in Figure 2.28. The precursor solution for the electrospinning process should have a high molecular weight. The electrospinning process of oxides involves the following steps. The first step is the mixing of molecules. Stoichiometric quantities of the precursor chemicals for the intended perovskite are mixed in one or more solvents, and a gel-forming medium (usually a polymer) is added to form a spinnable sol. In the second step, the above polymeric sol is shaped to xerogel nanofibers by electrospinning. Finally, the xerogel fibers are heated to a high temperature, in an air furnace or a furnace filled with oxygen (vacuum furnace is not workable), at a slow ramp, to remove the volatile and organic phases from the xerogel nanofibers to achieve the ceramic material of high purity and required morphology.

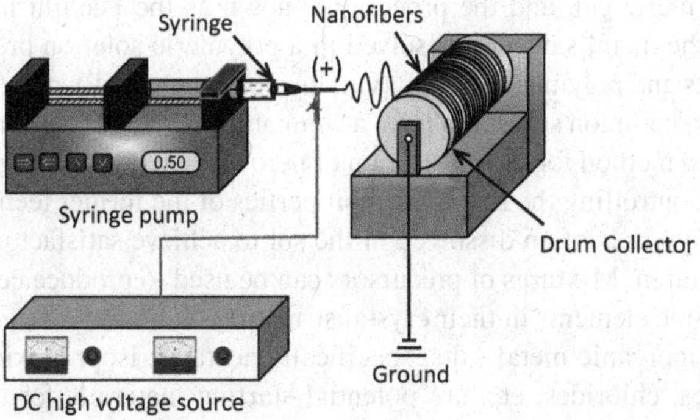

FIGURE 2.27 Schematic of an electrospinning unit.

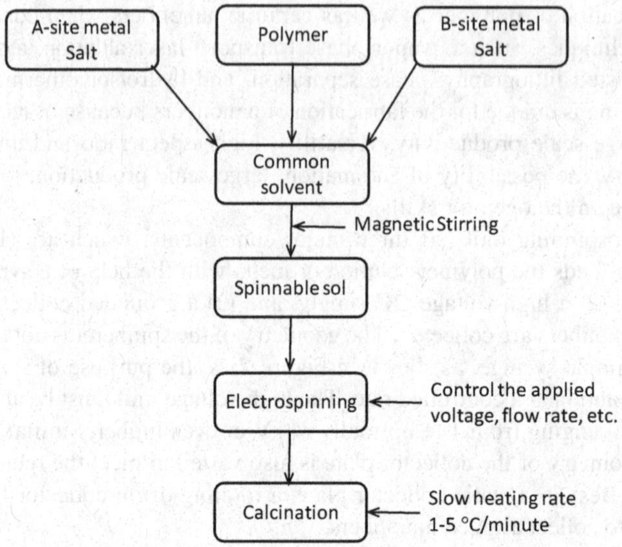

FIGURE 2.28 Flow chart of the electrospinning process for the fabrication of perovskite nanofibers.

Sol–gel assisted electrospinning has been successful in preparing perovskite oxide nanofibers with homogeneity and purity for a large number of applications. Despite the nanoscale dimensions of the oxides, the processing temperatures are lower than that of the usual solid-state processes. Additionally, the reactants used in the sol–gel electrospinning process are available in analytical grades, which allow the formation of high purity products. Moreover, the metal salts readily convert to the oxide by thermal or oxidative decomposition, and they are completely soluble in many common solvents.

The synthesis of perovskite oxides using the electrospinning process can be carried out in two different sol–gel routes based on the gel structure. In the first route, the chemistry of the sol–gel process is based on the complexing of cationic precursors in the presence of a complexing agent (e.g., EDTA, citric acid, etc.) and alcohol to form a polymeric gel, and the process is known as the Pechini method. In the second route, the metal salts are dissolved in a polymeric solution prepared by dissolving the salts and polymer (e.g., poly(vinylpyrrolidone) (PVP), poly(vinyl alcohol) (PVA), etc.) in a common solvent to form a spinnable precursor solution, which is the widely accepted method for the fabrication of perovskite oxides through electrospinning process. Controlling the rheological properties of the former technique is hard; therefore, a polymer is often dissolved in the sol to achieve satisfactory qualities of a spinnable solution. Mixtures of precursors can be used to produce ceramic nanofibers with multiple elements in their crystal structure.

Organic or inorganic metal salts, specifically acetates, isopropoxides, formates, citrates, nitrates, chlorides, etc. are potential starting materials for the perovskite oxide fiber processing. Nitrates and chlorides are widely used metal salts because of their availability, and thermally stable salts such as sulfates are generally avoided because of the difficulty in removing the anions. The homogeneity of the spinnable solutions near the molecular level is achieved by continuous mixing, and the movement of metal ions is expected to be prevented in the sol. Highly volatile solvents, in which the metal salt is sparingly soluble, are avoided for the electrospinning process as the sudden evaporation of the solvent can crystallize the metal salt and ultimately destroy the homogeneity of the precursor fibers. Maintaining an acidic pH may reduce the tendency of crystallization of metal salts; otherwise, a solvent with high solubility and boiling point is preferred. Such precautions avoid the rapid gelation of fibers during the electrospinning process, which otherwise results in large fibers.

During the electrospinning process, the viscous polymer solution is uninterruptedly supplied at a controlled rate with the help of a spinneret. When the spinneret is given a high positive charge, a droplet of the polymer solution dangles at the tip of the spinneret (Li and Xia 2004). As the applied voltage is increased, electrostatic charges are accumulated on the surface of the spherical precursor droplet, and the droplet transforms to a cone called *Taylor cone*, as shown in Figure 2.29. Subsequently, at a specific threshold voltage, the electrostatic field overcomes the surface tension of the polymer solution, and nanosized charged jets will be ejected from the tip of the Taylor cone geometry, which is dried during its travel from the spinneret to the collector to form high aspect ratio nanofibers.

The polycrystalline ceramic fibers obtained by the calcination of suitable precursors composite fibers usually give oxides with pseudomorphs of the starting materials. A large amount of thermal energy is required for the decomposition of the precursors; it also facilitates the sintering of the particles to form the required morphology. Sol–gel combined electrospinning method usually produces amorphous precursor fibers, and calcination of the precursor fibers is required to obtain polycrystalline products.

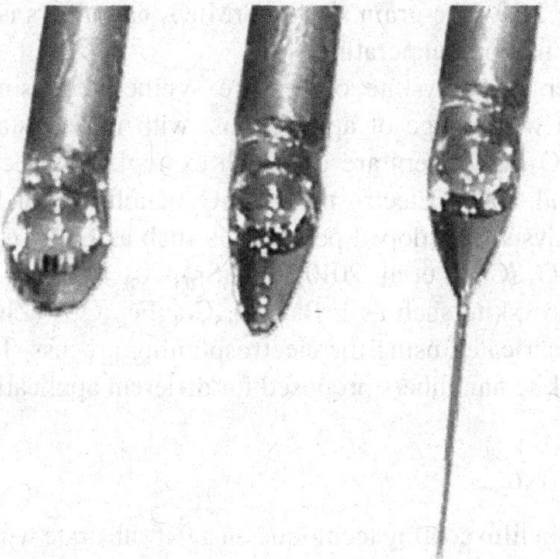

FIGURE 2.29 Photographs of the development of Taylor cone with an increasing electric field and eventual fiber ejection as the electric field increases, reproduced with permission of the Springer Nature (Laudenslager and Sigmund 2012).

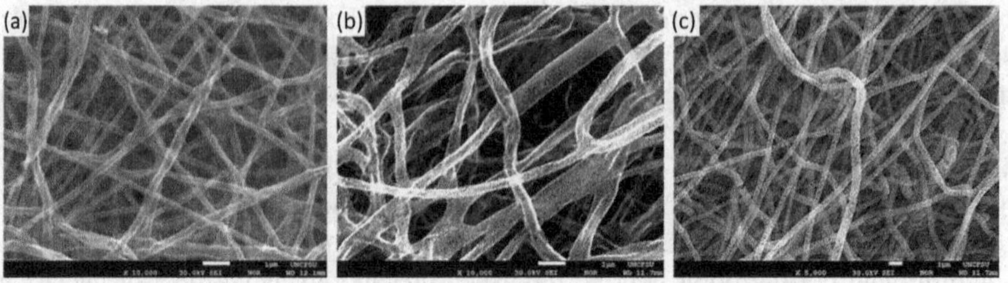

FIGURE 2.30 (a) Electrospun $SrMnO_3$ nanofibers obtained after calcining (a) 600, (b) 700, and 800°C, reproduced with permission of the Elsevier (George et al. 2018).

During the calcination of ceramic fibers, a low heating rate is often used, succeeded by a certain dwell period. A slow heating rate ensures the removal of organic components, without destroying the nanofibrillar morphological features in the end products, and also avoids the disintegration of ceramic fibers. The minimum calcination temperature is determined by the pyrolysis temperature of the polymer base used. Thermogravimetric analysis (TGA) is widely used to identify the degradation point of the polymer; it is also important to note that the calcination temperature should be well above the oxidation threshold of the metal salt. The calcination of the composite samples is performed at different high temperatures and dwell periods because the calcination temperature and the dwell period have a significant role in controlling the grain size of the fibers, which in turn reflects in the performance of the final products. If a dense rather than a nanoporous material is desired, sintering for an extended period is recommended. The high surface area of the particles leads to rapid densification due to the significant grain growth rate at high temperatures. Electrospun $SrMnO_3$ nanofibers obtained at different calcination temperatures is shown in Figure 2.30. The grain size of $SrMnO_3$ nanofibers is increased with an increase in the calcination temperature.

A large number of perovskite oxides are synthesized using the electrospinning process for a wide range of applications, with the assistance of a polymer. Ferroelectric $BaTiO_3$ nanofibers are one such example for electrospun perovskite nanofibers (Yuh et al. 2005). Electrospun $LaNiO_3$ nanofibers can be used for glucose sensing/electrocatalysis. The doped perovskites such as $La_{1-x}Ce_xCoO_{3\pm\delta}$ (Luo et al. 2015), $La_{0.8}Sr_{0.2}CoO_3$ (Chen et al. 2010), $Ba_{0.5}Sr_{0.5}Co_{0.8}Fe_{0.2}O_{3-\delta}$ (Hieu et al. 2012), etc. and double perovskite such as $PrBa_{0.5}Sr_{0.5}Co_{1.5}Fe_{0.5}O_{5+\delta}$ (Zhao et al. 2017) are also successfully fabricated using the electrospinning process. Table 2.9 lists some electrospun perovskite nanofibers proposed for different applications.

2.4.4 Spin Coating

Spin coating is a thin film coating technique on a flat substrate with a thickness at the nanoscale. The various stages of a spin coating technique are shown in Figure 2.31. The functional thin films are coated on glass or single-crystal substrates. To fabricate oxide layers, the precursors are prepared through the sol–gel route, and after spin coating, the substrates are heated at a high temperature to obtain the oxide layer.

TABLE 2.9
Electrospun Perovskite Nanofibers

Perovskite	Polymer	Solvent	Application	Reference
$BaTiO_3$	Poly(vinyl pyrrolidone) (PVP)	Ethanol	Ferroelectric material	Yuh et al. (2005)
$BaZrO_3$	PVP	Acetic acid	-	Calleja et al. (2011)
$Ba_{0.6}Sr_{0.4}TiO_3$	PVP	Ethanol	-	Maensiri et al. (2006)
$Ba_{0.5}Sr_{0.5}Co_{0.8}Fe_{0.2}O_{3-\delta}$	PVP	Ethanol and DMF	Cathode for low-temperature solid oxide fuel cells	Hieu et al. (2012)
$BiFeO_3$	PVP	DMF/acetone	Photocatalyst	Wang et al. (2013c)
	PVP	Acetone/water	Ferroelectric and photovoltaic devices	Fei et al. (2015)
$CoTiO_3$	PVP	Methanol/acetic acid	Catalyst for activating peroxymonosulfate in water	Lin et al. (2017)
$GdBaCo_2O_{5+\delta}$	PVP	Ethanol/water	Cathode for oxygen reduction reaction (ORR)	Jiang et al. (2013)
$LaNiO_3$	PVP	DMF	Electrochemical sensing of glucose and hydrogen peroxide	Wang et al. (2013a)
$La_{0.5}Sr_{0.5}Co_{0.8}Fe_{0.2}O_3$	PVP	H_2O, C_2H_5OH and DMF	Catalyst for metal air batteries	Park et al. (2014)
$La_{1-x}Ce_xCoO_{3\pm\delta}$	PVP	DMF	Catalytic oxidation of benzene	Luo et al. (2015)
$LaCoO_3$, $LaMnO_3$, $LaFeO_3$, $La_{0.8}Sr_{0.2}CoO_3$, and $La_{0.9}Ce_{0.1}CoO_3$	Citric acid/PVP	Ethanol/H_2O	Combustion of methane	Chen et al. (2010)
$La_xSr_{1-x}FeO_3$	PVP	DMF	Supercapacitor	Wang et al. (2019)
$La_{0.8}Sr_{0.2}MnO_3$	Polyacrylonitrile	DMF	Carbon monoxide sensor	Zhi et al. (2012)
$LaCoO_3$	Citric acid/PVP/SBA-15	DMF	Benzene oxidation catalyst	Luo et al. (2017)
	PVP	Water	Rechargeable Zn–air batteries	Shim et al. (2015)
$La_{0.6}Sr_{0.4}Co_{1-y}Fe_yO_3$	Polyvinyl butyral	Ethanol/DMF	Semiconductor-metal transition	Lubini et al. (2016)

(Continued)

TABLE 2.9 (*Continued*)
Electrospun Perovskite Nanofibers

Perovskite	Polymer	Solvent	Application	Reference
$LaMnO_3$	PVA	Water	-	Zhou et al. (2008)
$LaFeO_3$	PVP	Ethanol/acetic acid	Photocatalyst	Li and Wang (2015)
$La_{0.5}Sr_{0.5}CoO_{3-x}$	PVP	DMF	Rechargeable lithium oxygen batteries	Liu et al. (2015)
$La_{0.5}Sr_{0.5}TiO_3$	PVP	Ethanol	Ferromagnetism	Ponhan et al. (2014)
$L_{a0.88}Sr_{0.12}MnO_3$	PVP	DMF	Electrochemical detection of glucose	Xu et al. (2015)
$La_{1-x}K_xFeO_{3-\delta}$	PVP	DMF	Catalyst for soot combustion	Fang et al. (2018)
$Ag/LaFeO_3$	PVP	Ethanol and acetic acid	Photocatalyst	Li et al. (2016)
$La_{0.75}Sr_{0.25}MnO_3$	PVP	DMF	Rechargeable lithium–oxygen batteries	Xu et al. (2013)
$LaMnO_3$	PVP	Ethanol	Gas sensing	Chen and Yi (2018)
$La_{1-x}Sr_xCo_{0.2}Fe_{0.8}O_{3-\delta}$	PVP	DMF/ethanol	Soot oxidation	Lee et al. (2016)
$La_{0.75}Sr_{0.25}MnO_3$	PVP	DMF	Oxidation of CO and CH_4	Huang et al. (2014)
$LaMnO_3$	PVP	DMF	Fructose sensor	Xu et al. (2014)
$La_{0.6}Sr_{0.4}Co_{1-x}Fe_xO_{3-\delta}$	PVP and polyacrylonitrile (PAN)	DMF	Oxygen evolution reaction catalyst	Zhen et al. (2017)
$Pr_{0.5}Ba_{0.5}MnO_{3-\delta}$	PVP	DMF	Oxygen reduction reaction and oxygen evolution reaction	Zhang et al. (2016)
$PrBa_{0.5}Sr_{0.5}Co_{1.5}Fe_{0.5}O_{5+\delta}$	PVP	DMF	Oxygen evolution	Zhao et al. (2017)
$SrMnO_3$	PVP	Acetic acid/DMF	Supercapacitor	George et al. (2018)
$SrTiO_3$	PVP	Methyl alcohol and acetic acid	Photocatalyst	Yang et al. (2014)
$SrTi_{0.65}Fe_{0.35}O_{3-\delta}$	PVP	DMF/acetic acid	Oxygen sensing	Choi et al. (2013)

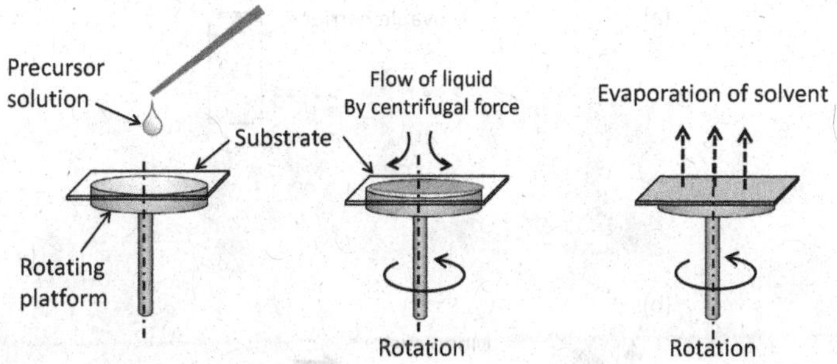

FIGURE 2.31 Spin coating process.

In spin coating, thin films are achieved by centrifugal force applied to the precursor solution. In spin coating, a liquid precursor droplet is placed at the center of the substrate, and the substrate is rapidly rotated until the excess solution spins off the substrate to form thin films with preferred thickness. The spin coating technique is comparatively easier and cheaper than CVD and PVD techniques; however, the thin oxide films obtained through the spin coating are polycrystalline in nature rather than single crystals in CVD or PVD. Unlike PVD or CVD, the selection of materials for spin coating is abundant. The major controlling factors of spin coating are rotation speed and the viscosity of the solution. However, factors such as time, the volatility of the solvent, and the wettability of the substrate play a vital role in determining the quality of the films.

$NiTiO_3$ (Phani and Santucci 2001), $CaZrO_3$ (Yu et al. 2004), and $Pb_{1/2}Sr_{1/2}Ti_{1-x}Fe_xO_3$ (Pontes et al. 2015) are some examples for the spin-coated thin films of perovskite structured oxide with the assistance of sol–gel technique and subsequent calcination at a high temperature (700°C–1,000°C). The spin coating can be performed using the suspension of perovskite oxides as well. $Ca(Ti, Fe)O_3$ film is obtained by spin coating of $Ca(Ti, Fe)O_3$ particle suspension with a particle size of 2–3 μm. A continuous dense film of 30–50 μm thickness is obtained at 1,235°C, which is higher than the sol-assisted technique (Itoh et al. 1997).

2.4.5 Langmuir–Blodgett Technique

Langmuir–Blodgett (LB) technique is designed for the deposition of monolayer of atoms. The alternating multiple layers are also possible by this technique. In the LB technique, as shown in Figure 2.32, a monomolecular layer of an amphiphilic precursor substance is developed on the water surface, and then it is transferred to a solid substrate. The precursor molecules are arranged on the air-water interface. Initially, the precursor solution containing the amphiphilic molecules is added slowly to a stagnant water surface. The molecules are arranged in such a way that the hydrophilic part is oriented towards the water. The monolayer on the water surface is then compressed using a movable barrier, as shown in Figure 2.32a, to form a continuous layer. The continuous monolayer film is then transferred to the required solid substrate by immersing the flat substrate to water and then extracted with the film adsorbed on the surface, as shown in Figure 2.32b. The transferring process

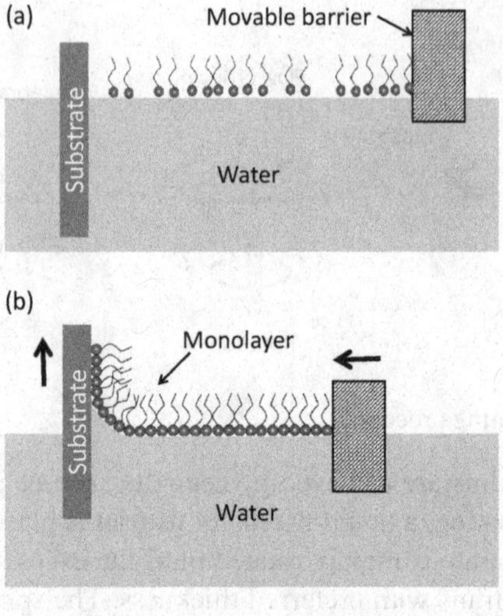

FIGURE 2.32 LB film technique (a) loosely packed precursor molecules by compression using a movable barrier across the surface of the water (b) transfer of Langmuir monolayer to a substrate by dip coating.

must be very slow in order to retain the continuity of the films. By repeating the process, one can obtain multi-layered films. The layer of molecules on the surface of the water is termed a Langmuir monolayer, and the film deposited on the substrate is called LB film.

In the reported technique for the fabrication of $PbTiO_3$ ultrathin layers, the LB film is fabricated by depositing a monolayer prepared from a solution containing stearic acid ($C_{17}H_{35}COOH$) and lead chloride ($PbCl_2$) and titanium potassium oxalate ($K_2TiO(C_2O_4)_2$). The LB film containing lead and titanium is then converted to an inorganic film, followed by ultraviolet/ozone (UVO) treatment and subsequent thermal annealing. The crystallographic orientation of $PbTiO_3$ thin films was controlled by conditions of precursor preparation such as the molecular ratio of lead and titanium, pH value, and/or temperature in the subphase and surface pressure (Sugai et al. 1999).

2.4.6 Spray Pyrolysis

Spray pyrolysis is a thin film deposition process by spraying a precursor solution on a heated surface. The process is similar to spray painting; here the paint is replaced by a precursor solution, and the coating is done on a heated surface where the precursors react to form the final compound. The reactants with volatile components upon heating are selected to result in the desired products. The reaction happens in two steps in a spray pyrolysis process. In the first step, the precursor in the solution form reaches the surface, and later, the reaction occurs at the heated surface. The controlling factors of the spray pyrolysis process are substrate temperature, carrier gas flow rate, nozzle-to-substrate distance, and the solution content and concentration. The substrate temperature plays the most significant role among the above parameters, since drying, decomposition, crystallization, and grain growth are dependent on the substrate temperature.

The typical spray pyrolysis process is shown in Figure 2.33. The substrate holder is heated by an electric heater whose temperature can be accurately controlled. At the spraying nozzle, the precursor solution is atomized with the help of compressed carrier gas, which is air in the case of oxide films. The pressure of the carrier gas is controlled by a pressure regulating valve, and if needed, the carrier gas is supplied in pulse mode. The flow of the precursor solution is also controlled by adjusting the flow rate of the pump. Porous $LaMnO_{3+\delta}$ (Kuai et al. 2018), Sr-doped $SmCoO_3$ (Shimada et al. 2016), $LaFe_xCo_{1-x}O_3$ (Dervishogullari et al. 2017), $La_{0.6}Sr_{0.4}Co_{0.2}Fe_{0.8}O_{3-\delta}$ (dos Santos-Gómez et al. 2018), $La_{1-x}Sr_xCoO_3$ (Viskadourakis et al. 2020), $NiSnO_3$ (Dridi et al. 2016), etc. are some examples of perovskite materials obtained by spray pyrolysis. Representative SEM images of the perovskite films obtained using spray pyrolysis are shown in Figure 2.34.

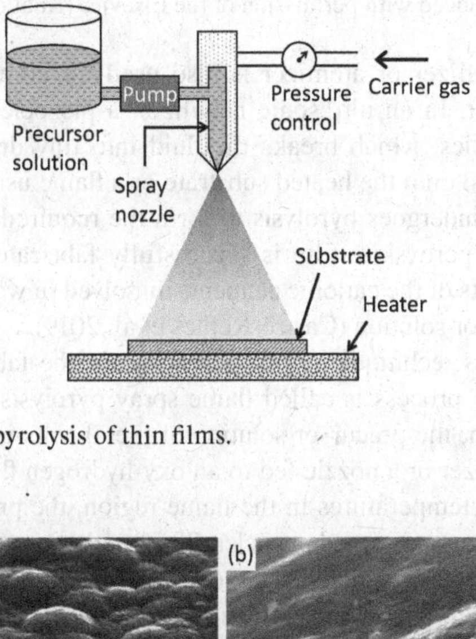

FIGURE 2.33 Spray pyrolysis of thin films.

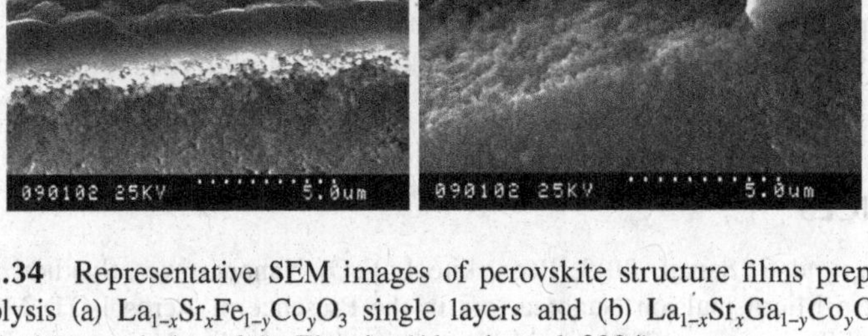

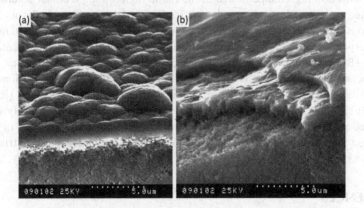

FIGURE 2.34 Representative SEM images of perovskite structure films prepared using spray pyrolysis (a) $La_{1-x}Sr_xFe_{1-y}Co_yO_3$ single layers and (b) $La_{1-x}Sr_xGa_{1-y}Co_yO_3$ bilayers, reproduced with permission of the Elsevier (Abrutis et al. 2004).

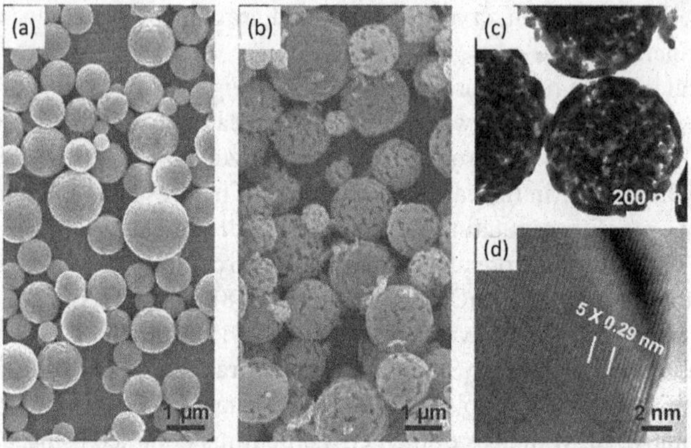

FIGURE 2.35 (a, b) SEM, (c) transmission electron microscope (TEM), and (d) high-resolution TEM (HRTEM) images of mesoporous $LaMnO_{3+\delta}$ particles obtained using a spray pyrolysis process, reproduced with permission of the Elsevier (Kuai et al. 2018).

An ultrasonic nebulizer or atomizer is also used for converting the precursor solution to a fine mist. In an ultrasonic nebulizer, a piezoelectric crystal vibrates at very high frequencies, which breaks the fluid into tiny droplets (aerosol). The aerosol is then pumped onto the heated substrate or a flame using a carrier gas (air), where the precursor undergoes pyrolysis to form the required oxide film/particles. $La_{1-x}Sr_xCo_{0.2}Fe_{0.8}O_{3-\delta}$ perovskite film is successfully fabricated through this technique using nitrate salts of the cationic elements dissolved in water/ethanol (75/25 by weight) as the precursor solution (Castro-Robles et al. 2019).

The spray pyrolysis technique can be extended for the fabrication of nanoparticles as well, and the process is called flame spray pyrolysis or flame hydrolysis. During the fabrication, the precursor solution of metal salts is atomized by means of an ultrasonic atomizer or a nozzle fed to an oxy-hydrogen flame. Due to the high drying rates and high temperatures in the flame region, the precursor solution will be instantly converted to nanosized particles. This method is suitable for precursors in gaseous, liquid, and solid states. Such a technique has a high production rate and yield particles with high purity and uniform size distribution. Nanoparticulates of $BaTiO_3$ (Brewster and Kodas 1997), La(Fe, Co)O_3 (Angel et al. 2020), $La_{1-x}A_xBO_3$ (A = vacancy, Ca, Sr, Y, Ce; B = Mn, Fe, Co, Ni, $Co_{0.5}Fe_{0.5}$, $Co_{0.2}Fe_{0.8}$) (Simmance et al. 2019), $LaCoO_3$ (Chiarello et al. 2005), $LaBO_{3\pm\delta}$ (B = Co, Mn, Fe) (Rossetti and Forni 2001), etc. are obtained by flame spray pyrolysis or flame hydrolysis. The atomized precursor compositions can also be dried with the assistance of a furnace, and then calcined to form the perovskite oxides with mesoporous structures, as shown in Figure 2.35.

REFERENCES

Abazari, R., and S. Sanati. 2013. Perovskite $LaFeO_3$ Nanoparticles Synthesized by the Reverse Microemulsion Nanoreactors in the Presence of Aerosol-OT: Morphology, Crystal Structure, and Their Optical Properties. *Superlattices Microstruct.* 64:148–157.

Abdel-Khalek, E. K., and H. M. Mohamed. 2013. Synthesis, Structural and Magnetic Properties of $La_{1-x}Ca_xFeO_3$ Prepared by the Co-Precipitation Method. *Hyperfine Interact.* 222:57–67.

Afsharikia, A., M. R. Dehghani, and M. Rezaei. 2017. Preparation of Vanadium-Based Perovskite by the Effective Method of Microemulsion on Enhanced Surface Area and Activity: Environmental Applications. *Mater. Chem. Phys.* 196:177–185.

Aimable, A., B. Xin, N. Millot, et al. 2008. Continuous Hydrothermal Synthesis of Nanometric $BaZrO_3$ in Supercritical Water. *J. Solid State Chem.* 181:183–189.

Alvarez-Galvan, C., A. Trunschke, H. Falcon, et al. 2018. Microwave-Assisted Coprecipitation Synthesis of $LaCoO_3$ Nanoparticles and Their Catalytic Activity for Syngas Production by Partial Oxidation of Methane. *Front. Energy Res.* 6:18.

Aman, D., T. Zaki, S. Mikhail, et al. 2011. Synthesis of a Perovskite $LaNiO_3$ Nanocatalyst at a Low Temperature Using Single Reverse Microemulsion. *Catal. Today* 164:209–213.

Angel, S., J. Neises, M. Dreyer, et al. 2020. Spray-Flame Synthesis of La(Fe,Co)O_3 Nano-Perovskites from Metal Nitrates. *AIChE J.* 66:e16748.

Anh, D. T. V., W. Olthuis, and P. Bergveld. 2004. Sensing Properties of Perovskite Oxide $La_{0.5}Sr_{0.5}CoO_{3-\delta}$ Obtained by Using Pulsed Laser Deposition. *Sens. Actuators B Chem.* 103:165–168.

Arney, D., B. Porter, B. Greve, et al. 2008. New Molten-Salt Synthesis and Photocatalytic Properties of $La_2Ti_2O_7$ Particles. *J. Photochem. Photobiol. Chem.* 199:230–235.

Baghurst, D. R., A. M. Chippindale, and D. M. P. Mingos. 1988. Microwave Syntheses for Superconducting Ceramics. *Nature* 332:311–311.

Baghurst, D. R., and D. M. P. Mingos. 1988. Application of Microwave Heating Techniques for the Synthesis of Solid State Inorganic Compounds. *J. Chem. Soc. Chem. Commun.* 1988:829–830.

Baharuddin, N. A., A. Muchtar, M. R. Somalu, et al. 2019. Synthesis and Characterization of Cobalt-Free $SrFe_{0.8}Ti_{0.2}O_{3-\delta}$ Cathode Powders Synthesized through Combustion Method for Solid Oxide Fuel Cells. *Int. J. Hydrog. Energy* 44:30682–30691.

Basavalingu, B., H. N. Girish, K. Byrappa, et al. 2008. Hydrothermal Synthesis and Characterization of Orthorhombic Yttrium Aluminum Perovskites (YAP). *Mater. Chem. Phys.* 112:723–725.

Bhat, M. H., A. Miura, P. Vinatier, et al. 2003. Microwave Synthesis of Lithium Lanthanum Titanate. *Solid State Commun.* 125:557–562.

Biasotto, G., A. Z. Simões, C. R. Foschini, et al. 2011. Microwave-Hydrothermal Synthesis of Perovskite Bismuth Ferrite Nanoparticles. *Mater. Res. Bull.* 46:2543–2547.

Brewster, J. H., and T. T. Kodas. 1997. Generation of Unagglomerated, Dense, $BaTiO_3$ Particles by Flame-Spray Pyrolysis. *AIChE J.* 43:2665–2669.

Buniatyan, V. V., M. H. Abouzar, N. W. Martirosyan, et al. 2010. PH-Sensitive Properties of Barium Strontium Titanate (BST) Thin Films Prepared by Pulsed Laser Deposition Technique. *Phys. Status Solidi A* 207:824–830.

Cai, Z., X. Xing, L. Li, et al. 2008. Molten Salt Synthesis of Lead Lanthanum Zirconate Titanate Ceramic Powders. *J. Alloys Compd.* 454:466–470.

Cai, Z., X. Xing, R. Yu, et al. 2007. Morphology-Controlled Synthesis of Lead Titanate Powders. *Inorg. Chem.* 46:7423–7427.

Calleja, A., X. Granados, S. Ricart, et al. 2011. High Temperature Transformation of Electrospun $BaZrO_3$ Nanotubes into Nanoparticle Chains. *CrystEngComm.* 13:7224–7230.

Castro-Robles, J. D., N. Soltani, and J. Á. Chávez-Carvayar. 2019. Structural, Morphological and Transport Properties of Nanostructured $La_{1-x}Sr_xCo_{0.2}Fe_{0.8}O_{3-\delta}$ Thin Films, Deposited by Ultrasonic Spray Pyrolysis. *Mater. Chem. Phys.* 225:50–54.

Chae, N. S., K. S. Park, Y. S. Yoon, et al. 2008. Sr- and Mg-Doped $LaGaO_3$ Powder Synthesis by Carbonate Coprecipitation. *Colloids Surf. Physicochem. Eng. Asp. Nanosci. Nanotechnol.* 313–314:154–157.

Chang, H. W., F. T. Yuan, Y. C. Yu, et al. 2013. Photovoltaic Property of Sputtered $BiFeO_3$ Thin Films. *J. Alloys Compd.* 574:402–406.

Chaoui, N., E. Millon, J. F. Muller, et al. 1999. On the Role of Ambient Oxygen in the Formation of Lead Titanate Pulsed Laser Deposition Thin Films. *Appl. Surf. Sci.* 138–139:256–260.

Chen, C., J. Cheng, S. Yu, et al. 2006. Hydrothermal Synthesis of Perovskite Bismuth Ferrite Crystallites. *J. Cryst. Growth* 291:135–139.

Chen, C.-Q., W. Li, C.-Y. Cao, et al. 2010. Enhanced Catalytic Activity of Perovskite Oxide Nanofibers for Combustion of Methane in Coal Mine Ventilation Air. *J. Mater. Chem.* 20:6968–6974.

Chen, D., and J. Yi. 2018. One-Pot Electrospinning and Gas-Sensing Properties of $LaMnO_3$ Perovskite/SnO_2 Heterojunction Nanofibers. *J. Nanoparticle Res.* 20:65.

Chen, J., X. Xing, A. Watson, et al. 2007a. Rapid Synthesis of Multiferroic $BiFeO_3$ Single-Crystalline Nanostructures. *Chem. Mater.* 19:3598–3600.

Chen, J., R. Yu, L. Li, et al. 2008. Structure and Shape Evolution of $Bi_{1-x}La_xFeO_3$ Perovskite Microcrystals by Molten Salt Synthesis. *Eur. J. Inorg. Chem.* 2008:3655–3660.

Chen, K., and X. Zhang. 2010. Synthesis of Calcium Copper Titanate Ceramics via the Molten Salts Method. *Ceram. Int.* 36:1523–1527.

Chen, X., L. Rieth, M. S. Miller, et al. 2009. High Temperature Humidity Sensors Based on Sputtered Y-Doped $BaZrO_3$ Thin Films. *Sens. Actuators B Chem.* 137:578–585.

Chen, Y., H. Yuan, G. Li, et al. 2007b. Crystal Growth and Magnetic Property of Orthorhombic $RMnO_3$ (R = Sm–Ho) Perovskites by Mild Hydrothermal Synthesis. *J. Cryst. Growth* 305:242–248.

Chen, Y., H. Yuan, G. Tian, et al. 2007c. Mild Hydrothermal Synthesis and Magnetic Properties of the Manganates $Pr_{1-x}Ca_xMnO_3$. *J. Solid State Chem.* 180:167–172.

Chiarello, G. L., I. Rossetti, and L. Forni. 2005. Flame-Spray Pyrolysis Preparation of Perovskites for Methane Catalytic Combustion. *J. Catal.* 236:251–261.

Chiu, C. C., C. C. Li, and S. B. Desu. 1991. Molten Salt Synthesis of a Complex Perovskite, $Pb(Fe_{0.5}Nb_{0.5})O_3$. *J. Am. Ceram. Soc.* 74:38–41.

Cho, W.-S., and M. Yoshimura. 1997. Hydrothermal Synthesis of $PbTiO_3$ Films. *J. Mater. Res.* 12:833–839.

Choi, J. Y., C. H. Kim, and D. K. Kim. 1998. Hydrothermal Synthesis of Spherical Perovskite Oxide Powders Using Spherical Gel Powders. *J. Am. Ceram. Soc.* 81:1353–1356.

Choi, S.-H., S.-J. Choi, B. K. Min, et al. 2013. Facile Synthesis of p-Type Perovskite $SrTi_{0.65}Fe_{0.35}O_{3-\delta}$ Nanofibers Prepared by Electrospinning and Their Oxygen-Sensing Properties. *Macromol. Mater. Eng.* 298:521–527.

Chou, C.-C., C.-S. Hou, G.-C. Chang, et al. 1999. Pulsed Laser Deposition of Ferroelectric $Pb_{0.6}Sr_{0.4}TiO_3$ Thin Films on Perovskite Substrates. *Appl. Surf. Sci.* 142:413–417.

Choudhary, V. R., B. S. Uphade, and S. G. Pataskar. 1999. Low Temperature Complete Combustion of Methane over Ag-Doped $LaFeO_3$ and $LaFe_{0.5}Co_{0.5}O_3$ Perovskite Oxide Catalysts. *Fuel* 78:919–921.

Choudhary, V. R., B. S. Uphade, S. G. Pataskar, et al. 1996. Low-Temperature Total Oxidation of Methane over Ag-Doped $LaMO_3$ Perovskite Oxides. *Chem. Commun.* 1996:1021–1022.

Christen, H. M., and G. Eres. 2008. Recent Advances in Pulsed-Laser Deposition of Complex Oxides. *J. Phys. Condens. Matter* 20:264005.

Cillessen, J. F. M., R. M. Wolf, and A. E. M. De Veirman. 1993. Hetero-Epitaxial Oxidic Conductor $La_{1-x}Sr_xCoO_3$ Prepared by Pulsed Laser Deposition. *Appl. Surf. Sci.* 69:212–215.

Civera, A., M. Pavese, G. Saracco, et al. 2003. Combustion Synthesis of Perovskite-Type Catalysts for Natural Gas Combustion. *Catal. Today* 83:199–211.

Combemale, L., G. Caboche, and D. Stuerga. 2009. Flash Microwave Synthesis and Sintering of Nanosized $La_{0.75}Sr_{0.25}Cr_{0.93}Ru_{0.07}O_{3-\delta}$ for Fuel Cell Application. *J. Solid State Chem.* 182:2829–2834.

Dervishogullari, D., C. A. Sharpe, and L. R. Sharpe. 2017. $LaFe_xCo_{(1-x)}O_3$ Thin-Film Oxygen Reduction Catalysts Prepared Using Spray Pyrolysis without Conductive Additives. *ACS Omega* 2:7695–7701.

Dhal, G. C., S. Dey, D. Mohan, et al. 2017. Solution Combustion Synthesis of Perovskite-Type Catalysts for Diesel Engine Exhaust Gas Purification. *Mater. Today Proc.* 4:10489–10493.

Djoudi, L., and M. Omari. 2015. Synthesis and Characterization of Perovskite Oxides $LaAl_{1-x}Ni_xO_{3-\delta}$ $(0 \leq x \leq 0.6)$ via Co-Precipitation Method. *J. Inorg. Organomet. Polym. Mater.* 25:796–803.

Dong, D., Y. Wu, X. Zhang, et al. 2011. Eggshell Membrane-Templated Synthesis of Highly Crystalline Perovskite Ceramics for Solid Oxide Fuel Cells. *J. Mater. Chem.* 21:1028–1032.

dos Santos-Gómez, L., J. Hurtado, J. M. Porras-Vázquez, et al. 2018. Durability and Performance of CGO Barriers and LSCF Cathode Deposited by Spray-Pyrolysis. *J. Eur. Ceram. Soc.* 38:3518–3526.

Dridi, R., A. Mhamdi, A. Labidi, et al. 2016. Electrical Conductivity and Ethanol Sensing of a New Perovskite $NiSnO_3$ Sprayed Thin Film. *Mater. Chem. Phys.* 182:498–502.

Dutta, A., J. Mukhopadhyay, and R. N. Basu. 2009. Combustion Synthesis and Characterization of LSCF-Based Materials as Cathode of Intermediate Temperature Solid Oxide Fuel Cells. *J. Eur. Ceram. Soc.* 29:2003–2011.

Eaton, C., J. A. Moyer, H. M. Alipour, et al. 2015. Growth of $SrVO_3$ Thin Films by Hybrid Molecular Beam Epitaxy. *J. Vac. Sci. Technol. A* 33:061504.

Einaga, H., Y. Nasu, M. Oda, et al. 2016. Catalytic Performances of Perovskite Oxides for CO Oxidation under Microwave Irradiation. *Chem. Eng. J.* 283:97–104.

Fabian, F. A., P. P. Pedra, J. L. S. Filho, et al. 2015. Synthesis and Characterization of La(Cr,Fe,Mn)O_3 Nanoparticles Obtained by Co-Precipitation Method. *J. Magn. Magn. Mater.* 379:80–83.

Fang, F., N. Feng, L. Wang, et al. 2018. Fabrication of Perovskite-Type Macro/Mesoporous $La_{1-x}K_xFeO_{3-\delta}$ Nanotubes as an Efficient Catalyst for Soot Combustion. *Appl. Catal. B Environ.* 236:184–194.

Farhadi, S., Z. Momeni, and M. Taherimehr. 2009. Rapid Synthesis of Perovskite-Type $LaFeO_3$ Nanoparticles by Microwave-Assisted Decomposition of Bimetallic $La[Fe(CN)_6]\cdot 5H_2O$ Compound. *J. Alloys Compd.* 471:L5–L8.

Farhadi, S., and S. Sepahvand. 2010. Microwave-Assisted Solid-State Decomposition of $La[Co(CN)_6]\cdot 5H_2O$ Precursor: A Simple and Fast Route for the Synthesis of Single-Phase Perovskite-Type $LaCoO_3$ Nanoparticles. *J. Alloys Compd.* 489:586–591.

Fei, L., Y. Hu, X. Li, et al. 2015. Electrospun Bismuth Ferrite Nanofibers for Potential Applications in Ferroelectric Photovoltaic Devices. *ACS Appl. Mater. Interfaces* 7:3665–3670.

Fei, L., J. Yuan, Y. Hu, et al. 2011. Visible Light Responsive Perovskite $BiFeO_3$ Pills and Rods with Dominant $\{111\}_c$ Facets. *Cryst. Growth Des.* 11:1049–1053.

Frangini, S., A. Masci, and F. Zaza. 2011. Molten Salt Synthesis of Perovskite Conversion Coatings: A Novel Approach for Corrosion Protection of Stainless Steels in Molten Carbonate Fuel Cells. *Corros. Sci.* 53:2539–2548.

Fujimoto, K., Y. Kobayashi, and K. Kubota. 1989. Growth of $BaTiO_3$-$SrTiO_3$ Thin Films by r.f. Magnetron Sputtering. *Thin Solid Films* 169:249–256.

Fumo, D. A., JoséR. Jurado, A. M. Segadães, et al. 1997. Combustion Synthesis of Iron-Substituted Strontium Titanate Perovskites. *Mater. Res. Bull.* 32:1459–1470.

Gabal, M. A., F. Al-Solami, Y. M. Al Angari, et al. 2019. Auto-Combustion Synthesis and Characterization of Perovskite-Type $LaFeO_3$ Nanocrystals Prepared via Different Routes. *Ceram. Int.* 45:16530–16539.

Ganguly, K., P. Ambwani, P. Xu, et al. 2015. Structure and Transport in High Pressure Oxygen Sputter-Deposited $BaSnO_{3-\delta}$. *APL Mater.* 3:062509.

Garg, A., A. Snedden, P. Lightfoot, et al. 2004. Investigation of Structural and Ferroelectric Properties of Pulsed-Laser-Ablated Epitaxial Nd-Doped Bismuth Titanate Films. *J. Appl. Phys.* 96:3408–3412.

Gargori, C., S. Cerro, R. Galindo, et al. 2012. Iron and Chromium Doped Perovskite ($CaMO_3$ M = Ti, Zr) Ceramic Pigments, Effect of Mineralizer. *Ceram. Int.* 38:4453–4460.

George, G., S. L. Jackson, C. Q. Luo, et al. 2018. Effect of Doping on the Performance of High-Crystalline $SrMnO_3$ Perovskite Nanofibers as a Supercapacitor Electrode. *Ceram. Int.* 44:21982–21992.

Gersten, B. L., M. M. Lencka, and R. E. Riman. 2004. Low-Temperature Hydrothermal Synthesis of Phase-Pure $(Ba,Sr)TiO_3$ Perovskite Using EDTA. *J. Am. Ceram. Soc.* 87:2025–2032.

Giannakas, A. E., A. K. Ladavos, and P. J. Pomonis. 2004. Preparation, Characterization and Investigation of Catalytic Activity for NO+CO Reaction of $LaMnO_3$ and $LaFeO_3$ Perovskites Prepared via Microemulsion Method. *Appl. Catal. B Environ.* 49:147–158.

Gibbons, K. E., M. O. Jones, S. J. Blundell, et al. 2000. Rapid Synthesis of Colossal Magnetoresistance Manganites by Microwave Dielectric Heating. *Chem. Commun.* 2000:159–160.

Gonell, F., N. Alem, P. Dunne, et al. 2019. Versatile Molten Salt Synthesis of Manganite Perovskite Oxide Nanocrystals and Their Magnetic Properties. *ChemNanoMat* 5:358–363.

González, A., E. Martínez Tamayo, A. Beltrán Porter, et al. 1997. Synthesis of High Surface Area Perovskite Catalysts by Non-Conventional Routes. *Catal. Today* 33:361–369.

Gopalan, S., K. Mehta, and A. V. Virkar. 1996. Synthesis of Oxide Perovskite Solid Solutions Using the Molten Salt Method. *J. Mater. Res.* 11:1863–1865.

Gupta, S. K., N. Pathak, and R. M. Kadam. 2016. An Efficient Gel-Combustion Synthesis of Visible Light Emitting Barium Zirconate Perovskite Nanoceramics: Probing the Photoluminescence of Sm^{3+} and Eu^{3+} Doped $BaZrO_3$. *J. Lumin.* 169:106–114.

Habermeier, H.-U. 2007. Thin Films of Perovskite-Type Complex Oxides. *Mater. Today* 10:34–43.

Haron, W., A. Wisitsoraat, U. Sirimahachai, et al. 2018. A Simple Synthesis and Characterization of $LaMO_3$ (M = Al, Co, Fe, Gd) Perovskites via Chemical Co-Precipitation Method. *Songklanakarin J. Sci. Technol.* 40:484–491.

He, Y., Y. Zhu, and N. Wu. 2004. Mixed Solvents: A Key in Solvothermal Synthesis of $KTaO_3$. *J. Solid State Chem.* 177:2985–2990.

Hernandez, B. A., K.-S. Chang, E. R. Fisher, et al. 2002. Sol–Gel Template Synthesis and Characterization of $BaTiO_3$ and $PbTiO_3$ Nanotubes. *Chem. Mater.* 14:480–482.

Hieu, N. T., J. Park, and B. Tae. 2012. Synthesis and Characterization of Nanofiber-Structured $Ba_{0.5}Sr_{0.5}Co_{0.8}Fe_{0.2}O_{3-\delta}$ Perovskite Oxide Used as a Cathode Material for Low-Temperature Solid Oxide Fuel Cells. *Mater. Sci. Eng. B* 177:205–209.

Hirai, T., T. Goto, H. Matsuhashi, et al. 1993. Preparation of Tetragonal Perovskite Single Phase $PbTiO_3$ Film Using an Improved Metal-Organic Chemical Vapor Deposition Method Alternately Introducing Pb and Ti Precursors. *Jpn. J. Appl. Phys.* 32:4078.

Hong, M., S. H. Liou, D. D. Bacon, et al. 1988. Superconducting Tl-Ba-Ca-Cu-O Films by Sputtering. *Appl. Phys. Lett.* 53:2102–2104.

Hou, Y.-D., L. Hou, T.-T. Zhang, et al. 2007. $(Na_{0.8}K_{0.2})_{0.5}Bi_{0.5}TiO_3$ Nanowires: Low-Temperature Sol–Gel–Hydrothermal Synthesis and Densification. *J. Am. Ceram. Soc.* 90:1738–1743.

Hsieh, F.-F., N. Okinaka, and T. Akiyama. 2009. Combustion Synthesis of Doped $LaGaO_3$ Perovskite Oxide with Fe. *J. Alloys Compd.* 484:747–752.

Hu, L., C. Wang, S. Lee, et al. 2013. $SrTiO_3$ Nanocuboids from a Lamellar Microemulsion. *Chem. Mater.* 25:378–384.

Huang, K., X. Chu, W. Feng, et al. 2014. Catalytic Behavior of Electrospinning Synthesized $La_{0.75}Sr_{0.25}MnO_3$ Nanofibers in the Oxidation of CO and CH_4. *Chem. Eng. J.* 244:27–32.

Huang, K.-C., T.-C. Huang, and W.-F. Hsieh. 2009. Morphology-Controlled Synthesis of Barium Titanate Nanostructures. *Inorg. Chem.* 48:9180–9184.

Huang, S.-T., W. W. Lee, J.-L. Chang, et al. 2014. Hydrothermal Synthesis of $SrTiO_3$ Nanocubes: Characterization, Photocatalytic Activities, and Degradation Pathway. *J. Taiwan Inst. Chem. Eng.* 45:1927–1936.

Ifrah, S., A. Kaddouri, P. Gelin, et al. 2007. Conventional Hydrothermal Process versus Microwave-Assisted Hydrothermal Synthesis of $La_{1-x}Ag_xMnO_{3+\delta}$ (x = 0, 0.2) Perovskites Used in Methane Combustion. *Comptes Rendus Chim.* 10:1216–1226.

Itoh, H., H. Asano, K. Fukuroi, et al. 1997. Spin Coating of a Ca(Ti,Fe)O_3 Dense Film on a Porous Substrate for Electrochemical Permeation of Oxygen. *J. Am. Ceram. Soc.* 80:1359–1365.

Jalan, B., R. Engel-Herbert, N. J. Wright, et al. 2009. Growth of High-Quality $SrTiO_3$ Films Using a Hybrid Molecular Beam Epitaxy Approach. *J. Vac. Sci. Technol. A* 27:461–464.

Jan, D.-J., C.-T. Lin, and C.-F. Ai. 2008. Structural Characterization of $La_{0.67}Sr_{0.33}MnO_3$ Protective Coatings for Solid Oxide Fuel Cell Interconnect Deposited by Pulsed Magnetron Sputtering. *Thin Solid Films* 516:6300–6304.

Ji, Q., P. Xue, H. Wu, et al. 2019. Structural Characterizations and Dielectric Properties of Sphere- and Rod-Like $PbTiO_3$ Powders Synthesized via Molten Salt Synthesis. *Nanoscale Res. Lett.* 14:62.

Jia, X., H. Fan, X. Lou, et al. 2009. Synthesis and Gas Sensing Properties of Perovskite $CdSnO_3$ Nanoparticles. *Appl. Phys. A* 94:837–841.

Jiang, X., H. Xu, Q. Wang, et al. 2013. Fabrication of $GdBaCo_2O_{5+\delta}$ Cathode Using Electrospun Composite Nanofibers and Its Improved Electrochemical Performance. *J. Alloys Compd.* 557:184–189.

Johnson, C., R. Gemmen, and N. Orlovskaya. 2004. Nano-Structured Self-Assembled $LaCrO_3$ Thin Film Deposited by RF-Magnetron Sputtering on a Stainless Steel Interconnect Material. *Compos. Part B Eng.* 35:167–172.

Johnson, C., N. Orlovskaya, A. Coratolo, et al. 2009. The Effect of Coating Crystallization and Substrate Impurities on Magnetron Sputtered Doped $LaCrO_3$ Coatings for Metallic Solid Oxide Fuel Cell Interconnects. *Int. J. Hydrog. Energy* 34:2408–2415.

Joshi, U. A., and J. S. Lee. 2005. Template-Free Hydrothermal Synthesis of Single-Crystalline Barium Titanate and Strontium Titanate Nanowires. *Small* 1:1172–1176.

Junwu, Z., S. Xiaojie, W. Yanping, et al. 2007. Solution-Phase Synthesis and Characterization of Perovskite $LaCoO_3$ Nanocrystals via A Co-Precipitation Route. *J. Rare Earths* 25:601–604.

Kajiyoshi, K., Y. Sakabe, and M. Yoshimura. 1997. Electrical Properties of $BaTiO_3$ Thin Film Grown by the Hydrothermal-Electrochemical Method. *Jpn. J. Appl. Phys.* 36:1209.

Kajiyoshi, K., K. Tomono, Y. Hamaji, et al. 1994. Contribution of Electrolysis Current to Growth of $SrTiO_3$ Thin Film by the Hydrothermal-Electrochemical Method. *J. Mater. Res.* 9:2109–2117.

Kao, C.-F., and C.-L. Jeng. 2000. Preparation and Characterisation of Lanthanum Nickel Strontium Oxides by Combined Coprecipitation and Molten Salt Reactions. *Ceram. Int.* 26:237–243.

Karthäuser, S., E. Vasco, R. Dittmann, et al. 2004. Fabrication of Arrays of $SrZrO_3$ Nanowires by Pulsed Laser Deposition. *Nanotechnology* 15:S122–S125.

Kim, I. W., C. W. Ahn, J. S. Kim, et al. 2002. Low-Frequency Dielectric Relaxation and Ac Conduction of $SrBi_2Ta_2O_9$ Thin Film Grown by Pulsed Laser Deposition. *Appl. Phys. Lett.* 80:4006–4008.

Kim, S., Y. L. Yang, A. J. Jacobson, et al. 1998. Diffusion and Surface Exchange Coefficients in Mixed Ionic Electronic Conducting Oxides from the Pressure Dependence of Oxygen Permeation. *Solid State Ion.* 106:189–195.

Kim, Y. K., A. Sumi, K. Takahashi, et al. 2005. Metalorganic Chemical Vapor Deposition of Epitaxial Perovskite $SrIrO_3$ Films on (100)$SrTiO_3$ Substrates. *Jpn. J. Appl. Phys.* 45:L36.

Kingsley, J. J., and L. R. Pederson. 1993. Combustion Synthesis of Perovskite $LnCrO_3$ Powders Using Ammonium Dichromate. *Mater. Lett.* 18:89–96.

Klvana, D., J. Kirchnerová, J. Chaouki, et al. 1999. Fiber-Supported Perovskites for Catalytic Combustion of Natural Gas. *Catal. Today* 47:115–121.

Kojima, T., K. Nomura, Y. Miyazaki, et al. 2006. Synthesis of Various $LaMO_3$ Perovskites in Molten Carbonates. *J. Am. Ceram. Soc.* 89:3610–3616.

Kuai, L., E. Kan, W. Cao, et al. 2018. Mesoporous $LaMnO_{3+\delta}$ Perovskite from Spray–Pyrolysis with Superior Performance for Oxygen Reduction Reaction and Zn– Air Battery. *Nano Energy* 43:81–90.

Kuang, Q., and S. Yang. 2013. Template Synthesis of Single-Crystal-Like Porous $SrTiO_3$ Nanocube Assemblies and Their Enhanced Photocatalytic Hydrogen Evolution. *ACS Appl. Mater. Interfaces* 5:3683–3690.

Kulkarni, A. S., and R. V. Jayaram. 2003. Liquid Phase Catalytic Transfer Hydrogenation of Aromatic Nitro Compounds on Perovskites Prepared by Microwave Irradiation. *Appl. Catal. Gen.* 252:225–230.

Kuprenaite, S., V. Astié, S. Margueron, et al. 2019. Relationship Processing–Composition–Structure–Resistivity of $LaNiO_3$ Thin Films Grown by Chemical Vapor Deposition Methods. *Coatings* 9:35.

Kwak, B. S., K. Zhang, E. P. Boyd, et al. 1991. Metalorganic Chemical Vapor Deposition of $BaTiO_3$ Thin Films. *J. Appl. Phys.* 69:767–772.

Kwo, J., M. Hong, D. J. Trevor, et al. 1988. In Situ Epitaxial Growth of $Y_1Ba_2Cu_3O_{7-x}$ Films by Molecular Beam Epitaxy with an Activated Oxygen Source. *Appl. Phys. Lett.* 53:2683–2685.

Labhsetwar, N. K., A. Watanabe, and T. Mitsuhashi. 2003. New Improved Syntheses of $LaRuO_3$ Perovskites and Their Applications in Environmental Catalysis. *Appl. Catal. B Environ.* 40:21–30.

Lan, L., Q. Li, G. Gu, et al. 2015. Hydrothermal Synthesis of γ-MnOOH Nanorods and Their Conversion to MnO_2, Mn_2O_3, and Mn_3O_4 Nanorods. *J. Alloys Comp.* 644:430–437.

Langfeld, K., E. V. Kondratenko, O. Görke, et al. 2011. Microemulsion-Aided Synthesis of Nanosized Perovskite-Type $SrCoO_x$ Catalysts. *Catal. Lett.* 141:772–778.

Laudenslager, M. J., and W. M. Sigmund. 2012. Electrospinning. In *Encyclopedia of Nanotechnology*, ed. B. Bhushan, 769–775. Dordrecht: Springer Netherlands.

Laughlin, B., J. Ihlefeld, and J.-P. Maria. 2005. Preparation of Sputtered $(Ba_x,Sr_{1-x})TiO_3$ Thin Films Directly on Copper. *J. Am. Ceram. Soc.* 88:2652–2654.

Lee, C., Y. Jeon, S. Hata, et al. 2016. Three-Dimensional Arrangements of Perovskite-Type Oxide Nano-Fiber Webs for Effective Soot Oxidation. *Appl. Catal. B Environ.* 191:157–164.

Lee, M., M. Kawasaki, M. Yoshimoto, et al. 1995. Heteroepitaxial Growth of $BaTiO_3$ Films on Si by Pulsed Laser Deposition. *Appl. Phys. Lett.* 66:1331–1333.

Lee, Y., T. Watanabe, T. Takata, et al. 2005. Preparation and Characterization of Sodium Tantalate Thin Films by Hydrothermal–Electrochemical Synthesis. *Chem. Mater.* 17:2422–2426.

Lee, Y.-H., and J.-M. Wu. 2004. Epitaxial Growth of $LaFeO_3$ Thin Films by RF Magnetron Sputtering. *J. Cryst. Growth* 263:436–441.

Lee, Y. N., R. M. Lago, J. L. G. Fierro, et al. 2001. Hydrogen Peroxide Decomposition over $Ln_{1-x}A_xMnO_3$ (Ln = La or Nd and A = K or Sr) Perovskites. *Appl. Catal. Gen.* 215:245–256.

Lemée, N., C. Dubourdieu, G. Delabouglise, et al. 2002. Semiconductive Nb-Doped $BaTiO_3$ Films Grown by Pulsed Injection Metalorganic Chemical Vapor Deposition. *J. Cryst. Growth* 235:347–351.

Li, B., W. Shang, Z. Hu, et al. 2014a. Template-Free Fabrication of Pure Single-Crystalline $BaTiO_3$ Nano-Wires by Molten Salt Synthesis Technique. *Ceram. Int.* 40:73–80.

Li, D., and Y. Xia. 2004. Electrospinning of Nanofibers: Reinventing the Wheel? *Adv. Mater.* 16:1151–1170.

Li, H.-L., Z.-N. Du, G.-L. Wang, et al. 2010. Low Temperature Molten Salt Synthesis of $SrTiO_3$ Submicron Crystallites and Nanocrystals in the Eutectic NaCl–KCl. *Mater. Lett.* 64:431–434.

Li, L., J. Deng, J. Chen, et al. 2009. Wire Structure and Morphology Transformation of Niobium Oxide and Niobates by Molten Salt Synthesis. *Chem. Mater.* 21:1207–1213.

Li, L., J. Song, Q. Lu, et al. 2014b. Synthesis of Nano-Crystalline $Sm_{0.5}Sr_{0.5}Co(Fe)O_{3-\delta}$ Perovskite Oxides by a Microwave-Assisted Sol–Gel Combustion Process. *Ceram. Int.* 40:1189–1194.

Li, S., and X. Wang. 2015. Synthesis of Different Morphologies Lanthanum Ferrite ($LaFeO_3$) Fibers via Electrospinning. *Optik* 126:408–410.

Li, S., Y. Zhao, C. Wang, et al. 2016. Fabrication and Characterization Unique Ribbon-like Porous Ag/$LaFeO_3$ Nanobelts Photocatalyst via Electrospinning. *Mater. Lett.* 170:122–125.

Li, T., G. Wang, K. Li, et al. 2014c. Electrical Properties of Lead-Free KNN Films on SRO/STO by RF Magnetron Sputtering. *Ceram. Int.* 40:1195–1198.

Li, Z., W. E. Lee, and S. Zhang. 2007. Low-Temperature Synthesis of $CaZrO_3$ Powder from Molten Salts. *J. Am. Ceram. Soc.* 90:364–368.

Lim, C., C. Kim, O. Gwon, et al. 2018. Nano-Perovskite Oxide Prepared via Inverse Microemulsion Mediated Synthesis for Catalyst of Lithium-Air Batteries. *Electrochim. Acta* 275:248–255.

Lin, K.-Y. A., T.-Y. Lin, Y.-C. Lu, et al. 2017. Electrospun Nanofiber of Cobalt Titanate Perovskite as an Enhanced Heterogeneous Catalyst for Activating Peroxymonosulfate in Water. *Chem. Eng. Sci.* 168:372–379.

Lippert, T., M. J. Montenegro, M. Döbeli, et al. 2007. Perovskite Thin Films Deposited by Pulsed Laser Ablation as Model Systems for Electrochemical Applications. *Prog. Solid State Chem.* 35:221–231.

Liu, D., Y. Yan, and H. Zhou. 2007. Synthesis of Micron-Scale Platelet $BaTiO_3$. *J. Am. Ceram. Soc.* 90:1323–1326.

Liu, G., H. Chen, L. Xia, et al. 2015. Hierarchical Mesoporous/Macroporous Perovskite $La_{0.5}Sr_{0.5}CoO_{3-x}$ Nanotubes: A Bifunctional Catalyst with Enhanced Activity and Cycle Stability for Rechargeable Lithium Oxygen Batteries. *ACS Appl. Mater. Interfaces* 7:22478–22486.

Liu, J. W., G. Chen, Z. H. Li, et al. 2007. Hydrothermal Synthesis and Photocatalytic Properties of $ATaO_3$ and $ANbO_3$ (A = Na and K). *Int. J. Hydrog. Energy* 32:2269–2272.

Liu, S., X. Qian, and J. Xiao. 2007. Synthesis and Characterization of $La_{0.8}Sr_{0.2}Co_{0.5}Fe_{0.5}O_{3\pm\delta}$ Nanopowders by Microwave Assisted Sol–Gel Route. *J. Sol–Gel Sci. Technol.* 44:187–193.

Liu, Shaomin, X. Tan, K. Li, et al. 2002. Synthesis of Strontium Cerates-Based Perovskite Ceramics via Water-Soluble Complex Precursor Routes. *Ceram. Int.* 28:327–335.

Livage, J., M. Henry, and C. Sanchez. 1988. Sol–Gel Chemistry of Transition Metal Oxides. *Prog. Solid State Chem.* 18:259–341.

López-Trosell, A., and R. Schomäcker. 2006. Synthesis of Manganite Perovskite $Ca_{0.5}Sr_{0.5}MnO_3$ Nanoparticles in w/o-Microemulsion. *Mater. Res. Bull.* 41:333–339.
Lozano-Sánchez, L. M., S.-W. Lee, T. Sekino, et al. 2013. Practical Microwave-Induced Hydrothermal Synthesis of Rectangular Prism-like $CaTiO_3$. *CrystEngComm* 15:2359–2362.
Lu, H. B., N. Wang, W. Z. Chen, et al. 2000. Laser Molecular Beam Epitaxy of $BaTiO_3$ and $SrTiO_3$ Ultra Thin Films. *J. Cryst. Growth* 212:173–177.
Lu, W., and H. Schmidt. 2008. Synthesis of Tin Oxide Hydrate ($SnO_2 \cdot xH_2O$) Gel and Its Effects on the Hydrothermal Preparation of $BaSnO_3$ Powders. *Adv. Powder Technol.* 19:1–12.
Lu, Y., C. Le Paven, H. V. Nguyen, et al. 2013. Reactive Sputtering Deposition of Perovskite Oxide and Oxynitride Lanthanum Titanium Films: Structural and Dielectric Characterization. *Cryst. Growth Des.* 13:4852–4858.
Lu, Z., J. Liu, Y. Tang, et al. 2004. Hydrothermal Synthesis of $CaSnO_3$ Cubes. *Inorg. Chem. Commun.* 7:731–733.
Lubini, M., E. Chinarro, B. Moreno, et al. 2016. Electrical Properties of $La_{0.6}Sr_{0.4}Co_{1-y}Fe_yO3$ (y = 0.2–1.0) Fibers Obtained by Electrospinning. *J. Phys. Chem. C* 120:64–69.
Luo, Y., K. Wang, Q. Chen, et al. 2015. Preparation and Characterization of Electrospun $La_{1-x}Ce_xCoO_\delta$: Application to Catalytic Oxidation of Benzene. *J. Hazard. Mater.* 296:17–22.
Luo, Y., K. Wang, J. Zuo, et al. 2017. Enhanced Activity for Total Benzene Oxidation over SBA-15 Assisted Electrospun $LaCoO_3$. *Mol. Catal.* 436:259–266.
Luo, Y.-R., and J.-M. Wu. 2001. $BaPbO_3$ Perovskite Electrode for Lead Zirconate Titanate Ferroelectric Thin Films. *Appl. Phys. Lett.* 79:3669–3671.
Macario, L. R., M. L. Moreira, J. Andrés, et al. 2010. An Efficient Microwave-Assisted Hydrothermal Synthesis of $BaZrO_3$ Microcrystals: Growth Mechanism and Photoluminescence Emissions. *CrystEngComm* 12:3612–3619.
Madoui, N., and M. Omari. 2016. Synthesis and Electrochemical Properties of $LaCr_{1-x}Co_xO_3$ ($0 \leq x \leq 0.5$) via Co-Precipitation Method. *J. Inorg. Organomet. Polym. Mater.* 26:1005–1013.
Maensiri, S., W. Nuansing, J. Klinkaewnarong, et al. 2006. Nanofibers of Barium Strontium Titanate (BST) by Sol–Gel Processing and Electrospinning. *J. Colloid Interface Sci.* 297:578–583.
Makovec, D., T. Goršak, K. Zupan, et al. 2013. Hydrothermal Synthesis of $La_{1-x}Sr_xMnO_3$ Dendrites. *J. Cryst. Growth* 375:78–83.
Manoharan, S. S., and K. C. Patil. 1993. Combustion Route to Fine Particle Perovskite Oxides. *J. Solid State Chem.* 102:267–276.
Mao, Y. 2012. Facile Molten-Salt Synthesis of Double Perovskite La_2BMnO_6 Nanoparticles. *RSC Adv.* 2:12675–12678.
Mao, Y., S. Banerjee, and S. S. Wong. 2003. Hydrothermal Synthesis of Perovskite Nanotubes. *Chem. Commun.* 2003:408–409.
Marchin, L., S. Guillemet-Fritsch, and B. Durand. 2008. Soft Chemistry Synthesis of the Perovskite $CaCu_3Ti_4O_{12}$. *Prog. Solid State Chem.* 36:151–155.
Marozau, I., A. Shkabko, M. Döbeli, et al. 2011. Pulsed Laser Deposition and Characterisation of Perovskite-Type $LaTiO_{3-x}N_x$ Thin Films. *Acta Mater.* 59:7145–7154.
Martín, M. J., J. E. Alfonso, J. Mendiola, et al. 1997. Pulsed Laser Deposition of $KNbO_3$ Thin Films. *J. Mater. Res.* 12:2699–2706.
Marton, Z., S. S. A. Seo, T. Egami, et al. 2010. Growth Control of Stoichiometry in $LaMnO_3$ Epitaxial Thin Films by Pulsed Laser Deposition. *J. Cryst. Growth* 312:2923–2927.
Matei, C., D. Berger, P. Marote, et al. 2007. Lanthanum-Based Perovskites Obtained in Molten Nitrates or Nitrites. *Prog. Solid State Chem.* 35:203–209.
Matsumoto, Y., T. Morikawa, H. Adachi, et al. 1992a. A New Preparation Method of Barium Titanate Perovskite Film Using Electrochemical Reduction. *Mater. Res. Bull.* 27:1319–1327.

Matsumoto, Y., T. Sasaki, and J. Hombo. 1992b. A New Preparation Method of Lanthanum Cobalt Oxide, $LaCoO_3$, Perovskite Using Electrochemical Oxidation. *Inorg. Chem.* 31:738–741.

Maxim, F., P. Ferreira, P. M. Vilarinho, et al. 2008. Hydrothermal Synthesis and Crystal Growth Studies of $BaTiO_3$ Using Ti Nanotube Precursors. *Cryst. Growth Des.* 8:3309–3315.

Mendoza-Mendoza, E., K. P. Padmasree, S. M. Montemayor, et al. 2012. Molten Salts Synthesis and Electrical Properties of Sr- and/or Mg-Doped Perovskite-Type $LaAlO_3$ Powders. *J. Mater. Sci.* 47:6076–6085.

Meng, W., and A. V. Virkar. 1999. Synthesis and Thermodynamic Stability of $Ba_2B'B''O_6$ and $Ba_3B^*B''_2O_9$ Perovskites Using the Molten Salt Method. *J. Solid State Chem.* 148:492–498.

Mergen, A., and D. Sert. 2012. Production of $Ba(Zn_{1/3}Nb_{2/3})O_3$ Ceramic by Coprecipitation. *Mater. Charact.* 63:63–69.

Miao, H., Q. Zhang, G. Tan, et al. 2008. Co-Precipitation/Hydrothermal Synthesis of $BiFeO_3$ Powder. *J. Wuhan Univ. Technol.-Mater Sci Ed.* 23:507–509.

Milt, V. G., R. Spretz, M. A. Ulla, et al. 1995. Fast, Reproducible Explosion Method to Produce Crystalline Perovskite-Type Oxides. *J. Mater. Sci. Lett.* 14:428–430.

Minohara, M., R. Yukawa, M. Kitamura, et al. 2018. Growth of Antiperovskite Oxide Ca_3SnO Films by Pulsed Laser Deposition. *J. Cryst. Growth* 500:33–37.

Miotello, A., and R. Kelly. 1999. Laser-Induced Phase Explosion: New Physical Problems When a Condensed Phase Approaches the Thermodynamic Critical Temperature. *Appl. Phys. A* 69:S67–S73.

Mizusaki, J., Y. Mima, S. Yamauchi, et al. 1989. Nonstoichiometry of the Perovskite-Type Oxides $La_{1-x}Sr_xCoO_{3-\delta}$. *J. Solid State Chem.* 80:102–111.

Modeshia, D., R. J. Darton, S. E. Ashbrook, et al. 2009. Control of Polymorphism in $NaNbO_3$ by Hydrothermal Synthesis. *Chem. Commun.* 2009:68–70.

Moetakef, P., and T. A. Cain. 2015. Metal–Insulator Transitions in Epitaxial $Gd_{1-x}Sr_xTiO_3$ Thin Films Grown Using Hybrid Molecular Beam Epitaxy. *Thin Solid Films* 583:129–134.

Moetakef, P., D. G. Ouellette, J. Y. Zhang, et al. 2012. Growth and Properties of $GdTiO_3$ Films Prepared by Hybrid Molecular Beam Epitaxy. *J. Cryst. Growth* 355:166–170.

Montenegro, M. J., M. Döbeli, T. Lippert, et al. 2002. Pulsed Laser Deposition of $La_{0.6}Ca_{0.4}CoO_3$ (LCCO) Films. A Promising Metal-Oxide Catalyst for Air Based Batteries. *Phys. Chem. Chem. Phys.* 4:2799–2805.

Moreira, M. L., J. Andrés, J. A. Varela, et al. 2009. Synthesis of Fine Micro-Sized $BaZrO_3$ Powders Based on a Decaoctahedron Shape by the Microwave-Assisted Hydrothermal Method. *Cryst. Growth Des.* 9:833–839.

Moreira, M. L., E. C. Paris, G. S. do Nascimento, et al. 2009. Structural and Optical Properties of $CaTiO_3$ Perovskite-Based Materials Obtained by Microwave-Assisted Hydrothermal Synthesis: An Experimental and Theoretical Insight. *Acta Mater.* 57:5174–5185.

Moyer, J. A., C. Eaton, and R. Engel-Herbert. 2013. Highly Conductive $SrVO_3$ as a Bottom Electrode for Functional Perovskite Oxides. *Adv. Mater.* 25:3578–3582.

Muthuraman, M., N. A. Dhas, and K. C. Patil. 1994. Combustion Synthesis of Oxide Materials for Nuclear Waste Immobilization. *Bull. Mater. Sci.* 17:977–987.

Nakayama, S., M. Okazaki, Y. L. Aung, et al. 2003. Preparations of Perovskite-Type Oxides $LaCoO_3$ from Three Different Methods and Their Evaluation by Homogeneity, Sinterability and Conductivity. *Solid State Ion.* 158:133–139.

Ngamou, P. H. T., and N. Bahlawane. 2009. Chemical Vapor Deposition and Electric Characterization of Perovskite Oxides $LaMO_3$ (M = Co, Fe, Cr and Mn) Thin Films. *J. Solid State Chem.* 182:849–854.

Nguyen, S. V., V. Szabo, D. T. On, et al. 2002. Mesoporous Silica Supported $LaCoO_3$ Perovskites as Catalysts for Methane Oxidation. *Microporous Mesoporous Mater.* 54:51–61.

Niederberger, M., and N. Pinna. 2009. *Metal Oxide Nanoparticles in Organic Solvents: Synthesis, Formation, Assembly and Application*. Heidelberg, NY: Springer.

Novojilov, M. A., O. Y. Gorbenko, I. E. Graboy, et al. 2000. Perovskite Rare-Earth Nickelates in the Thin-Film Epitaxial State. *Appl. Phys. Lett.* 76:2041–2043.

Nuraje, N., X. Dang, J. Qi, et al. 2012. Biotemplated Synthesis of Perovskite Nanomaterials for Solar Energy Conversion. *Adv. Mater.* 24:2885–2889.

Ortiz-Vitoriano, N., I. Ruiz de Larramendi, I. Gil de Muro, et al. 2010. Nanoparticles of $La_{0.8}Ca_{0.2}Fe_{0.8}Ni_{0.2}O_{3-\delta}$ Perovskite for Solid Oxide Fuel Cell Application. *Mater. Res. Bull.* 45:1513–1519.

Panneerselvam, M., and K. J. Rao. 2003. Microwave Preparation and Sintering of Industrially Important Perovskite Oxides: $LaMO_3$ (M = Cr, Co, Ni). *J. Mater. Chem.* 13:596–601.

Papa, C. M., A. J. Cesnik, T. C. Evans, et al. 2015. Electrochemical Synthesis of Binary and Ternary Niobium-Containing Oxide Electrodes Using the p-Benzoquinone/Hydroquinone Redox Couple. *Langmuir ACS J. Surf. Colloids* 31:9502–9510.

Park, H. W., D. U. Lee, P. Zamani, et al. 2014. Electrospun Porous Nanorod Perovskite Oxide/Nitrogen-Doped Graphene Composite as a Bi-Functional Catalyst for Metal Air Batteries. *Nano Energy* 10:192–200.

Pelosato, R., C. Cristiani, G. Dotelli, et al. 2010. Co-Precipitation in Aqueous Medium of $La_{0.8}Sr_{0.2}Ga_{0.8}Mg_{0.2}O_{3-\delta}$ via Inorganic Precursors. *J. Power Sources* 195:8116–8123.

Pelosato, R., C. Cristiani, G. Dotelli, et al. 2013. Co-Precipitation Synthesis of SOFC Electrode Materials. *Int. J. Hydrog. Energy* 38:480–491.

Peng, C. H., and S. B. Desu. 1992. Low-Temperature Metalorganic Chemical Vapor Deposition of Perovskite $Pb(Zr_xTi_{1-x})O_3$ Thin Films. *Appl. Phys. Lett.* 61:16–18.

Phani, A. R., M. Passacantando, and S. Santucci. 2007. Synthesis of Nanocrystalline $ZnTiO_3$ Perovskite Thin Films by Sol–Gel Process Assisted by Microwave Irradiation. *J. Phys. Chem. Solids* 68:317–323.

Phani, A. R., and S. Santucci. 2001. Structural Characterization of Nickel Titanium Oxide Synthesized by Sol–Gel Spin Coating Technique. *Thin Solid Films* 396:1–4.

Piticescu, R. M., P. Vilarnho, L. M. Popescu, et al. 2006. Perovskite Nanostructures Obtained by a Hydrothermal Electrochemical Process. *J. Eur. Ceram. Soc.* 26:2945–2949.

Ponhan, W., V. Amornkitbamrung, and S. Maensiri. 2014. Room Temperature Ferromagnetism Observed in Pure $La_{0.5}Sr_{0.5}TiO_3$ Nanofibers Fabricated by Electrospinning. *J. Alloys Compd.* 606:182–188.

Pontes, F. M., D. S. L. Pontes, A. J. Chiquito, et al. 2015. Effect of Fe-Doping on the Structural, Microstructural, Optical, and Ferroeletric Properties of $Pb_{1/2}Sr_{1/2}Ti_{1-x}Fe_xO_3$ Oxide Prepared by Spin Coating Technique. *Mater. Lett.* 138:179–183.

Poterala, S. F., Y. Chang, T. Clark, et al. 2010. Mechanistic Interpretation of the Aurivillius to Perovskite Topochemical Microcrystal Conversion Process. *Chem. Mater.* 22:2061–2068.

Prado-Gonjal, J., R. Schmidt, J.-J. Romero, et al. 2013. Microwave-Assisted Synthesis, Microstructure, and Physical Properties of Rare-Earth Chromites. *Inorg. Chem.* 52:313–320.

Prakash, A., P. Xu, X. Wu, et al. 2017. Adsorption-Controlled Growth and the Influence of Stoichiometry on Electronic Transport in Hybrid Molecular Beam Epitaxy-Grown $BaSnO_3$ Films. *J. Mater. Chem. C* 5:5730–5736.

Prakash, A. S., A. M. A. Khadar, K. C. Patil, et al. 2002. Hexamethylenetetramine: A New Fuel for Solution Combustion Synthesis of Complex Metal Oxides. *J. Mater. Synth. Process.* 10:135–141.

Primo-Martín, V., and M. Jansen. 2001. Synthesis, Structure, and Physical Properties of Cobalt Perovskites: $Sr_3CoSb_2O_9$ and $Sr_2CoSbO_{6-}\delta$. *J. Solid State Chem.* 157:76–85.

Qiao, L., and X. Bi. 2008. Microstructure and Ferroelectric Properties of $BaTiO_3$ Films on $LaNiO_3$ Buffer Layers by Rf Sputtering. *J. Cryst. Growth* 310:2780–2784.

Rajmohan, S., A. Manikandan, V. Jeseentharani, et al. 2016. Simple Co-Precipitation Synthesis and Characterization Studies of $La_{1-x}Ni_xVO_3$ Perovskites Nanostructures for Humidity Sensing Applications. *J. Nanosci. Nanotechnol.* 16:1650–6555.

Ran, R., D. Weng, X. Wu, et al. 2007. Rapid Synthesis of $La_{0.7}Sr_{0.3}MnO_{3+\lambda}$ Catalysts by Microwave Irradiation Process. *Catal. Today* 126:394–399.

Rivas, I., J. Alvarez, E. Pietri, et al. 2010. Perovskite-Type Oxides in Methane Dry Reforming: Effect of Their Incorporation into a Mesoporous SBA-15 Silica-Host. *Catal. Today* 149:388–393.

Rørvik, P. M., K. Tadanaga, M. Tatsumisago, et al. 2009. Template-Assisted Synthesis of $PbTiO_3$ Nanotubes. *J. Eur. Ceram. Soc.* 29:2575–2579.

Rossetti, I., and L. Forni. 2001. Catalytic Flameless Combustion of Methane over Perovskites Prepared by Flame–Hydrolysis. *Appl. Catal. B Environ.* 33:345–352.

Sadakane, M., T. Asanuma, J. Kubo, et al. 2005. Facile Procedure To Prepare Three-Dimensionally Ordered Macroporous (3DOM) Perovskite-Type Mixed Metal Oxides by Colloidal Crystal Templating Method. *Chem. Mater.* 17:3546–3551.

Sánchez, F., C. Ferrater, X. Alcobé, et al. 2001. Pulsed Laser Deposition of Epitaxial $LaNiO_3$ Thin Films on Buffered Si(100). *Thin Solid Films* 384:200–205.

Sánchez-Rodríguez, D., H. Wada, S. Yamaguchi, et al. 2017. Synthesis of $LaFeO_3$ Perovskite-Type Oxide via Solid-State Combustion of a Cyano Complex Precursor: The Effect of Oxygen Diffusion. *Ceram. Int.* 43:3156–3165.

Santiso, J., A. Figueras, and J. Fraxedas. 2002. Thin Films of Sr_2FeMoO_6 Grown by Pulsed Laser Deposition: Preparation and Characterization. *Surf. Interface Anal.* 33:676–680.

Santulli, A. C., H. Zhou, S. Berweger, et al. 2010. Synthesis of Single-Crystalline One-Dimensional $LiNbO_3$ Nanowires. *CrystEngComm* 12:2675–2678.

Sardar, K., M. R. Lees, R. J. Kashtiban, et al. 2011. Direct Hydrothermal Synthesis and Physical Properties of Rare-Earth and Yttrium Orthochromite Perovskites. *Chem. Mater.* 23:48–56.

Sariah, S., R. Kancharla Rajendar, R. Kamatala Chinna, et al. 2012. Mortar-Pestle and Microwave Assisted Regioselective Nitration of Aromatic Compounds in Presence of Certain Group V and VI Metal Salts under Solvent Free Conditions. *Int. J. Org. Chem.* 2:233–247.

Sasaki, T., Y. Matsumoto, J. Hombo, et al. 1991. A New Preparation Method of $LaMnO_3$ Perovskite Using Electrochemical Oxidation. *J. Solid State Chem.* 91:61–70.

Sasaki, T., T. Morikawa, J. Hombo, et al. 1990. Electrochemical Synthesis of $La_{1-x}Sr_xMnO_3$ Perovskite Films. *Electrochem. Ind. Phys. Chem.* 58:567–568.

Sekar, M. M. A., and K. C. Patil. 1992. Combustion Synthesis and Properties of Fine-Particle Dielectric Oxide Materials. *J. Mater. Chem.* 2:739–743.

Sennu, P., V. Aravindan, K. S. Nahm, et al. 2017. Exceptional Catalytic Activity of Hollow Structured $La_{0.6}Sr_{0.4}CoO_{3-\delta}$ Perovskite Spheres in Aqueous Media and Aprotic Li–O_2 Batteries. *J. Mater. Chem. A* 5:18029–18037.

Shami, M. Y., M. S. Awan, and M. Anis-ur-Rehman. 2011. Phase Pure Synthesis of $BiFeO_3$ Nanopowders Using Diverse Precursor via Co-Precipitation Method. *J. Alloys Compd.* 509:10139–10144.

Sharif, M. K., M. A. Khan, A. Hussain, et al. 2016. Synthesis and Characterization of Zr and Mg Doped $BiFeO_3$ Nanocrystalline Multiferroics via Micro Emulsion Route. *J. Alloys Compd.* 667:329–340.

Shen, Y., K. Zhao, F. He, et al. 2016. Synthesis of Three-Dimensionally Ordered Macroporous $LaFe_{0.7}Co_{0.3}O_3$ Perovskites and Their Performance for Chemical-Looping Steam Reforming of Methane. *J. Fuel Chem. Technol.* 44:1168–1176.

Shibuya, K., S. Mi, C.-L. Jia, et al. 2008. Sr_2TiO_4 Layered Perovskite Thin Films Grown by Pulsed Laser Deposition. *Appl. Phys. Lett.* 92:241918.

Shim, J., K. J. Lopez, H.-J. Sun, et al. 2015. Preparation and Characterization of Electrospun $LaCoO_3$ Fibers for Oxygen Reduction and Evolution in Rechargeable Zn–Air Batteries. *J. Appl. Electrochem.* 45:1005–1012.

Shimada, H., T. Yamaguchi, T. Suzuki, et al. 2016. High Power Density Cell Using Nanostructured Sr-Doped $SmCoO_3$ and Sm-Doped CeO_2 Composite Powder Synthesized by Spray Pyrolysis. *J. Power Sources* 302:308–314.

Shlyakhtin, O. A., Y.-J. Oh, and Yu. D. Tretyakov. 2000. Preparation of Dense $La_{0.7}Ca_{0.3}MnO_3$ Ceramics from Freeze-Dried Precursors. *J. Eur. Ceram. Soc.* 20:2047–2054.

Simmance, K., D. Thompsett, W. Wang, et al. 2019. Evaluation of Perovskite Catalysts Prepared by Flame Spray Pyrolysis for Three-Way Catalyst Activity under Simulated Gasoline Exhaust Feeds. *Catal. Today* 320:40–50, SI:Vehicle Emissions Catalys.

Simon-Seveyrat, L., A. Hajjaji, Y. Emziane, et al. 2007. Re-Investigation of Synthesis of $BaTiO_3$ by Conventional Solid-State Reaction and Oxalate Coprecipitation Route for Piezoelectric Applications. *Ceram. Int.* 33:35–40.

Singh, V., S. Watanabe, T. K. Gundu Rao, et al. 2012. Synthesis, Characterisation, Luminescence and Defect Centres in Solution Combustion Synthesised $CaZrO_3$:Tb^{3+} Phosphor. *J. Lumin.* 132:2036–2042.

Slamovich, E. B., and I. A. Aksay. 1996. Structure Evolution in Hydrothermally Processed (<100°C) $BaTiO_3$ Films. *J. Am. Ceram. Soc.* 79:239–247.

Song, S., J. Zhou, S. Zhang, et al. 2018. Molten-Salt Synthesis of Porous $La_{0.6}Sr_{0.4}Co_{0.2}Fe_{0.8}O_{2.9}$ Perovskite as an Efficient Electrocatalyst for Oxygen Evolution. *Nano Res.* 11:4796–4805.

Specchia, S., A. Civera, and G. Saracco. 2004. In Situ Combustion Synthesis of Perovskite Catalysts for Efficient and Clean Methane Premixed Metal Burners. *Chem. Eng. Sci.* 59:5091–5098.

Sugai, H., T. Iijima, and H. Masumoto. 1999. Preparation of Lead Titanate Ultrathin Film Using Langmuir–Blodgett Film as Precursor. *Jpn. J. Appl. Phys.* 38:5322.

Sun, Y., X. Xiong, Z. Xia, et al. 2013. Study on Visible Light Response and Magnetism of Bismuth Ferrites Synthesized by a Low Temperature Hydrothermal Method. *Ceram. Int.* 39:4651–4656.

Suzuki, M., Y. Enomoto, T. Murakami, et al. 1981. Thin Film Preparation of Superconducting Perovskite-Type Oxides by Rf Sputtering. *Jpn. J. Appl. Phys.* 20:13.

Takahashi, J., T. Toyoda, T. Ito, et al. 1990. Preparation of $LaNiO_3$ Powder from Coprecipitated Lanthanum-Nickel Oxalates. *J. Mater. Sci.* 25:1557–1562.

Takalkar, G., R. Bhosale, and F. AlMomani. 2019. Combustion Synthesized $A_{0.5}Sr_{0.5}MnO_{3-\delta}$ Perovskites (Where, A = La, Nd, Sm, Gd, Tb, Pr, Dy, and Y) as Redox Materials for Thermochemical Splitting of CO_2. *Appl. Surf. Sci.* 489:80–91.

Takashima, H., K. Ueda, and M. Itoh. 2006. Red Photoluminescence in Praseodymium-Doped Titanate Perovskite Films Epitaxially Grown by Pulsed Laser Deposition. *Appl. Phys. Lett.* 89:261915.

Takeda, Y., R. Kanno, M. Noda, et al. 1987. Cathodic Polarization Phenomena of Perovskite Oxide Electrodes with Stabilized Zirconia. *J. Electrochem. Soc.* 134:2656–2661.

Takeda, Y., Y. Sakaki, T. Ichikawa, et al. 1994. Stability of $La_{1-x}A_xMnO_{3-z}$ (A = Ca, Sr) as Cathode Materials for Solid Oxide Fuel Cells. *Solid State Ion.* 72:257–264.

Tang, J., M. Zhu, T. Zhong, et al. 2007. Synthesis of Fine $Pb(Fe_{0.5}Nb_{0.5})O_3$ Perovskite Powders by Coprecipitation Method. *Mater. Chem. Phys.* 101:475–479.

Tang, P., Y. Tong, H. Chen, et al. 2013. Microwave-Assisted Synthesis of Nanoparticulate Perovskite $LaFeO_3$ as a High Active Visible-Light Photocatalyst. *Curr. Appl. Phys.* 13:340–343.

Tang, R., Y. Nie, J. K. Kawasaki, et al. 2016. Oxygen Evolution Reaction Electrocatalysis on $SrIrO_3$ Grown Using Molecular Beam Epitaxy. *J. Mater. Chem. A* 4:6831–6836.

Tao, J., J. Ma, Y. Wang, et al. 2006. Electrochemical Synthesis of Small $SrTiO_3$ Particles. *J. Am. Ceram. Soc.* 89:3554–3556.

Tawichai, N., W. Sittiyot, S. Eitssayeam, et al. 2012. Preparation and Dielectric Properties of Barium Iron Niobate by Molten-Salt Synthesis. *Ceram. Int.* 38:S121–S124.

Tetsuka, H., Y. J. Shan, K. Tezuka, et al. 2005. Transparent Conductive In-Doped Cd_3TeO_6 Thin Films with Perovskite Structure Deposited by Radio Frequency Magnetron Sputtering. *J. Mater. Res.* 20:2256–2260.

Therese, G. H. A., M. Dinamani, and P. Vishnu Kamath. 2005. Electrochemical Synthesis of Perovskite Oxides. *J. Appl. Electrochem.* 35:459–465.

Therese, G. H. A., and P. V. Kamath. 1998. Electrochemical Synthesis of $LaMnO_3$ Coatings on Conducting Substrates. *Chem. Mater.* 10:3364–3367.

Therese, G. H. A., and P. V. Kamath. 2000. Electrochemical Synthesis of Metal Oxides and Hydroxides. *Chem. Mater.* 12:1195–1204.

Thirumal, M., P. Jain, and A. K. Ganguli. 2001. Molten Salt Synthesis of Complex Perovskite-Related Dielectric Oxides. *Mater. Chem. Phys.* 70:7–11.

Thirumalairajan, S., K. Girija, I. Ganesh, et al. 2012. Controlled Synthesis of Perovskite $LaFeO_3$ Microsphere Composed of Nanoparticles via Self-Assembly Process and Their Associated Photocatalytic Activity. *Chem. Eng. J.* 209:420–428.

Thirumalairajan, S., K. Girija, V. R. Mastelaro, et al. 2014. Surface Morphology-Dependent Room-Temperature $LaFeO_3$ Nanostructure Thin Films as Selective NO_2 Gas Sensor Prepared by Radio Frequency Magnetron Sputtering. *ACS Appl. Mater. Interfaces* 6:13917–13927.

Thomas, R., S. Mochizuki, T. Mihara, et al. 2002. Effect of Substrate Temperature on the Crystallization of $Pb(Zr,Ti)O_3$ Films on Pt/Ti/Si Substrates Prepared by Radio Frequency Magnetron Sputtering with a Stoichiometric Oxide Target. *Mater. Sci. Eng. B* 95:36–42.

Toprak, M. S., M. Darab, G. E. Syvertsen, et al. 2010. Synthesis of Nanostructured BSCF by Oxalate Co-Precipitation – As Potential Cathode Material for Solid Oxide Fuels Cells. *Int. J. Hydrog. Energy* 35:9448–9454.

Traina, K., M. C. Steil, J. P. Pirard, et al. 2007. Synthesis of $La_{0.9}Sr_{0.1}Ga_{0.8}Mg_{0.2}O_{2.85}$ by Successive Freeze-Drying and Self-Ignition of a Hydroxypropylmethyl Cellulose Solution. *J. Eur. Ceram. Soc.* 27:3469–3474.

Tu, H. Y., Y. Takeda, N. Imanishi, et al. 1999. $Ln_{0.4}Sr_{0.6}Co_{0.8}Fe_{0.2}O_{3-\delta}$ (Ln = La, Pr, Nd, Sm, Gd) for the Electrode in Solid Oxide Fuel Cells. *Solid State Ion.* 117:277–281.

Uskoković, V., and M. Drofenik. 2006. Synthesis of Lanthanum–Strontium Manganites by Oxalate-Precursor Co-Precipitation Methods in Solution and in Reverse Micellar Microemulsion. *J. Magn. Magn. Mater.* 303:214–220.

Varandili, S. B., A. Babaei, and A. Ataie. 2018. Characterization of B Site Codoped $LaFeO_3$ Nanoparticles Prepared via Co-Precipitation Route. *Rare Met.* 37:181–190.

Vieten, J., B. Bulfin, M. Roeb, et al. 2018. Citric Acid Auto-Combustion Synthesis of Ti-Containing Perovskites via Aqueous Precursors. *Solid State Ion.* 315:92–97.

Viskadourakis, Z., C. N. Mihailescu, and G. Kenanakis. 2020. Spray-Pyrolysis Deposited $La_{1-x}Sr_xCoO_3$ Thin Films for Potential Non-Volatile Memory Applications. *Appl. Phys. A* 126:80.

Vradman, L., E. Friedland, J. Zana, et al. 2017. Molten Salt Synthesis of $LaCoO_3$ Perovskite. *J. Mater. Sci.* 52:11383–11390.

Vradman, L., J. Zana, A. Kirschner, et al. 2013. Synthesis of $LaMnO_3$ in Molten Chlorides: Effect of Preparation Conditions. *Phys. Chem. Chem. Phys.* 15:10914–10920.

Wachowski, L. 1986. Influence of the Method of Preparation on the Porous Structure of Perovskite Oxides. *Surf. Coat. Technol.* 29:303–311.

Wang, B., S. Gu, Y. Ding, et al. 2013a. A Novel Route to Prepare $LaNiO_3$ Perovskite-Type Oxide Nanofibers by Electrospinning for Glucose and Hydrogen Peroxide Sensing. *Analyst* 138:362–367.

Wang, H. B., G. Y. Meng, and D. K. Peng. 2000. Aerosol and Plasma Assisted Chemical Vapor Deposition Process for Multi-Component Oxide $La_{0.8}Sr_{0.2}MnO_3$ Thin Film. *Thin Solid Films* 368:275–278.

Wang, L. S., and S. A. Barnett. 1995. Ag-Perovskite Cermets for Thin Film Solid Oxide Fuel Cell Air-Electrode Applications. *Solid State Ion.* 76:103–113.

Wang, S., K. Huang, C. Hou, et al. 2015. Low Temperature Hydrothermal Synthesis, Structure and Magnetic Properties of $RECrO_3$ (RE = La, Pr, Nd, Sm). *Dalton Trans.* 44:17201–17208.

Wang, S., K. Huang, B. Zheng, et al. 2013b. Mild Hydrothermal Synthesis and Physical Property of Perovskite Sr Doped $LaCrO_3$. *Mater. Lett.* 101:86–89.

Wang, W., N. Li, Y. Chi, et al. 2013c. Electrospinning of Magnetical Bismuth Ferrite Nanofibers with Photocatalytic Activity. *Ceram. Int.* 39:3511–3518.

Wang, W., B. Lin, H. Zhang, et al. 2019. Synthesis, Morphology and Electrochemical Performances of Perovskite-Type Oxide $La_xSr_{1-x}FeO_3$ Nanofibers Prepared by Electrospinning. *J. Phys. Chem. Solids* 124:144–150.

Wang, Y., S. Xie, J. Deng, et al. 2014. Morphologically Controlled Synthesis of Porous Spherical and Cubic $LaMnO_3$ with High Activity

Wang, Y., G. Xu, L. Yang, et al. 2009. Hydrothermal Synthesis of Single-Crystal $BaTiO_3$ Dendrites. *Mater. Lett.* 63:239–241.

Wang, Y., J. Zhu, L. Zhang, et al. 2006. Preparation and Characterization of Perovskite $LaFeO_3$ Nanocrystals. *Mater. Lett.* 60:1767–1770.

Wang, Y., S. Xie, J. Deng, et al. 2014. Morphologically Controlled Synthesis of Porous Spherical and Cubic $LaMnO_3$ with High Activity for the Catalytic Removal of Toluene. *ACS Appl. Mater. Interfaces* 6:17394–17401.

Watanabe, Y., Y. Matsumoto, H. Kunitomo, et al. 1994. Crystallographic and Electrical Properties of Epitaxial $BaTiO_3$ Film Grown on Conductive and Insulating Perovskite Oxides. *Jpn. J. Appl. Phys.* 33:5182.

Wei, X., G. Xu, Z. Ren, et al. 2008. PVA-Assisted Hydrothermal Synthesis of $SrTiO_3$ Nanoparticles with Enhanced Photocatalytic Activity for Degradation of RhB. *J. Am. Ceram. Soc.* 91:3795–3799.

Weifan, C., L. Fengsheng, L. Leili, et al. 2006. One-Step Synthesis of Nanocrytalline Perovskite $LaMnO_3$ Powders via Microwave-Induced Solution Combustion Route. *J. Rare Earths* 24:782–787.

Wrobel, F., A. F. Mark, G. Christiani, et al. 2017. Comparative Study of $LaNiO_3/LaAlO_3$ Heterostructures Grown by Pulsed Laser Deposition and Oxide Molecular Beam Epitaxy. *Appl. Phys. Lett.* 110:041606.

Wu, L., X. Mei, and W. Zheng. 2006. Hydrothermal Synthesis and Characterization of Double Perovskite Ba_2YSbO_6. *Mater. Lett.* 60:2326–2330.

Wu, W., F. Lu, K. H. Wong, et al. 2000. Epitaxial and Highly Electrical Conductive $La_{0.5}Sr_{0.5}TiO_3$ Films Grown by Pulsed Laser Deposition in Vacuum. *J. Appl. Phys.* 88:700–704.

Wu, X.-H., Y.-D. Wang, H.-L. Liu, et al. 2002. Preparation and Gas-Sensing Properties of Perovskite-Type $MSnO_3$ (M = Zn, Cd, Ni). *Mater. Lett.* 56:732–736.

Xiang, W., R. Dong, D. Lee, et al. 2007. Heteroepitaxial Growth of Nb-Doped $SrTiO_3$ Films on Si Substrates by Pulsed Laser Deposition for Resistance Memory Applications. *Appl. Phys. Lett.* 90:052110.

Xu, C.-Y., L. Zhen, R. Yang, et al. 2007. Synthesis of Single-Crystalline Niobate Nanorods via Ion-Exchange Based on Molten-Salt Reaction. *J. Am. Chem. Soc.* 129:15444–15445.

Xu, D., L. Luo, Y. Ding, et al. 2014. A Novel Nonenzymatic Fructose Sensor Based on Electrospun $LaMnO_3$ Fibers. *J. Electroanal. Chem.* 727:21–26.

Xu, D., L. Luo, Y. Ding, et al. 2015. Sensitive Electrochemical Detection of Glucose Based on Electrospun $La_{0.88}Sr_{0.12}MnO_3$ Naonofibers Modified Electrode. *Anal. Biochem.* 489:38–43.

Xu, G., Z. H. Ren, W. J. Weng, et al. 2005. Synthesis of Perovskite $Pb(Zr_{0.52}Ti_{0.48})O_3$ (PZT) Powders by a Modified Coprecipitation Method. *Key Eng. Mater.* 280–283:627–630.

Xu, J., X. Jia, X. Lou, et al. 2006. One-Step Hydrothermal Synthesis and Gas Sensing Property of $ZnSnO_3$ Microparticles. *Solid-State Electron.* 50:504–507.

Xu, J., J. Liu, Z. Zhao, et al. 2010. Three-Dimensionally Ordered Macroporous $LaCo_xFe_{1-x}O_3$ Perovskite-Type Complex Oxide Catalysts for Diesel Soot Combustion. *Catal. Today* 153:136–142.

Xu, J.-J., D. Xu, Z.-L. Wang, et al. 2013. Synthesis of Perovskite-Based Porous $La_{0.75}Sr_{0.25}MnO_3$ Nanotubes as a Highly Efficient Electrocatalyst for Rechargeable Lithium–Oxygen Batteries. *Angew. Chem. Int. Ed.* 52:3887–3890.

Xu, J.-J., Z.-L. Wang, D. Xu, et al. 2014. 3D Ordered Macroporous $LaFeO_3$ as Efficient Electrocatalyst for Li–O_2 Batteries with Enhanced Rate Capability and Cyclic Performance. *Energy Environ. Sci.* 7:2213–2219.

Xue, P., Y. Hu, W. Xia, et al. 2017. Molten-Salt Synthesis of $BaTiO_3$ Powders and Their Atomic-Scale Structural Characterization. *J. Alloys Compd.* 695:2870–2877.

Yamamoto, O., Y. Takeda, R. Kanno, et al. 1987. Perovskite-Type Oxides as Oxygen Electrodes for High Temperature Oxide Fuel Cells. *Solid State Ion.* 22:241–246.

Yan, J., H. Matsumoto, T. Akbay, et al. 2006. Preparation of $LaGaO_3$-Based Perovskite Oxide Film by a Pulsed-Laser Ablation Method and Application as a Solid Oxide Fuel Cell Electrolyte. *J. Power Sources* 157:714–719, Selected papers presented at the Ninth Grove Fuel Cell Symposium.

Yang, G., W. Yan, J. Wang, et al. 2014. Fabrication and Photocatalytic Activities of $SrTiO_3$ Nanofibers by Sol–Gel Assisted Electrospinning. *J. Sol–Gel Sci. Technol.* 71:159–167.

Yang, J., R. Li, J. Zhou, et al. 2010. Synthesis of $LaMO_3$ (M = Fe, Co, Ni) Using Nitrate or Nitrite Molten Salts. *J. Alloys Compd.* 508:301–308.

Yang, L., G. Wang, C. Mao, et al. 2009. Orientation Control of $LaNiO_3$ Thin Films by RF Magnetron Sputtering with Different Oxygen Partial Pressure. *J. Cryst. Growth* 311:4241–4246.

Yang, R., S. Y. Shen, C. B. Wang, et al. 2008. Pulsed Laser Deposition of Stoichiometric $KNbO_3$ Films on Si (100). *Thin Solid Films* 516:8559–8563.

Yang, S. Y., F. Zavaliche, L. Mohaddes-Ardabili, et al. 2005. Metalorganic Chemical Vapor Deposition of Lead-Free Ferroelectric $BiFeO_3$ Films for Memory Applications. *Appl. Phys. Lett.* 87:102903.

Yang, Z., Q. Li, L. Wang, et al. 2011. The Characteristics of (Pb,La)(Zr,Sn,Ti)O_3 Ceramics Synthesized by Coprecipitation Method Compared to Conventional Mixed Oxide Method. *J. Mater. Sci. Mater. Electron.* 22:162–166.

Yi, N., Y. Cao, Y. Su, et al. 2005. Nanocrystalline $LaCoO_3$ Perovskite Particles Confined in SBA-15 Silica as a New Efficient Catalyst for Hydrocarbon Oxidation. *J. Catal.* 230:249–253.

Yoon, K. H., Y. S. Cho, and D. H. Kang. 1998. Molten Salt Synthesis of Lead-Based Relaxors. *J. Mater. Sci.* 33:2977–2984.

Yu, T., C. H. Chen, X. F. Chen, et al. 2004. Fabrication and Characterization of Perovskite $CaZrO_3$ Oxide Thin Films. *Ceram. Int.* 30:1279–1282, 3rd Asian Meeting on Electroceramics.

Yuan, L., K. Huang, C. Hou, et al. 2014. Hydrothermal Synthesis and Magnetic Properties of $REFe_{0.5}Cr_{0.5}O_3$ (RE = La, Tb, Ho, Er, Yb, Lu and Y) Perovskite. *New J. Chem.* 38:1168–1172.

Yuh, J., J. C. Nino, and W. M. Sigmund. 2005. Synthesis of Barium Titanate ($BaTiO_3$) Nanofibers via Electrospinning. *Mater. Lett.* 59:3645–3647.

Zhang, C., M. Shang, M. Liu, et al. 2016. Multiferroicity in $SmFeO_3$ Synthesized by Hydrothermal Method. *J. Alloys Compd.* 665:152–157.

Zhang, C., T. Zhang, L. Ge, et al. 2014. Hydrothermal Synthesis and Multiferroic Properties of Y_2NiMnO_6. *RSC Adv.* 4:50969–50974.

Zhang, H.-T., L. R. Dedon, L. W. Martin, et al. 2015. Self-Regulated Growth of $LaVO_3$ Thin Films by Hybrid Molecular Beam Epitaxy. *Appl. Phys. Lett.* 106:233102.

Zhang, Y., Y.-F. Sun, and J.-L. Luo. 2016. Developing Cobalt Doped $Pr_{0.5}Ba_{0.5}MnO_{3-\delta}$ Electrospun Nanofiber Bifunctional Catalyst for Oxygen Reduction Reaction and Oxygen Evolution Reaction. *ECS Trans.* 75:955–964.

Zhao, B., L. Zhang, D. Zhen, et al. 2017. A Tailored Double Perovskite Nanofiber Catalyst Enables Ultrafast Oxygen Evolution. *Nat. Commun.* 8:1–9.

Zhao, J., Y. Liu, X. Li, et al. 2013. Highly Sensitive Humidity Sensor Based on High Surface Area Mesoporous $LaFeO_3$ Prepared by a Nanocasting Route. *Sens. Actuators B Chem.* 181:802–809.

Zhao, S., J. Zheng, F. Jiang, et al. 2015. Co-Precipitation Synthesis and Microwave Absorption Properties of $CaMnO_3$ Doped by La and Co. *J. Mater. Sci. Mater. Electron.* 26:8603–8608.

Zhen, D., B. Zhao, H.-C. Shin, et al. 2017. Electrospun Porous Perovskite $La_{0.6}Sr_{0.4}Co_{1-x}Fe_xO_{3-\delta}$ Nanofibers for Efficient Oxygen Evolution Reaction. *Adv. Mater. Interfaces* 4:1700146.

Zheng, W., W. Pang, and G. Meng. 1998. Hydrothermal Synthesis and Characterization of Perovskite-Type Ba_2SbMO_6 (M = In, Y, Nd) Oxides. *Mater. Lett.* 37:276–280.

Zheng, W., W. Pang, G. Meng, et al. 1999. Hydrothermal Synthesis and Characterization of $LaCrO_3$. *J. Mater. Chem.* 9:2833–2836.

Zhi, M., A. Koneru, F. Yang, et al. 2012. Electrospun $La_{0.8}Sr_{0.2}MnO_3$ nanofibers for a High-Temperature Electrochemical Carbon Monoxide Sensor. *Nanotechnology* 23:305501.

Zhou, H., Y. Mao, and S. S. Wong. 2007. Probing Structure–Parameter Correlations in the Molten Salt Synthesis of $BaZrO_3$ Perovskite Submicrometer-Sized Particles. *Chem. Mater.* 19:5238–5249.

Zhou, X., Y. Zhao, X. Cao, et al. 2008. Fabrication of Polycrystalline Lanthanum Manganite ($LaMnO_3$) Nanofibers by Electrospinning. *Mater. Lett.* 62:470–472.

Zhu, C., A. Nobuta, I. Nakatsugawa, et al. 2013. Solution Combustion Synthesis of $LaMO_3$ (M = Fe, Co, Mn) Perovskite Nanoparticles and the Measurement of Their Electrocatalytic Properties for Air Cathode. *Int. J. Hydrog. Energy* 38:13238–13248.

Zhu, D., H. Zhu, and Y. Zhang. 2002. Hydrothermal Synthesis of $La_{0.5}Ba_{0.5}MnO_3$ Nanowires. *Appl. Phys. Lett.* 80:1634–1636.

Zhu, X., J. Wang, Z. Zhang, et al. 2008. Perovskite Nanoparticles and Nanowires: Microwave–Hydrothermal Synthesis and Structural Characterization by High-Resolution Transmission Electron Microscopy. *J. Am. Ceram. Soc.* 91:2683–2689.

Zhu, Y., R. Tan, T. Yi, et al. 2000. Preparation of Nanosized $LaCoO_3$ Perovskite Oxide Using Amorphous Heteronuclear Complex as a Precursor at Low Temperature. *J. Mater. Sci.* 35:5415–542

Zipprich, W., S. Waschilewski, F. Rocholl, et al. 1997. Improved Preparation of $La_{1-x}Me_xCoO_{3-\delta}$ (Me = Sr, Ca) and Analysis of Oxide Ion Conductivity with Ion Conducting Microcontacts. *Solid State Ion.* 101–103:1015–1023.

Zomorrodian, A., H. Salamati, Z. Lu, et al. 2010. Electrical Conductivity of Epitaxial $La_{0.6}Sr_{0.4}Co_{0.2}Fe_{0.8}O_{3-\delta}$ Thin Films Grown by Pulsed Laser Deposition. *Int. J. Hydrog. Energy* 35:12443–12448.

3 Crystal Structure of Simple Perovskites

The performance of a perovskite material is dominated by its structure, and therefore, understanding of the perovskite structure is crucial to link with its properties. In a structural point of view, the perovskite materials have unique structural features compared with other materials. In this chapter, we provide a description of the single perovskite structure and possible structural distortions caused by octahedral tilting, B-cation displacement, and Jahn–Teller (JT) distortion. The complex perovskites and related structures are described in the next chapter.

3.1 DESCRIPTION OF PEROVSKITE STRUCTURES

Since the A/B site can be occupied by element with oxidation states ranging from 1^+ to 5^+, a large different compositions and crystal structures are possible among perovskites. The perovskite structure should have neutral charge; therefore, the sum of the oxidation states of A and B ions should be equivalent to the sum of charges of the oxygen ions. Based on the oxidation state of A and B cations in single perovskite structured ABO_3 materials, they are divided into $A^{1+}B^{5+}O_3$, $A^{2+}B^{4+}O_3$, $A^{3+}B^{3+}O_3$, $A^{4+}B^{2+}O_3$, and $A^{5+}B^{1+}O_3$ groups. Theoretically, ~2,346 perovskite-structured oxides can exist; however, ~265 ABO_3 perovskite-structured single oxides are synthesized experimentally (Yin et al. 2019). Many perovskite oxides form at a high temperature and pressure, and the phase transition with temperature and pressure is common among perovskite-structured materials.

The ideal crystal structure of perovskites is cubic, as shown in Figure 3.1; however, the crystal structure varies depending on the size of the cations and their bond lengths. The cubic structured perovskites belong to the space group $Pm\overline{3}m$. In such a perovskite structure (refer to Figure 1.2a), the atomic positions are

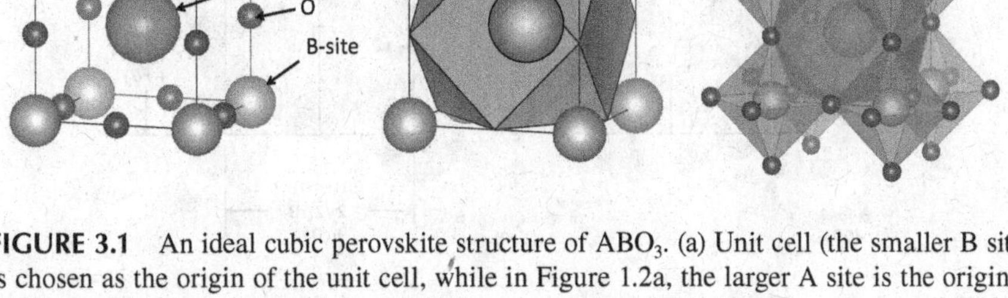

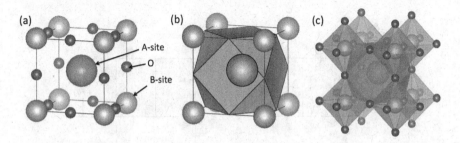

FIGURE 3.1 An ideal cubic perovskite structure of ABO_3. (a) Unit cell (the smaller B site is chosen as the origin of the unit cell, while in Figure 1.2a, the larger A site is the origin); (b) the coordination of A-site ion; and (c) the coordination of B-site ion.

A: 0,0,0; B: ½, ½, ½; and X: ½, ½, 0; ½, 0,½; 0,½, ½.

The structure is also represented using B cation in the origin (refer to Figure 3.1a) with the atomic positions as

B: 0,0,0; A: ½, ½, ½; X: ½, 0,0; 0,½, 0; 0,0,½.

In the ABO_3 perovskite cubic crystal structure shown in Figure 3.1a, the A-site cation occupies a 12-fold coordination site, and the B-site cation occupies a six-fold coordination site. The A-site cation is located at the body center of the cube, and the B cation is located at each of the eight corners, and the O anion is at each of the centers of the 12 edges. The A-site cation has a coordination number of 12, forming a cuboctahedral shape (Figure 3.1b). The unit cell comprises a three-dimensional network of corner-sharing BO_6 octahedra with linear O–B–O bonds (Figure 3.1c). In most of the cases, the B site is occupied by a transition-metal element, and the A site is an alkaline-earth metal element or a rare earth element. The ideal cubic structure deviates to tetragonal or rhombohedral or hexagonal phases depending on the difference in the size of cations occupying A and B sites.

For an ideal cubic structure, the following equation must be satisfied.

$$(r_A + r_O) = \frac{\sqrt{2}}{2} a_p = \sqrt{2}(r_B + r_O) \tag{3.1}$$

The above equation can be derived by considering (100) or (010) or (001) and (110) planes of the perovskite unit cell shown in Figure 3.2.

The lattice parameter a of the perovskite unit cell $a_p = 2(r_B + r_O)$ and therefore the diagonal of (001) plane is $2\sqrt{2}(r_B + r_O)$, which is also the measure of side length in (110) projection, and the same can be equal to $2(r_A + r_O)$.

where r_A, r_B, and r_O are the radii of A, B, and O ions in the lattice, and a_p is the lattice parameter of the perovskite unit cell.

Therefore,

$$\sqrt{2}(r_B + r_O) = (r_A + r_B) = a_p / \sqrt{2}\text{, or} \tag{3.2}$$

$$\frac{r_A + r_O}{r_B + r_O} = \sqrt{2} \tag{3.3}$$

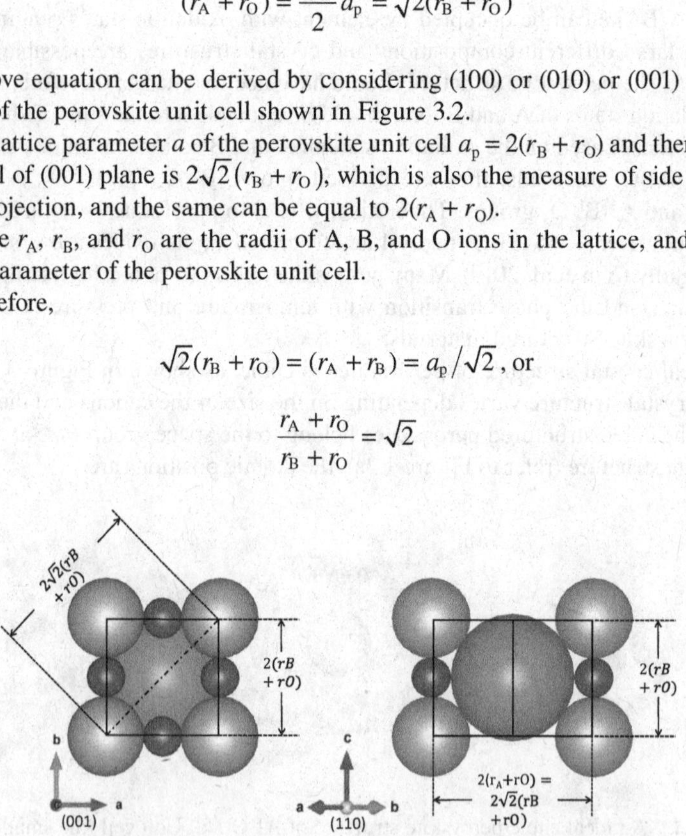

FIGURE 3.2 The projection of perovskite unit cell along (001) and (110) planes.

In terms of Goldschmidt tolerance factor,

$$t = \frac{(r_A + r_O)}{\sqrt{2}(r_B + r_O)} = 1\left(\text{for ideal structure}\right) \quad (3.4)$$

which can also represent the tolerance factor in terms of bond lengths (A – O) and (B – O) since the lattice parameter

$$a_P = 2(B - O) \quad \text{and} \quad \sqrt{2a_p} = 2(A - O) \quad (3.5)$$

$$\frac{(A-O)}{(B-O)} = \sqrt{2} \quad \text{or} \quad \frac{(A-O)}{\sqrt{2}(B-O)} = 1 \quad (3.6)$$

$SrTiO_3$ is an example of an ideal cubic perovskite structure, with $t = 1.00$ ($r_{Sr} = 1.44$ Å, $r_{Ti} = 0.605$ Å, and $r_O = 1.40$ Å). If the A-site ion is smaller than the ideal case or the difference between r_A and r_B is small, then the value of t is less than one. As a result [BO_6] octahedra will tilt in order to allow the dense packing, with a crystal structure symmetry lower than the cubic symmetry. In the ideal cubic perovskite structure, the value of t falls within 0.9–1. Apparently, if t is greater than one, due to a large A or a small B ion, then hexagonal variants of the perovskite structure are formed, e.g., $BaNiO_3$ type structures. In such a case, the close-packed layers are stacked in a hexagonal manner, resulting in a face sharing of the [NiO_6] octahedra. The t value for $BaNiO_3$ is 1.13 ($r_{Ba} = 1.61$ Å and $r_{Ni} = 0.48$ Å). If t falls between 0.71–0.9, the octahedral framework is rearranged to form a crystal structure with a lower symmetry than cubic, resulting in an orthorhombic or rhombohedral structure. If t is less than 0.71, the structure can be trigonal or orthorhombic or tetragonal with lower symmetries. The tolerance factor gives a rough estimate on the structure of perovskite crystals (Yin et al. 2019), and it is valuable in predicting the structure of perovskite oxides, as the ionic radius of the metallic ions is well-known. Moreover, a more accurate prediction of oxide structures is possible by Eq. 3.7 (Bartel et al. 2019) with an overall accuracy of 92%.

$$t = \frac{r_O}{r_B} - n_A\left(n_A - \frac{r_A / r_B}{\ln\left(\frac{r_A}{r_B}\right)}\right) \quad (3.7)$$

Here n_A is the oxidation state of A-site ions.

In the case of A or B site, partial substitution as in $A_{1-x}A'_xBO_3$ or $AB_{1-x}B'O_3$, the tolerance factor can be rewritten as

$$t = \frac{\left[(1-x)r_A + xr_{A'} + r_O\right]}{\sqrt{2(r_B + r_O)}} \quad (3.8)$$

or

$$t = \frac{(r_A + r_O)}{\sqrt{2((1-x)r_B + xr_{B'} + r_O)}} \quad (3.9)$$

respectively.

3.2 DISTORTIONS IN PEROVSKITE OXIDES

3.2.1 Octahedral Tilting

In general, the deviation from the ideal cubic structure due to a smaller or larger A-site ions is accommodated in the crystal structure through [BO_6] octahedral tilting, while maintaining their connectivity through corner-sharing. On a broad perspective [BO_6] octahedral tilting significantly affects the physical properties of perovskites. Due to the rigid nature of the perovskite oxide lattices, the possible compensation for the reduced size of A-site cuboctahedron cage (Figure 3.1b) is possible through the modification of the geometry of the six anions surrounding the B-site ions (Figure 3.1c). The tilt patterns of perovskite structures are often represented by Glazer notation, a technique to arrive at the nature of tilt in a distorted perovskite structured material as three-component tilts along the x, y, and z axes or a, b, and c direction of the ideal cubic structure, as shown in Figure 3.3a. Glazer notation is based on the relation between the lattice constants and a given distortion of an octahedron starting from a cubic perovskite structure. The angles of tilt in the tilted octahedra are measured along [001] direction denoted as φ, [110] direction as θ, [111] direction as ϕ, [0$\bar{1}$1] direction as ω, respectively (Lufaso and Woodward 2001), which is shown in Figure 3.3b.

The perovskite structure is composed of corner-sharing octahedral network, and the main axis of the tilt can be parallel to each crystallographic axis. However, the amplitude of tilt in each octahedron may be different from the others; additionally, two subsequent stacked layers along the tilt axis may be tilted in phase or anti-phase. The octahedral tilting is described by Glazer notation (Glazer 1972), which is defined as follows:

1. Start from the cubic (ideal perovskite) or pseudo cubic (non-cubic perovskite) structure.
2. Use three consecutive letters a, b, and c to represent the tilt along three orthogonal crystallographic axes along [100], [010], and [001] directions. If the tilt amplitude is the same, use the same letter; otherwise a different letter.
3. Use the superscripts +, −, or 0 above the letter to represent the same tilt, opposite tilt, or no tilt of two consecutive layers about that axis.

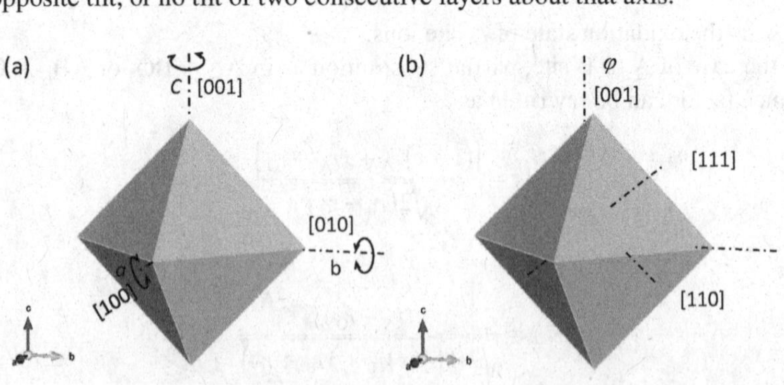

FIGURE 3.3 The rotational parameters for the octahedral tilting.

According to this definition, the ideal cubic aristotype lacking any octahedron tilting has the tilt symbol $a^0b^0c^0$ or $a^0a^0a^0$ and space group $Pm\bar{3}m$. According to Glazer there are 23 tilt systems (Glazer 1972) possible with respect to an ideal cubic perovskite structure with four octahedra unit cells, as listed in Table 3.1 (Woodward 1997), and the predicted symmetry of tilted structures with respect to ideal cubic structure is shown in Figure 3.4 (Howard and Stokes 2002).

Octahedral distortion is common in the case of perovskite structures with Goldschmidt number t being lower than 0.9, i.e., the A-site cations are smaller.

Let's use $NaTaO_3$ as an example to understand the Glazer notation. The structural parameters are listed in Table 3.2 (Ahtee and Darlington 1980). Start from a simple case, the structure at 893 K with a space group *P4/mbm* (127), and its Glazer tilt is denoted as $a^0b^0c^+$ with only one tilt, which is equivalent to No. 21 in Table 3.1. The structural model is shown in Figure 3.5a, where Ta is located at the center of octahedron formed by O (O is at the corners of the octahedra and is not shown in the figure). To make it simple, just consider the octahedra (Figure 3.5b). From this structure, we can identify the pseudo cubic structure along [$1\bar{1}0$], [110], and [001] direction, as indicated in the figure. The projections along these crystallographic directions are shown in Figure 3.5c–e. Around the [$1\bar{1}0$] axis (Figure 3.5c), no tilt occurs which is denoted as a^0; and so does along [110] axes (Figure 3.5d), which is denoted as b^0. However, along the third axis [001], an evident tilt is shown and the octahedra on the top and bottom layers have the same tilt, so the bottom layer does not appear (Figure 3.5e). This case is denoted as c^+, which only means the octahedra on consecutive layers have the same tilt, while within the layer, as shown in Figure 3.5e, neighboring octahedra have opposite tilts. Note that in Figure 3.5c and d, the center corner of the octahedra is slightly off the square center because of the tilt around [001], while the projected squares still show edges at nearly 90°.

The structure at 803 K is of a lower symmetry, with a space group of *Bmmb* (63) and Glazer notation of $a^0b^+c^-$ with two tilts (equivalent to No. 17 in Table 3.1). Its structural model is shown in Figure 3.6a and the octahedral model in Figure 3.6b. The pseudo cubic model is simply along the three axes of the structure, as indicated in Figure 3.6b, and their projections are shown in Figure 3.6c–e, respectively. Along [100] axis, no tilt occurs (Figure 3.6c), as evidenced from the project square patterns (denoted as a^0). While along [010] axis, same tilts occur between top and bottom layers (denoted as b^+), as shown in Figure 3.6d. However along [001] direction, the top layer and bottom layers have opposite tilts (denoted as c^-), so that the bottom layer appears along the projection, as shown in Figure 3.6e.

The structure at room temperature (RT) is of a *Pcmn* (62) space group and Glazer notation is $a^-b^+a^-$ with three tilts (equivalent to No. 10 in Table 3.1). Its structural and octahedral models are shown in Figure 3.7a and b, respectively. The pseudo cubic structure can be found along [101], [010], and [$10\bar{1}$] axes, as indicated in Figure 3.7b. The [101] projection is shown in Figure 3.7c with opposite tilts on top and bottom layers (denoted as a^-). The [010] projection is shown in Figure 3.7d, with the same tilt on top and bottom layers (denoted as b^+). Along the [$10\bar{1}$] projection, again the top and bottom layers have the same tilt and the tilt magnitude equals to that along the [101], so such a tilt is denoted as a^-.

TABLE 3.1
The Possible Phase Transformations from Ideal Cubic Perovskite Structures by BO_6 Octahedral Tilting

No.	Glazer Tilt	Space Group	Structure	Predicted Lattice Parameter (d = B–O Bond Length)
			Three tilts	
1	$a^+b^+c^+$	*Immm* (71)	Orthorhombic	–
2	$a^+b^+b^+$	*Immm* (71)	Orthorhombic	–
3	$a^+a^+a^+$	$Im\bar{3}$ (204)	Cubic	$a = d(8\cos + 4)/3$
4	$a^+b^+c^-$	*Pmmn* (59)	Orthorhombic	–
5	$a^+a^+c^-$	$P4_2/nmc$ (137)	Tetragonal	$a = 2d[\cos\varphi + \sin\varphi - \cos\theta(\sin\varphi - \cos\varphi)]$ $c = 4d\cos\theta$
6	$a^+b^+b^-$	*Pmmn* (59)	Orthorhombic	–
7	$a^+a^+a^-$	$P4_2/nmc$ (137)	Tetragonal	–
8	$a^+b^-c^-$	$P2_1/m$ (11)	Monoclinic	–
9	$a^+a^-c^-$	$P2_1/m$ (11)	Monoclinic	–
10	$a^+b^-b^-$	*Pmna* (62)	Orthorhombic	$a = d\left[8\left(2 + \cos^2\frac{\omega}{3}\right)\right]^{1/2}$, $b = d[48/(1 + 2\sec^2\omega)]^{1/2}, c = 2\sqrt{2}d\cos\omega$
11	$a^+a^-a^-$	*Pmna* (62)	Orthorhombic	–
12	$a^-b^-c^-$	$F\bar{1}$ (2)	Triclinic	–
13	$a^-b^-b^-$	*I2/a* (15)	Monoclinic	–
14	$a^-a^-a^-$	$R\bar{3}c$ (167)	Hexagonal	$a = 2\sqrt{2}d\cos, c = 4\sqrt{3}d$
			Two tilts	
15	$a^0b^+c^+$	*Immm* (71)	Orthorhombic	–
16	$a^0b^+b^+$	*I4/mmm* (139)	Tetragonal	$a = 2d(1+\cos\theta), c = 4d\cos\theta$
17	$a^0b^+c^-$	*Cmcm* (63)	Orthorhombic	$a = 4d\cos\theta, b = 2d(1+\cos\theta),$ $c = 2d(1+\cos\theta)$
18	$a^0b^+b^-$	*Cmcm* (63)	Orthorhombic	–
19	$a^0b^-c^-$	*I2/m* (12)	Monoclinic	–
20	$a^0b^-b^-$	*Imma* (74)	Orthorhombic	$a = 2\sqrt{2}d, b = 4d\cos\theta, c = 2\sqrt{2}d\cos\theta$
			One tilt	
21	$a^0a^0c^+$	*P4/mbm* (127)	Tetragonal	$= 2\sqrt{2}d\cos\varphi, c = 2d$
22	$a^0a^0c^-$	*I4/mcm* (140)	Tetragonal	$a = 2\sqrt{2}d\cos\varphi., c = 4d$
			Zero tilt	
23	$a^0a^0a^0$	$Pm\bar{3}m$ (221)	Cubic	$a_p = 2d$

Source: Reproduced with Permission from the International Union of Crystallography (Woodward 1997; Lufaso and Woodward 2001)

Notes: θ – the octahedral tilt about the cubic [001], ϕ is the octahedral tilt angle about the cubic [110], φ is the octahedral angle about the cubic [111], and ω is the octahedral tilt angle about the cubic [011].

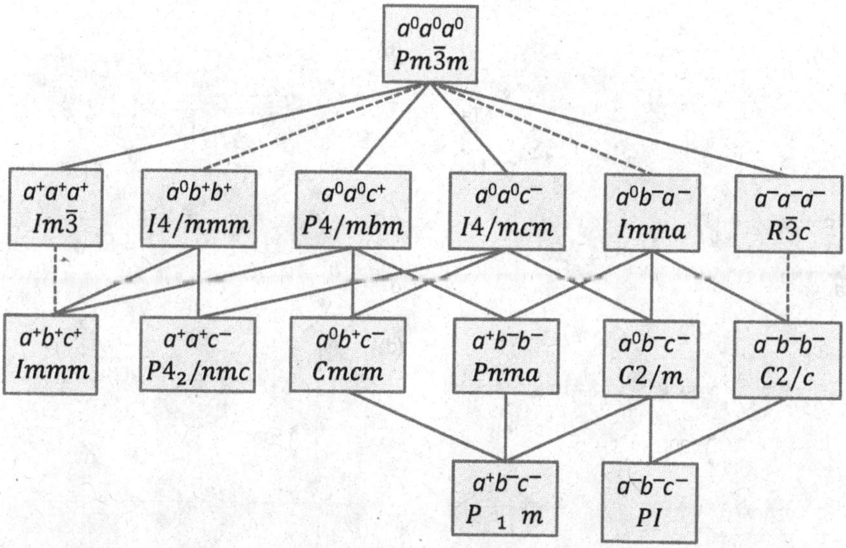

FIGURE 3.4 The space groups associated with the tilt systems for ABO_3 single perovskites starting from cubic aristotype and the dotted line indicate the possible phase transformation, reproduced with permission from the International Union of Crystallography (Howard and Stokes 2002).

TABLE 3.2
$NaTaO_3$ Phase Structures

Phase	Glazer Notation	Space Group/ Lattice Parameters	Atomic Site
$NaTaO_3$ at 893 K	$a^0b^0c^+$ (one tilt)	*P4/mbm* (127), tetragonal, $a = 0.55552$ nm, $c = 0.39338$ nm	Ta: 0, 0, 0 Na: 0, ½, ½ O(1): 0, 0, ½ O(2): ¼+0.0231, ¼–0.0231, 0
$NaTaO_3$ at 803 K	$a^0b^+c^-$ (two tilts)	*Bmmb* (63), orthorhombic, $a = 0.78453$ nm, $b = 0.78541$ nm, $c = 0.78633$ nm	Ta: ¼, 0, ¼ Na(1): 0, ¼, –0.021 Na(2): 0, ¼, ½+0.008 O(1): ¼+0.0185, 0, 0 O(2): 0, 0.0204, ¼–0.0318 O(3): ¼+0.0241, ¼, ¼–0.0023
$NaTaO_3$ at RT	$a^-b^+a^-$ (three tilts)	*Pcmn* (62), orthorhombic, $a = 0.54842$ nm, $b = 0.77952$ nm, $c = 0.55213$ nm	Ta: ½, 0, 0 Na: –0.0031, ¼, –0.0117 O(1): ½–0.0599, ¼, 0.0074 O(2): ¼+0.0357, –0.0295, ¼+0.0345

Source: Reproduced with permission from the International Union of Crystallography (Ahtee and Darlington 1980)

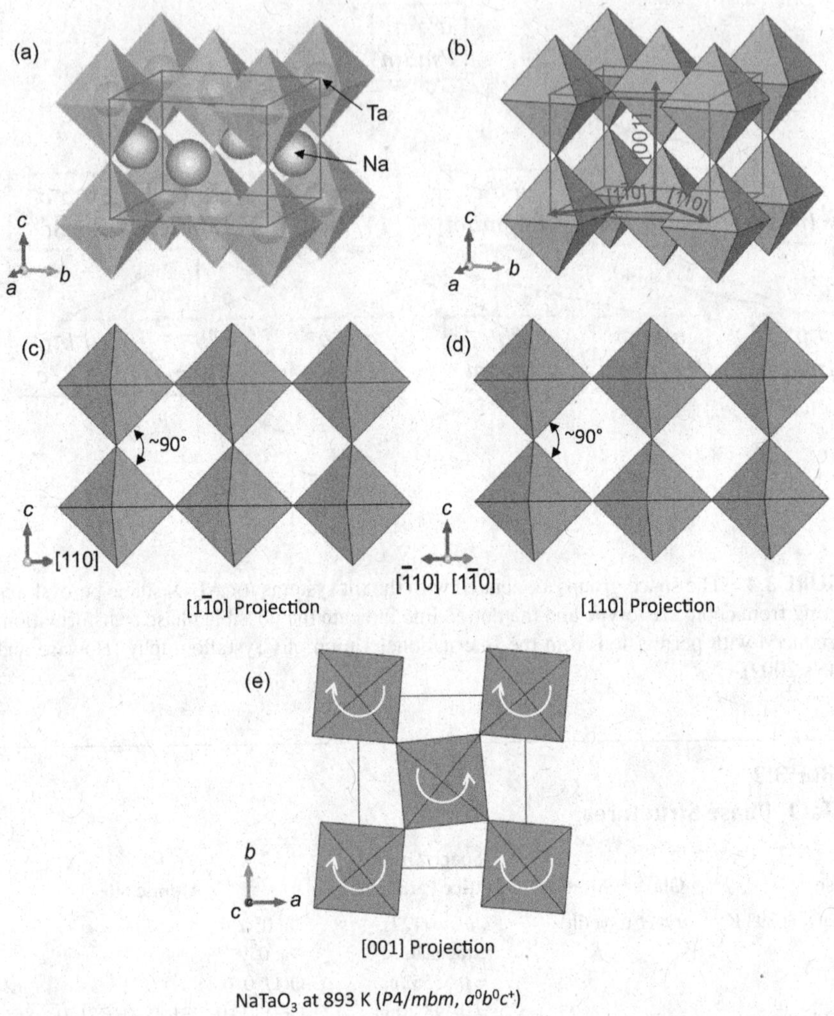

FIGURE 3.5 $NaTaO_3$ structure at 893 K with one tilt $a^0b^0c^+$. (a) Structural model (O atoms at the octahedral corners are not shown); (b) structural model showing octahedra only; (c) $[1\bar{1}1]$ projection with zero tilt; (d) [110] project with zero tilt; and (e) [001] projection showing the same tilts of the corresponding octahedra on the top and bottom layers so that the bottom layer does not appear.

3.2.2 B-Cation Displacement

If the B-site cations are very smaller than that of an ideal cubic structured perovskite with a tolerance factor considerably less than one, the resulting change in the lattice is through the displacement of B-site ion from the center of a perfect BO_6 octahedra. Such a displacement result in a significant change in the physical properties of the material. Off-center displacement occurs when the effective size of a cation B is such that the unstressed B–O bond length is less than $1/\sqrt{2}$ times the effective

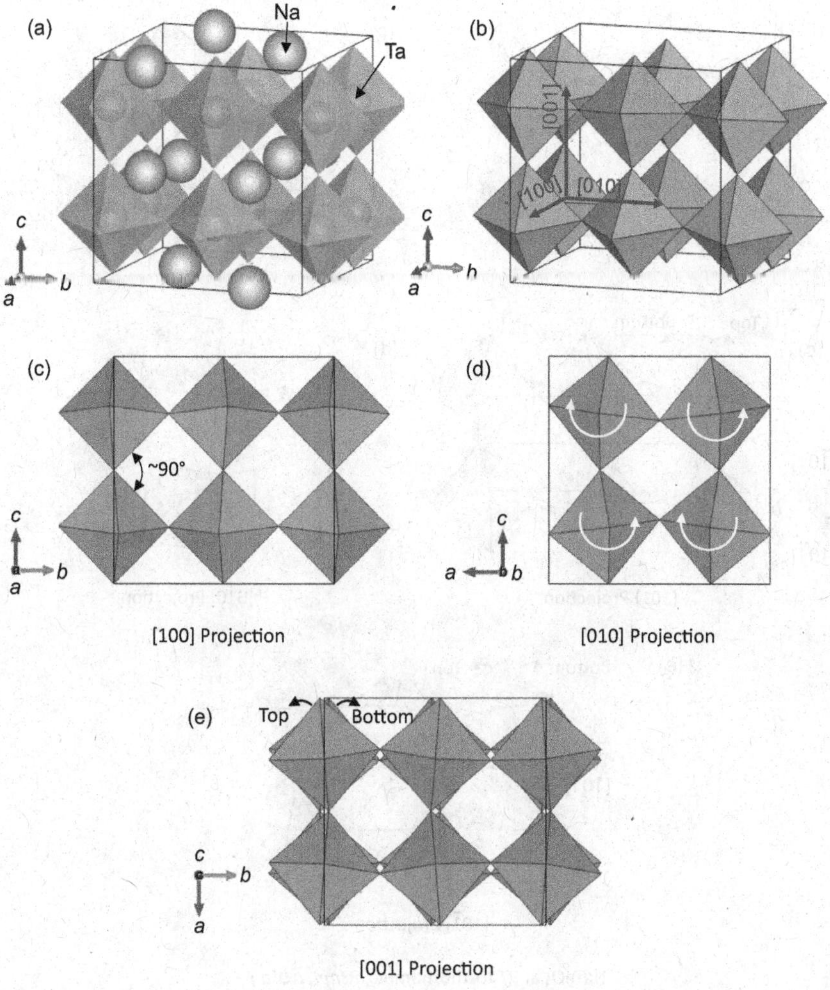

$NaTaO_3$ at 803 K (orthorhombic, *Bmmb*, $a^0b^+c^-$)

FIGURE 3.6 $NaTaO_3$ structure at 803 K with two tilts $a^0b^+c^-$. (a) Structural model (O atoms at the octahedral corners are not shown); (b) structural model showing octahedra only; (c) [100] projection with zero tilt; (d) [010] project with the same tilts of the corresponding octahedra on the top and bottom layers so that the bottom layer does not appear; and (e) [001] projection showing the opposite tilts of the corresponding octahedra along the top and bottom layers so that the bottom layer appears.

oxygen size (Megaw 1968). The cations displacements and octahedral distortions are sensitive to temperature and pressure. $BaTiO_3$ is extensively studied for the phase transition and property change with temperature and pressure. $BaTiO_3$ transforms from rhombohedral to orthorhombic at 183 K, orthorhombic to tetragonal at 263 K, and tetragonal to cubic at 393 K, due to cation displacement, and such displacements

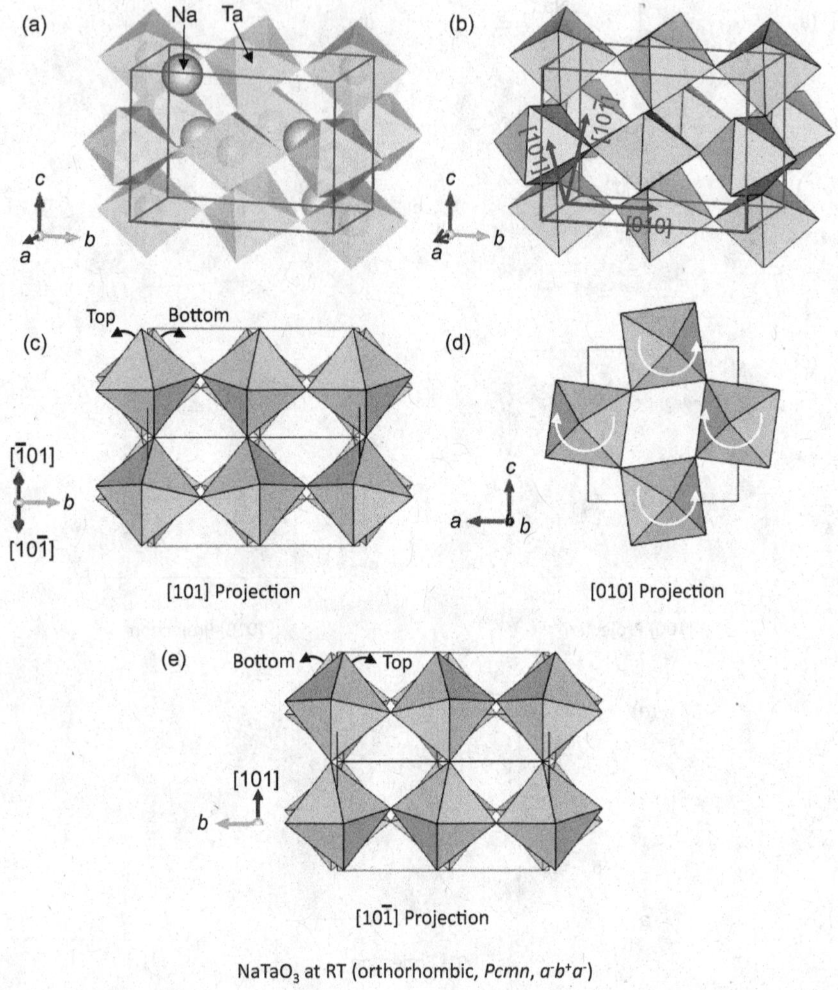

FIGURE 3.7 $NaTaO_3$ structure at RT with three tilts $a^-b^+a^-$. (a) Structural model (O atoms at the octahedral corners are not shown); (b) structural model showing octahedra only; (c) [101] projection showing the opposite tilts of the corresponding octahedra along the top and bottom layers so that the bottom layer appears; (d) [010] project with the same tilts of the corresponding octahedra on the top and bottom layers so that the bottom layer does not appear; and (e) $[10\bar{1}]$ projection showing the opposite tilts of the corresponding octahedra along the top and bottom layers so that the bottom layer appears.

occur due to the electronic instability of Ti^{4+} ions. Figure 3.8 shows the displacement of Ti^{4+} ions along different directions resulting in the phase transformation at different temperatures. The corresponding temperature-pressure phase transformation diagram of $BaTiO_3$ is shown in Figure 3.9. These distortions are considered as the second-order JT distortions.

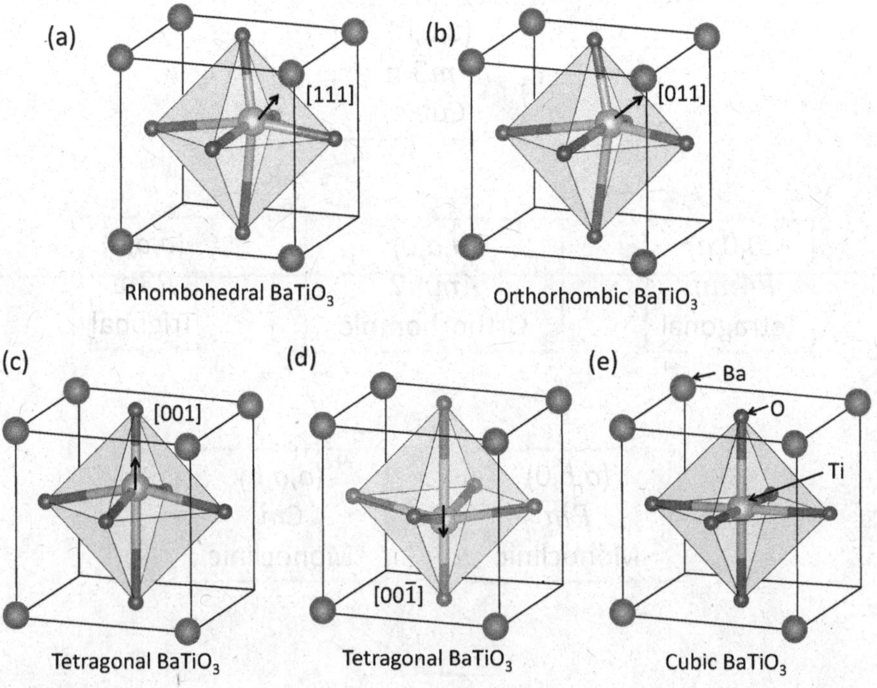

FIGURE 3.8 Crystal structure transformation by the displacement of Ti^{4+} in $BaTiO_3$ with respect to temperature.

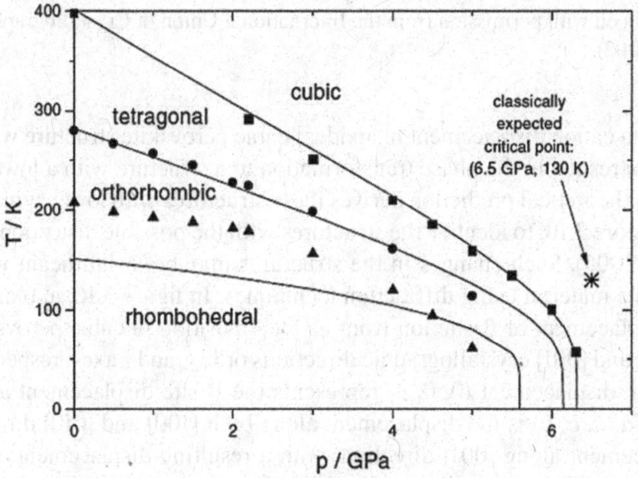

FIGURE 3.9 The pressure *vs.* temperature phase transformation diagram for $BaTiO_3$; the dots are the experimental results and the solid lines are the calculated results, reproduced with permission from the American Physical Society (Hayward and Salje 1997).

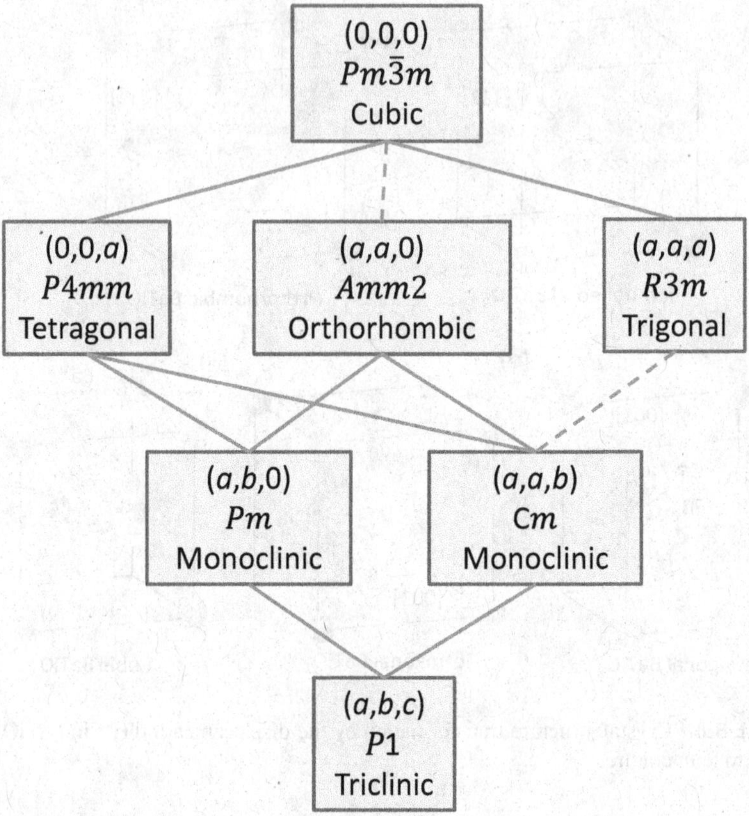

FIGURE 3.10 Theoretical prediction of phase transformation by B-site cation displacement, reproduced with permission from the International Union of Crystallography (Howard and Stokes 2002).

The B-site cation displacement in an ideal cubic perovskite structure with a space group $Pm\bar{3}m$ results in the phase transformation to a structure with a lower symmetry. A group theoretical prediction derives these structures with lower symmetries, as shown in Figure 3.10, to identify the structures with the possible distortions (Howard and Stokes 2005). Such changes in the structures may be insignificant while characterizing the material using diffraction techniques. In figure 3.10, a, b and c represent the displacement of B-site ion from its ideal position in cubic perovskite along [100], [010], and [001] crystallographic directions or *x*, *y*, and *z* axes, respectively. For example, the displacement (0, 0, *a*) represents the B-site displacement along [001] direction and (*a*, *a*, 0) is the displacement along both [100] and [010] direction, and zero displacement along [001] direction, with a resulting displacement along [110] direction if the displacements are equal in both the directions. If the displacement along the above three four-fold rotation axes are equal, then they are represented by the same letters, e.g., (*a*, *a*, *a*); however if the displacements are unequal, it is represented as (*a*, *b*, *c*).

3.2.3 Jahn–Teller (JT) Distortion

Another distortion in the ABO_3 perovskite is due to JT distortions of BO_6 octahedra. In the case of a transition metal ion in an octahedral symmetry system, the *d*-atomic orbitals are split into two degenerate sets, as per molecular orbital theory, as t_{2g} (d_{xz}, d_{yz}, and d_{xy}) and e_g (d_{z^2} and $d_{x^2-y^2}$). t_{2g} is at the lower energy level which splits into three orbitals, and the e_g at the upper energy level splits into two orbitals. JT distortions are dominant in perovskite materials with B-site cations and certain *d*-electron configurations, especially the ones with an odd number of electrons in the upper (e_g) pair of *d* orbitals. In general, JT distortions are geometrical distortions observed in non-linear molecules with a degenerated electronic ground state in such a way that the degeneracy is removed, as the distortion lowers the energy and symmetry (Alonso et al. 2000). JT distortion is encountered in octahedral complexes where axial bonds are elongated or shortened when compared with the equatorial bonds, as shown in Figure 3.11. Upon comparing the undistorted with JT distorted octahedra, the distortion occurs along the [001] direction; however, the position of A- and B-site ions and the coordination remain unchanged. In the case of elongation apical B–O bonds are longer than the lateral B–O bonds, and *vice versa* in compression (Tilley 2016).

In the case of B-site ions such as high-spin Cr^{2+}, Mn^{3+} $\left(d^4, t_{2g}^3 e_g^1\right)$; low-spin Co^{2+}, Ni^{3+} (d^7, $t_{2g}^6 e_g^1$) and Cu^{2+} (d^9, $t_{2g}^6 e_g^3$) (Halperin and Englman 1971) exhibit a degenerated ground state resulting from the two identical electron distributions. If t_{2g}^3 levels (d_{xz}, d_{yz}, and d_{xy}) occupy an equal number of electrons, the e_g higher energy level of d^4 orbital splits into either $d^1_{x^2-y^2}, d^0_{z^2}$ or $d^0_{x^2-y^2}, d^1_{z^2}$ degenerated energy configurations creating an energy imbalance, or *vice versa*. Therefore such degeneracy of the electronic state is removed by distortion of axial bonds in octahedral complexes either by elongation or compression to attain a lower energy due to the lower symmetry, as shown in Figure 3.12. If the splitting energy, Δ, is larger than the energy required to pair electrons, electrons will be paired in t_{2g} before occupying e_g, and *vice versa* if Δ is smaller. As a result, JT effect is particularly strong for high-spin d^4 and in d^9 configuration, giving rise to elongation or compression of the octahedra. The JT distortions due to the degeneracy of t_{2g} levels are insignificant, but the distortions by

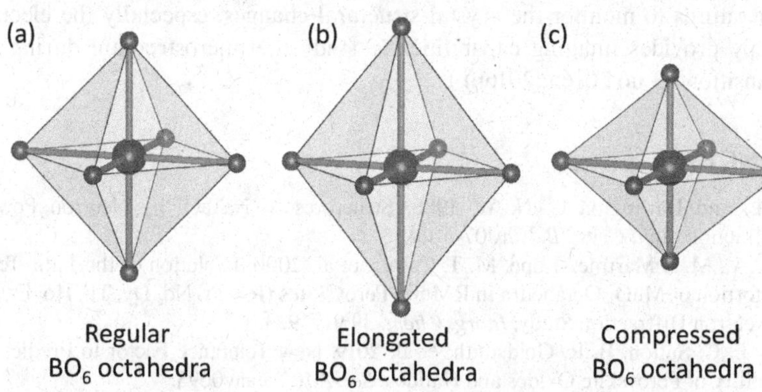

FIGURE 3.11 JT distorted octahedra.

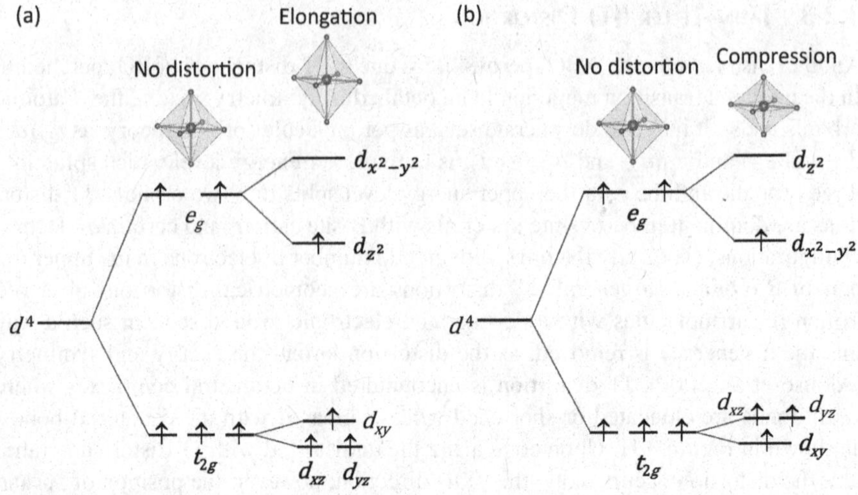

FIGURE 3.12 JT (a) elongation and (b) compression in high-spin d^4 orbital in an octahedral site due to degeneracy of ground and excited states.

the degeneracy of e_g orbitals are predominant (Lalena and Cleary 2010). JT distortions are sensitive to pressure and temperature, and the distortions will be reverted to a high symmetry state at high temperature or pressure.

The structure of perovskite often deviates from the cubic structure based on several parameters such as the size and oxidation state of the cations, temperature, pressure, and synthetic route. Therefore, materials with the general formula ABO_3 adopts different structures belonging to different symmetries; however, they are considered as perovskites or perovskite derivatives. Some major synthetic perovskites with ABO_3 formula and their symmetry, crystal structure, and lattice parameters are listed in Appendix A1.

The crystal structure of perovskites can be well studied by neutron, X-ray (Ahtee and Darlington 1980), or electron diffraction (Luo et al. 2009) techniques at different temperatures to monitor the crystal structural changes, especially the electron microscopy provides imaging capabilities to study the microstructure during the phase transitions (Luo 2016a; 2016b).

REFERENCES

Ahtee, M., and Darlington C. N. W. 1980. Structures of $NaTaO_3$ by Neutron Powder Diffraction. *Acta Cryst. B* 36:1007–1014.

Alonso, J. A., M. J. Martínez-Lope, M. T. Casais, et al. 2000. Evolution of the Jahn–Teller Distortion of MnO_6 Octahedra in $RMnO_3$ Perovskites (R = Pr, Nd, Dy, Tb, Ho, Er, Y): A Neutron Diffraction Study. *Inorg. Chem.* 39:917–923.

Bartel, C. J., C. Sutton, B. R. Goldsmith, et al. 2019. New Tolerance Factor to Predict the Stability of Perovskite Oxides and Halides. *Sci. Adv.* 5:eaav0693.

Glazer, A. M. 1972. The Classification of Tilted Octahedra in Perovskites. *Acta Cryst. B* 28:3384–3392.

Halperin, B., and R. Englman. 1971. Cooperative Dynamic Jahn–Teller Effect. II. Crystal Distortion in Perovskites. *Phys. Rev. B* 3:1698–1708.

Hayward, S. A., and E. K. H. Salje. 2002. The Pressure Temperature Phase Diagram of $BaTiO_3$: A Macroscopic Description of the Low-Temperature Behaviour. *J. Phys. Condens. Matter* 14:L599–L604.

Howard, C. J., and H. T. Stokes. 2002. Group-Theoretical Analysis of Octahedral Tilting in Perovskites. Erratum. *Acta Cryst. B* 58:565–565.

Howard, C. J., and H. T. Stokes. 2005. Structures and Phase Transitions in Perovskites – A Group-Theoretical Approach. *Acta Cryst. A* 61:93–111.

Lalena, J. N., and D. A. Cleary. 2010. *Principles of Inorganic Materials Design*. Hoboken, NJ: John Wiley & Sons.

Lufaso, M. W., and P. M. Woodward. 2001. Prediction of the Crystal Structures of Perovskites Using the Software Program *SPuDS*. *Acta Cryst. B* 57:725–738.

Luo, Z. P., D. J. Miller, J. F. Mitchell. 2009. Structure and Charge Ordering Behavior of the Colossal Magnetoresistive Manganite $Nd_{0.5}Sr_{0.5}MnO_3$. *J. Appl. Phys.* 105:07D528.

Luo, Z. 2016a. *A Practical Guide to Transmission Electron Microscopy, Volume I: Fundamentals*. New York: Momentum Press, pp. 94–95.

Luo, Z. 2016b. *A Practical Guide to Transmission Electron Microscopy, Volume II: Advanced Microscopy*. New York: Momentum Press, pp. 114–115.

Megaw, H. D. 1968. A Simple Theory of the Off-Centre Displacement of Cations in Octahedral Environments. *Acta Cryst. B* 24:149–153.

Tilley, R. J. D. 2016. *Perovskites: Structure-Property Relationships*. 1st edition. Chichester, UK: Wiley.

Woodward, P. M. 1997. Octahedral Tilting in Perovskites. I. Geometrical Considerations. *Acta Cryst. B* 53:32–43.

Yin, W.-J., B. Weng, J. Ge, et al. 2019. Oxide Perovskites, Double Perovskites and Derivatives for Electrocatalysis, Photocatalysis, and Photovoltaics. *Energy Environ. Sci.* 12:442–462.

4 Structural Variants of Perovskite Oxides

The materials with a certain systematic ordering of BO_6 octahedra are also considered under the class of perovskites with complex structures, which are different from the single perovskites as described in the previous chapter. The unit cell of such a structure often comprises more than one layer of A-site and/or B-site polyhedral slabs. Depending on the coordination of B-site cations, the shape of the BO_6 octahedra in an ideal cubic perovskite can transform into cubic, tetrahedral, pyramidal, planar, or linear arrangements, ultimately resulting in crystal structures with lower symmetries than that of a cubic perovskite. Some most commonly encountered perovskite-type derived structures are discussed in this chapter.

4.1 DOUBLE PEROVSKITES

The double perovskites have the closest analogy to single perovskite structures. Double perovskite unit cells are composed of two layers of BO_6 slabs along three axes. In general, double perovskites are perovskites with A- and B-lattice sites occupied by more than two different types ions, with general formulas $A'A''B_2O_6$ (double A-site), $A_2B'B''O_6$ (double B-site), or $A'A''B'B''O_6$ (mixed site). In all the above structures, the A- and B-site ions balance the charge of oxygen for a neutral charge according to $2q_A + q_{B'} + q_{B''} = -6$, $q_{A'} + q_{A''} + 2q_B = -6$ and $q_{A'} + q_{A''} + q_{B'} + q_{B''} = -6$, respectively, concerning $A_2B'B''O_6$, $A'A''B_2O_6$, and $A'A''B'B''O_6$ double perovskites. Here, q represents the charge of the specific cation A, A′, A″, B, B′, or B″.

In the double perovskite structures, A and B sites are occupied by ions with different oxidation states to accomplish the charge balance, as listed in Table 4.1. For instance, the double perovskite with the general formula $A_2B'B''O_6$ can be rewritten based on the charge of the cation as eight classes. The A-site and B-site cations with a range of oxidation states can form double perovskite structures with the defined stoichiometry. Considering $A_2B'B''O_6$ double perovskites as an example, if the average oxidation state for B-site is 4^+ when the oxidation state of A-site ion is 2^+, which is accomplished by a combination of B'^{4+}/B''^{4+}, B'^{3+}/B''^{5+}, B'^{2+}/B''^{6+}, or B'^{1+}/B''^{7+} cations. Similarly, for an A^{3+} cation, the average B-site cation oxidation state is 3^+, and the combination of B-site ions are B'^{3+}/B''^{3+}, B'^{2+}/B''^{4+}, or B'^{1+}/B''^{5+} cations. A^{2+} (Ca, Sr, Ba, etc.) and A^{3+} (La, Bi, etc.) cations are the common A-site cations for $A_2B'B''O_6$ perovskites. A^{1+} ions are also potential A-site ions, with B-site ions B'^{5+}/B''^{5+}, B'^{4+}/B''^{6+}, and B'^{3+}/B''^{7+}.

The crystal structure of double perovskites depends on the arrangement of A′, A″ or B′ and B″ cations in the respective lattices. Such an arrangement is significant if the size difference between A′ and A″ or B′ and B″ cations is substantial; thus, the structure of the double perovskites is adjusted to reduce the lattice strains through a certain ordering of $B'O_6$ or $B''O_6$ octahedra. These arrangements of $B'O_6$ or $B''O_6$

TABLE 4.1
The Possible Oxidation State of Elements in Double Perovskite Oxides

Type of Double Perovskite	Formulae Based on the Oxidation State	Examples for A-Site Cations	Examples for B-Site Cations
$A_2B'B''O_6$	$A_2^{(2+)}B'^{(2+)}B''^{(6+)}O_6$	Ba^{2+}, Sr^{2+}, Ca^{2+}, Mg^{2+}, Zn^{2+}, Pb^{2+}	Zn^{2+}, Mg^{2+}, Sn^{2+}, Cu^{2+}, Cr^{6+}, Fe^{6+}, Mo^{6+}, W^{6+}, U^{6+}
	$A_2^{(2+)}B'^{(3+)}B''^{(5+)}O_6$		Co^{3+}, Fe^{3+}, Ni^{3+}, Mn^{3+}, V^{5+}, Nb^{5+}, Ta^{5+}, Sb^{5+}
	$A_2^{(2+)}B'^{(4+)}B''^{(4+)}O_6$		Ti^{4+}, Sn^{4+}, Zr^{4+}, Si^{4+}, Ge^{4+}
	$A_2^{(1+)}B'^{(5+)}B''^{(5+)}O_6$	Cs^{+}, Rb^{+}, K^{+}, Na^{+}, etc.	V^{5+}, Nb^{5+}, Ta^{5+}, Sb^{5+}
	$A_2^{(1+)}B'^{(4+)}B''^{(6+)}O_6$		Si^{4+}, Sn^{4+}, Ge^{4+}, Ti^{4+}, Zr^{4+}, Mo^{6+}, W^{6+}
	$A_2^{(3+)}B'^{(3+)}B''^{(3+)}O_6$	Bi^{3+}, Dy^{3+}, Er^{3+}, Eu^{3+}, La^{3+}, Gd^{3+}, Tb^{3+}, Y^{3+}	Bi^{3+}, Tl^{3+}, Al^{3+}, Ga^{3+}, Co^{3+}, Fe^{3+}, Ni^{3+}
	$A_2^{(3+)}B'^{(2+)}B''^{(4+)}O_6$		Zn^{2+}, Mg^{2+}, Si^{4+}, Ge^{4+}, Sn^{4+}, Ti^{4+}, Ir^{4+}, Ru^{4+}
	$A_2^{(3+)}B'^{(1+)}B''^{(5+)}O_6$		V^{5+}, Nb^{5+}, Ta^{5+}, Sb^{5+}
$A'A''B_2O_6$	$A'^{(1+)}A''^{(1+)}B_2^{(5+)}O_6$	Cs^{+}, K^{+}, Na^{+}	V^{5+}, Nb^{5+}, Ta^{5+}, Sb^{5+}
	$A'^{(2+)}A''^{(2+)}B_2^{(4+)}O_6$	Ba^{2+}, Ca^{2+}, Sr^{2+}, Pb^{2+}	Si^{4+}, Ge^{4+}, Sn^{4+}, Ti^{4+}, Ir^{4+}, Ru^{4+}
	$A'^{(3+)}A''^{(3+)}B_2^{(3+)}O_6$	Bi^{3+}, Dy^{3+}, Er^{3+}, Eu^{3+}, La^{3+}, Nd^{3+}, Pr^{3+}, Gd^{3+}, Tb^{3+}, Y^{3+}	Al^{3+}, Bi^{3+}, Ga^{3+}, Co^{3+}, Fe^{3+}, Ni^{3+}
$A'A''B'B''O_6$	$A'^{(1+)}A''^{(1+)}B'^{(3+)}B''^{(7+)}O_6$	Na^{+}, K^{+}, Cs^{+}, Rb^{+}	Co^{3+}, Fe^{3+}, Ni^{3+}, Mn^{7+}, Tc^{7+}
	$A'^{(1+)}A''^{(1+)}B'^{(4+)}B''^{(6+)}O_6$		Si^{4+}, Ge^{4+}, Sn^{4+}, Ti^{4+}, Ir^{4+}, Ru^{4+}, W^{6+}, Mo^{6+}
	$A'^{(1+)}A''^{(1+)}B'^{(5+)}B''^{(5+)}O_6$		Nb^{5+}, Ta^{5+}, Sb^{5+}
	$A'^{(1+)}A''^{(2+)}B'^{(2+)}B''^{(7+)}O_6$	Na^{+}, K^{+}, Cs^{+}, Rb^{+}, Ba, Sr, Ca, Pb	Mg^{2+}, Zn^{2+}, Mn^{7+}, Tc^{7+}
	$A'^{(1+)}A''^{(2+)}B'^{(3+)}B''^{(6+)}O_6$	Na^{+}, K^{+}, Cs^{+}, Rb^{+}, Ba, Sr, Ca, Pb	Co^{3+}, Fe^{3+}, Ni^{3+}, Mo^{6+}, W^{6+}, U^{6+}

(Continued)

TABLE 4.1 (*Continued*)
The Possible Oxidation State of Elements in Double Perovskite Oxides

Type of Double Perovskite	Formulae Based on the Oxidation State	Examples for A-Site Cations	Examples for B-Site Cations
	$A'^{(1+)}A''^{(2+)}B'^{(4+)}B''^{(5+)}O_6$	Na^+, K^+, Cs^+, Rb^+, Ba^+, Sr, Ca, Pb	Si^{4+}, Ge^{4+}, Sn^{4+}, Ti^{4+}, Ir^{4+}, Ru^{4+}, Nb^{5+}, Ta^{5+}, Sb^{5+}
	$A'^{(1+)}A''^{(3+)}B'^{(1+)}B''^{(7+)}O_6$	Na^+, K^+, Cs^+, Rb^+, Bi^{3+}, Dy^{3+}, Er^{3+}, Eu^{3+}, La^{3+}, Gd^{3+}, Tb^{3+}, Y^{3+}	Li^+, Na^+, K^+, Mn^{7+}, Tc^{7+}
	$A'^{(1+)}A''^{(3+)}B'^{(2+)}B''^{(6+)}O_6$		Zn^{2+}, Mg^{2+}, Sn^{2+}, Cu^{2+}, Cr^{6+}, Fe^{6+}, Mo^{6+}, W^{6+}, U^{6+}
	$A'^{(1+)}A''^{(3+)}B'^{(3+)}B''^{(5+)}O_6$		Co^{3+}, Fe^{3+}, Ni^{3+}, Nb^{5+}, Ta^{5+}, Sb^{5+}
	$A'^{(1+)}A''^{(3+)}B'^{(4+)}B''^{(4+)}O_6$		Si^{4+}, Ge^{4+}, Sn^{4+}, Ti^{4+}, Ir^{4+}, Ru^{4+}
	$A'^{(2+)}A''^{(2+)}B'^{(1+)}B''^{(7+)}O_6$	Ca^{2+}, Sr^{2+}, Ba^{2+}, Pb^{2+}	Li^+, Na^+, K^+, Mn^{7+}, Tc^{7+}
	$A'^{(2+)}A''^{(2+)}B'^{(2+)}B''^{(6+)}O_6$		Zn^{2+}, Mg^{2+}, Sn^{2+}, Cu^{2+}, Cr^{6+}, Fe^{6+}, Mo^{6+}, W^{6+}, U^{6+}
	$A'^{(2+)}A''^{(2+)}B'^{(3+)}B''^{(5+)}O_6$		Co^{3+}, Fe^{3+}, Ni^{3+}, Nb^{5+}, Ta^{5+}, Sb^{5+}
	$A'^{(2+)}A''^{(2+)}B'^{(4+)}B''^{(4+)}O_6$		Si^{4+}, Ge^{4+}, Sn^{4+}, Ti^{4+}, Ir^{4+}, Ru^{4+}
	$A'^{(2+)}A''^{(3+)}B'^{(1+)}B''^{(6+)}O_6$	Ca^{2+}, Sr^{2+}, Ba^{2+}, Pb^{2+}, Bi^{3+}, Dy^{3+}, Er^{3+}, Eu^{3+}, La^{3+}, Gd^{3+}, Tb^{3+}, Y^{3+}	Li^+, Na^+, K^+, Mo^{6+}, W^{6+}, U^{6+}
	$A'^{(2+)}A''^{(3+)}B'^{(2+)}B''^{(5+)}O_6$		Zn^{2+}, Mg^{2+}, Sn^{2+}, Cu^{2+}, Nb^{5+}, Ta^{5+}, Sb^{5+}
	$A'^{(2+)}A''^{(3+)}B'^{(3+)}B''^{(4+)}O_6$		Co^{3+}, Fe^{3+}, Ni^{3+}, Si^{4+}, Ge^{4+}, Sn^{4+}, Ti^{4+}, Ir^{4+}, Ru^{4+}
	$A'^{(3+)}A''^{(3+)}B'^{(1+)}B''^{(5+)}O_6$	Bi^{3+}, Dy^{3+}, Er^{3+}, Eu^{3+}, La^{3+}, Gd^{3+}, Tb^{3+}, Y^{3+}	Li^+, Na^+, K^+, Nb^{5+}, Ta^{5+}, Sb^{5+}
	$A'^{(3+)}A''^{(3+)}B'^{(2+)}B''^{(4+)}O_6$		Zn^{2+}, Mg^{2+}, Sn^{2+}, Cu^{2+}, Si^{4+}, Ge^{4+}, Sn^{4+}, Ti^{4+}, Ir^{4+}, Ru^{4+}
	$A'^{(3+)}A''^{(3+)}B'^{(3+)}B''^{(3+)}O_6$		Al^{3+}, Bi^{3+}, Ga^{3+}, Co^{3+}, Fe^{3+}, Ni^{3+}

octahedra, in general, result in a random distribution $B'O_6$ or $B''O_6$ octahedra forming a cubic structure (e.g. Ba_2FeNbO_6) or an orthorhombic structure (e.g. $SrLaCuRuO_6$), with a rocksalt (NaCl) type (e.g. Ba_2FeMoO_6) or a layered (e.g. La_2CuSnO_6) ordering, mostly determined by the charge difference between B′ and B″ ions.

There are three types of ordering observed in typical double perovskite structures based on the A- or B-site cations. The rocksalt structure is the most symmetric one which is formed when the B′ and B″ cations are occupied in identical positions. In rocksalt structure, all the B′ and B″ octahedra are alternatively isolated from each other, which may be considered as zero-dimensional ordering, as shown in Figure 4.1a. The cations also adopt columnar or layered order, as shown in Figure 4.1b and c, corresponding to one- or two-dimensional ordering, respectively. In columnar ordering, $B'O_6$ or $B''O_6$ octahedra are interconnected to form columns as in Figure 4.1b. In layered ordering, each octahedral slab is formed by either $B'O_6$ or $B''O_6$ octahedral layers. Similarly, A-site ion ordering can be observed in $A'A''BO_6$ perovskite structures, as shown in Figure 4.2. Such ordering takes place in double perovskite structures which are dependent on the difference in the charge of B-site cations, in general, and the size of A-site ions in $A'A''BO_6$ perovskites. However, the above ordering of B-site ions is more prone as compared to the A-site ions.

Rocksalt ordering is dominant if the cations (B″ or A″) are highly charged than the other (B′ or A′) cations, and the rocksalt ordering separates the B″ ions with high charges to the largest distance. Some examples of rocksalt-ordered perovskites are listed in Table 4.2. In columnar-ordered structures, a B″ ion has two B″ and

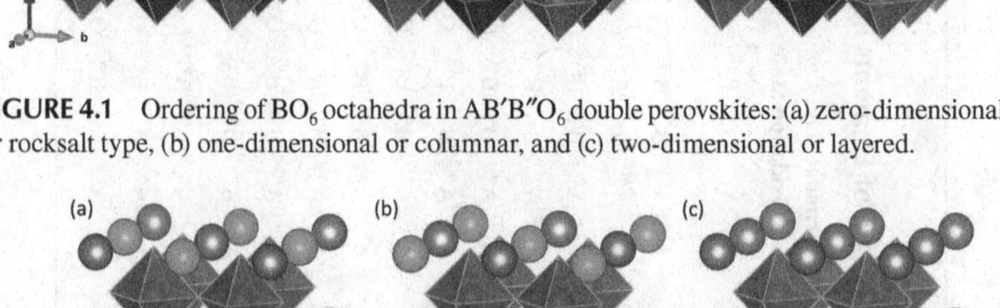

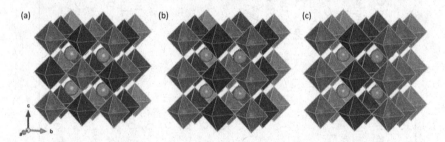

FIGURE 4.1 Ordering of BO_6 octahedra in $AB'B''O_6$ double perovskites: (a) zero-dimensional or rocksalt type, (b) one-dimensional or columnar, and (c) two-dimensional or layered.

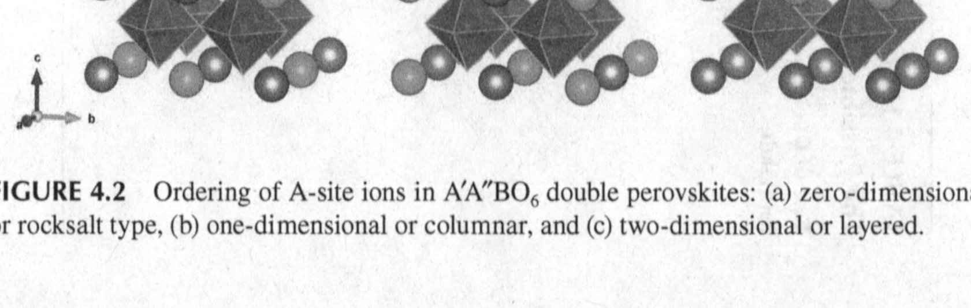

FIGURE 4.2 Ordering of A-site ions in $A'A''BO_6$ double perovskites: (a) zero-dimensional or rocksalt type, (b) one-dimensional or columnar, and (c) two-dimensional or layered.

TABLE 4.2
List of Double Perovskites with Rocksalt Ordering

Double Perovskite	Space Group	Lattice Parameters: Length (Å) a	b	c	Angle α	β	γ	Reference
Ba_2CaIrO_6	$Fm\bar{3}m$	8.364			90	90	90	Singh and Pulikottil (2019)
Ba_2CaMoO_6	$Fm\bar{3}m$	8.376			90	90	90	Nguyen et al. (2019)
Ba_2CeBiO_6	$Fm\bar{3}m$	8.754			90	90	90	Hatakeyama et al. (2010)
Ba_2CeIrO_6	$P2_1/n$	5.966	5.957	8.427	90	89.98	90	Revelli et al. (2019)
Ba_2CoMoO_6	$Fm\bar{3}m$	8.086			90	90	90	Martínez-Lope et al. (2002)
Ba_2CuWO_6	$I4/m$	5.564		8.636	90	90	90	Persson (2016a)
Ba_2EuNbO_6	$P2_1/n$	6.003	6.004	8.524	90	90.11	90	Persson (2014f)
Ba_2FeMoO_6	$Fm\bar{3}m$	8.058			90	90	90	Sahnoun et al. (2017)
Ba_2InBiO_6	$I4/mcm$	5.978		8.475	90	90	90	Fua et al. (2000)
Ba_2LaIrO_6	$P2_1/n$	6.051	6.060	8.574	90	90.29	90	Zhou et al. (2009b)
Ba_2NaOsO_6	$Fm\bar{3}m$	8.287			90	90	90	Erickson et al. (2007)
Ba_2NdMoO_6	$I4/m$	6.014		8.575	90	90	90	Persson (2016b)
Ba_2SmNbO_6	$P2_1/n$	6.019	6.016	8.535	90	90.2	90	Khandy and Gupta (2020)
Ba_2YIrO_6	$Fm\bar{3}m$	8.350			90	90	90	Fuchs et al. (2018)
Ba_2ZnMoO_6	$Fm\bar{3}m$	8.103			90	90	90	Li and Liu (2015)
Ca_2AlNbO_6	$P2_1/n$	5.377	5.415	7.626	90	89.95	90	Chan et al. (2000)
Ca_2FeSbO_6	$P2_1/n$	5.439	5.538	7.740	90	89.97	90	Persson (2014e)
Ca_2GdTaO_6	$P2_1/n$	5.576	5.848	8.084	90	90.3	90	Wang and Yao (2019)
Ca_2MgWO_6	$P2_1/n$	5.425	5.551	7.722	90	90.09	90	Yang et al. (2003)

(Continued)

TABLE 4.2 (*Continued*)
List of Double Perovskites with Rocksalt Ordering

Double Perovskite	Space Group	Lattice Parameters: Length (Å) a	b	c	Angle α	β	γ	Reference
Ca_2TiSiO_6	$Fm\bar{3}m$	7.410			90	90	90	Persson (2014b)
Er_2NiRuO_6	$P2_1/n$	5.240	5.632	7.534	90	89.88	90	Kayser et al. (2017)
Eu_2CoMnO_6	$P2_1/n$	5.340	5.592	7.598	90	89.93	90	Krishnamurthy and Venimadhav (2020)
In_2NiMnO_6	$P2_1/n$	5.135	5.337	7.546	90	90.13	90	Yi et al. (2013)
La_2CoIrO_6	$P2_1/n$	5.582	5.658	7.908	90	89.98	90	Song et al. (2017)
Lu_2CoMnO_6	$P2_1/n$	5.164	5.547	7.415	90	89.57	90	Lee et al. (2014)
Nd_2CoPtO_6	$P2_1/n$	5.442	5.697	7.749	90	90	90	Ouchetto et al. (1997)
Pb_2ScSbO_6	$Fm\bar{3}m$	8.105			90	90	90	Larrégola et al. (2009)
Sm_2CoRuO_6	$P2_1/n$	5.379	5.739	7.671	90	90	90	Das et al. (2019)
Sr_2AlTaO_6	$Fm\bar{3}m$	7.791			90	90	90	Takahashi et al. (2001)
Sr_2CaIrO_6	$P2_1/n$	5.780	5.830	8.200	90	90.2	90	Kayser et al. (2014)
Sr_2CdWO_6	$P2_1/n$	5.750	5.817	8.152	90	90.07	90	Gateshki et al. (2007)
Sr_2CeIrO_6	$P2_1/n$	5.834	5.844	8.256	90	90.19	90	Panda et al. (2013)
Sr_2CoFeO_6	$Fm\bar{3}m$	7.732			90	90	90	Pan et al. (2015)
Sr_2CoMoO_6	$I4/m$	5.565		7.948	90	90	90	Sereda et al. (2018)
Sr_2CoNbO_6	$I4/m$	5.585		7.907	90	90	90	Rendón Ramírez et al. (2013)
Sr_2CoOsO_6	$I4/m$	5.548		7.956	90	90	90	Morrow et al. (2013)
Sr_2CoReO_6	$I4/m$	5.563		7.952	90	90	90	Persson (2016c)
Sr_2CoSbO_6	$P2_1/n$	5.582	5.601	7.898	90	89.84	90	Faik et al. (2008)
Sr_2CoTaO_6	$\bar{3}$	7.928			90	90	90	Azcondo et al. (2015)

(*Continued*)

TABLE 4.2 (*Continued*)
List of Double Perovskites with Rocksalt Ordering

Double Perovskite	Space Group	Lattice Parameters: Length (Å) *a*	*b*	*c*	Angle *α*	*β*	*γ*	Reference
Sr_2CoTiO_6	*I4/m*	5.516		7.796	90	90	90	Azcondo et al. (2019)
Sr_2CrReO_6	*I4/mmm*	5.520		7.820	90	90	90	Hauser et al. (2012)
Sr_2CrSbO_6	$P2_1/n$	5.576	5.355	7.847	90	89.98	90	Faik et al. (2009)
Sr_2CrTaO_6	$Fm\bar{3}m$	7.884			90	90	90	Haid et al. (2019)
Sr_2CuMoO_6	*I4/m*	5.426		8.372	90	90	90	Vasala et al. (2014)
Sr_2DyMoO_6	$P2_1/n$	5.801	5.837	8.217	90	90.24	90	Vasala and Karppinen (2015)
Sr_2FeCoO_6	*I4/m*	5.461		7.711	90	90	90	Pradheesh et al. (2012)
Sr_2FeMoO_6	$Fm\bar{3}m$	7.888			90	90	90	Cernea et al. (2014)
Sr_2FeReO_6	*I4/mmm*	5.561		7.900	90	90	90	Blasco et al. (2009)
Sr_2FeWO_6	$P2_1/n$	5.651	5.614	7.942	90	89.99	90	Persson (2014g)
Sr_2GaSbO_6	*I4/m*	5.546		7.908	90	90	90	Lufaso et al. (2006)
Sr_2GdBiO_6	$P2_1/n$	5.913	6.029	8.439	90	90.15	90	Horyń et al. (1996)
Sr_2YIrO_6	$\bar{3}$	8.182			90	90	90	Corredor et al. (2017)
Sr_2ZrMnO_6	$Fm\bar{3}m$	7.840			90	90	90	Landínez et al. (2013)
Tb_2CoMnO_6	$P2_1/n$	5.294	5.601	7.542	90	90.01	90	Mandal et al. (2020)
Tb_2LiRuO_6	$P2_1/n$	5.308	5.684	7.536	90	90.71	90	Makowski et al. (2009)
Tm_2NiRuO_6	$P2_1/n$	5.230	5.656	7.523	90	90	90	Seinen et al. (1987)
Y_2AlCrO_6	$P2_1/n$	5.215	5.429	7.455	90	90.01	90	Das et al. (2016)
Y_2CoMnO_6	$P2_1/n$	5.233	5.593	7.468	90	89.76	90	Blasco et al. (2016)
Y_2NiMnO_6	$P2_1/n$	5.228	5.552	7.487	90	89.74	90	Su et al. (2015)
Yb_2CoMnO_6	$P2_1/n$	5.194	5.568	7.440	90	90.40	90	Blasco et al. (2015)

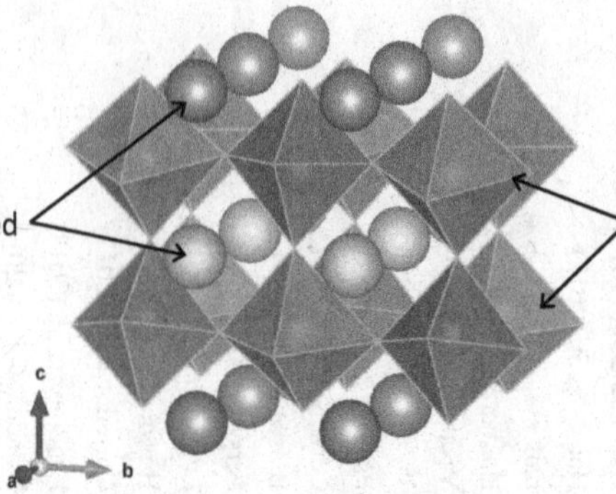

FIGURE 4.3 Double ordering of A- and B-site elements in A′A″B′B″O_6 double perovskite structures.

four B′ neighbors, which may be observed in the systems with charge ordering of mixed-valence elements in the B-site (e.g. $LaCaMn^{3+}Mn^{4+}O_6$), which may also arise due to the size difference of A-site cations, especially in A′A″BO_6 perovskites. The layered ordering is the least type observed, and one can expect this ordering when a B″ cation has two B′ and four B″ ions as the nearest neighbors (e.g. La_2CuZrO_6). It is important to note that the anion environment is subsequently changed with both A-site and B-site ordering except in the rocksalt structure.

Some double perovskite structures, often the ones with A′A″B′B″O_6 composition, in which B′/B″ octahedra are rocksalt ordered and A′/A″ ions ordered as layers called as double ordering as shown in Figure 4.3. They represent the major class of perovskites with such compositions, and the preferential ordering of A-site ions is dependent on the B-site ordering, which intern depends on the difference in the charges of B-site ions and sizes of A-site ions. Overall the unique ordering in A′A″B′B″O_6 perovskite results in lower space group symmetries than those of the AB′B″O_6 or A′A″BO_6 counterparts.

When the B-site ions are randomly ordered in a double perovskite, the resulting unit cell has the cell parameter two times of that of undistorted cubic perovskite (a_p). However, the octahedral titling may reduce the symmetry in these perovskite structures. With a unit cell twice of the size of the single $Pm\bar{3}m$ perovskite, the double perovskite with rocksalt ordering of B-site ions possesses a space group of $Fm\bar{3}m$. As in the case of single perovskites, a smaller A-site cation in double perovskites can result in octahedral tilting. The permissible tilts using the Glazer notation (Glazer 1972) in rocksalt-structured double perovskites are listed in Table 4.3 (Howard and Stokes 2005). The theoretical prediction of possible phase transformation in rocksalt-ordered double perovskite from ideal $Fm\bar{3}m$ space group is shown in Figure 4.4 (Howard and Stokes 2004).

4.2 LAYERED PEROVSKITES

In general, layered perovskites are composed of A-site cationic layers that separate the BO_6 octahedral slabs, forming 2D slabs of ideal perovskite structures separated by 2D A-site cation layers. Based on the structure of A-cation layers and BO_6

TABLE 4.3
The Plausible Tilting of Octahedra in Double Perovskites Using the Glazer Notation in Rocksalt-Ordered Double Perovskites

No.	Tilt Configuration	Space Group	Structure
1	$a^-b^-c^-$	$I\bar{1}$(No. 2)	Triclinic
2	$a^-a^-b^+$	$P2_1/n$(No. 14)	Tetragonal
3	$a^0b^-b^-$	$I2/m$(No. 12)	Orthorhombic
4	$a^0a^0c^+$	$P4/mnc$(No. 128)	Tetragonal
5	$a^0a^0c^-$	$I4/m$(No. 87)	Tetragonal
6	$a^-a^-a^-$	$R\bar{3}$(No. 148)	Trigonal
7	$a^0a^0a^0$	$Fm\bar{3}m$(No. 225)	Cubic

Source: Reproduced with permission of the International Union of Crystallography (Howard and Stokes 2004).

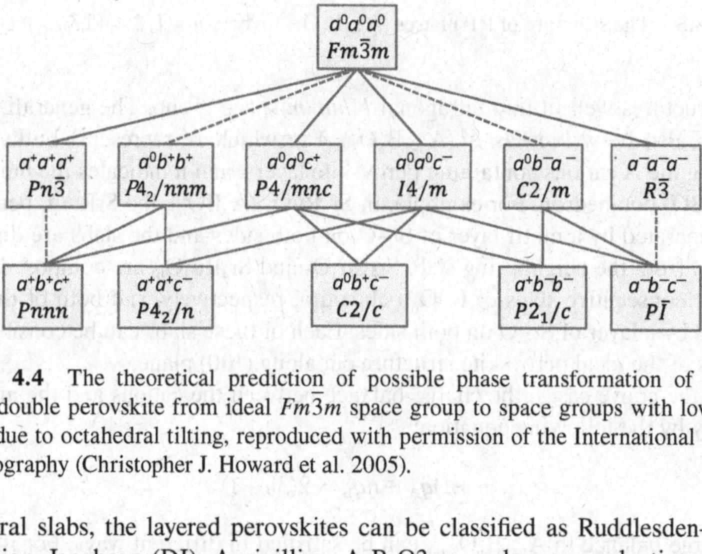

FIGURE 4.4 The theoretical prediction of possible phase transformation of rocksalt-ordered double perovskite from ideal $Fm\bar{3}m$ space group to space groups with lower symmetries due to octahedral tilting, reproduced with permission of the International Union of Crystallography (Christopher J. Howard et al. 2005).

octahedral slabs, the layered perovskites can be classified as Ruddlesden–Popper (RP), Dion–Jacobson (DJ), Aurivillius, $A_nB_nO3_{n+2}$, and superconducting cuprate-layered phases. The differentiating characteristics for the layered perovskites are (1) the motif which separates the layers and (2) the offsetting of the layers from each other (McKigney et al. 2007).

4.2.1 Ruddlesden–Popper Phase

RP perovskites have a general formula $A_{n+1}B_nO_{3n+1}$, ($n \geq 1$) comprising n-layers of the perovskite structure stacked in between rocksalt-structured A–O layers. A-site cations have a coordination number of 9, and B-site cations are coordinated to six oxygen anions. The common RP perovskite structures adopt the layered structures with the general formulae A_2BO_4, $A_3B_2O_7$ and $A_4B_3O_{10}$, as shown in Figure 4.5.

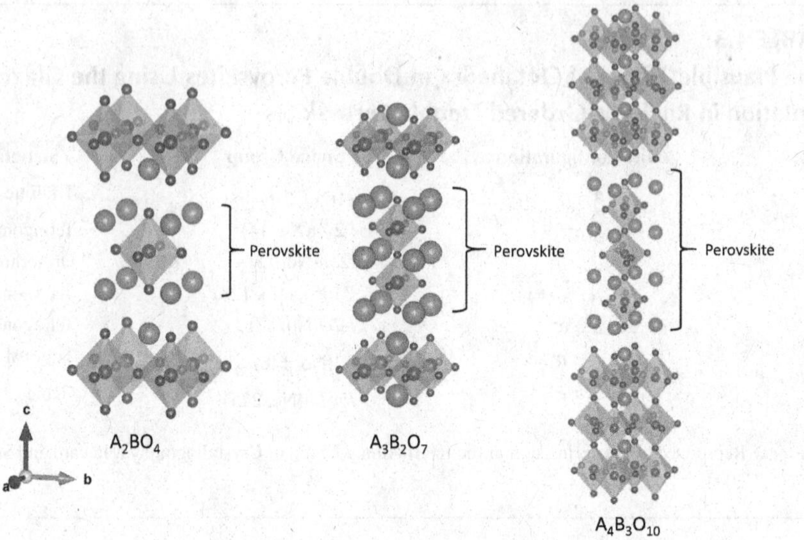

FIGURE 4.5 The structure of RP phases ($A_{n+1}B_nO_{3n+1}$ where $n = 1$, 2, and 3).

These structures well fit into tetragonal *I4/mmm* space group. The generalized formula can also be written as $A'_2(A_{n-1}B_nO_{3n+1})$ in which A′ represents bulky cation separating the A cations containing perovskite layers, and n indicates the number of slabs of BO_6 octahedron. For example, in Sr_2RuO_4 (A_2BO_4), the $SrRuO_3$ perovskite slab is separated by a motif layer of Sr–O on both sides and the slabs are displaced by (½, ½) from the neighboring slab. $Sr_3Ru_2O_7$ and $Sr_4Ru_3O_{10}$ are composed of two and three consecutive slabs of RuO_6 octahedra, respectively, and both of them are separated by a layer of Sr–O on both sides. Each of these slabs can be considered as segments of the ideal perovskite structure cut along (100) plane.

In all the above cases, the charge balance between the cations and the anions is necessary by the following equation:

$$(n+1)q_A + nq_B = 2(3n+1) \tag{4.1}$$

This charge balance in $A_{n+1}B_nO_{3n+1}$ can be satisfied in different ways. For instance, in $A_3B_2O_7$ structure $La_2SrAl_2O_7$, $q_A = 3$ (La) or $q_A = 2$ (Sr) and $q_B = 3$ (Al), and in $Sr_3Ru_2O_7$, $q_A = 2$ (Sr) and $q_B = 4$ (Ru), where q_A and q_B are the charges of A- and B-site ions, respectively. Table 4.4 lists some of the possible A_2BO_4 structures with respect to the oxidation state of A- and B-site cations.

The variation from RP structure is observed in A_2BO_4-type compounds based on the coordination of B-site ions with oxygen atoms. The three phases partially related to each other since the positions of cations in the lattice are identical; however, the position of oxygen is different. The formation of RP *T* phases is dependent on the effective atomic radii of the A- and B-site ions, and one can find the effective atomic radius from the article by Shannon (Shannon 1976). If the ratio r_A/r_B falls between 1.9 and 2.3, the formation of *T* phases occurs (Ganguli 1979). Depending on the

TABLE 4.4
The A_2BO_4-Layered Perovskite Oxides with RP-Type Structures with Possible Occupants of A- and B-Sites

Composition	Examples for A-Site Cations	Examples for B-Site Cations
$A_2^{2+}B^{4+}O_4$	Ba^{2+}, Sr^{2+}, Ca^{2+}, Pb^{2+}	Ti^{4+}, Sn^{4+}, Zr^{4+}, Si^{4+}, Ge^{4+}, Mn^{4+}, Ru^{4+}, Ir^{4+}
$A_2^{3+}B^{2+}O_4$	La^{3+}, Dy^{3+}, Gd^{3+}, Nd^{3+}, Y^{3+}	Zn^{2+}, Mg^{2+}, Sn^{2+}, Fe^{2+}, Cu^{2+}
$A_2^{1+}B^{6+}O_4$	Na^{+}, Cs^{+}, Rb^{+},	W^{6+}, Mo^{6+}, U^{6+}, Np^{6+}
$A^{2+}A^{3+}B^{3+}O_4$	Ba^{2+}, Sr^{2+}, Ca^{2+}, Pb^{2+}, La^{3+}, Dy^{3+}, Gd^{3+}, Nd^{3+}	Co^{3+}, Fe^{3+}, Ni^{3+}, Al^{3+}, Ga^{3+}, Mn^{3+}, Cr^{3+}

above ratio, these oxides exist as three different phases: *T*, *T** and *T′*, in which B-site ions are coordinated with six, five, and four oxygen atoms (BO_6, BO_5, and BO_4), respectively, as shown in Figure 4.6. Since the cation positions in this structure are identical to those in the RP perovskites, they are considered as layered perovskites. The examples for *T* phases are Sr_2RuO_4 and Sr_2TiO_4, whereas $LaSrCuO_4$, (Ce, Nd, Sr)$_2CuO_4$, $CaLaGaO_4$, etc. crystallize to *T** phase with BO_5 coordination. In RE_2CuO_4 (RE = Nd and Pr) of *T′* phase, the octahedrally coordinated CuO_6 slabs are replaced by CuO_2 layers with square planar coordination, and these planes are separated by Nd_2O_2 layers which have a fluorite structure. *T** and *T′* phases exhibit abundant oxygen vacancies; therefore, they are superconducting in nature.

Since the RP phases are composed of two different structural units, i.e. ABO_3 and AO, within the same lattice, the A_2BO_4 structure exhibits strong anisotropic features. Moreover, the B-site cations are coordinated by six oxygen anions, but the B–O bond lengths to "apical" and "equatorial" oxygen are different because of the J-T effect, resulting in two types of oxygen species in the BO_6 octahedra

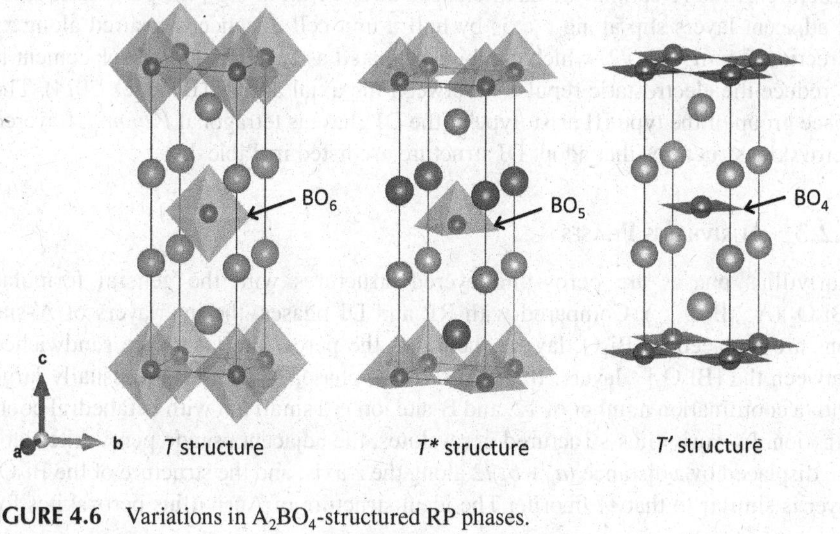

FIGURE 4.6 Variations in A_2BO_4-structured RP phases.

(Gopalakrishnan et al. 1977). Additionally, A_2BO_4 oxides can accommodate excess oxygen as interstitial oxygen defects in the A–O interstitial sites. By appropriate doping, oxygen vacancies can be induced into A_2BO_4 system, which may lead to lattice expansions. Therefore, A_2BO_4 oxides exhibit superior properties in the applications that involve oxygen transport, e.g. catalytic oxidation, electrochemical energy devices, chemical sensors, and oxygen permeation membranes (Zhu et al. 2014). Table 4.5 lists some layered perovskites that adopt RP structure.

4.2.2 Dion–Jacobson Phases

In DJ phase, a single layer of univalent alkali atoms often separates the perovskite-structured slabs apart. The DJ series of perovskites can be represented by the formula $A'(A_{n-1}B_nO_{3n+1})$ ($n > 1$). Similar to RP structures, a large number of cation mixtures are possible in the DJ phase to accomplish the charge balance. For example, when $n = 2$, A′ and A ions have an oxidation states of 1^+ and 3^+, respectively, and B-site ion has 5^+ state (Figure 4.7a). Similarly, when $n = 3$, A^{+1} and A^{+2} are paired with B^{5+} (Figure 4.7b).

Three aristotype structures are identified in the DJ phases, depending on the size and coordination of the A and A′ cations, which may be designated as type I, type II, and type III, respectively, as shown in Figure 4.7b. In the case of large A′ cations (e.g. Cs^+ or Rb^+), the perovskite slabs in the neighboring layers are aligned, with a tetragonal crystal structure with a space group of *P4/mmm* and $a = b \approx a_p$, which can be classified as type I. In the type I structures, there is no displacement of BO_6 octahedra or ions, and the neighboring layers are symmetric with respect to the axial anions in [100] direction. In other words, the atoms in the same plane have the same x and y coordinates but different z coordinates.

Type II structures are observed in perovskites with smaller A′ site ions (e.g. K^+). The perovskite blocks may "slip" by half a unit cell along the a or b axis only, in this case the space group of the aristotype structure is *Cmcm*, an orthorhombic crystal structure. If the A′ cations are smaller, such as Li^+, Na^+, or Ag^+, the perovskite slabs in adjacent layers slip along c axis by half a unit cell distance measured along xy direction, i.e. $(a_p + b_p)/2$, which can be designated as type III. This displacement is to reduce the electrostatic repulsion between the axial anions (Benedek 2014). The space group of the type III aristotype of the DJ phase is tetragonal *I4/mmm*. Layered perovskite structures that adopt DJ structure are listed in Table 4.6.

4.2.3 Aurivillius Phases

Aurivillius phases are perovskite-layered structures with the general formulae $(Bi_2O_2)(A_{n-1}B_nO_{3n+1})$. Compared with RP and DJ phases, the interlayers of A′-site ions are replaced by Bi_2O_3 layers; therefore, the perovskite layers are sandwiched between the $[Bi_2O_2]^{2+}$ layers. In an Aurivillius phase, A-site ions are usually large with a coordination number of 12, and B-site ion is a small ion with octahedral coordination. In Aurivillius-structured perovskites, the adjacent pseudo perovskite slabs are displaced by a distance $(a_p + b_p)/2$ along the z-axis, and the structure of the Bi_2O_2 layer is similar to that of fluorite. The ideal structure of Aurivillius perovskites for

TABLE 4.5
Examples of Perovskite Oxides with RP-Type Structures with Single and Multiple Layers

Perovskite	Space Group	Crystal System	Lattice Parameter (Å)			Reference
			a	*b*	*c*	
$n = 1$, A_2BO_4						
Ba_2CoO_4	*Pnma*	Orthorhombic	7.650	5.850	10.337	Zhang et al. (2019)
Ca_2RuO_4	*Pbca*	Orthorhombic	5.394	5.599	11.765	Braden et al. (1998)
$La_2CoO_{4+\delta}$	*I4/mmm*	Tetragonal	5.527		12.881	Brunelli and Ceretti (2015)
La_2CuO_4	*Fmmm*	Orthorhombic	5.365	5.409	13.170	Reehuis et al. (2006)
La_2NiO_4	*I4/mmm*	Orthorhombic	3.869	3.869	12.646	Zhou et al. (2009a)
$La_{2-x}Sr_xNiO_4$	*I4/mmm*	Tetragonal	5.397		12.733	Freeman et al. (2018)
$LaLiTiO_4$	*P4/nmm*	Tetragonal	3.772		12.083	Chaudhry et al. (2011)
$LaSrCoO_4$	*I4/mmm*	Tetragonal	3.824		12.659	Yang et al. (2005)
$LaSrVO_4$	*I4/mmm*	Tetragonal	3.869		12.652	Longo et al. (1973)
$LiEuTiO_4$	*Pbcm*	Orthorhombic	11.415	5.355	5.353	Huang et al. (2017)
Sm_2CoO_4	*Fmm2*	Orthorhombic	5.312	5.363	11.869	Lehmann and Müller-Buschbaum (1980)
Sr_2IrO_4	*I4₁/acd*	Tetragonal	5.493		25.777	Bhandari et al. (2019)
Sr_2MnO_4	*I4/mmm*	Tetragonal	3.787		12.496	Broux et al. (2013)
Sr_2PbO_4	*Pbam*	Orthorhombic	6.155	10.060	3.498	Zhao et al. (2015)
Sr_2TiO_4	*I4/mmm*	Tetragonal	3.884		12.600	Zhang et al. (2018)
$SrLaAlO_4$	*I4/mmm*	Tetragonal	3.754		12.649	Pan et al. (2018)
$SrLaFeO_4$	*I4/mmm*	Tetragonal	3.885		12.784	Xu et al. (2017)
$SrRuO_4$	*I4/mmm*	Tetragonal	3.873		12.732	Mackenzie and Maeno (2003)

(Continued)

TABLE 4.5 (*Continued*)
Examples of Perovskite Oxides with RP-Type Structures with Single and Multiple Layers

Perovskite	Space Group	Crystal System	Lattice Parameter (Å) a	b	c	Reference
$n = 2$, $A_3B_2O_7$						
$Ca_3Mn_2O_7$	*I4/mmm*	Tetragonal	3.696		19.487	Liu et al. (2018b)
$K_2SrTa_2O_7$	*I4/mmm*	Tetragonal	3.977		21.706	Kodenkandath et al. (2000)
$La_2SrAl_2O_7$	*I4/mmm*	Tetragonal	3.771		20.197	Zvereva et al. (2018)
$La_3Ni_2O_{7-\delta}$	*Fmmm*	Orthorhombic	5.393	5.436	20.516	Zhang et al. (1994)
$LaCa_2Mn_2O_7$	*Cmcm*	Orthorhombic	5.465	5.413	19.272	Green and Neumann (2000)
$LaSr_2MnCrO_7$	*I4/mmm*	Tetragonal	3.853		20.071	Singh and Singh (2010)
$Li_2LaTa_2O_7$	*I4/mmm*	Tetragonal	3.925		19.089	Persson (2014h)
$Li_2SrNb_2O_7$	*Fmmm*	Orthorhombic	5.594	5.601	18.011	Nagai et al. (2019)
$Sr_3Mn_2O_{6+\delta}$	*P4/mbm*	Tetragonal	10.835		20.168	Hadermann et al. (2005)
$Sr_3MnTiO_{7-\delta}$	*I4/mmm*	Tetragonal	3.849		20.163	Chowki et al. (2016)
$Sr_3Ru_2O_7$	*I4/mmm*	Tetragonal	3.890		20.552	Autieri et al. (2014)
$Sr_3Ti_2O_7$	*I4/mmm*	Tetragonal	3.899		20.399	Hungría et al. (2002)
$n = 3$, $A_4B_3O_{10}$						
$Ca_2La_2CuTi_2O_{10}$	*I4/mmm*	Tetragonal	3.884		27.727	Singh et al. (2012)
$Ca_4Mn_3O_{10-\delta}$	*Pbca*	Orthorhombic	5.260	5.256	26.835	Battle et al. (1998)
$K_2La_2Ti_3O_{10}$	*I4/mmm*	Tetragonal	3.877		29.824	Huang et al. (2010)

(*Continued*)

TABLE 4.5 (*Continued*)
Examples of Perovskite Oxides with RP-Type Structures with Single and Multiple Layers

Perovskite	Space Group	Crystal System	Lattice Parameter (Å) *a*	*b*	*c*	Reference
$K_2Nd_2Ti_3O_{10}$	*I4/mmm*	Tetragonal	3.859		29.656	Amow and Greedan (1998)
$La_4Co_3O_{10-\delta}$	*Pnma*	Orthorhombic	5.457	28.553	5.6542	Hansteen and Fjellvåg (1998)
$La_4Ni_3O_{10}$	*Cmca*	Orthorhombic	5.413	28.033	5.441	Kumar et al. (2020)
$Li_2Ca_2Ta_3O_{10}$	*I4/mmm*	Tetragonal	3.852		28.339	Toda et al. (1999)
$Na_2Ca_2Ta_3O_{10}$	*I4/mmm*	Tetragonal	3.887		28.655	Toda et al. (1999)
$Sr_4Mn_3O_{10}$	*Cmca*	Orthorhombic	5.470	12.380	12.510	González-Jiménez et al. (2018)
$Sr_4Ru_3O_{10}$	*Pbam*	Orthorhombic	5.528	5.526	28.651	Crawford et al. (2002)
$n = 4$, $A_5B_4O_{13}$						
$La_{2.25}Li_2Nb_{1.25}Ti_{2.75}O_{13}$	*I4/mmm*	Tetragonal	3.880		32.399	Tilley (2016)
$K_{2.5}Bi_{2.5}Ti_4O_{13}$	$Pb2_1m$	Orthorhombic	5.515	5.545	37.781	Liu et al. (2016)
$n = 5$, $A_6B_5O_{16}$						
$Ba_6Mn_5O_{16}$	*Cmca*	Orthorhombic	5.707	13.186	19.927	Grimaud et al. (2013)
$n = 6$, $A_7B_6O_{19}$						
$Li_4Sr_3Nb_6O_{20}$	*I4/mmm*	Tetragonal	3.953		26.041	Bhuvanesh et al. (1999)

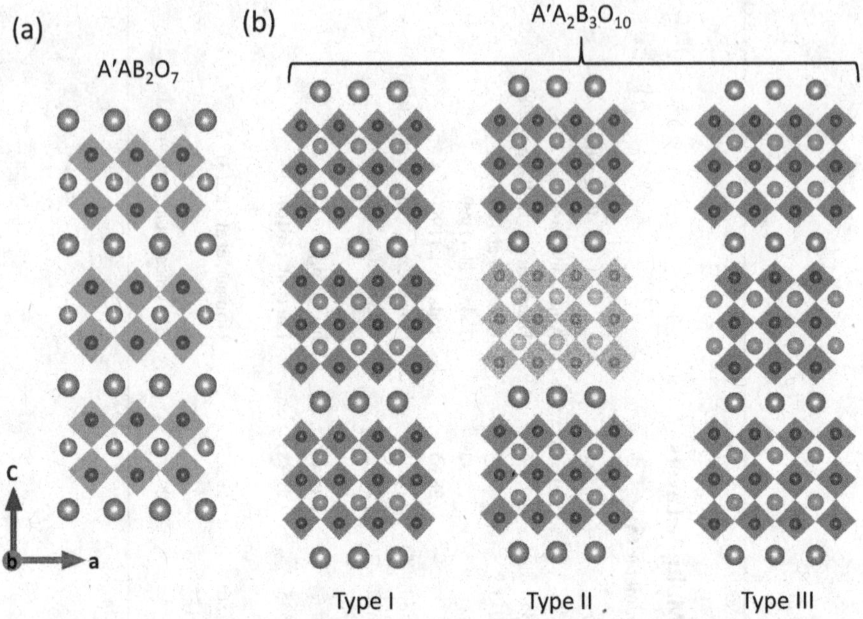

FIGURE 4.7 DJ layered perovskites when (a) $n = 1$ and (b) $n = 3$ with different aristotypes.

$n = 1 - 5$ is shown in Figure 4.8. The cations that can occupy A site are Na, K, Ca, Sr, Ba, etc. and B-site are Fe, Cr, Ti, Ga, Nb, V, Mo, W, etc. (Kendall et al. 1996).

The actual structure of Aurivillius phases is deviated from the ideal structure tremendously for higher values of n, especially when $n > 2$, due to distortions of the metal-oxygen octahedra in the perovskite slabs and alterations in the cation arrangements in the inter-perovskite layers. One such example is $Bi_2LaTiNbO_9$, where the B-site Ti or Nb ions occupy the octahedra in the perovskite layer with the interruption of the O–Ti/Nb–O chain along the z-axis by the $[Bi_2O_2]^{2+}$ layers, as in Figure 4.9a. The ordered intergrowth of slabs of more than one layer thickness is also common among these structures as shown in Figure 4.9b–d. Examples of Aurivillius perovskites with multiple layers are listed in Table 4.7.

4.2.4 $A_NB_NO_{3N+2}$-Layered Structures

$A_nB_nO_{3n+2}$-type layered perovskite oxides are formed by Ca, La, or Sr in the A-site and Ti, Nb, or Ta in the B-site. Figures 4.10 shows perovskite-layered crystal structures adopt the formula $A_nB_nO_{3n+2}$. Similar to the RP phases, these perovskite series have a layered structure stalked along c-axis, with n represents the thickness of octahedra layers. This structure is analogous to ABO_3 perovskites when n is approaching infinity. Each octahedral slab in the structure is displaced by the distance of $(a_p + a_p)/2$ with respect to the neighboring layer. Unlike the RP phases, these layered structures have crystallographic anisotropy as the BO_6 octahedra are often distorted or tilted.

TABLE 4.6
Examples of Perovskite Oxides with DJ-Type Structures with Multiple Layers

DJ Phase	Space Group	Crystal Structure	Lattice Parameter (Å)			Reference
$n = 2$						
$AgLaNb_2O_7$	*I4₁/acd*	Tetragonal	7.776		42.587	Sato et al. (1993)
$CsBiNb_2O_7$	$P2_1am$	Tetragonal	5.495		11.376	Autieri et al. (2019)
$KlaNb_2O_7$	*C222*	Orthorhombic	3.906	21.603	3.888	Sato et al. (1992b)
$NaLaNb_2O_7$	*I4/mmm*	Tetragonal	3.902		21.182	Sato et al. (1992a)
$BaSrTa_2O_7$	*Immm*	Orthorhombic	3.994	7.843	20.161	Le Berre et al. (2004)
$RbBiNb_2O_7$	$Pmc2_1$	Orthorhombic	11.232	5.393	5.463	Li et al. (2012)
$n = 3$						
$AgCa_2Ta_3O_{10}$	*I4/mmm*	Tetragonal	3.869		29.370	Toda et al. (1996)
$BaNd_2Ti_3O_{10}$	*Cmcm*	Orthorhombic	3.865	28.156	7.6221	Kolar et al. (1981)
$CsBa_2Ta_3O_{10}$	*P4/mmm*	Tetragonal	3.965		15.739	Hojamberdiev et al. (2016)
$CsBi_2Ti_2TaO_{10}$	$P\bar{4}$	Tetragonal	3.861		15.589	Kim et al. (2016)
$CsCa_2Nb_3O_{10}$	*Pnam*	Orthorhombic	7.740	7.746	30.185	Persson (2014c)
$CsCa_2Ta_3O_{10}$	*P4/mmm*	Tetragonal	3.866		15.254	Persson (2014i)
$CsCaLaNb_2TiO_{10}$	*P4/mmm*	Tetragonal	3.869		15.235	Hong et al. 2000
$CsNd_2Ti_2TaO_{10}$	*P4/mmm*	Tetragonal	3.832		15.276	Tilley (2016)
$Kca_2Nb_3O_{10}$	$P12_1/m1$	Orthorhombic	7.742	7.771	14.859	Fukuoka et al. (2000)
$Kca_2Ta_3O_{10}$	*C222*	Orthorhombic	3.866	29.777	3.852	Jiang et al. (2018)
$KSr_2Nb_3O_{10}$	$P2_1/m$	Orthorhombic	7.821	7.765	29.989	Kawaguchi et al. (2018)

(*Continued*)

TABLE 4.6 (*Continued*)
Examples of Perovskite Oxides with DJ-Type Structures with Multiple Layers

DJ Phase	Space Group	Crystal Structure	Lattice Parameter (Å)			Reference
$LiCa_2Ta_3O_{10}$	*I4/mmm*	Tetragonal	3.852		28.339	Mitsuyama et al. (2008)
$NaCa_2Nb_3O_{10}$	*P4₂/ncm*	Tetragonal	5.473		29.014	Thangadurai and Weppner (2002)
$RbBi_2Ti_2NbO_{10}$	*Ima2*	Orthorhombic	5.417	5.459	30.491	Kim et al. (2016)
$RbCa_2Nb_3O_{10}$	*P4/mmm*	Tetragonal	3.859		14.911	Liang et al. (2009)
$RbCa_2Ta_3O_{10}$	*P4/mmm*	Tetragonal	3.857		15.044	Persson (2014a)
$RbSr_2Nb_3O_{10}$	*P4/mmm*	Tetragonal	3.899		15.283	Tilley (2016)
$n = 4$						
$Ca_2Na_2Nb_4O_{13}$	*I4/mmm*	Tetragonal	3.872		36.937	Bharathy et al. (2008a)
$RbCa_2NaNb_4O_{13}$	*P4/mmm*	Tetragonal	3.870		18.894	Sato et al. (1993b)
$Kca_2NaNb_4O_{13}$	–	Orthorhombic	3.861	3.878	37.23	Knyazev et al. (2019)
$Kca_3Nb_3TiO_{13}$	–	Orthorhombic	3.842	3.861	37.27	Ram and Clearfield (1994)
$Kca_2Sr_{0.5}Nb_3Ti_{0.5}O_{11.5}$	–	Orthorhombic	3.863	3.878	37.39	Ram and Clearfield (1994)
$n = 5$						
$Kca_2Na_2Nb_5O_{16}$	–	Orthorhombic	3.863	3.882	45.00	Woo et al. (2017)
$Kca_4Nb_3Ti_2O_{16}$	–	Orthorhombic	3.833	3.856	45.00	Ram and Clearfield (1994)
$n = 6$						
$Kca_2Na_3Nb_6O_{19}$	–	Orthorhombic	3.869	3.883	52.80	Woo et al. (2017)

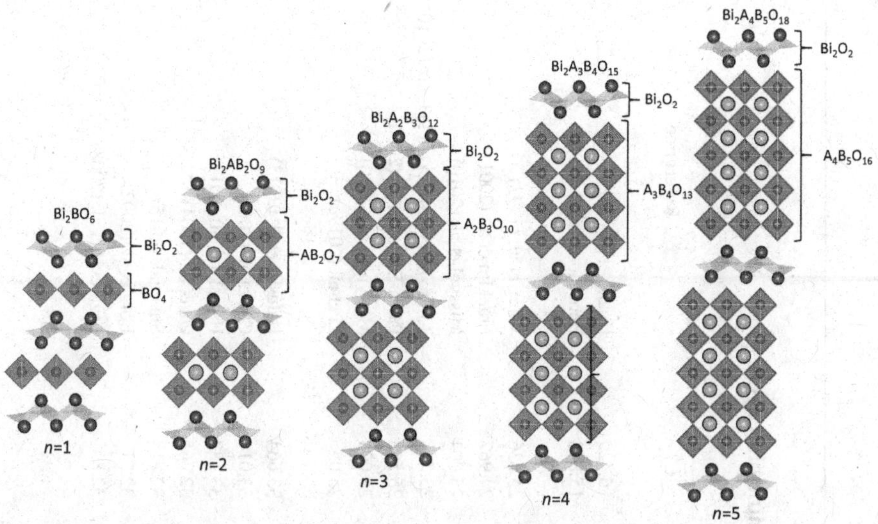

FIGURE 4.8 Structure of perovskites with Aurivillius phases.

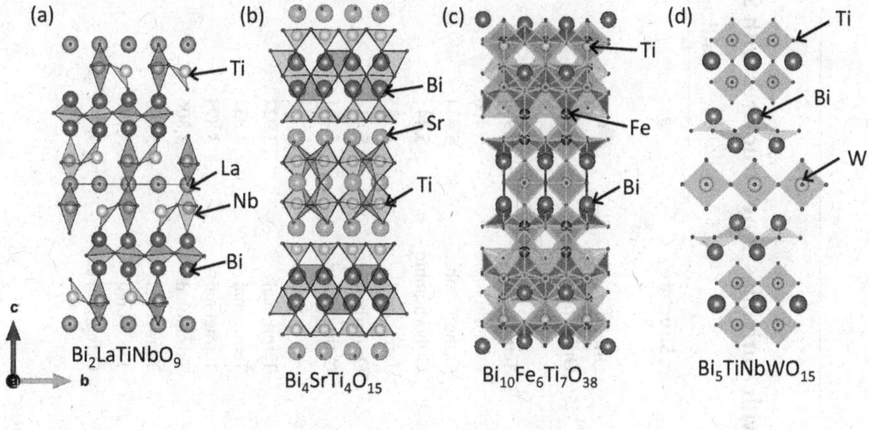

FIGURE 4.9 Structures that deviate from Aurivillius phases.

The interlayers between the octahedral blocks of $A_nB_nO_{3n+2}$-layered perovskites are formed, when the two adjacent octahedra do not share the oxygen atoms. There is an ambiguity on the number of possible octahedral layers as the repeating sequences involving two values n_1 and n_2 in the $A_nB_nO_{3n+2}$ structure, as in the following cases, (4,4,4,5), (4,4,5), (4,5), (4,5,5), etc., forming a non-centrosymmetric perovskite layers. Such systems can be represented by non-integer values based on the number of octahedral slabs in the blocks with alternating numbers of layers, e.g. $n = 4.5$, corresponding to alternating blocks of four and five layers (Lichtenberg et al. 2001).

TABLE 4.7
Examples of Perovskite Oxides with Aurivillius-Type Structures with Single and Multiple Layers

Phase	Space Group	Structure	Lattice Parameter (Å) a	b	C	Reference
$n = 1$						
Bi_2WO_6	$P2_1ab$	Orthorhombic	5.456	5.436	16.429	Huang et al. (2019)
$Bi_{0.7}Yb_{1.3}WO_6$	$A12_1$	Orthorhombic	8.107	3.705	15.838	Berdonosov et al. (2006)
$n = 2$						
$Bi_2BaNb_2O_9$	$I4/mmm$	Tetragonal	3.939		25.636	Wang et al. (1993)
$Bi_2CaTa_2O_9$	$A2_1am$	Orthorhombic	5.466	5.432	24.962	Franklin et al. (2001)
$Bi_2LaNbTiO_9$	$Pbam$	Orthorhombic	5.434	5.442	25.004	Missyul et al. (2010)
$Bi_2W_2O_9$	$Pna2_1$	Orthorhombic	5.44	5.413	23.74	Alfaro and Martínez-de la Cruz (2010)
$Bi_2BaTa_2O_9$	$I4/mmm$	Tetragonal	3.935		25.568	Paiva-Santos et al. (2000)
$Bi_2LaNbTiO_9$	$Pbam$	Orthorhombic	5.434	5.442	25.004	Missyul et al. (2010)
Bi_3NbTiO_9	$A2_1am$	Orthorhombic	5.431	5.389	25.05	Li et al. (2019c)
$n = 3$						
$Bi_2BaSrNb_2TiO_{12}$	$I4/mmm$	Tetragonal	3.922		33.664	Haluska et al. (2013)
$Bi_2La_2Ti_3O_{12}$	$I4/mmm$	Tetragonal	3.832		33.014	Chu et al. (2003)
$Bi_2Sr_2Nb_2TiO_{12}$	$I4/mmm$	Tetragonal	3.893		33.188	Haluska et al. (2013)
$Bi_2Nd_2Ti_3O_{12}$	$I4/mmm$	Tetragonal	3.806		32.765	Gu et al. (2004)
$Bi_2Pr_2Ti_3O_{12}$	$I4/mmm$	Tetragonal	3.809		32.814	Shi et al. (2015)
$Bi_2Sm_2Ti_3O_{12}$	$I4/mmm$	Tetragonal	3.796		32.71	Borg et al. (2002)
$Bi_4Ti_3O_{12}$	$B2cb$	Orthorhombic	5.448	5.411	32,83	Lazarević et al. (2008)

(Continued)

TABLE 4.7 (*Continued*)
Examples of Perovskite Oxides with Aurivillius-Type Structures with Single and Multiple Layers

Phase	Space Group	Structure	Lattice Parameter (Å) a	b	C	Reference
$n = 4$						
$BaBi_4Ti_4O_{15}$	*I4/mmm*	Tetragonal	3.864		41.760	Khokhar et al. (2013)
$Bi_4BaTi_4O_{15}$	$A2_1am$	Orthorhombic	5.443	5.432	41.694	Hutchison et al. (1977)
$Bi_4CaTi_4O_{15}$	$A2_1am$	Orthorhombic	5.423	5.402	40.593	Yokosuka et al. (2002)
$Bi_5Ti_3FeO_{15}$	$A2_1am$	Orthorhombic	5.468	5.440	41.247	Nakashima et al. (2010)
$n = 5$						
$Bi_4Ba_2Ti_5O_{18}$	*I4/mmm*	Tetragonal	3.88		50.3	Shi (2015)
$Bi_5LaTi_3Fe_2O_{18}$	*B2cb*	Orthorhombic	5.445	5.453	49.469	Naresh and Mandal (2015)
$Bi_6Ti_3Fe_2O_{18}$	*B2cb*	Orthorhombic	5.465	5.454	49.332	García-Guaderrama et al. (2006)

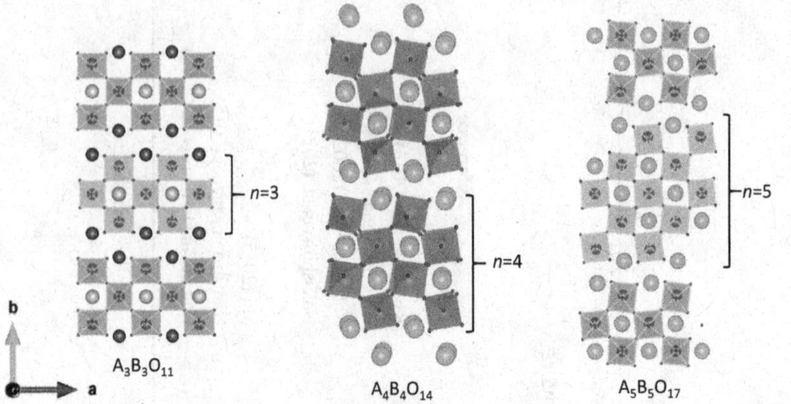

FIGURE 4.10 The structure of $A_nB_nO_{3n+2}$-layered perovskites.

4.2.5 Superconducting Cuprates

The layered perovskite-structured superconductors with $ACuO_3$ perovskite slabs are replacing conventional metallic alloys due to their high superconducting temperatures. The superconductivity of such materials is originated from the CuO_2 sheets present in them. The cationic layers separate the subsequent CuO_2 sheets, resulting in a general formula CuO_2_$(ACuO_2)_{n-1}$, where n is the number of CuO_2 layers, which is also an equivalent number of Cu atoms present in the system. The separating layers of the cations between the CuO_2 sheets adopt a rocksalt or fluorite structure, which often results in an overall T' or T^* structures as discussed before.

The simplest superconducting layered structure is observed in $A'_{2-x}A_xCuO_4$ (A′ = La, Nd, and Sm and A = Ca, Sr, and Ba) with either a T or T' structure. However, layered $LnBa_2Cu_3O_7$ (Ln = Y, Nd, Sm, Eu, Gd, Dy, Ho, Er, Tm, or Yb) materials are identified as superior materials as superconductors with the superconducting temperatures ~90 K above the boiling point of liquid nitrogen (−196°C or 77 K). $YBa_2Cu_3O_{7-y}$ (Y-123) belongs to the above class of superconducting materials, reported as the first superconductor with T_c above the liquid nitrogen temperature. The highest T_c of 133 K was recorded in Hg-based superconductor $HgBa_2Ca_{n-1}Cu_nO_{2n+2+\delta}$ when $n = 2$ (Schilling et al. 1993).

$YBa_2Cu_3O_7$ is composed of three perovskite-type unit cells stacked along the c-axis. The bottom and top of the unit cell are with Ba as the A cation, and the middle is with Y as the cation, as in Figure 4.11. Since the structure lacks oxygen atoms as compared to the idealized $A_3B_3O_9$ perovskite structure, the Cu atoms are coordinated with 5 and 4 oxygen atoms in the structure. They were thus forming square CuO_5 pyramids (the base perpendicular to c-axis) and square CuO_4 sheets (plane perpendicular to c-axis), replacing the CuO_6 octahedra. As a result, the Ba atoms are coordinated to a layer of BO_4 square sheets and BO_5 square pyramids. The Y atom at the center is coordinated to two CuO_5 square pyramidal layers, as in Figure 4.11a. The Cu atoms exist in the above structure as Cu^{2+} and Cu^{3+} ions. Therefore, the above structure can adopt $YBa_2Cu_3O_6$ and $YBaCu_3O_{7+\delta}$ formulas

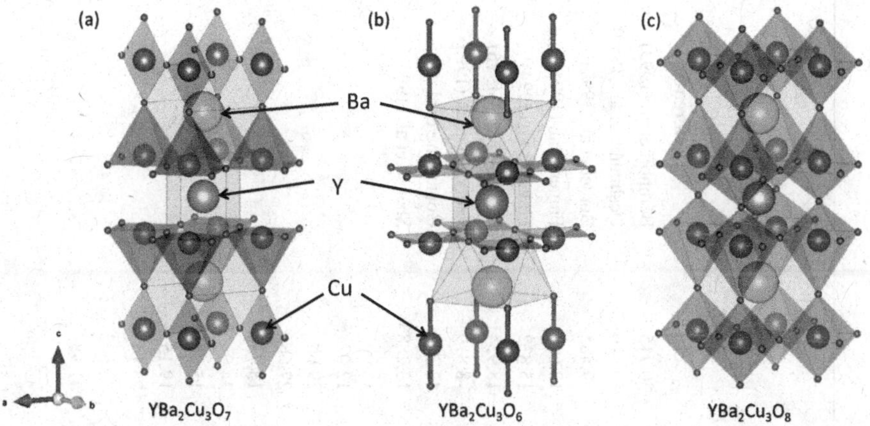

FIGURE 4.11 The perovskite-type structures observed in $YBa_2Cu_3O_{7\pm\delta}$ superconducting cuprates with (a) T^*, (b) T', and (c) distorted BO_6 structures with respect to the number of oxygen present.

depending on the number of oxygen atoms in the structure. In the structure of $YBa_2Cu_3O_6$, the CuO_5 pyramids in $YBa_2Cu_3O_7$ are replaced by CuO_4 planes with the planes oriented perpendicular to *c*-axis and the CuO_4 planes by linear coordination, as in Figure 4.11b. As the number of oxygen atom increases, as in $YBaCu_3O_{7+\delta}$, the structure approaches the structure of ideal perovskite with distorted octahedra, as shown in Figure 4.11c.

At several instances, the replacement of cations with Bi_2O_2 or Tl_2O_2, or with other cations, results in higher T_c. Many of these structures adopt a perovskite-type structure. The popular ones adopt a general formula $A_2Ca_{n+1-x}BxCu_nO_{2n+4}$, where A = Y, Bi, Tl, or Hg (Luo et al. 1999; Luo et al. 2004) and B = Sr or Ba. Representative perovskite-related structured superconducting oxides are presented in Table 4.8 with their unit cell characteristics.

4.3 ANION-DEFICIENT PEROVSKITES

A perovskite structure is said to be anion deficient when the sum of the charges of the A and B cations in the oxide perovskites is below six, which is equal to six for ideal perovskites. The ability of B-site transition metals to form different oxidation states reduces the number of oxygen surrounding the B-site ions to compensate the charge imbalance. In such structures, the number of surrounding oxygen atoms can reduce to 5, 4, 3, or 2, resulting in the transformation of BO_6 octahedron into a BO_5 tetragonal pyramid or a BO_4 square or a BO_2 linear perovskite layer. One can distinguish such changes in the perovskite structures in superconducting cuprates as discussed in the previous section.

The anion deficiency ranges from small δ values to a large value, resulting in the formulation from $ABO_{3-\delta}$ to ABO_2 structures in single perovskites. ABO_2 compounds are considered as an infinite layer compound since the A-site cation is occupied in

TABLE 4.8
Examples of Superconducting Cuprates with Layered Structures

Composition	Space Group	Crystal Structure	Lattice Parameter (Å) a	b	c	Reference
$Bi_2CaSr_2Cu_2O_8$	*Amaa*	Orthogonal	5.405	5.402	30.715	Bernhoff et al. (1993)
$Bi_2CaSr_2Cu_2O_9$	*Bbmb*	Orthogonal	5.396	5.397	30.649	Deshmukh et al. (1989)
$HgBa_2CuO_{4+\delta}$	*P4/mmm*	Tetragonal	3.8815		9.485	Bordet et al. (1996)
$HgBa_2CaCuO_{6+\delta}$	*P4/mmm*	Tetragonal	3.8554		12.693	Finger et al. (1994)
$HgBa_2Ca_2Cu_3O_{8+\delta}$	*P4/mmm*	Tetragonal	3.842		15.832	Finger et al. (1994)
$HgBa_2Ca_3Cu_4O_{10+\delta}$	*P4/mmm*	Tetragonal	3.8498		18.9341	Loureiro et al. (1996)
$HgBiSr_7Cu_2SbO_{15}$	*Pmam*	Orthorhombic	7.680	11.555	8.879	Pelloquin et al. (1995)
$HgTl_2Ba_4Cu_2O_{10}$	*I4/mmm*	Tetragonal	3.858		42.203	Goutenoire et al. (1994)
$(La,Ba)_2CuO_4$	*I4/mmm*	Tetragonal	3.787		13.2883	Shaked et al. (1994)
$(La,Sr,Ca)_3Cu_2O_6$	*I4/mmm*	Tetragonal	3.821		19.60	
$(Nd,Ce)_2CuO_4$	*I4/mmm*	Tetragonal	3.95		12.07	
$(Pb,Cu)(Eu,Ce)_2(Sr,Eu)_2Cu_2O_9$	*I4/mmm*	Tetragonal	3.80		29.60	
$PbBaYSrCu_3O_8$	*I4/mmm*	Tetragonal	3.842		27.66	
$Pb_2Sr_2Ycu_3O_8$	*Cmmm*	Orthorhombic	5.390	5.430	15.730	
$(Sr,Ca)_5Cu_4O_{10}$	*I4/mmm*	Tetragonal	3.86		34.0	
$TlBa_2Ca_2Cu_3O_{9+\delta}$	*P4/mmm*	Tetragonal	3.843		15.871	
$Tl_2Ba_2Ca_3Cu_4O_{12}$	*P4/mmm*	Tetragonal	3.85		19.15	
$Tl_2Ba_2CuO_6$	*I4/mmm*	Tetragonal	3.864		23.14	
$Tl_2Ba_2Ca_2Cu_3O_{10}$	*I4/mmm*	Tetragonal	3.8487		35.662	
$Tl_2Ba_2Ca_3Cu_4O_{12}$	*I4/mmm*	Tetragonal	3.85		41.98	

(*Continued*)

TABLE 4.8 (*Continued*)
Examples of Superconducting Cuprates with Layered Structures

Composition	Space Group	Crystal Structure	Lattice Parameter (Å) a	b	c	Reference
$Tl_2Ba_2CaCu_2O_8$	*I4/mmm*	Tetragonal	3.856		29.312	Chaplot et al. (1991)
$Tl_2Ba_2Ca_4Cu_5O_{13}$	*P4/mmm*	Tetragonal	3.85		22.25	Shaked et al. (1994)
$Tl_2Ba_2Ca_4Cu_5O_{14}$	*I4/mmm*	Tetragonal	3.85		48.2	
$Tl_5Ba_3Sr_5Cu_3O_{19}$	*A2mm*	Orthorhombic	3.754	30.631	9.219	Letouzé et al. (1997)
$TlBa_2CaCu_2O_7$	*P4/mmm*	Tetragonal	3.833		12.680	Zhou (2019)
$TlLaSrCuO_5$	*P4/mmm*	Tetragonal	3.779		8.847	Nagashima et al. (1990)
$Y_2Ba_4Cu_7O_{14+\delta}$	*Ammm*	Orthorhombic	3.851	3.869	50.290	Gupta and Gupta (1993)
Y_2BaCuO_5	*Pnma*	Orthorhombic	12.188	5.662	7.132	Kamiya et al. (1994)
$Yba_2CoCu_2O_{7.25}$	*P4/mmm*	Tetragonal	3.891		11.674	Koren and Polturak (1994)
$Y_4Ba_2CaCu_2O_{7-\delta}$	*P4/mmm*	Tetragonal	3.856		12.754	Shaked et al. (1994)
$Yba_2Cu_3O_7$	*Pmmm*	Orthorhombic	3.813	3.881	11.630	She and Liu (2008)
$Yba_2Cu_4O_8$	*Ammm*	Orthorhombic	3.841	3.871	27.240	Shaked et al. (1994)
$Yba_4Cu_3O_9$	$Pm\bar{3}$	Cubic	8.117			Kalanda et al. (2002)
$YbaCuFeO_5$	*P4mm*	Tetragonal	3.867		7.656	Chen et al. (2019)
$YbaSrCu_3O_{7.03}$	*Pmmm*	Orthorhombic	3.852	3.788	11.552	Nagarajan and Rao (1993)

between the BO_4 planes as shown in Figure 4.12, which may be assumed that A-site ions completely lost the oxygen atoms (Li et al. 2019a).

Anion-deficient perovskite with the general formula $A_nB_nO_{3n-1}$ adopt numerous different structures; a few common ones are shown in Figure 4.13. Figure 4.13a shows a typical structure adopted by compounds with Mn^{3+} or Cu^{2+} ions in the

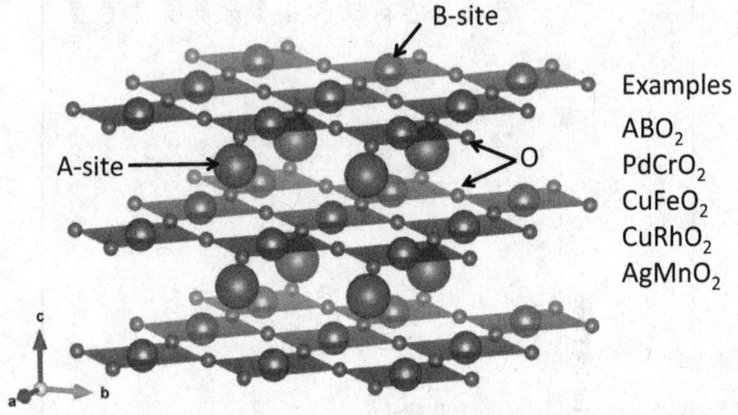

FIGURE 4.12 A representative structure of the infinite-layer ABO_2 compound.

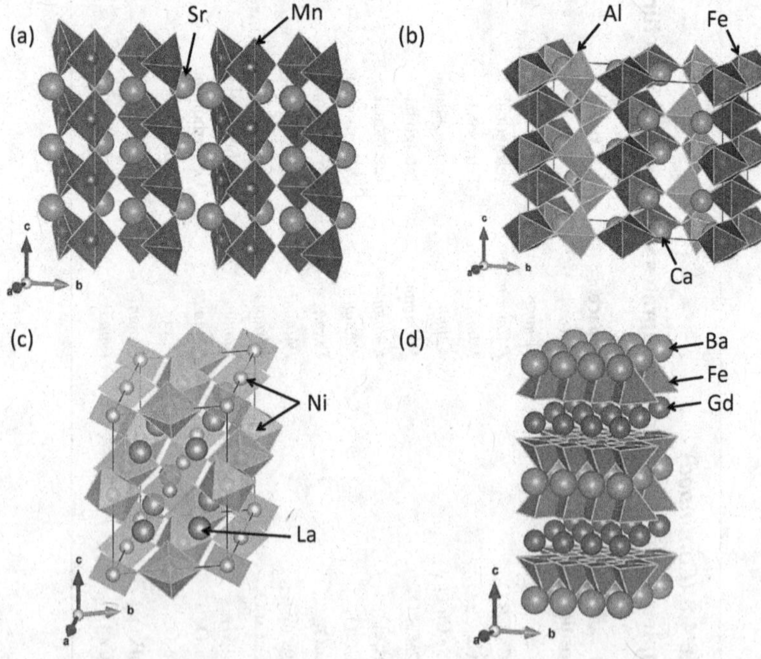

FIGURE 4.13 A list of anion-deficient perovskite structures.

B-site, which adopts an ordered structure of tetragonal pyramids, which is formed by the removal of apical oxygen ions from J-T active ions (e.g. $A_2Mn_2O_5$ (A = Ca, Sr)) (Abakumov et al. 2008). Figure 4.13b shows a brownmillerite structure, which is the structure of a naturally existing mineral brownmillerite with the formula Ca_2FeAlO_5. The perovskites with brownmillerite structures are composed of alternating BO_6 octahedra and BO_4 tetrahedra layers. BO_4 tetrahedra are formed due to the missing oxygen atom in the equatorial position (Colville and Geller 1971). The brownmillerite-structured perovskites are the most common anion-deficient perovskite structures, which can be viewed as an ideal perovskite with oxygen vacancies ordered along the [101] direction in alternate layers. Thus, the unit cell structure is enlarged along *a*- and *c*-axes as compared to the ideal perovskite. In the structure of brownmillerite perovskites, the apex-linked chains of tetrahedra are not regular; therefore, with respect to the orientation of chains, they adopt different symmetries with the same crystal structure, as listed in Table 4.9.

Figure 4.13c shows the structure of $La_2Ni_2O_5$, in which the B-site Ni^{2+} adopts mixed coordination. In the structure, half of Ni^{2+} ions exhibit an octahedral coordination, and the other half exhibit a square-planar coordination. The anion vacancies are often coupled with A-site ordering, if the A-site is occupied with ions in different oxidation states and ionic sizes. One such example is $GdBaFe_2O_5$, in which Ba and Gd atom layers are separated by ordered BO_5 square pyramids (Figure 4.13d). In $GdBaFe_2O_5$, the oxygen vacancies lie within the Gd layers; as a result, the unit cell doubles along the *c*-axis, lowering the symmetry to tetragonal.

4.4 CATION-DEFICIENT PEROVSKITES

The structure of perovskite-type materials allows a significant amount of A- and B-site cation vacancies; however, A-site deficient structures are very common. If the A-site is empty, the materials adopt structures similar to that of WO_3 or ReO_3. The structure of ReO_3 is shown in Figure 4.14. ReO_3 has a structure similar to a perovskite structure, in which the apex sharing octahedra form the backbone of the structure. However, cation displacement or octahedral tilting can encounter in these systems, and the crystal structure is sensitive to temperature. For instance, WO_3 exhibits monoclinic (~15 K) → triclinic (298 K) → monoclinic (573 K) → orthorhombic (823 K) → tetragonal (1123 K) structures with the increase in temperature (Vogt et al. 1999). Distortion and tilting of the WO_6 octahedra occur with the phase transitions, and the respective tilt systems are $(a^-b^-c^-)$ in triclinic, $(a^-b^+c^-)$ in monoclinic, $(a^0a^0c^-)$ in tetragonal, and $(a^0b^+c^-)$ in orthorhombic as compared to ideal cubic WO_3 structure with $(a^0a^0c^0)$ tilt, similar to ReO_3.

Perovskite structures with partially occupied A-site exhibit different crystal structures. If the empty A-site of WO_3 is occupied, a cation-deficient perovskite with the formula A_xWO_3 is formed. Li, Na, Ca, La, Ce, Tb, Gd, Nd, etc. are potential A-site occupants in WO_3 structures, and the possible A-site occupants are highlighted in the periodic table in Figure 4.15. Such structures are commonly called as tungsten bronzes. Based on the percentage of A-site present in the structure, A_xWO_3 adopts different structures. In general, with very low A-content, the structure is orthorhombic, but the pure WO_3 is monoclinic. As the percentage of A-site ions increases, the

TABLE 4.9
Anion-Deficient Brownmillerite-Structured Perovskites

Formula	Space Group	Structure	Lattice Parameter (Å) *a*	*b*	*c*	Reference
$Ba_2In_2O_5$	*Ima*2	Orthorhombic	16.719	6.083	5.956	Ito et al. (2017)
$Ba_2In_{1.9}Ti_{0.1}O_{5.05}$	*Icmm*	Orthorhombic	6.053	16.829	5.962	Ito et al. (2017)
Ba_2GaInO_5	*Icmm*	Orthorhombic	6.086	16.790	5.970	Tilley (2016)
$Ca_2Al_2O_5$	*I*2*mb*	Orthorhombic	5.228	14.469	5.400	Kahlenberg et al. (2000)
Ca_2AlFeO_5	*Pcmn*	Orthorhombic	5.580	14.500	5.340	Li et al. (2019b)
Ca_2AlMnO_5	*Ibm*2	Orthorhombic	5.463	14.9530	5.231	Motohashi et al. (2013)
$Ca_2Co_2O_5$	*Pcmb*	Orthorhombic	5.304	14.752	10.988	Zhang et al. (2014)
Ca_2CoFeO_5	*Pbcm*	Orthorhombic	5.367	11.107	14.779	Grosvenor and Greedan (2009)
$Ca_2Fe_2O_5$	*Pmna*	Orthorhombic	5.425	14.784	5.593	Ismail et al. (2016)
Ca_2FeMnO_5	*Pmna*	Orthorhombic	5.330	15.386	5.479	Nakahara et al. (1997)
Ca_2MnAlO_5	*I*2*mb*	Orthorhombic	5.230	14.953	5.463	Palmer et al. (2006)
$CaSrGaMnO_{5.035}$	*I*2*mb*	Orthorhombic	5.310	15.7705	5.480	Tilley (2016)
$Ca_4Al_2Fe_2O_{10}$	*Pcmn*	Orthorhombic	5.580	14.500	5.340	de la Torre et al. (2001)
$La_{0.6}Sr_{0.4}MnO_{2.5}$	*Pnma*	Orthorhombic	5.420	16.622	5.517	Parsons et al. (2009)
$Sr_2Al_{1.07}Mn_{0.93}O_5$	*Imma*	Orthorhombic	5.436	15.623	5.607	Hadermann et al. (2007)
Sr_2GaMnO_5	*Icmm*	Orthorhombic	5.489	16.226	5.354	Pomjakushin et al. (2004)
$SrCoO_{2.5}$	*Ima*2	Orthorhombic	15.778	5.574	5.470	Muñoz et al. (2008)
Sr_2GaMnO_5	*Ima*2	Orthorhombic	16.124	5.563	5.403	Pomjakushin et al. (2002)

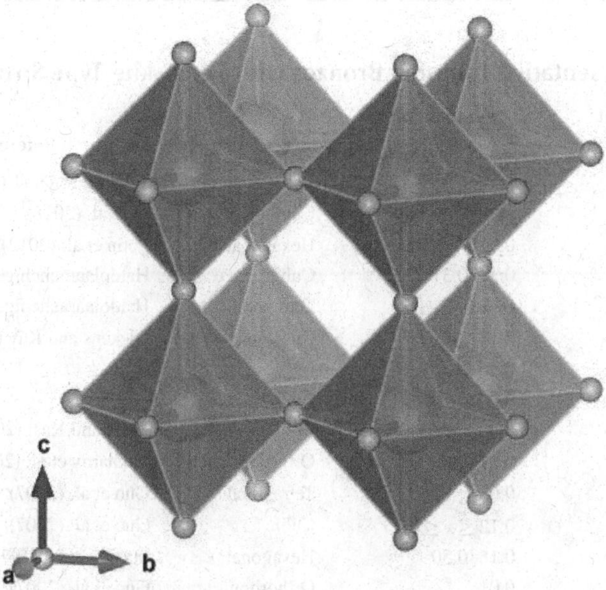

FIGURE 4.14 A unit cell of ReO_3.

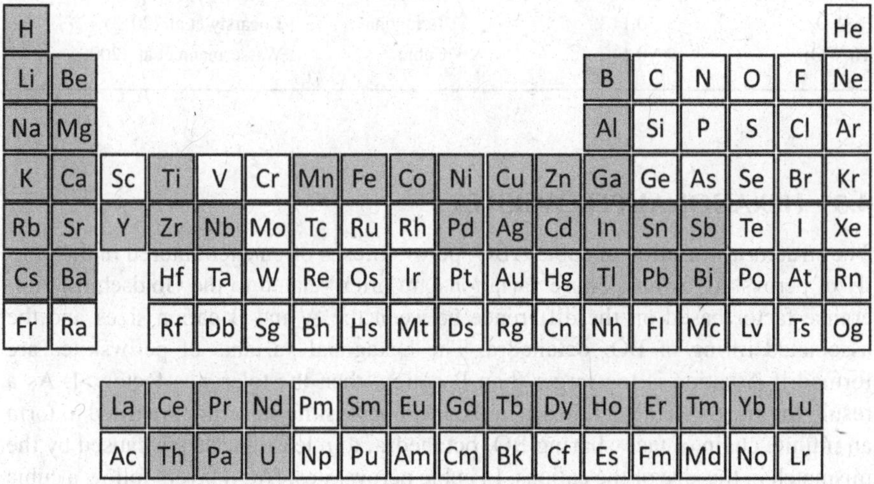

FIGURE 4.15 Possible elements in the periodic table to form perovskite-structured cation-deficient tungsten bronzes.

structure transforms to tetragonal and finally to a cubic structure. The partial occupancy of A-site ions is possible in the case of perovskites with Ti, Nb, and Ta at the B-site forming the respective bronzes. A few A-site cation-deficient perovskites type tungsten bronzes are listed in Table 4.10.

TABLE 4.10
List of Representative Tungsten Bronzes with Perovskite-Type Structures

Anion-Deficient Perovskite Structure	Mol. Fraction of A-Site	Structure	Reference
Na_xWO_3	($x = 0.25$–0.35)	Tetragonal	Takusagawa and Jacobson (1976)
	($x = 0.35$–1)	Cubic	Li et al. (2018)
K_xWO_3	0.2167–0.26	Hexagonal	Potin et al. (2012)
	0.27–0.31	Cubic	Haldolaarachchige et al. (2014)
	($x = 0.37$–1)	Tetragonal	Haldolaarachchige et al. (2014)
Li_xWO_3	0.14	Cubic	Dickens and Kay (1983)
	0.36	Cubic	
	0.93	Cubic	
La_xWO_3	$0.086 \leq x \leq 0.150$	Cubic	Hoch and Kasl (2005)
Ca_xWO_3	$0.01 \leq x \leq 0.02$	Orthorhombic	Zakharov et al. (2000)
	$0.03 \leq x \leq 0.11$	Tetragonal	Guo et al. (2007)
	$0.12 \leq x \leq 0.15$	Cubic	Guo et al. (2007)
Rb_xWO_3	0.16–0.50	Hexagonal	Stanley et al. (1978)
B_xWO_3	0.01	Orthorhombic	Guo et al. (2007)
	$x = 0.48$–0.08	Tetragonal	Guo et al. (2007)
Y_xWO_3	$0.08 < x < 0.13$	Cubic	Hoch and Kasl (2005)
Cs_xWO_3	0.32	Hexagonal	Liu et al. (2018a)
Ba_xWO_3	0.14	Tetragonal	Juan et al. (2010)
In_xWO_3	0.11	Tetragonal	Bocarsly et al. (2013)
Th_xWO_3	0.1	Cubic	Wassermann et al. (2000)

4.5 HEXAGONAL PEROVSKITES

The structural variation of cubic ABO_3 perovskites is often encountered in the family of perovskite oxides. These variations are often related to the Goldschmidt tolerance factor based on the difference between the A and B cation sizes and the associated tilting of BO_6 octahedra. The hexagonal variants of perovskites are formed if A cation is too larger than B cation, then the tolerance factor >1. As a result, the apex-sharing octahedra in the ideal perovskite structure are tilted to form an infinite chain of face-sharing BO_6 octahedra, to release the strain caused by the mismatch in the size of the cations. In cubic perovskites, [AO_3] layers follow a cubic close packing with an *ABC* stacking sequence, whereas *AB* stacking sequence in hexagonal perovskites.

An example of ABO_3 perovskite that adopts hexagonal structure is $BaNiO_3$. The Ni^{4+} ions in the structure are coordinated to oxygen atom octahedrally (Figure 4.16a), and these octahedra are occupied at the corners of the unit cell. The NiO_6 octahedra share three oxygen atoms, thus sharing a face of the octahedra to the nearest octahedra; as a result, the structure can be viewed as a column of face-sharing

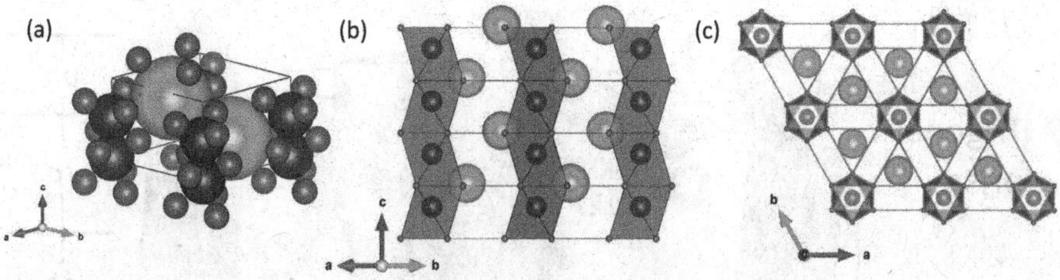

FIGURE 4.16 Hexagonal-structured ABO_3 perovskite: (a) unit cell, (b) unit cell projected on (110) plane, and (c) perpendicular to *c*-axis.

NiO_6 octahedra separated by Ba cations, as in Figure 4.16b. Figure 4.16c reveals the hexagonal stalking in the structure.

Structures that slightly deviate from the hexagonal structures are also identified as hexagonal structures. Such structures are also composed of BO_6 octahedral chains separated by A-site cations layers. Considering the stacking in ideal hexagonal perovskites along an axis perpendicular to *c*-axis, one can observe layers with $\cdots AB \cdots$ stacking of A_3O_9 ($A = Ca^{2+}$, Sr^{2+}, or Ba^{2+}) polyhedral, with octahedral interstices occupied by B-site ions forming a chain of octahedra (*–oh–oh–oh–oh–*) (oh represents octahedron), as in Figure 4.16a and c. The structures that deviate from the hexagonal structures often occupy a second A-site anion and oxygen vacancy in their lattice. The latter structure is formed by the stacking of $A_3A'O_6$ layers, which results in a trigonal prism (*tp*) of A′ ions, and octahedral interstices occupied by B cations. Therefore, the columns or the chains are composed of alternating face-sharing BO_6 octahedra and $A'O_6$ trigonal prisms (*–oh–tp–oh–tp–oh–tp–*), as shown in Figure 4.17. Such structures adopt a generalized formula $A_{3n+3m}A'_nB_{3m+n}O_{9m+6n}$ which is composed of chains of trigonal prisms and face-shared octahedra, where *m* is the (A_3O_9) sheets and *n* is ($A_3A'O_6$) sheets.

The values of *m* and *n* in the structure $A_{3n+3m}A'_nB_{3m+n}O_{9m+6n}$ can result in endless composition and structures. Depending on the values of *m* and *n*, the structure adopts a difference in the stalking of A_3O_9 and $A_3A'O_6$ layers. In the ideal hexagonal perovskite, $m = 1$ and $n = 0$. The majority of these perovskite-structured compounds are observed when $m = 0$ and $n = 1$, in which chains of alternating trigonal prisms and octahedra compose the structure; therefore, the trigonal prisms and octahedra are in equal number. Let $m = 0$ and $n = 1$, and then the general formula $A_{3n+3m}A'_nB_{3m+n}O_{9m+6n}$ reduces to $A_3A'BO_6$, the A′ cation occupies the trigonal prismatic site, and the B cation occupies the octahedral site. The A-site is occupied by alkaline earth metals such as Ca, Sr, and Ba, and A′- and B-sites are occupied with the following combination of oxidation states: $1^+/5^+$, $2^+/4^+$, $3^+/3^+$, and $4^+/2^+$.

The $A_3A'BO_6$ structure with alternating units of octahedra and trigonal prisms in the columns is shown in Figure 4.17. Based on the values of *m* and *n*, one can calculate the ratio of the alternating trigonal prism and octahedra. For instance, ($m = 1$, $n = 0$) has 2*H*—ABO_3-type structure with $tp/oh = 0$, and for ($m = 0$, $n = 1$) $A_3A'BO_6$

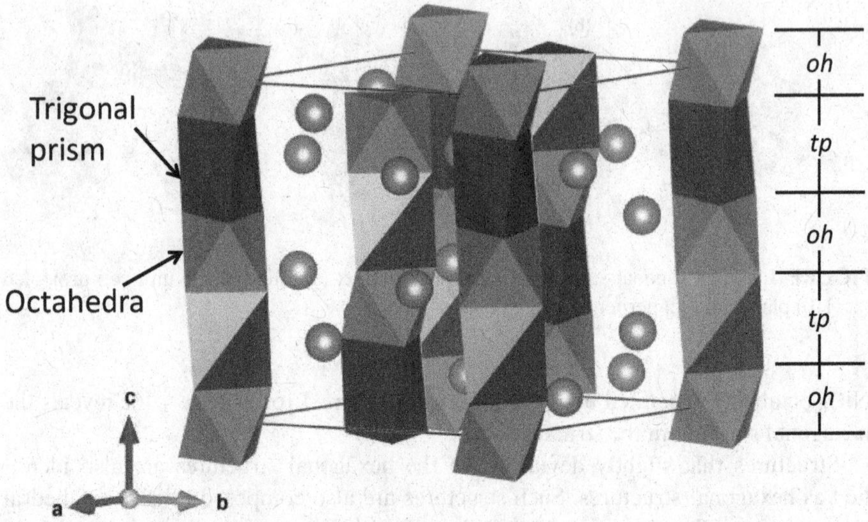

FIGURE 4.17 The hexagonal perovskite structure with alternating octahedral and tetrahedral units, when $m = 0$ and $n = 1$.

($tp/oh = 1$). Therefore, the generalized formula for the ratio of trigonal prism to octahedra, *tp/oh*, can be written as follows:

$$\frac{tp}{oh} = \frac{n}{(3m+n)} \tag{4.2}$$

A systematic approach to represent the $A_{3n+3m}A'_nB_{3m+n}O_{9m+6n}$ structures is $A_{1+x}\left(A'_xB_{1-x}\right)O_3$ comparable to an ideal perovskite structure.

Where

$$x = \frac{n}{(3m+n)} \tag{4.3}$$

and therefore

$$\frac{tp}{oh} = \frac{x}{(1-x)} \tag{4.4}$$

Based on the values of *m* and *n*, the number and order of tetragonal prism to ocahedra vary. The representative images of the pillers of hexagonal perovskites with different values of *m* and *n* are shown in Figure 4.18. Table 4.11 lists the possible combinations of *m* and *n* in $A_{3n+3m}A'_nB_{3m+n}O_{9m+6n}$ structure and the *–oh–tp–* stalking sequence.

The perovskite structures so far discussed belong to *2H* structures with *ABAB* close packing. Many complex hexagonal perovskites with the close packing mixed with that of cubic *ABCABC* close packing. Therefore, it is more convenient to

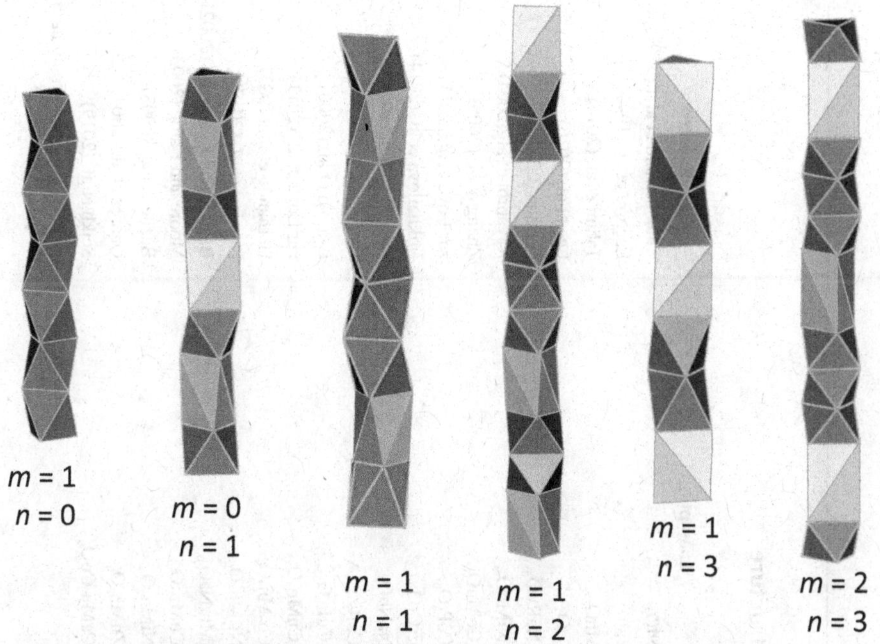

FIGURE 4.18 The hexagonal perovskite structure with alternating octahedral and tetrahedral units.

represent them as layers than the stalking sequence followed by the structure of the primitive unit cell. Such a notation is called the Ramsdell notation, in which *H*, *R*, and *C* indicate the hexagonal, rhombohedral, and cubic lattice types, respectively, and the number in the front, e.g. *nH* or *mR*, represents the number of layers with hexagonal structure. The representative hexagonal perovskite structures with mixed cubic/hexagonal packing are shown in Figure 4.19.

The above complex hexagonal structures with mixed close packing exists in polytypes. The term "polytype" is used to represent the crystal structures with the same composition but with different stalking sequences of cubic-packed layer (denoted as *c*) and hexagonal-packed layer (denoted as *h*). For instance, 4*H* structure has a common stalking sequence of (*chch*), which can also be arranged as (*cchh*), (*chhc*), (*hhcc*), (*hchc*), or (*hcch*) sequences. Therefore, 4*H* hexagonal structure can have six polytypes. In more complex hexagonal structures with several mixed close packing, it is impractical to represent them as stacking sequence in terms of *c* and *h*; hence, the sequence is often represented as $c_p h_q$, where *p* and *q* indicate the number of cubic-packed layers and the number of hexagonal-packed layers. Based on the values of *p* and *q*, the number of polytypes can be easily calculated using the combination formula:

$$^{p+q}C_{p \text{ or } q} = (p+q)!/(p! \times q!) \tag{4.5}$$

TABLE 4.11
The Representative Combinations of *m* and *n* in $A_{3n+3m}A'_nB_{3m+n}O_{9m+6n}$ Structure

Formula	Value of *m* and *n*	Stalking Sequence	Example	Reference
ABO_3	$m = 1; n = 0$	*–oh–oh–oh–oh–*	$BaNiO_3$	Chamberland et al. (1970)
			$PrNiO_3$	Lacorre et al. (1991)
			$BaMnO_3$	Tahini et al. (2016)
			$SrCoO_3$	Lee et al. (2016)
$A_3A'BO_6$	$m = 0; n = 1$	*–oh–tp–oh–tp–*	Ba_3NaRuO_6,	Persson (2014d)
			Ca_3NiMnO_6,	Rayaprol et al. (2004)
			Ca_3CoMnO_6,	Zhang et al. (2009)
			Sr_3CoPtO_6,	zur Loye et al. (2011)
			Sr_3ErCrO_6,	Mikhailova et al. (2014)
			Sr_3NaNbO_6	Bharathy et al. (2008b)
$A_6A'_1B_4O_{15}$	$m = 1; n = 1$	*–4oh–tp–4oh–tp–*	$Sr_6Co_5O_{15}$ (A′ = B = Co)	Battle et al. (1997)
			$Ba_6CuIr_4O_{15}$,	Iwasaki et al. (2006)
			$Ba_6CuMn_4O_{15}$	zur Loye et al. (2011)
$A_9A'_2B_5O_{21}$	$m = 1; n = 2$	*–3oh–tp–2oh–tp–*	$Sr_9Co_2Mn_5O_{21}$,	Hernando et al. (2003)
			$Sr_9Ni_2Mn_5O_{21}$	Demirel et al. (2017)
$A_{12}A'_3B_6O_{27}$ ($A_4A'B_2O_9$)	$m = 1; n = 3$	*–2oh–tp–2oh–tp–*	$Ba_4Mn_2NaO_9$,	Bazuev and Kellerman (2002)
			$Sr_4CuMn_2O_9$,	Moore and Battle (2003)
			$Sr_4NiMn_2O_9$,	Bazuev et al. (2003)
			$Sr_4ZnMn_2O_9$	Quarez et al. (2004)
			$Sr_3CaMn_2CoO_9$	Seikh et al. (2019)

(Continued)

TABLE 4.11 (*Continued*)
The Representative Combinations of *m* and *n* in $A_{3n+3m}A'_nB_{3m+n}O_{9m+6n}$ Structure

Formula	Value of *m* and *n*	Stalking Sequence	Example	Reference
$A_{15}A'_4B_7O_{33}$	$m = 1; n = 4$	–2*oh*–*tp*–2*oh*–*tp*–	$(Sr_{0.5}Ca_{0.5})_{15}Mn_7Co_4O_{33}$	Boulahya et al. (2007)
$A_9A'B_7O_{24}$	$m = 2, n = 1$	–7*oh*–*tp*–7*oh*–*tp*–	$Ba_9Rh_8O_{24}$ (A′ = B = Rh)	Stitzer et al. (2001)
$A_{15}A'_3B_9O_{36}$	$m = 2; n = 3$	–3*oh*–*tp*–3*oh*–*tp*–	$Ba_5CuIr_3O_{12}$	Blake et al. (1999)
$(A_5A'B_3O_{12})$			$(Sr_{0.75}Ba_{0.25})_5NiMn_3O_{12}$	Hernando et al. (2003)
$A_{12}A'B_{10}O_{33}$	$m = 3, n = 1$	–10*oh*–*tp*–10*oh*–*tp*–	$Ba_{12}Rh_{9.25}Ir1_{.75}O_{33}$ (A′ = B = Rh)	Stitzer et al. (2004)
$A_{42}A'_9B_{24}O_{99}$	$m = 5; n = 9$	–3*oh*–*tp*–2*oh*–*tp*–	$Sr_{14}Cu_3Mn_8O_{33}$	Abed et al. (2003)
$(A_{14}A'_3B_8O_{33})$				
$A_{48}A'_9B_{30}O_{117}$	$m = 7, n = 9$	–4*oh*–*tp*–4*oh*–*tp*–	$Ba_{16}Cu_3Ir_{10}O_{39}$	Blake et al. (1999)
$(A_{16}A'_3B_{10}O_{39})$				
$A_{33}A'_3B_{27}O_{90}$	$m = 8, n = 3$	–9*oh*–*tp*–9*oh*–*tp*–	$Ba_{11}Rh_{10}O_{30}$	Stitzer et al. (2003a)
$(A_{11}A'B_9O_{30})$			(A′ = B = Rh)	

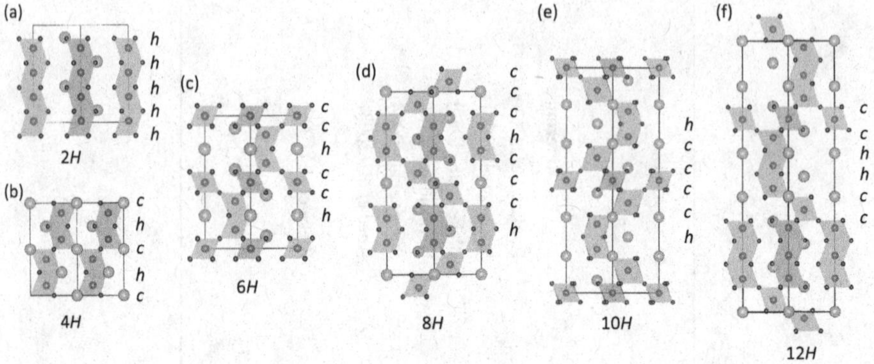

FIGURE 4.19 The hexagonal perovskite structure with cubic (*c*) and hexagonal (*h*) packing.

In reality, the number of polytypes formed is far less than the theoretically predicted numbers, as the crystal structure adopts the most favorable stalking sequences with a low energy state. In many instances of very complex hexagonal perovskite structures, the polytypes exist in different symmetries. For example, 12*L* hexagonal structure, representing 12 close-packed layers in perovskite structure, can exist in both 12*H* and 12*R* symmetries, as in Figure 4.20. Representative perovskites with mixed cubic and hexagonal stalking are presented in Table 4.12.

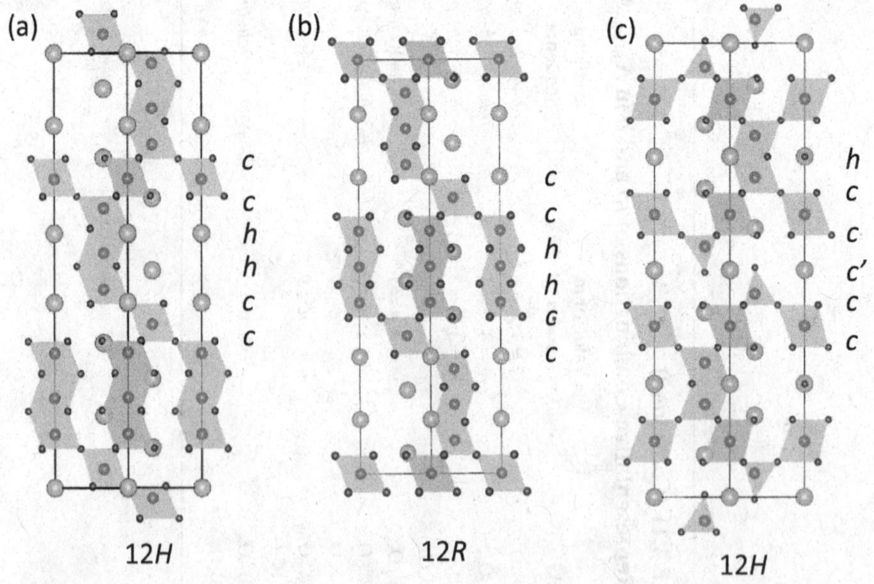

FIGURE 4.20 Polytypes of 12*H* hexagonal perovskite structure with different symmetries.

TABLE 4.12
The Examples of Hexagonal Perovskites with Different Stalking Sequences

Stalking	Sequence	Space Group	Example	Reference
4*H*	$(ch)_2$	$P6_3/mmc$	$SrMnO_3$,	Hong and Sleight (1997)
			$BaRuO_3$,	Adkin and Hayward (2007)
			$BaMnO_{2.75}$	Aich et al. (2019)
5*H*	(*hhccc*)	*P*3*m*1	$Ba_5Nb_4O_{15}$,	Abakumov et al. (1998)
			$Ba_5Ta_4O_{15}$	Kim et al. (2006)
6*H*	$(cch)_2$	$P6_3/mmc$	$Ba_4Ru_3LiO_{12}$	Battle et al. (1992)
	$(h)_6$	$P6_3/mmc$	$Ba_3Te_2O_9$,	Jacobson et al. (1981)
			$Ba_3W_2O_9$	Urushihara et al. (2017)
	$(hcc)_2$	$P6_3/mmc$	$Ba_3NaIr_2O_9$,	zur Loye et al. (2009)
			$Ba_3LnIr_2O_9$ (Y, Ce, Pr, and Tb),	Doi and Hinatsu (2004)
			$Ba_3NiRu_2O_9$,	Lightfoot and Battle (1990)
	(*hchhhc*)	$P\bar{6}m2$	$BaMn_{0.6}Fe_{0.4}O_{2.72}$	Miranda et al. (2009)
7*H*	*cchcch*	$R\bar{3}m$	$Ba_7Li_3Ru_4O_{20}$	Stitzer et al. (2003b)
8*H*	$(chhh)_2$	$P6_3/mmc$	$BaMnO_{2.875}$	Miranda et al. (2009)
	$(ccch)_2$	$P6_3mc$	$Ba_4Ru_3NaO_{12}$	Battle et al. (1992)
	$(cchc)_2$	*P*63*cm*	$Ba_8Ta_6NiO_{24}$	Abakumov et al. (1996)
9*R*	$(chh)_3$	$R\bar{3}m$	$BaRuO_3$, $BaMnO_3$	Kozlenko et al. (2017)
10*H*	$(chchh)_2$	$P6_3/mmc$	$BaMnO_{2.80}$	Miranda et al. (2009)
	$(cchhh)_2$	$P6_3/mmc$	$Ba_5Ln_{1-x}Mn_{4+y}O_{15-\delta}$	Yin et al. (2017)

(*Continued*)

TABLE 4.12 (*Continued*)
The Examples of Hexagonal Perovskites with Different Stalking Sequences

Stalking	Sequence	Space Group	Example	Reference
12H	$(hhcc)_3$	$P6_3/mmc$	$Ba_3LaNb_3O_{12}$,	Fang et al. (2008)
			$Ba_3LaTa_3O_{12}$,	Bezjak et al. (2009)
			`$Ba_4Nb_2WO_{12}$,	Hu et al. (2009);
			$Ba_4Ta_2WO_{12}$	Kim et al. (2015)
	$(c'cchcc)_2$	$P6_3/mmc$	$Ba_6Ru_2Na_2V_2O_{17}$	Quarez et al. (2003)
			$Ba_6Ru_2Na_2Mn_2O_{17}$	Yang et al. (2007)
12*R*	$(cchh)_3$	$R\bar{3}m$	$Ba_4Ti_2Mn_2O_{12}$	Kim et al. (2015)
			$La_4Ti_3O_{12}$	
15*R*	$(chhhh)_3$	$R\bar{3}m$	$BaMnO_{2.90}$	Miranda et al. (2009)
	$(ccc'cc)_3$	$R3m$	$Ba_5MnNa_2V_2O_{13}$	Bendraoua et al. (2004)
18*H*	$(hhcccc)_3$	–	$Ba_6GdW_3O_{18}$, $Ba_6YW_3O_{18}$	Tilley (2016)
18*R*	$(ccchh)_3$	$R\bar{3}m$	$Ba_5KNb_5O_{18}$	Polubinskii et al. (2014)
21*R*	$(chhhhhh)_3$	$R\bar{3}m$	$BaMnO_{2.92}$	Miranda et al. (2009)
24*H*	$(hhhhchhc)_3$	–	$Ba_8ReW_3O_{24}$	Tilley (2016)
27*H*	$(hhccchhcc)_3$	$R\bar{3}m$	$Ba_9Nb_6WO_{27}$	Bezjak et al. (2009)

Note: c' represents the anion-deficient layer of cubic close packing.

REFERENCES

Abakumov, A. M., G. V. Tendeloo, A. A. Scheglov, et al. 1996. The Crystal Structure of $Ba_8Ta_6NiO_{24}$: Cation Ordering in Hexagonal Perovskites. *J. Solid State Chem.* 125:102–107.

Abakumov, A. M., R. V. Shpanchenko, E. V. Antipov, et al. 1998. Synthesis and Structural Study of Hexagonal Perovskites in the $Ba_5Ta_4O_{15}$–$MZrO_3$(M = Ba, Sr) System. *J. Solid State Chem.* 141:492–499.

Abakumov, A. M., J. Hadermann, G. V. Tendeloo, et al. 2008. Chemistry and Structure of Anion-Deficient Perovskites with Translational Interfaces. *J. Am. Ceram. Soc.* 91:1807–1813.

Abed, A., E. Gaudin, H.-C. Loye, et al. 2003. (3+1)D Superspace Structural Determination of Two New Modulated Composite Phases: $Sr_{1+x}(CuxMn_{1-x})O_3$; $x = 3/11$ and $x = 0.3244$. *Solid State Sci.* 5:59–71.

Adkin, J. J., and M. A. Hayward. 2007. $BaMnO_{3-x}$ Revisited: A Structural and Magnetic Study. *Chem. Mater.* 19:755–762.

Aich, P., C. Meneghini, L. Tortora, et al. 2019. Fluorinated Hexagonal 4*H* $SrMnO_3$: A Locally Disordered Manganite. *J. Mater. Chem. C* 7:3560–3568.

Alfaro, S. O., and A. Martínez-de la Cruz. 2010. Synthesis, Characterization and Visible-Light Photocatalytic Properties of Bi_2WO_6 and $Bi_2W_2O_9$ Obtained by Co-Precipitation Method. *Appl. Catal. Gen.* 383:128–133.

Amow, G., and J. E. Greedan. 1998. The Layered Perovskite $K_2Nd_2Ti_3O_{10}$. *Acta Crystallogr. Sect. C* 54:1053–1055.

Autieri, C., M. Cuoco, and C. Noce. 2014. Structural and Electronic Properties of Sr_2RuO_4-$Sr_3Ru_2O_7$ Heterostructure. *Phys. Rev. B* 89:075102.

Autieri, C., P. Barone, J. Slawinska, et al. 2019. Persistent Spin Helix in Rashba-Dresselhaus Ferroelectric $CsBiNb_2O_7$. *Phys. Rev. Mater.* 3:084416.

Azcondo, M. T., J. R. de Paz, K. Boulahya, et al. 2015. Complex Magnetic Behaviour of $Sr_2CoNb_{1-x}Ti_xO_6$ ($0 \leq x \leq 0.5$) as a Result of a Flexible Microstructure. *Dalton Trans.* 44:3801–3810.

Azcondo, M. T., M. Orfila, J. Marugán, et al. 2019. Novel Perovskite Materials for Thermal Water Splitting at Moderate Temperature. *ChemSusChem* 12:4029–4037.

Battle, P. D., S. H. Kim, and A. V. Powell. 1992. The Crystal Structure and Electronic Properties of $Ba_4Ru_3MO_{12}$(M = Li, Na, Mg, Zn). *J. Solid State Chem.* 101:161–172.

Battle, P. D., G. R. Blake, J. Darriet, et al. 1997. Modulated Structure Of$Ba_6ZnIr_4O_{15}$; a Comparison With $Ba_6CuIr_4O_{15}$ AndSrMn_{1-x}CoxO_3-$_y$. *J. Mater. Chem.* 7:1559–1564.

Battle, P. D., M. A. Green, J. Lago, et al. 1998. Crystal and Magnetic Structures of $Ca_4Mn_3O_{10}$, an $n = 3$ Ruddlesden–Popper Compound. *Chem. Mater.* 10:658–664.

Bazuev, G., and D. G. Kellerman. 2002. Incommensurate Complex Oxides $Sr_4NiMn_2O_9$ and $Sr_3NiMnO_{6.36}$. *Russ.J. Inorg. Chem.* 47:1772–1782.

Bazuev, G., V. Krasil'nikov, and D. G. Kellerman. 2003. Synthesis and Magnetic Properties of Incommensurate Phases $A_4CuMn_2O_9$ (A = Ca, Sr). *J. Alloys Compd.* 352:190–196.

Bendraoua, A., E. Quarez, F. Abraham, et al. 2004. Electrosynthesis and Crystal Structure of the New 15*R* Hexagonal Perovskite $Ba_5MnNa_2V_2O_{13}$. *J. Solid State Chem.* 177:1416–1424.

Benedek, N. A. 2014. Origin of Ferroelectricity in a Family of Polar Oxides: The Dion–Jacobson Phases. *Inorg. Chem.* 53:3769–3777.

Berdonosov, P., D. Charkin, K. Knight, et al. 2006. Phase Relations and Crystal Structures in the Systems (Bi, Ln)$_2WO_6$ and (Bi, Ln)$_2MoO_6$ (Ln = lanthanide). *J. Solid State Chem.* 179:3437–3444.

Bernhoff, H., M. Qvarford, S. Söderholm, et al. 1993. Photoemission Study of the $Bi_2CaSr_2Cu_2O_8$ Superconductor with Cu, Ag and Au Overlayers. *Phys. C Supercond.* 218:103–108.

Bezjak, J., B. Jančar, P. Boullay, et al. 2009. Hexagonal Perovskite-Type Phases in the BaO-Rich Part of the BaO–WO_3–Nb_2O_5 System. *J. Am. Ceram. Soc.* 92:3022–3032.

Bhandari, C., Z. S. Popović, and S. Satpathy. 2019. Electronic Structure and Optical Properties of Sr_2 IrO_4 under Epitaxial Strain. *New J. Phys.* 21:013036.

Bharathy, M., V. A. Rassolov, S. Park, et al. 2008a. Crystal Growth of Two New Photoluminescent Oxides: $Sr_3Li_6Nb_2O_{11}$ and $Sr_3Li_6Ta_2O_{11}$. *Ionrg. Chem.* 21:9941–9945.

Bharathy, M., V. A. Rassolov, and H. Loye. 2008b. Crystal Growth of Sr_3NaNbO_6 and Sr_3NaTaO_6: New Photoluminescent Oxides. *Chem. Mater.* 20:2268–2273.

Bhuvanesh, N. S. P., M. P. Crosnier-Lopez, O. Bohnke, et al. 1999. Synthesis, Crystal Structure, and Ionic Conductivity of Novel Ruddlesden–Popper Related Phases, $Li_4Sr_3Nb_{5.77}Fe_{0.23}O_{19.77}$ and $Li_4Sr_3Nb_6O_{20}$. *Chem. Mater.* 11:634–641.

Blake, G. R., P. D. Battle, J. Sloan, et al. 1999. Neutron Diffraction Study of the Structures of $Ba_5CuIr_3O_{12}$ and $Ba_{16}Cu_3Ir_{10}O_{39}$. *Chem. Mater.* 11:1551–1558.

Blasco, J., J. A. Rodríguez-Velamazán, C. Ritter, et al. 2009. Electron Doping Effects on Sr_2FeReO_6. *Solid State Sci.* 11:1535–1541.

Blasco, J., J. L. García-Muñoz, J. García, et al. 2015. Evidence of Large Magneto-Dielectric Effect Coupled to a Metamagnetic Transition in Yb_2CoMnO_6. *Appl. Phys. Lett.* 107:012902.

Blasco, J., J. García, G. Subías, et al. 2016. Magnetoelectric and Structural Properties of Y_2CoMnO_6: The Role of Antisite Defects. *Phys. Rev. B* 93:214401.

Bocarsly, J. D., D. Hirai, M. N. Ali, et al. 2013. Superconducting Phase Diagram of In_xWO_3 Synthesized by Indium Deintercalation. *EPL Europhys. Lett.* 103:17001.

Bordet, P, F. Duc, S. LeFloch, J. J. Capponi, et al. 1996. Single Crystal X-Ray Diffraction Study of the $HgBa_2CuO_{4+\delta}$ Superconducting Compound. *Physica C* 271:189–196.

Borg, S., G. Svensson, and J.-O. Bovin. 2002. Structure Study of $Bi_{2.5}Na_{0.5}Ta_2O_9$ and $Bi_{2.5}Na_{m-1.5}Nb_mO_{3m+3\,(M}$ = 2–4) by Neutron Powder Diffraction and Electron Microscopy. *J. Solid State Chem.* 167:86–96.

Boulahya, K., M. Hernando, M. Parras, et al. 2007. New Stabilized Phases in the Sr/Ça–Mn–Co–O System: Structural–Magnetic Properties Relationship. *J Mater Chem.* 17:1620–1626.

Braden, M., G. André, S. Nakatsuji, et al. 1998. Crystal and Magnetic Structure of Ca_2RuO_4 Magnetoelastic Coupling and the Metal-Insulator Transition. *Phys. Rev. B* 58:847–861.

Broux, T., M. Bahout, O. Hernandez, et al. 2013. Reduction of Sr_2MnO_4 Investigated by High Temperature in Situ Neutron Powder Diffraction under Hydrogen Flow. *Inorg. Chem.* 52:1009–1017.

Brunelli, M., and M. Ceretti. 2015. Exploring Local Disorder in Fast Oxygen Ion Conductors by Atomic Pair Distribution Function Analysis. *EPJ Web Conf.* 104:01009. EDP Sciences.

Cernea, M., F. Vasiliu, C. Bartha, et al. 2014. Characterization of Ferromagnetic Double Perovskite Sr_2FeMoO_6 Prepared by Various Methods. *Ceram. Int.* 40:11601–11609.

Chamberland, B. L., A. W. Sleight, and J. F. Weiher. 1970. Preparation and Characterization of $BaMnO_3$ and $SrMnO_3$ Polytypes. *J. Solid State Chem.* 1:506–511.

Chan, J. Y., T. A. Vanderah, R. S. Roth, et al. 2000. Structure and Microwave Dielectric Properties in the $x\,Ca_2AlNbO_6{:}_{(1\text{-}x)}Ca_3Nb_2O_8$ System. https://www.nist.gov/publications/structure-and-microwave-dielectric-properties-x-$Ca_2AlNbO_{6\text{-}1\text{-}x}$-$Ca_3Nb_2O_8$-system.

Chaplot, S. L., B. A. Dasannacharya, R. Mukhopadhyay, et al. 1991. Inelastic Neutron Scattering from $Tl_2CaBa_2Cu_2O_8$. *Bull. Mater. Sci.* 14:603–605.

Chaudhry, A., A. Canning, R. Boutchko, et al. 2011. First-Principles Studies of Ce-Doped $RE_2M_2O_7$ (RE = Y, La;M = Ti, Zr, Hf): A Class of Non-Scintillators. *J. Appl. Phys.* 109:083708.

Chen, H. W., Y.-W. Chen, J.-L. Kuo, et al. 2019. Spin-Charge-Lattice Coupling in $YBaCuFeO_5$: Optical Properties and First-Principles Calculations. *Sci. Rep.* 9:1–9. Group.

Chowki, S., B. Sahu, A. K. Singh, et al. 2016. Dielectric Response of Double Layered Perovskite Sr_3MnTiO_7. *AIP Conf. Proc.* 1731:090040.

Chu, M.-W., M.-T. Caldes, L. Brohan, et al. 2003. Bulk and Surface Structures of the Aurivillius Phases: $Bi_{4-x}La_xTi_3O_{12}$ ($0 \leq x \leq 2.00$). *Chem. Mater.* 1:31–42.

Colville, A. A., and S. Geller. 1971. The Crystal Structure of Brownmillerite, Ca_2FeAlO_5. *Acta Crystallogr. B* 27:2311–2315.

Corredor, L. T., G. Aslan-Cansever, M. Sturza, et al. 2017. Iridium Double Perovskite Sr_2YIrO_6: A Combined Structural and Specific Heat Study. *Phys. Rev. B* 95:064418.

Crawford, M. K., R. L. Harlow, W. Marshall, et al. 2002. Structure and Magnetism of Single Crystal $Sr_4Ru_3O_{10}$: A Ferromagnetic Triple-Layer Ruthenate. *Phys. Rev. B* 65:214412.

Das, I., S. Chanda, S. Saha, et al. 2016. Electronic Structure and Transport Properties of Antiferromagnetic Double Perovskite Y_2AlCrO_6. *RSC Adv.* 6:80415–80423.

Das, M., P. Dutta, S. Giri, et al. 2019. Octahedral Tilting and Emergence of Ferrimagnetism in Cobalt-Ruthenium Based Double Perovskites. *J. Phys. Condens. Matter* 31:385801.

de la Torre, A. G., A. Cabeza, A. Calvente, et al. 2001. Full Phase Analysis of Portland Clinker by Penetrating Synchrotron Powder Diffraction. *Anal. Chem.* 73:151–156.

Demirel, S., E. Oz, S. Altin, et al. 2017. Structural, Magnetic, Electrical and Electrochemical Properties of $SrCoO_{2.5}$, $Sr_9Co_2Mn_5O_{21}$ and $SrMnO_3$ Compounds. *Ceram. Int.* 43:14818–14826.

Deshmukh, S., E. W. Rothe, and G. P. Reck. 1989. Spectral Analysis of Fluorescence of Excited Species from 193 Nm Photoablation of Polycarbonate and the High Tc Superconductors $Bi_2CaSr_2Cu_2O_9$ and $YBa_2Cu_3O_{7-d}$. *AIP Conf. Proc.* 191:430–432.

Dickens, P. G., and S. A. Kay. 1983. Thermochemistry of the Cubic and Hexagonal Lithium Tungsten Bronze Phases Li_xWO_3. *Solid State Ion.* 8:291–295.

Doi, Y., and Y. Hinatsu. 2004. The Structural and Magnetic Characterization of 6*H*-Perovskite-Type Oxides $Ba_3LnIr_2O_9$ (Ln = Y, Lanthanides). *J. Phys. Condens. Matter* 16:2849.

Erickson, A. S., S. Misra, G. J. Miller, et al. 2007. Ferromagnetism in the Mott Insulator Ba_2NaOsO_6. *Phys. Rev. Lett.* 99:016404.

Faik, A., M. Gateshki, J. M. Igartua, et al. 2008. Crystal Structures and Cation Ordering of Sr_2AlSbO_6 and Sr_2CoSbO_6. *J. Solid State Chem.* 181:1759–1766.

Faik, A., J. M. Igartua, M. Gateshki, et al. 2009. Crystal Structures and Phase Transitions of Sr_2CrSbO_6. *J. Solid State Chem.* 182:1717–1725.

Fang, L., Y. Wu, C. Hu, et al. 2008. $Ba_3LaTa_3O_{12}$: A New Microwave Dielectric of $A_4B_3O_{12}$-Type Cation-Deficient Perovskites. *Mater. Lett.* 62:594–596.

Finger, L. W., R. M. Hazen, R. T. Downs, et al. 1994. Crystal Chemistry of $HgBa_2CaCu_2O_{8+\delta}$ and $HgBa_2Ca_2Cu_3O_{8+\delta}$ Single-Crystal X-Ray Diffraction Results. *Physica C* 226:216–221.

Franklin, J. R. 2001. In-Situ Synthesis of Piezoelectric-Reinforced Metal Matrix Composites. https://www.semanticscholar.org/paper/In-situ-Synthesis-of-Piezoelectric-Reinforced-Metal-Franklin/8107c64a5aea3a9ac1829497de7e40856c0655e3.

Freeman, P. G., R. A. Mole, N. B. Christensen, et al. 2018. Stability of Charge-Stripe Ordered $La_{2-x}Sr_xNiO_{4+\delta}$ at One Third Doping. *Phys. B Condens. Matter* 536:720–725.

Fua, W. T., M. J. Polderman, and F. M. Mulder. 2000. Structural and Transport Properties of the $BaBi_{1-x}In_xO_3$ System. *Mater. Res. Bull.* 35:1205–1211.

Fuchs, S., T. Dey, G. Aslan-Cansever, et al. 2018. Unraveling the Nature of Magnetism of the $5d^4$ Double Perovskite Ba_2YIrO_6. *Phys. Rev. Lett.* 120:237204.

Fukuoka, H., T. Isami, and S. Yamanaka. 2000. Crystal Structure of a Layered Perovskite Niobate $KCa_2Nb_3O_{10}$. *J. Solid State Chem.* 151:40–45.

Ganguli, D. 1979. Cationic Radius Ratio and Formation of K_2NiF_4-Type Compounds. *J. Solid State Chem.* 30:353–356.

García-Guaderrama, M., L. Fuentes-Montero, A. Rodriguez, et al. 2006. Structural Characterization of $Bi_6Ti_3Fe_2O_{18}$ Obtained by Molten Salt Synthesis. *Integr. Ferroelectr.* 83:41–47.

Gateshki, M., J. M. Igartua, and A. Faik. 2007. Crystal Structure and Phase Transitions of Sr_2CdWO_6. *J. Solid State Chem.* 180:2248–2255.

Glazer, A. M. 1972. The Classification of Tilted Octahedra in Perovskites. *Acta Crystallogr. Sect. B* 28:3384–3392.

González-Jiménez, I. N., A. Torres-Pardo, S. Rano, et al. 2018. Multicationic $Sr_4Mn_3O_{10}$ Mesostructures: Molten Salt Synthesis, Analytical Electron Microscopy Study and Reactivity. *Mater. Horiz.* 5:480–485.

Gopalakrishnan, J., G. Colsmann, and B. Reuter. 1977. Studies on the $La_{2-x}Sr_xNiO_4$ $(0 \leqslant x \leqslant 1)$ System. *J. Solid State Chem.* 22:145–149.

Goutenoire, F., M. Hervieu, C. Martin, et al. 1994. Cationic Substitutions in the "2201-1201" Intergrowth $HgT_{12}Ba_4Cu_2O_{10}$. *Chem. Mater.* 6:1654–1658.

Green, M. A., and D. A. Neumann. 2000. Synthesis, Structure, and Electronic Properties of $LaCa_2Mn_2O_7$. *Chem. Mater.* 12:90–97.

Grimaud, A., C. E. Carlton, M. Risch, et al. 2013. Oxygen Evolution Activity and Stability of $Ba_6Mn_5O_{16}$, $Sr_4Mn_2CoO_9$, and $Sr_6Co_5O_{15}$: The Influence of Transition Metal Coordination. *J. Phys. Chem. C* 117:25926–25932.

Grosvenor, A. P., and J. E. Greedan. 2009. Analysis of Metal Site Preference and Electronic Structure of Brownmillerite-Phase Oxides ($A_2B'_xB_{2-x}O_5$; A = Ca, Sr; B′/B = Al, Mn, Fe, Co) by X-Ray Absorption Near-Edge Spectroscopy. *J. Phys. Chem. C* 113:11366–11372.

Gu, B., Y.-H. Wang, X.-C. Peng, et al. 2004. Giant Optical Nonlinearity of a $Bi_2Nd_2Ti_3O_{12}$ Ferroelectric Thin Film. *Appl. Phys. Lett.* 85:3687–3689.

Guo, J., C. Dong, L. Yang, et al. 2007. Crystal Structure and Electrical Properties of New Tungsten Bronzes: B_xWO_3 $(0.01 \leq x \leq 0.08)$. *Mater. Res. Bull.* 42:1384–1389.

Gupta, R. P., and M. Gupta. 1993. Distribution of the Hole-Carrier Density in $Y_2Ba_4Cu_7O_{14+\delta}$ $(\delta = 0.0, 0.5, 1.0)$ Superconductors. *Phys. Rev. B Condens. Matter* 48:16068–16077.

Hadermann, J., G. Van Tendeloo, and A. M. Abakumov. 2005. Transmission Electron Microscopy and Structural Phase Transitions in Anion-Deficient Perovskite-Based Oxides. *Acta Crystallogr. A* 61:77–92.

Hadermann, J., A. M. Abakumov, H. D'Hondt, et al. 2007. Synthesis and Crystal Structure of the $Sr_2Al_{1.07}Mn_{0.93}O_5$ Brownmillerite. *J. Mater. Chem.* 17:692–698.

Haid, S., W. Benstaali, A. Abbad, et al. 2019. Thermoelectric, Structural, Optoelectronic and Magnetic Properties of Double Perovskite Sr_2CrTaO_6: First Principle Study. *Mater. Sci. Eng. B* 245:68–74.

Haldolaarachchige, N., Q. Gibson, J. Krizan, et al. 2014. Superconducting Properties of the K_xWO_3 Tetragonal Tungsten Bronze and the Superconducting Phase Diagram of the Tungsten Bronze Family. *Phys. Rev. B* 89:104520.

Haluska, M., S. Speakman, and S. T. Misture. 2013. Crystal Structure Determinations of The Three-Layer Aurivillius Ceramics Using A New High-Resolution X- Ray Powder Diffractometer. *Adv. X-Ray Anal.* 46:192–197.

Hansteen, O. H., and H. Fjellvåg. 1998. Synthesis, Crystal Structure, and Magnetic Properties of $La_4Co_3O_{10+\delta}$ $(0.00 \leq \delta \leq 0.30)$. *J. Solid State Chem.* 141:212–220.

Hatakeyama, T., S. Takeda, F. Ishikawa, et al. 2010. Photocatalytic Activities of Ba_2RBiO_6 (R = La, Ce, Nd, Sm, Eu, Gd, Dy) under Visible Light Irradiation. *J. Ceram. Soc. Jpn.* 118:91–95.

Hauser, A. J., J. R. Soliz, M. Dixit, et al. 2012. Fully Ordered $Sr_2CrReO_6$$ Epitaxial Films: A High-Temperature Ferrimagnetic Semiconductor. *Phys. Rev. B* 85:161201.

Hernando, M., K. Boulahya, M. Parras, J. Gonzalez-Calbet, et al. 2003. Synthesis and Microstructural Characterization of Two New One-Dimensional Members of the $(A_3NiMnO_6)_\alpha(A_3Mn_3O_9)_\beta$ Homologous Series (A: Ba, Sr). *Cheminform* 2003:2419–2425.

Hoch, M., and C. Kasl. 2005. The Metal-Insulator Transition in La_xWO_3 and Y_xWO_3. *APS March Meeting*, March 2005, Los Angeles, CA.

Hojamberdiev, M., M. F. Bekheet, E. Zahedi, et al. 2016. New Dion–Jacobson Phase Three-Layer Perovskite $CsBa_2Ta_3O_{10}$ and Its Conversion to Nitrided $Ba_2Ta_3O_{10}$ Nanosheets via a Nitridation–Protonation–Intercalation–Exfoliation Route for Water Splitting. *Cryst. Growth Des.* 16:2302–2308.

Hong, S.-T., and A. W. Sleight. 1997. Crystal Structure of 4*H* $BaRuO_3$: High Pressure Phase Prepared at Ambient Pressure. *J. Solid State Chem.* 128:251–255.

Hong, Y.-S., S.-J. Kim, S.-J. Kim, et al. 2000. B-Site Cation Arrangement and Crystal Structure of Layered Perovskite Compounds $CsLn_2Ti_2NbO_{10}$ (Ln = La, Pr, Nd, Sm) and $CsCaLaTiNb_2O_{10}$. *J. Mater. Chem.* 10:1209–1214.

Horyń, R., M. Wolcyrz, A. Wojakewski, et al. 1996. Synthesis and Characterization of the $BiRESr_2O_6$-Type Ternaries (RE = La and Lanthanides). *J. Alloys Compd.* 242:35–40.

Howard, C. J., and H. T. Stokes. 2004. Octahedral Tilting in Cation-Ordered Perovskites – A Group-Theoretical Analysis. *Acta Crystallogr.* B 60:674–684.

Howard, C. J., and H. T. Stokes. 2005. Structures and Phase Transitions in Perovskites – A Group-Theoretical Approach. *Acta Crystallogr. A* 61:93–111.

Hu, C., L. Fang, H. Su, et al. 2009. Effects of Sr Substitution on Microwave Dielectric Properties of $Ba_3LaNb_3O_{12}$ Ceramics. *J. Alloys Compd.* 487:504–506.

Huang, Y., Y. Wei, S. Cheng, et al. 2010. Photocatalytic Property of Nitrogen-Doped Layered Perovskite $K_2La_2Ti_3O_{10}$. *Sol. Energy Mater. Sol. Cells* 94:761–766.

Huang, J., K. Yang, Z. Zhang, et al. 2017. Layered Perovskite $LiEuTiO_4$ as a 0.8 V Lithium Intercalation Electrode. *Chem. Commun.* 53:7800–7803.

Huang, C., L. Chen, H. Li, et al. 2019. Synthesis and Application of Bi_2WO_6 for the Photocatalytic Degradation of Two Typical Fluoroquinolones under Visible Light Irradiation. *RSC Adv.* 9:27768–27779.

Hungría, T., J. G. Lisoni, and A. Castro. 2002. $Sr_3Ti_2O_7$ Ruddlesden–Popper Phase Synthesis by Milling Routes. *Chem. Mater.* 14:1747–1754.

Hutchison, J. L., J. S. Anderson, and C. N. R. Rao. 1977. Electron Microscopy of Ferroelectric Bismuth Oxides Containing Perovskite Layers. *Proc. R. Soc. Lond. Ser. Math. Phys. Sci.* 355:301–312.

Ismail, M., W. Liu, M. S. C. Chan, et al. 2016. Synthesis, Application, and Carbonation Behavior of $Ca_2Fe_2O_5$ for Chemical Looping H_2 Production. *Energy Fuels* 30:6220–6232.

Ito, S., T. Mori, P. Yan, et al. 2017. High Electrical Conductivity in $Ba_2In_2O_5$ Brownmillerite Based Materials Induced by Design of a Frenkel Defect Structure. *RSC Adv.* 7:4688–4696.

Iwasaki, K., T. Ito, T. Matsui, et al. 2006. Synthesis of an Oxygen Nonstoichiometric $Sr_6Co_5O_{15}$ Phase. *Mater. Res. Bull.* 41:732–739.

Jacobson, A. J., J. C. Scanlon, K. R. Poeppelmeier, et al. 1981. The Preparation and Characterization of $Ba_3Te_2O_9$; a New Oxide Structure. *Mater. Res. Bull.* 16:359–367.

Jiang, D., W. Ma, Y. Yao, et al. 2018. Dion–Jacobson-Type Perovskite $KCa_2Ta_3O_{10}$ Nanosheets Hybridized with g-C_3N_4 Nanosheets for Photocatalytic H_2 Production. *Catal. Sci. Technol.* 8:3767–3773.

Juan, G., Z. Zhili, G. Chaojun, et al. 2010. Crystal Structure and Superconductivity of Ba_xWO_3 Prepared by Microwave Method. *Diwen Wuli Xuebao* 32:436–440.

Kahlenberg, V., R. X. Fischer, and C. S. J. Shaw. 2000. Rietveld Analysis of Dicalcium Aluminate ($Ca_2Al_2O_5$)—A New High Pressure Phase with the Brownmillerite-Type Structure. *Am. Mineral.* 85:1061–1065.

Kalanda, N. A., V. M. Trukhan, and S. F. Marenkin. 2002. Phase Transformations in the Systems Y_2BaCuO_5–"$Ba_3Cu_5O_8$" and Y_2BaCuO_5–$BaCuO_2$. *Inorg. Mater.* 38:597–603.

Kamiya, H., A. Kondo, T. Yokoyama, et al. 1994. Effect of Y_2BaCuO_5 Particle Size on the Properties of $YBa_2Cu_3O_{7-x}$ Superconductor. *Adv. Powder Technol.* 5:339–351.

Kawaguchi, T., K. Horigane, Y. Itoh, et al. 2018. Crystal Structure and Superconducting Properties of $KSr_2Nb_3O_{10}$. *Phys. B Condens. Matter* 536:830–832.

Kayser, P., M. J. Martínez-Lope, J. A. Alonso, et al. 2014. Crystal and Magnetic Structure of Sr_2MIrO_6 (M = Ca, Mg) Double Perovskites – A Neutron Diffraction Study. *Eur. J. Inorg. Chem.* 2014:178–185.

Kayser, P., J. A. Alonso, A. Muñoz, et al. 2017. Structural and Magnetic Characterization of the Double Perovskites R_2NiRuO_6 (R = Pr-Er): A Neutron Diffraction Study. *Acta Mater.* 126:114–123.

Kendall, K. R., C. Navas, J. K. Thomas, et al. 1996. Recent Developments in Oxide Ion Conductors: Aurivillius Phases. *Chem. Mater.* 8:642–649.

Khandy, S. A., and D. C. Gupta. 2020. Magneto-Electronic, Mechanical, Thermoelectric and Thermodynamic Properties of Ductile Perovskite Ba_2SmNbO_6. *Mater. Chem. Phys.* 239:121983.

Khokhar, A., M. L. V. Mahesh, A. R. James, et al. 2013. Sintering Characteristics and Electrical Properties of $BaBi_4Ti_4O_{15}$ Ferroelectric Ceramics. *J. Alloys Compd.* 581:150–159.

Kim, J.-R., D.-W. Kim, H. S. Jung, et al. 2006. Low-Temperature Sintering and Microwave Dielectric Properties of $Ba_5Nb_4O_{15}$ with ZnB_2O_4 Glass. *J. Eur. Ceram. Soc.* 26:2105–2109.

Kim, S. W., R. Zhang, P. S. Halasyamani, et al. 2015. $K_4Fe_3F_{12}$: An Fe^{2+}/Fe^{3+} Charge-Ordered, Ferrimagnetic Fluoride with a Cation-Deficient, Layered Perovskite Structure. *Inorg. Chem.* 54:6647–6652.

Kim, H. G., T. T. Tran, W. Choi, et al. 2016. Two New Non-Centrosymmetric $n = 3$ Layered Dion–Jacobson Perovskites: Polar $RbBi_2Ti_2NbO_{10}$ and Nonpolar $CsBi_2Ti_2TaO_{10}$. *Chem. Mater.* 28:2424–2432.

Knyazev, A. V., O. V. Krasheninnikova, E. V. Syrov, et al. 2019. Thermodynamic and X-Ray Studies of Layered Perovskite $KCa_2NaNb_4O_{13}$. *J. Chem. Thermodyn.* 138:255–261.

Kodenkandath, T. A., and J. B. Wiley. 2000. Synthesis and Structure of a Double-Layered Perovskite and Its Hydrate, $K_2SrTa_2O_7{\cdot}mH_2O$ ($m = 0, 2$). *Mater. Res. Bull.* 35:1737–1742.

Kolar, D., S. Gaberscek, B. Volavsek, et al. 1981. Synthesis and Crystal Chemistry of $BaNd_2Ti_3O_{10}$, $BaNd_2Ti_5O_{14}$, and $Nd_4Ti_9O_{24}$. *J. Solid State Chem.* 38:158–164.

Koren, G., and E. Polturak. 1994. Properties of $YBa_2Cu_3O_{7-\delta}/YBa_2Co_xCu_{3-x}O_y/YBa_2Cu_3O_{7-\delta}$ Josephson Edge Junctions with $0.1 \leqslant x \leqslant 0.3$ and the Effect of Flux Flow on Their Normal Resistance. *Phys. C Supercond.* 230:340–348.

Kozlenko, D. P., N. Dang, T. Phan, et al. 2017. The Structural, Magnetic and Vibrational Properties of Ti-Doped $BaMnO_3$. *J. Alloys Compd.* 695:2539–2548.

Krishnamurthy, J., and A. Venimadhav. 2020. Magnetic Field-Induced Metamagnetic, Magnetocaloric and Pyrocurrent Behaviors of Eu_2CoMnO_6. *J. Magn. Magn. Mater.* 500:166387.

Kumar, S., Ø. Fjellvåg, A. O. Sjåstad, et al. 2020. Physical Properties of Ruddlesden–Popper ($N = 3$) Nickelate: $La_4Ni_3O_{10}$. *J. Magn. Magn. Mater.* 496:165915.

Lacorre, P., J. B. Torrance, J. Pannetier, et al. 1991. Synthesis, Crystal Structure, and Properties of Metallic $PrNiO_3$: Comparison with Metallic $NdNiO_3$ and Semiconducting $SmNiO_3$. *J. Solid State Chem.* 91:225–237.

Larrégola, S. A., J. A. Alonso, J. C. Pedregosa, et al. 2009. The Role of the Pb^{2+} 6*s* Lone Pair in the Structure of the Double Perovskite Pb_2ScSbO_6. *Dalton Trans.* 2003:5453–5459.

Lazarević, Z. Ž., B. D. Stojanović, C. O. Paiva-Santos, et al. 2008. Study of Structure and Properties of $Bi_4Ti_3O_{12}$ Prepared by Mechanochemical Syntheses. *Ferroelectrics* 368:154–162.

Le Berre, F., M.-P. Crosnier-Lopez, and J.-L. Fourquet. 2004. Cationic Ordering in the New Layered Perovskite $BaSrTa_2O_7$. *Solid State Sci.* 6:53–59.

Lee, N., H. Y. Choi, Y. J. Jo, et al. 2014. Strong Ferromagnetic-Dielectric Coupling in Multiferroic Lu_2CoMnO_6 Single Crystals. *Appl. Phys. Lett.* 104:112907.

Lee, J. G., H. J. Hwang, O. Kwon, et al. 2016. Synthesis and Application of Hexagonal Perovskite $BaNiO_3$ with Quadrivalent Nickel under Atmospheric and Low-Temperature Conditions. *Chem. Commun.* 52:10731–10734.

Lehmann, U., and H. Müller-Buschbaum. 1980. Ein Beitrag zur Chemie der Oxocobaltate(II): La_2CoO_4, Sm_2CoO_4. *Z. Für Anorg. Allg. Chem.* 470:59–63.

Letouzé, F., C. Martin, M. Hervieu, et al. 1997. A New Structure Related to the Layered Cuprates: The "1201" Shear-Like Phase $Tl_5Ba_3Sr_5Cu_3O_{19}$, Third Member of the Series $(TlA_2CuO_5)M\cdot Tl_2A_2O_4$. *J. Solid State Chem.* 128:150–155.

Li, B.-W., M. Osada, T. C. Ozawa, et al. 2012. $RbBiNb_2O_7$: A New Lead-Free High-T_c Ferroelectric. *Chem. Mater.* 24:3111–3113.

Li, Y., and X. Liu. 2015. Sol–Gel Synthesis, Structure and Luminescence Properties of Ba_2ZnMoO_6:Eu^{3+} Phosphors. *Mater. Res. Bull.* 64:88–92.

Li, X., R. Xie, X. Cao, et al. 2018. Synthesis of Cubic Sodium Tungsten Bronze Na_xWO_3 in Air. *J. Am. Ceram. Soc.* 101:4458–4462.

Li, D., K. Lee, B. Y. Wang, et al. 2019a. Superconductivity in an Infinite-Layer Nickelate. *Nature* 572:624–627.

Li, Z., Y. Yin, J. D. Rumney, et al. 2019b. High-Pressure in-Situ X-Ray Diffraction and Raman Spectroscopy of Ca_2AlFeO_5 Brownmillerite. *High Press. Res.* 39:92–105.

Li, Z., W. Qi, J. Cao, et al. 2019c. Multiferroic Properties of Single Phase Bi_3NbTiO_9 Based Textured Ceramics. *J. Alloys Compd.* 788:701–704.

Liang, Z.-H., K.-B. Tang, Q.-W. Chen, et al. 2009. $RbCa_2Nb_3O_{10}$ from X-Ray Powder Data. *Acta Crystallogr. Sect. E Struct. Rep. Online* 65:i44.

Lichtenberg, F., A. Herrnberger, K. Wiedenmann, et al. 2001. Synthesis of Perovskite-Related Layered $AnB_nO_{3n+2} = ABO_x$ Type Niobates and Titanates and Study of Their Structural, Electric and Magnetic Properties. *Prog. Solid State Chem.* 29:1–70.

Lightfoot, P., and P. D. Battle. 1990. The Crystal and Magnetic Structures of $Ba_3NiRu_2O_9$, $Ba_3CoRu_2O_9$, and $Ba_3ZnRu_2O_9$. *J. Solid State Chem.* 89:174–183.

Liu, S., M. Avdeev, Y. Liu, et al. 2016. A New n = 4 Layered Ruddlesden–Popper Phase $K_{2.5}Bi_{2.5}Ti_4O_{13}$ Showing Stoichiometric Hydration. *Inorg. Chem.* 55:1403–1411.

Liu, J., B. Chen, C. Fan, et al. 2018a. Controllable Synthesis of Small Size Cs_xWO_3 Nanorods as Transparent Heat Insulation Film Additives. *CrystEngComm* 20:1509–1519.

Liu, M., Y. Zhang, L.-F. Lin, et al. 2018b. Direct Observation of Ferroelectricity in $Ca_3Mn_2O_7$ and Its Prominent Light Absorption. *Appl. Phys. Lett.* 113:022902.

Longo, J. M., and P. M. Raccah. 1973. The Structure of La_2CuO_4 and $LaSrVO_4$. *J. Solid State Chem.* 6:526–531.

Loureiro, S. M., E. V. Antipov, E. M. Kopnin, et al. 1996, Structure and Superconductivity of the $HgBa_2Ca_3Cu_4O_{10+\delta}$ Phase. *Physica C* 257:117–124.

Lufaso, M. W., R. B. Macquart, Y. Lee, et al. 2006. Structural Studies of Sr_2GaSbO_6, Sr_2NiMoO_6, and Sr_2FeNbO_6 Using Pressure and Temperature. *J. Phys. Condens. Matter* 18:8761–8780.

Luo, Z. P., H. Hashimoto, H. Ihara, et al. 1999. Transmission Electron Microscopy Characterization of the High-T_c Superconductor $HgBa_2Ca_3Cu_4O_{10+\Delta}$. *Philos. Mag. Lett.* 79:429–439.

Luo, Z. P., Y. Li, H. Hashimoto, et al. 2004. Defective Structure in the High-Superconductor Hg-1234. *Physica C* 408–410:50–51.

Mackenzie, A. P., and Y. Maeno. 2003. The Superconductivity of Sr_2RuO_4 and the Physics of Spin-Triplet Pairing. *Rev. Mod. Phys.* 75:657–712.

Makowski, S. J., J. A. Rodgers, P. F. Henry, et al. 2009. Coupled Spin Ordering in the Ln_2LiRuO_6 Double Perovskites. *Chem. Mater.* 21:264–272.

Mandal, R., M. Chandra, V. Roddatis, et al. 2020. Magneto-Dielectric Effect in Relaxor Dipolar Glassy Tb_2CoMnO_6 Film. *Phys. Rev. B* 101:094426.

Martínez-Lope, M. J., J. A. Alonso, M. T. Casais, et al. 2002. Preparation, Crystal and Magnetic Structure of the Double Perovskites Ba_2CoBO_6 (B = Mo, W). *Eur. J. Inorg. Chem.* 2002:2463–2469.

McKigney, E. A., R. E. Del Sesto, L. G. Jacobsohn, et al. 2007. Nanocomposite Scintillators for Radiation Detection and Nuclear Spectroscopy. *Nucl. Instrum. Methods Phys. Res. Sect. Accel. Spectrometers Detect. Assoc. Equip.* 579:15–18.

Mikhailova, D., C.-Y. Kuo, P. Reichel, et al. 2014. Structure, Magnetism, and Valence States of Cobalt and Platinum in Quasi-One-Dimensional Oxides A_3CoPtO_6 with A = Ca, Sr. *J. Phys. Chem. C* 118:5463–5469.

Miranda, L., D. C. Sinclair, M. Hernando, et al. 2009. Mn-Rich $BaMn_{1-x}FexO_{3-\delta}$ Perovskites Revisited: Structural, Magnetic, and Electrical Properties of Two New 6*H′* Polytypes. *Chem. Mater.* 21:5272–5283.

Missyul, A. B., I. A. Zvereva, T. T. M. Palstra, et al. 2010. Double-Layered Aurivillius-Type Ferroelectrics with Magnetic Moments. *Mater. Res. Bull.* 45:546–550.

Mitsuyama, T., A. Tsutsumi, T. Hata, et al. 2008. Enhanced Photocatalytic Water Splitting of Hydrous $LiCa_2Ta_3O_{10}$ Prepared by Hydrothermal Treatment. *Bull. Chem. Soc. Jpn.* 81:401–406.

Moore, C. A., and P. D. Battle. 2003. Crystal and Magnetic Structures of $Sr_4MMn_2O_9$ (M = Cu or Zn). *Cryst. Magn. Struct.* 176:88–96.

Morrow, R., R. Mishra, O. D. Restrepo, et al. 2013. Independent Ordering of Two Interpenetrating Magnetic Sublattices in the Double Perovskite Sr_2CoOsO_6. *J. Am. Chem. Soc.* 135:18824–18830.

Motohashi, T., Y. Hirano, Y. Masubuchi, et al. 2013. Oxygen Storage Capability of Brownmillerite-Type $Ca_2AlMnO_{5+\delta}$ and Its Application to Oxygen Enrichment. *Chem. Mater.* 25:372–377.

Muñoz, A., C. de la Calle, J. A. Alonso, et al. 2008. Crystallographic and Magnetic Structure of $SrCoO_{2.5}$ Brownmillerite: Neutron Study Coupled with Band-Structure Calculations. *Phys. Rev. B* 78:054404.

Nagai, T., H. Shirakuni, A. Nakano, et al. 2019. Weak Ferroelectricity in $n = 2$ Pseudo Ruddlesden–Popper-Type Niobate $Li_2SrNb_2O_7$. *Chem. Mater.* 31:6257–6261.

Nagarajan, R., and C. N. R. Rao. 1993. Structure and Superconducting Properties of Ga-Substituted $YBa_2Cu_3O_{7-\delta}$ and $YBaSrCu_3O_{7-\delta}$ Systems. *J. Mater. Chem.* 3:969–973.

Nagashima, T., M. Watahiki, Y. Fukai, et al. 1990. Synthesis and Crystal Structure of a New Family of Superconductors (Tl, Pb) (R, Sr)$_2$CuO$_5$(R = La, Nd). In *Advances in Superconductivity II*, ed. T. Ishiguro, and K. Kajimura, 83–86. Tokyo: Springer Japan.

Nakahara, Y., S. Kato, M. Sugai, et al. 1997. Synthesis and Crystal Structure of $(Sr_{1-x}Ca_x)2FeMnO_y$ ($x = 0$–1.0). *Mater. Lett.* 30:163–167.

Nakashima, S., H. Fujisawa, S. Ichikawa, et al. 2010. Structural and Ferroelectric Properties of Epitaxial $Bi_5Ti_3FeO_{15}$ and Natural-Superlattice-Structured $Bi_4Ti_3O_{12}$–$Bi_5Ti_3FeO_{15}$ Thin Films. *J. Appl. Phys.* 108:074106.

Naresh, G., and T. K. Mandal. 2015. Efficient COD Removal Coinciding with Dye Decoloration by Five-Layer Aurivillius Perovskites under Sunlight-Irradiation. *ACS Sustain. Chem. Eng.* 3:2900–2908.

Nguyen, L. T., R. J. Cava, and A. M. Fry-Petit. 2019. Low Temperature Structural Phase Transition in the Ba_2CaMoO_6 Perovskite. *J. Solid State Chem.* 277:415–421.

Ouchetto, K., F. Archaimbault, J. Choisnet, et al. 1997. New Ordered and Distorted Perovskites: The Mixed Platinates Ln_2MPtO_6 (Ln = La, Pr, Nd, Sm, Eu, Gd; M = Mg, Co, Ni, Zn). *Mater. Chem. Phys.* 51:117–124.

Paiva-Santos, C. O., T. Mazon, M. A. Zaghete, et al. 2000. Crystal Structure of $BaBi_2Ta_2O_9$. *Powder Diffr.* 15:134–138.

Palmer, H. M., A. Snedden, A. J. Wright, et al. 2006. Crystal Structure and Magnetic Properties of $Ca_2MnAlO_{5.5}$, an $n = 3$ Brownmillerite Phase. *Chem. Mater.* 18.:1130–1133.

Pan, Y.-W., P.-W. Zhu, and X. Wang. 2015. High-Pressure Synthesis, Characterization, and Equation of State of Double Perovskite Sr_2CoFeO_6. *Chin. Phys. B* 24:017503.

Pan, Z., X. Dai, Y. Lei, et al. 2018. Crystal Growth and Properties of the Disordered Crystal Yb:$SrLaAlO_4$: A Promising Candidate for High-Power Ultrashort Pulse Lasers. *CrystEngComm* 20:3388–3395.

Panda, S. K., and I. Dasgupta. 2013. Electronic Structure and Magnetism in Ir-Based Doble-Perovskite Sr_2CeIrO_6. *Mod. Phys. Lett. B* 2:1350041.

Parsons, T. G., H. D'Hondt, J. Hadermann, et al. 2009. Synthesis and Structural Characterization of $La_{1-x}A_xMnO_{2.5}$ (A = Ba, Sr, Ca) Phases: Mapping the Variants of the Brownmillerite Structure. *Chem. Mater.* 21:5527–5538.

Pelloquin, D., M. Hervieu, C. Michel, et al. 1995. Double Cationic Ordering in the "1201" Substituted Type Cuprate $HgBiSr_7Cu_2SbO_{15}$. *J. Solid State Chem.* 116:53–60.

Persson, K. 2014a. *Materials Data on $RbCa_2Ta_3O_{10}$ (SG:123) by Materials Project.* Berkeley, CA: LBNL Materials Project; Lawrence Berkeley National Laboratory (LBNL).

Persson, K. 2014b. *Materials Data on Ca_2TiSiO_6 (SG:225) by Materials Project.* mp-9413. Berkeley, CA: LBNL Materials Project; Lawrence Berkeley National Laboratory (LBNL).

Persson, K. 2014c. *Materials Data on $CsCa_2Nb_3O_{10}$ (SG:62) by Materials Project.* mp-581330. Berkeley, CA: LBNL Materials Project; Lawrence Berkeley National Laboratory (LBNL).

Persson, K. 2014d. *Materials Data on Ba_3NaRuO_6 (SG:167) by Materials Project.* mp-9745. Berkeley, CA: LBNL Materials Project; Lawrence Berkeley National Laboratory (LBNL).

Persson, K. 2014e. *Materials Data on Ca_2FeSbO_6 (SG:11) by Materials Project.* mvc–4117. Berkeley, CA: LBNL Materials Project; Lawrence Berkeley National Laboratory (LBNL).

Persson, K. 2014f. *Materials Data on $EuBa_2NbO_6$ (SG:87) by Materials Project.* mp-13323. Berkeley, CA: LBNL Materials Project; Lawrence Berkeley National Laboratory (LBNL).

Persson, K. 2014g. *Materials Data on Sr_2FeWO_6 (SG:14) by Materials Project.* mp-19266. Berkeley, CA: LBNL Materials Project; Lawrence Berkeley National Laboratory (LBNL).

Persson, K. 2014h. *Materials Data on $Li_2LaTa_2O_7$ (SG:139) by Materials Project.* mp-15901. Berkeley, CA: LBNL Materials Project; Lawrence Berkeley National Laboratory (LBNL).

Persson, K. 2014i. *Materials Data on $CsCa_2Ta_3O_{10}$ (SG:123) by Materials Project.* mp-10347. Berkeley, CA: LBNL Materials Project; Lawrence Berkeley National Laboratory (LBNL).

Persson, K. 2016a. *Materials Data on Ba_2CuWO_6 (SG:87) by Materials Project.* mp-505618. Berkeley, CA: LBNL Materials Project; Lawrence Berkeley National Laboratory (LBNL).

Persson, K. 2016b. *Materials Data on Ba_2NdMoO_6 (SG:225) by Materials Project.* mp-18904. Berkeley, CA: LBNL Materials Project; Lawrence Berkeley National Laboratory (LBNL).

Persson, K. 2016c. *Materials Data on Sr_2CoReO_6 (SG:87) by Materials Project.* mp-31515. Berkeley, CA: LBNL Materials Project; Lawrence Berkeley National Laboratory (LBNL).

Polubinskii, V. V., Y. Titov, N. M. Belyavina, et al. 2014. Synthesis and Crystal Structure of the $A_6B_5O_{18}$ Perovskite-Like Compounds. *J. Solid Stat. Sci.* 29:1–5.

Pomjakushin, V. Y., A. M. Balagurov, T. V. Elzhov, et al. 2002. Atomic and Magnetic Structures, Disorder Effects, and Unconventional Superexchange Interactions in $A_2MnGaO_{5+\delta}$ (A = Ca) Oxides of Layered Brownmillerite-Type Structure. *Phys. Rev. B* 66:184412.

Pomjakushin, V., D. Sheptyakov, P. Fischer, et al. 2004. Atomic and Magnetic Structures, and Unconventional Superexchange Interactions in Sr_2GaMnO_{5+x} ($0<x<0.5$) and $Sr_2GaMn(O,F)_6$. *J. Magn. Magn. Mater.* 272–276:820–822.

Potin, V., S. Bruyere, M. Gillet, et al. 2012. Growth, Structure, and Stability of K_xWO_3 Nanorods on Mica Substrate. *J. Phys. Chem. C* 116:1921–1929.

Pradheesh, R., H. S. Nair, V. Sankaranarayanan, et al. 2012. Large Magnetoresistance and Jahn-Teller Effect in Sr_2FeCoO_6. *Eur. Phys. J. B* 85:260

Quarez, E., F. Abraham, and O. Mentré. 2003. Synthesis, Crystal Structure and Characterization of New 12*H* Hexagonal Perovskite-Related Oxides $Ba_6M_2Na_2X_2O_{17}$ (M = Ru, Nb, Ta, Sb; X = V, Cr, Mn, P, As). *J. Solid State Chem.* 176:137–150.

Quarez, E., P. Roussel, O. Pérez, et al. 2004. Crystal Structure of the Mixed Mn^{4+}/Mn^{5+} 2H-Perovskite-Type $Ba_4Mn_2NaO_9$ Oxide. *Solid State Sci.* 6:931–938.

Ram, R. A. M., and A. Clearfield. 1994. Pillaring Studies on Some Layered Oxides with Ruddlesden–Popper Related Structures. *J. Solid State Chem.* 112:288–294.

Rayaprol, S., K. Sengupta, E. V. Sampathkumaran, et al. 2004. Magnetic Behavior of Spin-Chain Compounds, Sr_3ZnRhO_6 and Ca_3NiMnO_6, from Heat Capacity and Ac Susceptibility Studies. *J. Solid State Chem.* 177:3270–3273.

Reehuis, M., C. Ulrich, K. Prokes, et al. 2006. Crystal Structure and High-Field Magnetism of La_2CuO_4. *Phys. Rev. B* 73:144513.

Rendón Ramírez, J. M., O. A. Almanza M., R. Cardona, et al. 2013. Structural, Magnetic and Electronic Properties of The Sr_2CoNbO_6 Complex Perovskite. *Int. J. Mod. Phys. B* 27 :1350171.

Revelli, A., C. C. Loo, D. Kiese, et al. 2019. Spin-Orbit Entangled $j = 1/2$ Moments in Ba_2CeIrO_6: A Frustrated Fcc Quantum Magnet. *Phys. Rev. B* 100:085139.

Sahnoun, O., H. Bouhani-Benziane, M. Sahnoun, et al. 2017. Magnetic and Thermoelectric Properties of Ordered Double Perovskite Ba_2FeMoO_6. *J. Alloys Compd.* 714:704–708.

Sato, M., J. Abo, and T. Jin. 1992a. Structure Examination of $NaLaNb_2O_7$ Synthesized by Soft Chemistry. *Solid State Ion.* 57:285–293.

Sato, M., J. Abo, T. Jin, et al. 1992b. Structure Determination of $KLaNb_2O_7$ Exhibiting Ion Exchange Ability by X-Ray Powder Diffraction. *Solid State Ion.* 51:85–89.

Sato, M., J. Watanabe, and K. Uematsu. 1993a. Crystal Structure and Ionic Conductivity of a Layered Perovskite, $AgLaNb_2O_7$. *J. Solid State Chem.* 107:460–470.

Sato, M., Y. Kono, and T. Jin. 1993b. Structural Characterization and Ion Conductivity of $MCa_2NaNb_4O_{13}$ (M = Rb, Na) with Four Units of Perovskite Layer. *J. Ceram. Soc. Jpn.* 101:980–984.

Schilling, A., M. Cantoni, J. D. Guo, et al. 1993. Superconductivity above 130 K in the Hg–Ba–Ca–Cu–O system. *Nature* 363:56–58.

Seikh, M. M., V. Caignaert, N. Sakly, et al. 2019. Effect of Thermal Treatment upon the Structure Incommensurability and Magnetism of the Spin Chain Oxide $Sr_3CaMn_2CoO_{9+\delta}$. *J. Alloys Compd.* 790:572–576.

Seinen, P. A., F. P. F. van Berkel, W. A. Groen, et al. 1987. The Ordered Perovskite System Ln_2NiRuO_2. *Mater. Res. Bull.* 22:535–542.

Sereda, V. V., D. S. Tsvetkov, A. L. Sednev, et al. 2018. Thermodynamics of Sr_2NiMoO_6 and Sr_2CoMoO_6 and Their Stability under Reducing Conditions. *Phys. Chem. Chem. Phys.* 20:20108–20116.

Shaked, H., P. M. Keane, J. C. Rodriguez, et al. 1994. *Crystal Structures of the High-Tc Superconducting Copper-Oxides*. Amsterdam, The Netherlands: Elsevier Science B.V.

Shannon, R. D. 1976. Revised Effective Ionic Radii and Systematic Studies of Interatomic Distances in Halides and Chalcogenides. *Acta Crystallogr. A* 32:751–767.

She, J.-L., and R.-S. Liu. 2008. A Simplified Synthetic Experiment of $YBa_2Cu_3O_{7-x}$ Superconductor for First-Year Chemistry Laboratory. *J. Chem. Educ.* 85:825.

Shi, J. 2015. Crystal Structure Studies, Electrical and Magnetic Properties of 2, 3, 4, 5-Layer Aurivillius Oxides. Thesis, New York State College of Ceramics at Alfred University. Kazuo Inamori School of Engineering. https://aura.alfred.edu/handle/10829/7311.

Singh, D., and R. Singh. 2010. Synthesis and Characterization of Ruddlesden–Popper (RP) Type Phase $LaSr_2MnCrO_7$. *J. Chem. Sci.* 122:807–811.

Singh, V., J. Dwivedi, and I. Sharma. 2012. Synthesis, Structure and Electric Transport Properties of Sr_3NbCrO_7. *J. Mater. Sci. Res.* 2:p148.

Singh, V., and J. J. Pulikkotil. 2019. Double Perovskite Ba_2CaIrO_6: A Slater-Type Antiferromagnet System. *J. Magn. Magn. Mater.* 475:550–553.

Song, J., B. Zhao, L. Yin, et al. 2017. Reentrant Spin Glass Behavior and Magnetodielectric Coupling of an Ir-Based Double Perovskite Compound, La_2CoIrO_6. *Dalton Trans.* 46:11691–11697.

Stanley, R. K., R. C. Morris, and W. G. Moulton. 1978. Possible New Phase Transitions in Hexagonal Rb_xWO_3. *Solid State Commun.* 27:1277–1280.

Stitzer, K. E., M. D. Smith, J. Darriet, et al. 2001. Crystal Growth, Structure Determination and Magnetism of a New Hexagonal Rhodate: $Ba_9Rh_8O_{24}$. *Chem. Commun. Camb. Engl.*:1680–1681.

Stitzer, K. E., A. El Abed, J. Darriet, et al. 2003a. Crystal Growth and Structure Determination of Barium Rhodates: Stepping Stones toward 2*H*–$BaRhO_3$. *J. Am. Chem. Soc.* 3:856–864.

Stitzer, K. E., W. R. Gemmill, M. D. Smith, et al. 2003b. Crystal Growth of a Novel Oxygen-Deficient Layered Perovskite: $Ba_7Li_3Ru_4O_{20}$. *J. Solid State Chem.* 175:39–45.

Stitzer, K., A. El Abed, J. Darriet, et al. 2004. Crystal Growth, Structure Determination and Magnetism of a New M = 3, N = 1 Member of the $A_{3n+3m}A'_nB_{3m+n}O_{9m+6n}$ Family of Oxides: 12*R*-$Ba_{12}Rh_{9.25}Ir_{1.75}O_{33}$. *J. Solid State Chem.* 117:1405–1411.

Su, J., Z. Z. Yang, X. M. Lu, et al. 2015. Magnetism-Driven Ferroelectricity in Double Perovskite Y_2NiMnO_6. *ACS Appl. Mater. Interfaces* 7:13260–13265.

Tahini, H. A., X. Tan, U. Schwingenschlögl, et al. 2016. Formation and Migration of Oxygen Vacancies in $SrCoO_3$ and Their Effect on Oxygen Evolution Reactions. *ACS Catal.* 6:5565–5570.

Takahashi, Y., H. Zama, T. Morishita, et al. 2001. O_2-Annealing Effects on Dielectric Properties of Sr_2AlTaO_6/YBa_2Cu_3Oy Films. *Phys. C Supercond.* 357–360:1364–1367.

Takusagawa, F., and R. A. Jacobson. 1976. Crystal Structure Studies of Tetragonal Sodium Tungsten Bronzes, Na_xWO_3. I. $Na_{0.33}WO_3$ and $Na_{0.48}WO_3$. *J. Solid State Chem.* 18:163–174.

Téllez, D. L., C. D. Toro, A.V.G. Rebaza, et al. 2013. Structural, Magnetic, Multiferroic and Electronic Properties of Sr_2ZrMnO_6 Double Perovskite. *J. Mol. Struct.* 1034:233–237.

Thangadurai, V., and W. Weppner. 2002. Determination of the Sodium Ion Transference Number of the Dion–Jacobson-Type Layered Perovskite $NaCa_2Nb_3O_{10}$ Using Ac Impedance and Dc Methods. *Chem. Mater.* 3:1136–1143.

Tilley, R. J. D. 2016. *Perovskites: Structure-Property Relationships*. 1st edition. Chichester, UK: John Wiley & Sons, Ltd.

Toda, K., T. Suzuki, and M. Sato. 1996. Synthesis and High Ionic Conductivity of New Layered Perovskite Compounds, $AgLaTa_2O_7$ and $AgCa_2Ta_3O_{10}$. *Solid State Ion.* 93:177–181.

Toda, K., M. Takahashi, T. Teranishi, et al. 1999. Synthesis and Structure Determination of Reduced Tantalates, $Li_2LaTa_2O_7$, $Li_2Ca_2Ta_3O_{10}$ and $Na_2Ca_2Ta_3O_{10}$, with a Layered Perovskite Structure. *J. Mater. Chem.* 9:799–803.

Urushihara, D., T. Asaka, K. Fukuda, et al. 2017. Discovery of the High-Pressure Phase of $Ba_3W_2O_9$ and Determination of Its Crystal Structure. *Inorg. Chem.* 56:13007–13013.
Vasala, S., H. Saadaoui, E. Morenzoni, et al. 2014. Characterization of Magnetic Properties of $CuWO_6$ and Sr_2CuMoO_6. *Phys. Rev. B* 89:134419.
Vasala, S., and M. Karppinen. 2015. $A_2B'B''O_6$ Perovskites: A Review. *Prog. Solid State Chem.* 43:1–36.
Vogt, T., P. M. Woodward, and B. A. Hunter. 1999. The High-Temperature Phases of WO_3. *J. Solid State Chem.* 144:209–215.
Wang, X., and X. Yao. 1993. Dielectric Relaxation in $Bi_2BaNb_2O_9$ Ceramic. *J. Mater. Sci. Technol.* 9:461–463.
Wang, S., Q. Sun, B. Devakumar, et al. 2019. Novel Highly Efficient and Thermally Stable $Ca_2GdTaO_6:Eu^{3+}$ Red-Emitting Phosphors with High Color Purity for UV/Blue-Excited WLEDs. *J. Alloys Compd.* 804:93–99.
Wassermann, K., M. T. Pope, M. Salmen, et al. 2000. Thermal Degradation of Polyoxotungstates—An Effective Method for the Preparation of Tungsten Bronzes. *J. Solid State Chem.* 149:378–383.
Woo, J. U., S. H. Kweon, M. Im, et al. 2017. Synthesis and Dielectric Properties of Layered-Perovskite $KCa_2Na_{n-3}NbnO3_{n+1}$ Ceramics. *Ceram. Int.* 43:15089–15094. Elsevier Limited.
Xu, L., Y.-M. Yin, N. Zhou, et al. 2017. Sulfur Tolerant Redox Stable Layered Perovskite $SrLaFeO_{4-\delta}$ as Anode for Solid Oxide Fuel Cells. *Electrochem. Commun.* 76:51–54.
Yang, J. H., W. K. Choo, and C. H. Lee. 2003. Ca_2MgWO_6 from Neutron and X-Ray Powder Data. *Acta Crystallogr. C* 59:i86–88.
Yang, X., L. Luo, and H. Zhong. 2005. Preparation of $LaSrCoO_4$ Mixed Oxides and Their Catalytic Properties in the Oxidation of CO and C_3H_8. *Catal. Commun.* 6:13–17.
Yang, H., Y. Tang, L. D. Yao, et al. 2007. Synthesis, Structure and Phase Separation of a New 12 *R*-Type Perovskite-Related Oxide $Ba_3NdMn_2O_9$. *J. Alloys Comp.* 432:283–288.
Yi, W., Q. Liang, Y. Matsushita, et al. 2013. High-Pressure Synthesis, Crystal Structure, and Properties of In_2NiMnO_6 with Antiferromagnetic Order and Field-Induced Phase Transition. *Inorg. Chem.* 52:14108–14115.
Yin, C., G. Tian, G. Li, et al. 2017. New 10H Perovskites $Ba5Ln_{1-x}Mn_{4+y}O1_{5-\delta}$ with Spin Glass Behaviour. *RSC Adv.* 7:33869–33874.
Yokosuka, M. 2002. Dielectric and Piezoelectric Properties of Mn-Modified $Bi_4CaTi_4O1_5$ Based Ceramics. *Jpn. J. Appl. Phys.* 41:7123–7126.
Zakharov, N., P. Werner, I. Zibrov, et al. 2000. Structural Studies of Calcium Tungsten Bronzes, Ca_xWO_3, Formed at High Pressure. *Cryst. Res. Technol.* 35:713–720.
Zhang, Z., M. Greenblatt, and J. B. Goodenough. 1994. Synthesis, Structure, and Properties of the Layered Perovskite $La_3Ni_2O_{7-\delta}$. *J. Solid State Chem.* 108:402–409.
Zhang, Y., H. J. Xiang, and M.-H. Whangbo. 2009. Interplay between Jahn-Teller Instability, Uniaxial Magnetism and Ferroelectricity in Ca_3CoMnO_6. *Phys. Rev. B* 79:054432.
Zhang, J., H. Zheng, C. D. Malliakas, et al. 2014. Brownmillerite $Ca_2Co_2O_5$: Synthesis, Stability, and Re-Entrant Single-Crystal-to-Single-Crystal Structural Transitions. *Chem. Mater.* 26:7172–7182.
Zhang, H., S. Ni, Y. Mi, et al. 2018. Ruddlesden–Popper Compound Sr_2TiO_4 Co-Doped with La and Fe for Efficient Photocatalytic Hydrogen Production. *J. Catal.* 359:112–121.
Zhang, Q., G. Cao, F. Ye, et al. 2019. Anomalous Magnetic Behavior in Ba_2CoO_4 with Isolated CoO_4 Tetrahedra. *ArXiv190302490 Cond-Mat.* http://arxiv.org/abs/1903.02490.
Zhao, D., J.-F. Han, J.-Y. Cui, et al. 2015. A New Pb(IV)-Based Photocathode Material Sr_2PbO_4 with Good Light Harvesting Ability. *J. Mater. Chem. A* 3:12051–12058.
Zhou, N., G. Chen, H. J. Zhang, et al. 2009a. Synthesis and Transport Properties of La_2NiO_4. *Phys. B Condens. Matter* 404:4150–4154.

Zhou, Q., B. J. Kennedy, M. Avdeev, et al. 2009b. Structural Studies of the Phases in Ba_2LaIrO_6–New Light on an Old Problem. *J. Solid State Chem.* 182:3195–3200.

Zhou, T. 2019. Slow Change of the Electron Clouds of Ions: A Common Medium for Electron Pairing in LaOFeP, Nb and $TlBa_2CaCu_2O_7$. DOI: 10.13140/RG.2.2.32188.18564

Zhu, J., H. Li, L. Zhong, et al. 2014. Perovskite Oxides: Preparation, Characterizations, and Applications in Heterogeneous Catalysis. *ACS Catal.* 4:2917–2940.

zur Loye, H.-C., S.-J. Kim, R. Macquart, et al. 2009. Low Temperature Structural Phase Transition of $Ba_3NaIr_2O_9$. *Solid State Sci.* 11:608–613.

zur Loye, H.-C., Q. Zhao, D. E. Bugaris, et al. 2011. 2H-Perovskite Related Oxides: Synthesis, Structures, and Predictions. *CrystEngComm* 14:23–39.

Zvereva, I. A., E. A. Tugova, V. F. Popova, et al. 2018. The Impact of Nd^{3+}/La^{3+} Substitution on the Cation Distribution and Phase Diagram in the $La_2SrAl_2O_7$-$Nd_2SrAl_2O_7$ System. *Chim. Techno Acta* 5:80–85.

5 Magnetic Properties of Perovskite Oxides

The magnetic properties of perovskites are associated with the presence of incompletely filled *d*- or *f*-block elements at A- or B-sites, and they are often transition metals (*d*-block) and lanthanides (*f*-block). The crystallographic geometries of these cations from *d*- and *f*-block metals are important, because they govern the orbital energy levels, which influence the crystal field–ligand interactions. Notably, the magnetic nature in 3*d*-block elements arises from the spin state (S) of electrons, which depends on the crystal field interactions. Therefore, the magnetic behavior of the perovskite is a resultant of spin states of the cations (low, intermediate, or high spin) and magnetic moments. At high temperatures, the perovskites often show paramagnetic nature due to the disorders arising with temperature-dependent magnetic susceptibility as per the Curie–Weiss law. However, at lower temperatures, magnetic moments can interact with each other, resulting in antiferromagnetic, ferromagnetic, or ferrimagnetic behavior. The magnetic properties of the perovskites are influenced by external factors such as temperature, pressure, or dopants. As the magnetic property varies under any of the above circumstances, it subsequently changes the electronic properties.

Many times, the magnetic property of the perovskite is determined by the characteristics of B-site ions; for example, if the B-site of the perovskite is occupied by a paramagnetic ion, then the perovskite exhibits paramagnetic behavior at normal conditions. Additionally, if the perovskites are composed of a nonmagnetic A-site ion and Mn or Co occupies the B-site, then the perovskites potentially exhibit inherent ferromagnetism. The B-site ions in the octahedral site have the electronic band with multiplet structure, originated due to the intra-atomic exchange and correlation. According to Hund's rule, the lowest-energy configuration of electrons corresponds to the state of maximum multiplicity or maximum spin and orbital angular momentum. Hund's rule qualitatively determines the spin state of localized *d* electrons in the perovskites based on the ionic state of atoms.

In actual scenario, there is a remarkable difference in the atomic theory based on free ions and the cations occupying the perovskite lattice. They are as follows: (1) the degeneracy of *d* states into the e_g and t_{2g} energy levels by a ligand-field energy $10Dq$; (2) the energy differences between electronic states of cations are not as wide as free ions, due to polarization and Coulomb interactions of electrons; (3) the perovskite is not purely ionic as there is covalent bonding between the *d*-orbitals of B-site ions and the neighboring oxygen *p*-orbitals. Therefore, the electronic configurations of cations must be considered in terms of occupancy of electrons in e_g and t_{2g} levels. Since the energy between the electronic configurations of different ions in the perovskite is closer than the free ions, Hund's rule must be applied on perovskite cation considering the ligand-field splitting of *d*-orbitals. Sometimes Hund's rule is not valid, especially if the number of *d* electrons is between 4 and 7, where the ligand-field splitting is greater than the intra-atomic exchange energy. For instance, in $LaMnO_3$, Mn^{3+} is a Jahn–Teller ion; as per Hund's rule, the ions exist in high-spin state t_{2g} (3*e*↑) and e_g (↑). Due to ligand-field splitting, the occupancy of the e_g state requires an energy equivalent to the ligand-field splitting. In Mn^{3+}, the ligand-field splitting energy is higher than the intra-atomic exchange energy; therefore, the low-spin configuration is more favored.

In perovskites, transition metals as the B-site cations with 3*d* shells can lead to multiple charge states for the metal cations. For example, $LaMnO_3$ has the La^{3+} charge state and Mn^{3+} with $3d^4$ configuration, and $CaMnO_3$ has Ca^{2+} charge state and Mn^{4+} with $3d^3$ configuration. In the 50–50 solid solutions of the above compounds as in $La_{0.5}Ca_{0.5}MnO_3$, both Mn^{3+} and Mn^{4+} ions are present (Coey 2004). On cooling, $La_{0.5}Ca_{0.5}MnO_3$ undergoes ferromagnetic conducting state at 220 K and insulating antiferromagnetic state near 160 K. The change in magnetic phase transformation and the corresponding change in resistivity for $La_{0.5}Ca_{0.5}MnO_{3-\delta}$ are shown in Figure 5.1 (Zhao et al. 2002). The above temperature-dependent transitions occur due to the charge ordering of Mn^{3+} and Mn^{4+} ions through the intervening oxygen ions.

Often in perovskite lattices, the cations exhibit localized electrons and spins; therefore, a long-range magnetic ordering takes place through spin–spin interactions. Such spin–spin interaction between the neighboring ions through the separating oxygen, but overlapping orbitals, is called superexchange, which is a mode of charge ordering in perovskites. Superexchange leads to antiferromagnetic or ferromagnetic ordering, depending on the coupling between nearest-neighbor (NN) cations. If the superexchange interaction takes place between the NN cations with antiparallel spin, then it is antiferromagnetic ordering and *vice versa* in ferromagnetic ordering, as shown in Figure 5.2a and b. Superexchange is prominent between two singly occupied orbitals of metals through a 180° B–O–B bond, and the electrons move back and forth from occupied to unoccupied states.

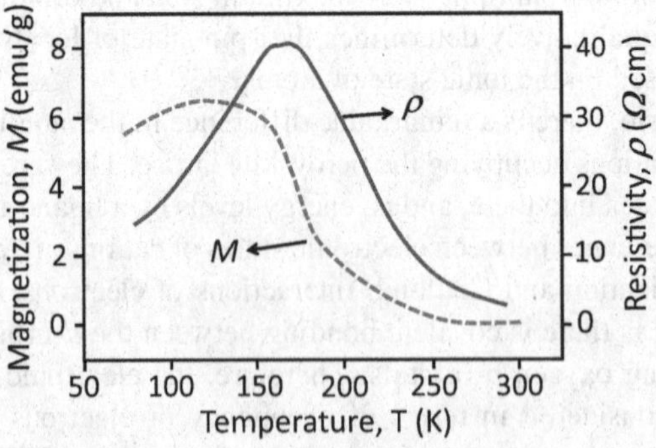

FIGURE 5.1 The dependence of temperature on magnetization and resistivity of $La_{0.5}Ca_{0.5}MnO_{3-\delta}$.

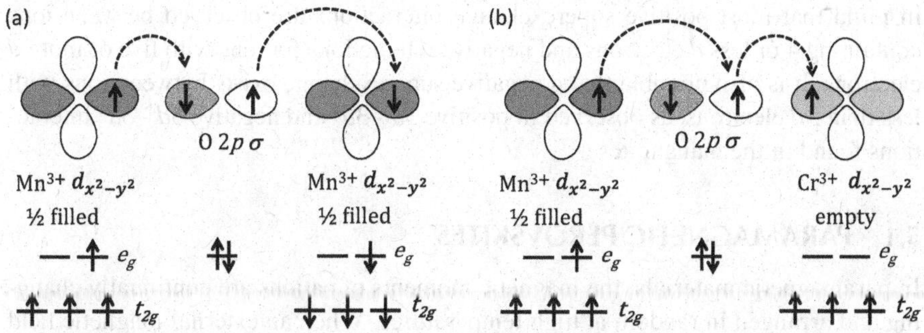

FIGURE 5.2 Superexchange interactions: (a) the covalent interaction through the O 2*p*-orbital leading to antiferromagnetic coupling and (b) Mn^{3+}–O–Cr^{3+} interaction leading to ferromagnetic coupling.

The change in the magnetic behavior of the perovskite with respect to temperature by superexchange mechanism can be predicted by the Goodenough–Kanamori–Anderson (GKA) rules (Kundu 2016). According to the GKA rules, if the angle of alignment of cation–anion–cation link is 180°, then two paramagnetic nearest neighboring ions order in an antiferromagnetic fashion. If the angle of alignment is 90°, then the resulting magnetic moment is ferromagnetic. A schematic of the possible alignment of Mn^{3+}/Mn^{4+} ions in perovskite resulting in antiferromagnetic and ferromagnetic ordering is shown in Figure 5.3. The superexchange model is applicable for both octahedral and tetrahedral coordinated cations and other coordinated polyhedrals such as square pyramids in brownmillerites and other crystal systems. Even though the superexchange is successful in explaining the phase change in magnetic materials, there is no actual electron transfer taking place. Finally, one must keep

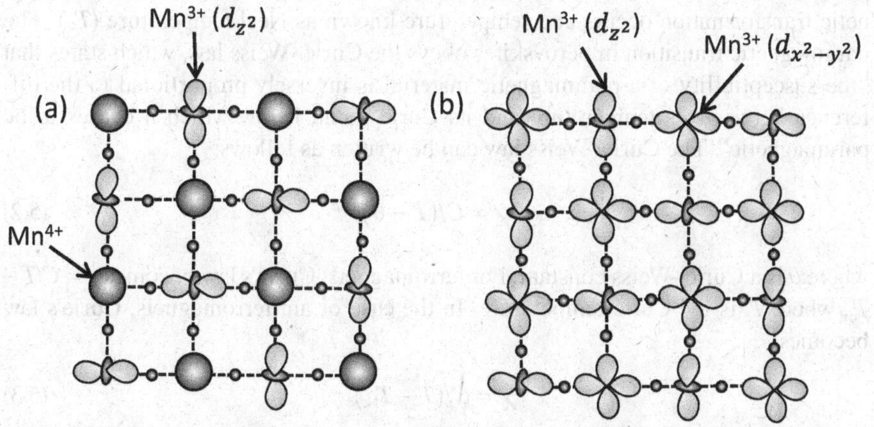

FIGURE 5.3 The alignment of orbitals for (a) antiferromagnetic and (b) ferromagnetic ordering.

in mind that most positive superexchange interactions are observed between ions containing 4 or less d electrons and negative interactions for ions with five or more d electrons. It is also possible that a negative superexchange exists between ions with less than 5 d electrons, as observed in positive $3d^3$–$3d^4$ and negative $3d^4$–$3d^4$ interactions found in the manganites.

5.1 PARAMAGNETIC PEROVSKITES

In paramagnetic materials, the magnetic moments of cations are continually changing and arranged in random at high temperatures. When an external magnetic field H is applied, the magnetic dipoles in the paramagnetic phases will try to align themselves parallel to the magnetic flux of perovskite; nonetheless, the thermal agitation opposes it. The paramagnetic susceptibility (χ) is a measure of the ability of a material to be magnetized in an applied magnetic field, which is the ratio of magnetization M to the applied magnetic field intensity H, $\chi = M/H$. The χ of paramagnetic materials is a temperature-dependent parameter, which is governed by Curie's law. According to Curie's law, the magnetization in a paramagnetic material is directly proportional to the applied magnetic field and perceived as inversely proportional to the temperature if the material is heated. A representative plot of temperature dependence on susceptibility and inverse susceptibility of a paramagnetic material is shown in Figure 5.4a, in comparison to a ferromagnetic (Figure 5.4b) and antiferromagnetic (Figure 5.4c) materials.

According to Curie's law, the magnetic susceptibility χ is

$$\chi = C/T \tag{5.1}$$

where C is a Curie constant and T is the temperature.

Perovskite oxides exhibit transformation from a magnetic phase to another with a change in temperature. The transformation from ferromagnetic to paramagnetic perovskites occurs at Curie temperature (T_c) and antiferromagnetic to paramagnetic transformation occurs at a temperature known as Néel temperature (T_N). The paramagnetic transition in perovskites obeys the Curie–Weiss law, which states that "the susceptibility of a paramagnetic material is inversely proportional to the difference between its temperature and its Curie point, below which it ceases to be paramagnetic". The Curie–Weiss law can be written as follows:

$$\chi = C/(T - \theta) \tag{5.2}$$

where θ is a Curie–Weiss constant. For ferromagnets, Curie's law becomes $\chi = C/T - T_c$, where T_c is the Curie temperature. In the case of antiferromagnets, Curie's law becomes

$$\chi = C/(T - T_N) \tag{5.3}$$

where T_N is the Néel temperature.

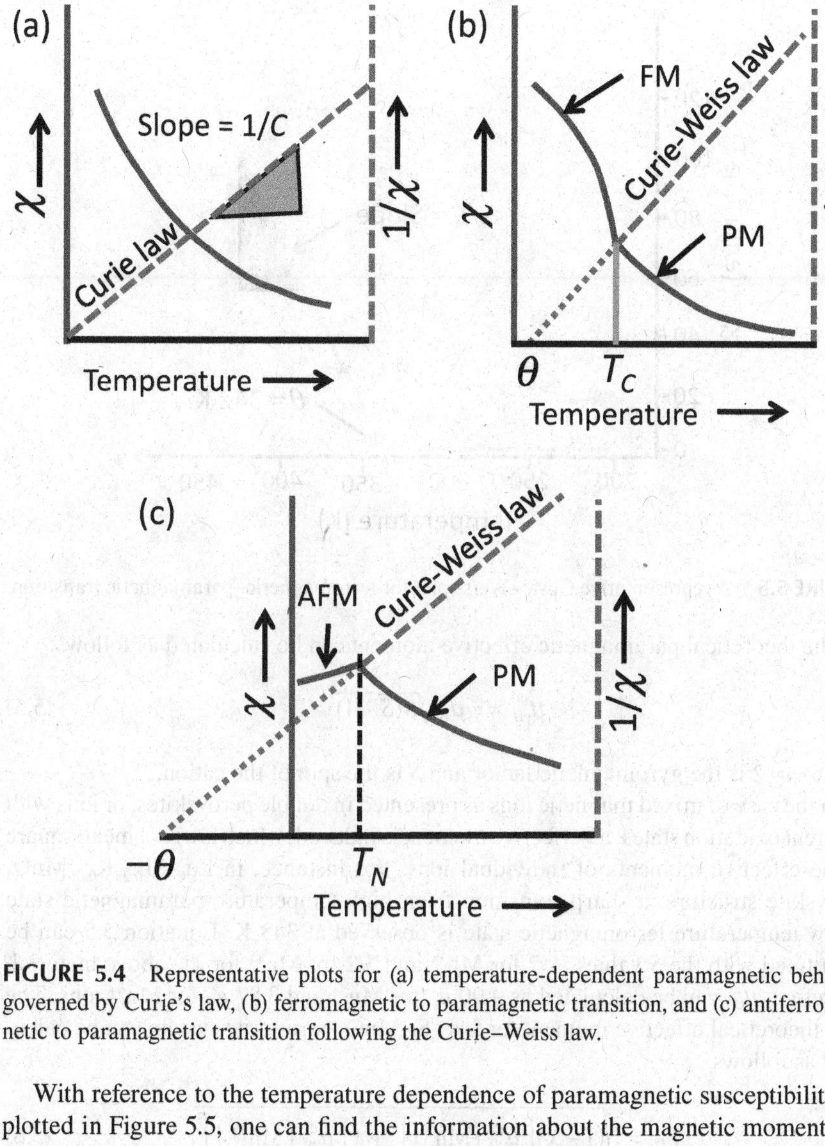

FIGURE 5.4 Representative plots for (a) temperature-dependent paramagnetic behavior governed by Curie's law, (b) ferromagnetic to paramagnetic transition, and (c) antiferromagnetic to paramagnetic transition following the Curie–Weiss law.

With reference to the temperature dependence of paramagnetic susceptibility, as plotted in Figure 5.5, one can find the information about the magnetic moment and state of the cations by using Curie's law. The molar Curie constant (C_m) is related to the effective paramagnetic moment as follows:

$$C_m = \frac{N_A \mu_B^2}{3k_B} \mu_{eff}^2 \tag{5.4}$$

where N_A is Avogadro's constant (6.023×10^{23} mol^{-1}), μ_B is the Bohr magneton (9.274×10^{-21} emu), k_B is the Boltzmann constant (1.38016×10^{-16} erg K^{-1}), and μ_{eff} is the effective paramagnetic moment.

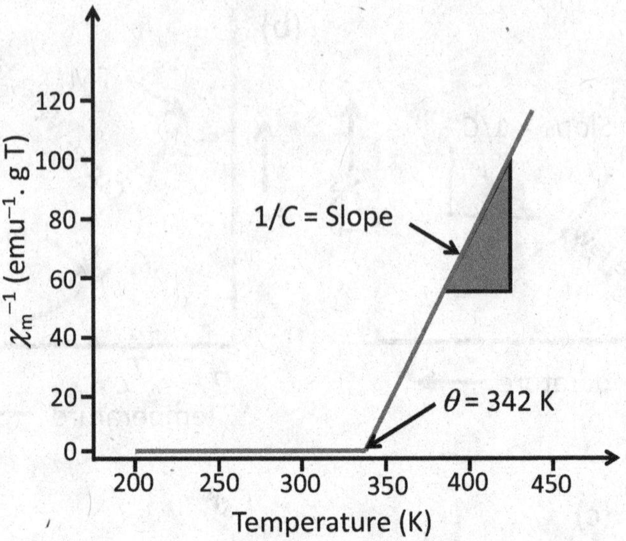

FIGURE 5.5 A representative Curie–Weiss plot for ferromagnetic–paramagnetic transition.

The theoretical paramagnetic effective moment can be calculated as follows:

$$\mu_{\text{eff}}^{\text{th}} = g\mu_{\text{B}}\sqrt{S(S+1)} \tag{5.5}$$

where $g = 2$ is the gyromagnetic factor and S is the spin of the cation.

In the case of mixed magnetic ions as presented in double perovskites, or ions with different oxidation states, an effective moment is induced, which is a root mean square of the effective moments of individual ions. For instance, in $La_{0.7}Sr_{0.25}K_{0.05}MnO_3$ perovskite structure, a sharp transition from high-temperature paramagnetic state to low-temperature ferromagnetic state is observed at 343 K. Equation 5.5 can be substituted with the S values as 2 for Mn^{3+} and 3/2 for Mn^{4+} for the above material. Therefore, $\mu_{\text{eff}}^{\text{th}}$ can be calculated as 4.90 μ_{B} for Mn^{3+} and 3.87 μ_{B} for Mn^{4+} ions. The total theoretical effective moment for $La_{1-x}Sr_xMnO_3$ perovskite system can be calculated as follows:

$$\mu_{\text{eff}} = \sqrt{(1-x)\left[\mu_{\text{eff}}^{\text{th}}\left(\text{Mn}^{3}\right)\right]^2 + x\left[\mu_{\text{eff}}^{\text{th}}\left(\text{Mn}^{4+}\right)\right]^2} \tag{5.6}$$

where x is the fraction of Mn^{4+} ions.

One can also estimate the magnetic moments from the Curie–Weiss plot (with reference to the representative plot as shown in Figure 5.5) which can be obtained using the following equation (Jonker and Van Santen 1953):

$$\mu_{\text{eff}} = 2.84\sqrt{\chi_{\text{m}}(T-\theta)}\mu_{\text{B}} = 2.84\sqrt{C\mu_{\text{B}}} \tag{5.7}$$

where χ_{m} is χ per mole and μ_{B} is the Bohr magneton.

Since for ferromagnets, $\chi_m = C/T - T_c$. In Figure 5.5, T_c is 342 K:

$$C = \frac{\Delta T}{\Delta(1/\chi)} = \frac{450-342}{120} = 0.9 \text{ g K m}^{-3} \tag{5.8}$$

$$\mu_{\text{eff}} = 2.84\sqrt{C} = 2.69\mu_B \tag{5.9}$$

By knowing the moments of the cations from the Curie–Weiss plots, one can estimate the average valence of the cations in different oxidation states, since the magnetic moments are constant for ions with a fixed valence. For instance, the valence of Tb ions in superconducting $La_{2-x-y}Tb_yBa_xCuO_4$ is Tb^{3+} and Tb^{4+} (Bao et al. 1995). The contribution to susceptibility from Tb ions is increasing with the concentration of Tb in the lattice at a constant temperature, due to the change in the valence of Tb ions. To calculate the valence, initially, one has to find the effective magnetic moment using Eq. 5.4. Then, the theoretical values of moments must be calculated using Eq. 5.5. In order to evaluate the fraction of Tb^{3+} ions, Eq. 5.4 is modified as

$$C_m = \frac{N_A \mu_B^2}{3k_B}\left(x\mu_{\text{eff}}^2\left(Tb^{3+}\right) + (1-x)\mu_{\text{eff}}^2\left(Tb^{4+}\right)\right) \tag{5.10}$$

where x is the total fraction of Tb ions in the lattice.

Finally, the average valence can be estimated as

$$v = 3x + 4(1-x) \tag{5.11}$$

This method can be extended to estimate the nominal valences of cations present in paramagnetic perovskites. For example, $SrCrO_3$ prepared at high temperature and pressure displays cubic structure and shows an effective magnetic moment of the cations as 2.98 μ_B estimated from the plot of $1/\chi$ versus T at above 150 K. This value is in accordance with the expected spin-only value for Cr^{4+} (d^2) cations. Another example is Mn cation in $AA'_3B_4O_{12}$ structure, and the effective magnetic moment calculated from the inverse susceptibility plot is 5.51 μ_B. However, the theoretical value for high-spin Mn^{3+} cation is 4.90 μ_B. This difference indicates that some Mn^{3+} cations disproportionate into Mn^{4+} and Mn^{2+} ions and the spin-only magnetic moment of Mn^{2+} cations is 5.92 μ_B.

The Curie temperature of the perovskite materials is dependent on the crystal structure which can be realized in the case of double perovskites with the general formula $A_2B'B''O_6$ (A = Ca, Sr, and Ba; B′ = Mn and Cr; and B″ = Mo, Re, and W). A perovskite with a more symmetrical structure, such as cubic or tetragonal phases, exhibits a higher T_c than an orthorhombic or hexagonal phase. In general, with reference to Figure 5.6, the highest Curie temperatures are observed for the systems having a tolerance factor, $t \approx 1$. As the symmetry lowers, the ferromagnetic–paramagnetic interaction is significantly suppressed (Philipp et al. 2003).

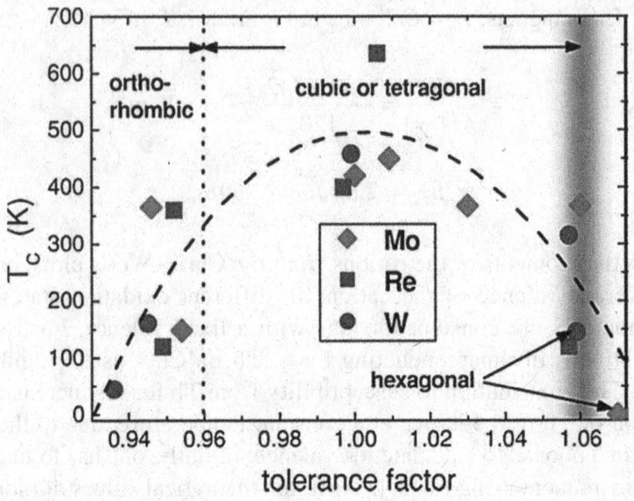

FIGURE 5.6 Curie temperatures of different double-perovskite materials versus their tolerance factors, reproduced with permission of the American Physical Society (Philipp et al. 2003).

5.2 ANTIFERROMAGNETIC PEROVSKITES

The majority of the perovskites are antiferromagnetic (magnetic moments align antiparallel to the neighboring moments), and ideal cubic perovskites are more likely to exhibit this type of magnetic behavior. The ideal cubic phase contains B-site cations inside the apex-shared octahedra and A-site cations in between the octahedra; this arrangement is more favorable for the superexchange mechanism. The lanthanoid perovskite manganites, $LnMnO_3$ (Ln = La, Pr, and Nd), are insulators at normal-temperature and high-temperature regimes. However, they undergo the antiferromagnetic transition at a low critical temperature, known as the Néel temperature, T_N. The above antiferromagnetic ordering is A-type, in which the ferromagnetically aligned *ab*-layers are coupled with a ferromagnetically aligned neighboring layer in the opposite direction along the *c*-axis, as shown in Figure 5.7. Since Mn^{3+} is often subjected to the Jahn–Teller distortion, an anisotropic magnetic exchange interaction arises in the above perovskites due to the presence of long and short Mn–O distances. At high temperatures, due to the lattice expansion, the Jahn–Teller distortions disappear, and at the same time destroying the antiferromagnetic order.

Antiferromagnetic ordering is classified into four basic types: A, C, E, and G types. The subscripts *x*, *y*, or *z* represents the orientation of spin corresponding to *a*-, *b*-, or *c*-axis in the unit cell. A-type ordering is frequently observed in ideal cubic perovskites with antiferromagnetic ordering along $(001)_p$ plane, and the magnetic moments in $(001)_p$ planes are aligned opposite to each other. Therefore, the B-site cation is surrounded by neighbors with four parallel spins and two antiparallel spins. C-type ordering is similar to A-type but along $(110)_p$ plane. The E-type order comprises a more complex arrangement of cation ordering. In this ordering, each cation has two neighbors with parallel and four with antiparallel spins. G-type ordering has

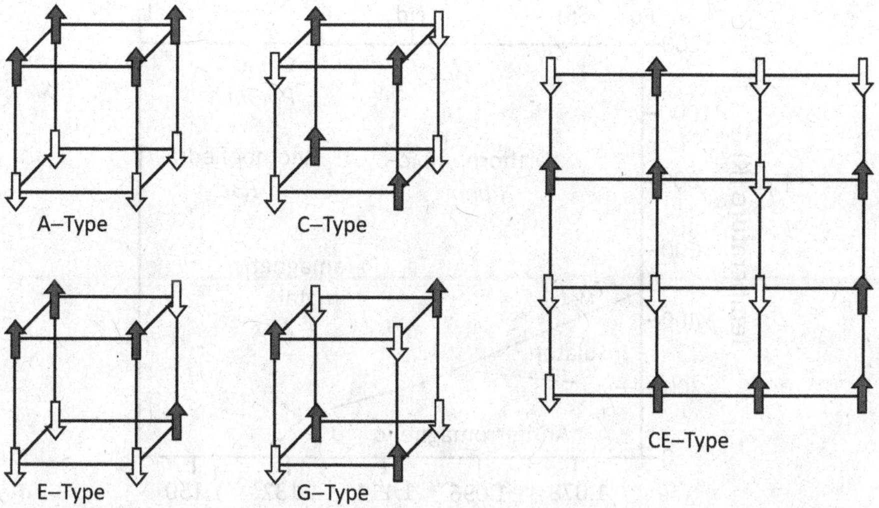

FIGURE 5.7 Antiferromagnetic ordering classified into A, C, E, G, and CE types.

ferromagnetic ordering along with $(111)_p$ and each cation surrounded by six antiparallel spins.

$BiCoO_3$ shows C-type antiferromagnetic ordering below Néel temperature T_N ~470 K, comprising Bi^{3+} and Co^{3+} cations in a high spin with $S = 2$, and coordinating with four oxygen atoms forming a square pyramid. Above the Néel temperature, this perovskite structure adopts a tetragonal symmetry belonging to the space group *P*4*mm*. Below the Néel temperature, the C-type antiferromagnetic ordering of Co^{3+} cations doubles the *a*- and *b*-axis parameters; hence, it becomes a trigonal unit cell. An example of G-type antiferromagnetic order is $CaMnO_3$ phase with T_N ~125 K. The most complex ordering is CE, which is observed in manganite perovskite with the composition $La_{0.5}A_{0.5}MnO_3$ (A = alkaline earth metal) (Figure 5.7). In these perovskites, Mn exists as Mn^{3+} and Mn^{4+} ions, and interactions between them result in antiferromagnetic ordering, which consists of both C- and E-type alignments.

The structural, magnetic, and electronic behaviors of perovskites are dependent on the A-site cations as well. For instance, $LnNiO_3$ (Ln = lanthanides) perovskites exhibit a significant change in the physical properties with a decrease in the ionic size of the lanthanide ions, as shown in the phase diagram in Figure 5.8. The parameter that controls the metal-insulator transition temperature T_{M-I}, the Néel temperature T_N, and the stability of the different crystallographic structures is Ni–O–Ni superexchange angle θ. With respect to the Ni–O–Ni angle, the degree of overlap of O 2*p*- and Ni 3*d*-orbitals increases as moving from Lu ($r_a = 0.977$ Å) to La ($r_a = 1.160$ Å) in the periodic table. The increase in overlap of orbitals, results in enhancement of the metallic state (T_{M-I} decreases), as well as the antiferromagnetic state (T_N increases). In the case of $NdNiO_3$ and $PrNiO_3$, T_N is close to T_{M-I}; therefore, these materials remain metallic at all the temperature regimes. The representative antiferromagnetic ordering of $LnNiO_3$ perovskites is shown in Figure 5.9. The magnetic behavior of $LnNiO_3$ depends on the magnetic nature of A-site ions as well. For instance, in the

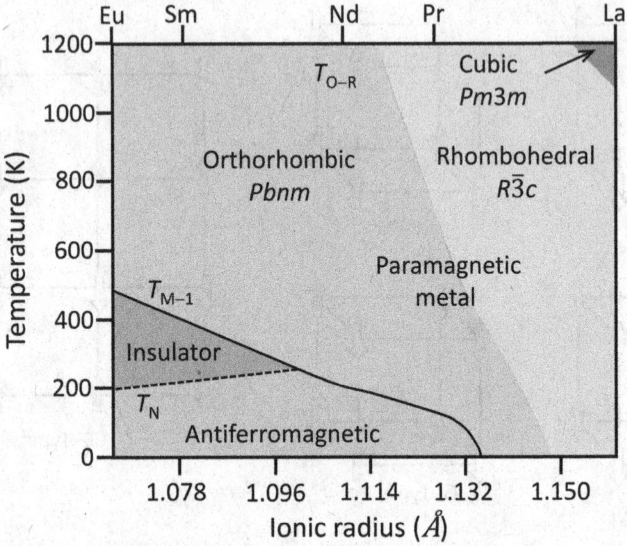

FIGURE 5.8 Phase diagram for the $RNiO_3$ perovskites, reproduced with permission of the Royal Society of Chemistry (Medarde 1997).

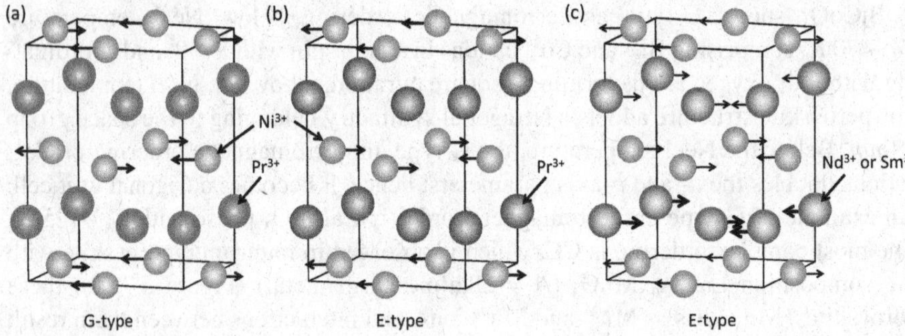

FIGURE 5.9 (a) G-type and (b) E-type orientations of antiferromagnetic ordering in $PrNiO_3$, and (c) E-type ordering in $NdNiO_3$ and $SmNiO_3$.

case of $NdNiO_3$, due to E_x-type arrangement of the Ni magnetic moments, Nd^{3+} ions exhibit two different values for the magnetic moments, as shown in Figure 5.9c as large and small arrows, which are 2 μ_B ("large" Nd) and 0.8 μ_B ("small" Nd) measured at 200 mK (Medarde 1997). Therefore, these compounds exhibit two T_N values.

The series of lanthanide perovskites, $LnCoO_3$, $LnFeO_3$, $LnVO_3$, $LnMnO_3$, and $LnTiO_3$, exhibit G-type antiferromagnetic ordering in general, and the magnetic properties are rare earth ionic size-dependent moving from La to Lu.

The magnetic properties of $A_2B'B''O_6$ perovskites are interesting, because various combinations of paramagnetic cations are possible at the three cation sites, resulting in unique magnetic properties. Most of the double perovskites exhibit antiferromagnetic properties at very high Curie temperatures (T_c). When the $A_2B'B''O_6$ forms an ideal cubic lattice, three types of ordering are observed, I, II and III, as shown in Figure 5.10.

Type I is favored when the 90° B′–O–B″–O–B′ NN interactions are stronger than the 180° B′–O–B″–O–B′ (NNN) interactions. As a result, eight of the twelve NN cations are ordered antiferromagnetic, while NNN cations all ordered ferromagnetically, which is commonly observed in double perovskites with a B′/B″ combination of a rare earth and a *4d*/*5d* element. Type II ordering observed in perovskite with NNN interactions is stronger than NN interactions; therefore, all six NNN cations are ordered antiferromagnetic. Type II ordering is frequently observed in perovskites with two transition metals in the B-site. Type III ordering takes place if both NN is slightly stronger than NNN interactions; as a result, eight of the twelve NN cations are ordered antiferromagnetic, and two of the six NNN cations are ordered antiferromagnetic. The above ordering is dependent on the site symmetry as the NN and NNN interactions become strong or weak with respect to the site symmetry. For instance, if the A-site is occupied by a large cation, then the symmetry of the lattice decreases due to the octahedral tilting. In A_2LaRuO_6, the magnetic ordering changes from type I to type III when A-site atom changes from Ca to Ba, during which the NN interactions are weakened relative to the NNN interactions (Battle et al. 1983). The magnetic behavior of type I antiferromagnetic ordering of Ba_2PrRuO_6 is shown in Figure 5.11 (Izumiyama et al. 2001).

Similarly, in the case of $A_2B'B'O_6$ perovskites, a distinguishable T_N can be observed in perovskites with different B′-site elements. For instance, perovskites with the general formula A_2LnRuO_6 (A = Ca, Sr, and Ba; and Ln = rare earths) exhibit different T_N values for every exchange in the A or Ln ions. Table 5.1 compiles

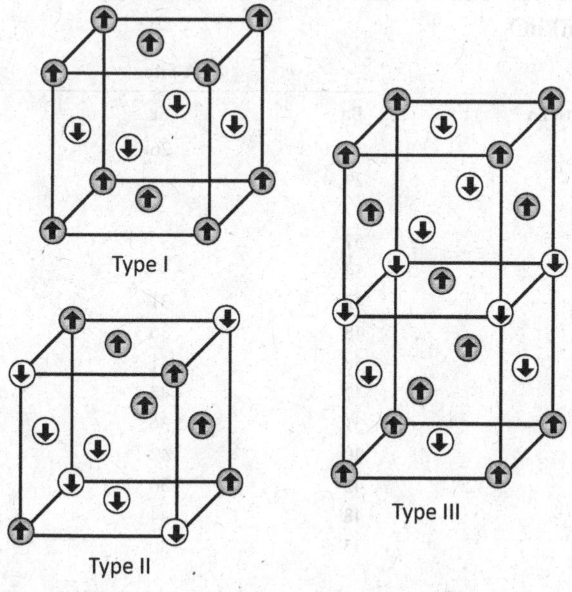

FIGURE 5.10 Common antiferromagnetic orderings found in the $A_2B'B''O_6$ perovskites: (a) type I, (b) type II, and (c) type III (only magnetic ions are shown in the figure).

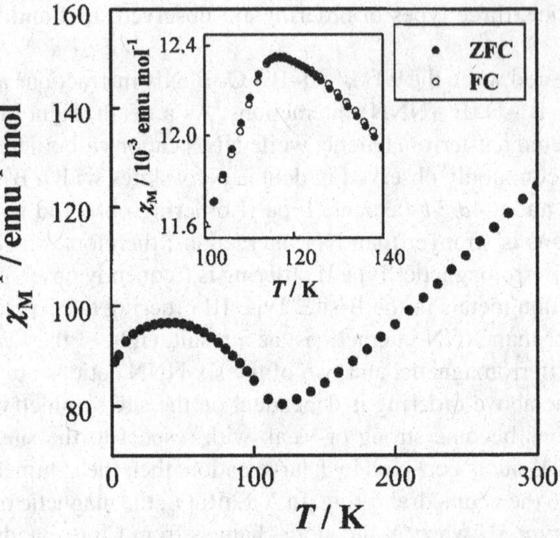

FIGURE 5.11 The reciprocal magnetic susceptibilities of Ba_2PrRuO_6 against temperature. Inset shows the temperature dependence of the magnetic susceptibilities, where the ZFC (filled symbols) and FC (open symbols) are molar magnetic susceptibilities in the applied field of 0.1 T, reproduced with permission of the IOP Publishing (Izumiyama et al. 2001).

TABLE 5.1
The Néel Temperatures for Double Perovskites A_2LnRuO_6

	A-Site		
B′-Site Ln	**Ba**	**Sr**	**Ca**
Y	37	26	14
La	29.5	-	11.5
Pr	117	-	10
Nd	57	-	11.5
Sm	54	-	10
Eu	42	31	18
Gd	48	36	10
Tb	58	41	11
Dy	47	44	11
Ho	51	36	12
Er	40	42	16
Tm	42	36	17
Yb	48	44	34
Lu	35	30	14

Reproduced with Permission of the American Institute of Physics (Sakai et al. 2005).

the influence of A (Ba, Sr, and Ca) and rare earth ions on the antiferromagnetic transition temperatures (T_N). One can distinguish that if the A-site ion is large, as in the case of Ba, T_N is high. For smaller cations, here Ca, the T_N significantly drops.

Assume a double perovskite with a single magnetic cation; then, the magnetic behavior of the perovskite can be compared to that of a simple perovskite. In such a case, the antisite disorder (a defect that affects the spin polarization and the Néel temperature) between two B-cations can complicate the resultant magnetic ordering. The Sr_2YRuO_6 shows type I antiferromagnetic behavior below $T_N = 26$ K, which has Ru^{5+} with the paramagnetic nature. In the case of two paramagnetic B-site ions, the magnetic ordering of B′ and B″ cations can take place either at the same temperature or at different temperatures. For instance, A_2LnRuO_6 (A = Sr and Ba; and Ln-paramagnetic rare earth) double perovskites show two interpenetrating type I antiferromagnetic ordering, whereas Sr_2CoOsO_6 shows two interpenetrating type II antiferromagnetic ordering.

The double perovskites with anion vacancy commonly display G_z-type antiferromagnetic ordering below T_N. In the case of brownmillerite with a single magnetic ion as in the case of Sr_2MnGaO_5, the magnetic ordering takes place by the superexchange interactions between two consecutive MnO_2 layers along the *b* direction, even though the Mn atoms along the *b*-axis are approximately displaced twice as compared to the *ac* plane (Figure 5.12). The magnetic structure of $Sr_2MnGaO_{5.5}$ is also antiferromagnetic; however, it is different from Sr_2MnGaO_5. In $Sr_2MnGaO_{5.5}$, the MnO_2 layers (in which manganese atoms are ordered antiferromagnetically) are coupled ferromagnetically, which corresponds to C-type magnetic structures (Figure 5.12b). The change in the magnetic structure from Sr_2MnGaO_5 to $Sr_2MnGaO_{5.5}$ occurs due to the appearance of extra oxygen atoms in the Ga layers, which additionally stabilize

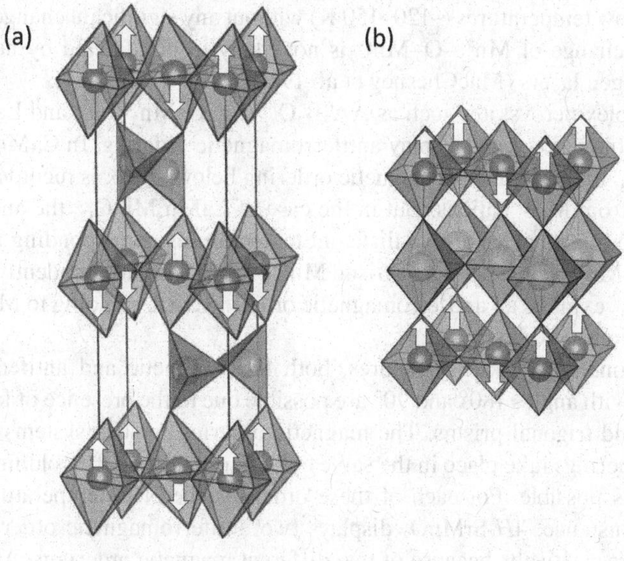

FIGURE 5.12 The magnetic structure of Sr_2MnGaO_5 (G-type) and Sr_2MnGaO_5 (C-type).

the C-type magnetic structure due to more distant diagonal superexchange interactions through the GaO_6 octahedra (Abakumov et al. 2004).

Brownmillerites $A_2Fe_2O_5$ (A = Ca and Sr) with single magnetic Fe cation show G_z-type antiferromagnetic ordering at all temperatures. Since Fe^{3+} ions are situated in octahedral and tetrahedral environments in $A_2Fe_2O_5$ (A = Ca and Sr), the net magnetic moments are the sum of spins arising from the Fe^{3+} ions with octahedral (Fe_I) and tetrahedral (Fe_{II}) coordination with the oxygen atoms. The existence of covalent spin transfer between Fe^{3+} ions in the different coordination environments through the oxygen atoms results in a high Néel temperature, T_N ~700 and 720 K, when A = Sr and Ca, respectively (Takeda et al. 1968; Greaves et al. 1975). Brownmillerites with two magnetic cations display temperature-dependent G_y-type and G_x-type magnetic ordering. Ca_2FeMnO_5 exhibits G_y-type antiferromagnetic ordering up to a temperature of 407 K and G_x-type ordering in between 407 and 500 K, and above 500 K, which falls between the antiferromagnetic ordering of the corresponding pure phases $Ca_2Fe_2O_5$ and $Ca_2Mn_2O_5$, with T_N at 725 and 350 K, respectively (Ganguli and Gopalakrishnan 1986). Ca_2FeMnO_5 is paramagnetic (Rykov et al. 2008). The variation in the ratio of Fe and Mn in Ca_2FeMnO_5 perovskite influences the antiferromagnetic ordering transition temperature. In the presence of Mn and Fe, the additional Mn^{3+}–O_2–Fe^{3+} interaction in the solid solutions is likely to lower the Néel temperature, since one can observe a change in T_N as the ratio of Mn/Fe is varied as in $Ca_2Fe_{1.5}Mn_{0.5}O_5$ which shows the G_y-type ordering below 400 K and G_y-type ordering above 465 K. However, $Ca_2Fe_{1.33}Mn_{0.67}O_5$ displays only G_x-type antiferromagnetic ordering at all the temperatures.

In the case of layered perovskite structures, the nonmagnetic A-site-oxygen layers are likely to disrupt the magnetic exchange of the B^{n+}–O–B^{n+} ions, and therefore one can expect a reduced T_N as the number of layers increases. On the contrary, the series of $CaMnO_3$, $Ca_4Mn_3O_{10}$, $Ca_3Mn_2O_7$, and Ca_2MnO_4 exhibit antiferromagnetic ordering at low temperatures (~120–150 K) without any significant change in T_N. The magnetic exchange of Mn^{4+}–O–Mn^{4+} is not likely to be affected by nonmagnetic calcium-oxygen layers (MacChesney et al. 1967).

The complex perovskites such as $AA'_3B_4O_{12}$ (e.g. $CaMn_3V_4O_{12}$ and $LaMn_3V_4O_{12}$) family of cubic perovskites display antiferromagnetic ordering. In $CaMn_3V_4O_{12}$ and $LaMn_3V_4O_{12}$, the low antiferromagnetic ordering below ~54 K is recorded, which is originating from Mn^{2+} cations. But in the case of $LaMn_3Mn_4O_{12}$, the antiferromagnetic ordering takes place at two different temperatures corresponding to the magnetic A′-site Mn^{3+} (T_N = 21 K) and B-site Mn^{4+} (T_N = 78 K) independently. Similarly, $LaMn_3Cr_4O_{12}$ exhibits an antiferromagnetic ordering corresponding to Mn^{3+} (~50 K) and Cr^{3+} cations (150 K).

In hexagonal perovskite structures, both ferromagnetic and antiferromagnetic alignments with angles 180° and 90° are possible due to the presence of face-sharing octahedra and trigonal prisms. The magnetic ordering in such systems is complex since the orderings take place in the same column or between the columns, or A-site polyhedral is possible. For each of these ordering, the Néel temperatures are different. For instance, 4*H*-$SrMnO_3$ displays two antiferromagnetic orderings at 270 and 350 K, respectively, because of two different magnetic orderings. At 270 K, the magnetic order is due to coupling of face-centered units via apex-sharing octahedra, whereas at 370 K, the transition is due to aligning magnetic moments of face-sharing octahedra. The net magnetic moment is a cumulative of different magnetic ordering discussed above. As a result, the resultant magnetic moment in 2H-$BaCoO_3$ is antiferromagnetic, since the ferromagnetic superexchange geometry of columns (inter-columnar) contributes less than the antiferromagnetic ordering from adjacent columns (intra-columnar). Overall, the magnetic ordering of perovskite structures is sensitive to minute changes in the crystal structures, such as symmetry, sizes of cations, cation or anion deficiency, and distortions of BO_6 octahedra, as we have seen in the magnetic ordering in anion-deficient structures, due to changes in the stoichiometry of ions.

5.3 FERRIMAGNETIC PEROVSKITES

Ferrimagnetic (FiM) ordering is similar to antiferromagnetic ordering, but the magnetic moments are unbalanced in ferrimagnets due to the difference in magnetic moments, which is present in all the directions. The ferrimagnets exhibit a weak ferromagnetic behavior due to the unbalance moments; therefore, the ferrimagnetic ordering is often confused with weak ferromagnetic ordering. A comparison of magnetic susceptibility curves of para-, ferro-, antiferro-, and ferrimagnetic ordering with temperature is shown in Figure 5.13. Ferrimagnetic ordering is observed in complex perovskite structures, due to the interactions between the multiple magnetic cations in their lattice. Y_2CoRuO_6 with a combination of 3*d* and 4*d* metals at *B*-sites, with Co–O–Ru angles (141.7°) significantly deviated from ideal 180°, results in a robust Co↑–O–Ru↓ antiferromagnetic interactions with a FiM Curie temperature (T_c) close to 82 K. In Y_2CoRuO_6, the magnetic moments of Co atoms (1.71 μ_B) are aligned opposite to the magnetic moments of the Ru (0.51 μ_B) ions. The two antiparallel

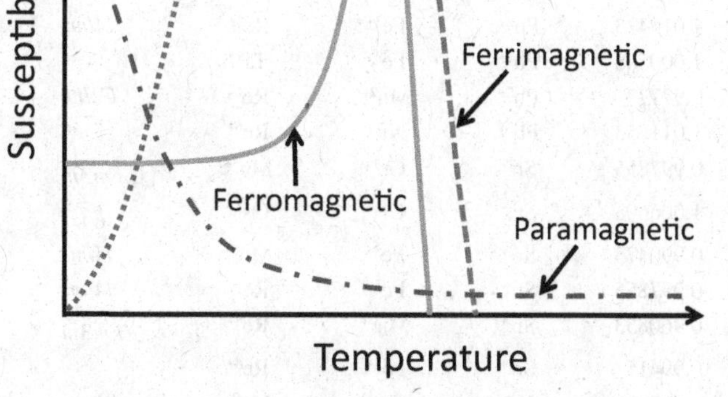

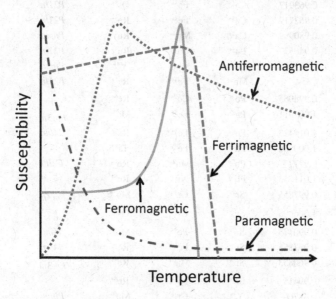

FIGURE 5.13 Comparison of different magnetic orderings with respect to temperature.

ferromagnetic moments of Co and Ru spins stacked along [101] planes are with different magnitudes, resulting in ferrimagnetic behavior (Deng et al. 2018). More representative, ferrimagnetically ordered double perovskites are listed in Table 5.2.

Complex perovskites with the general formula $AA'_3B_4O_{12}$ are also potential candidates for ferrimagnetic ordering because of possible multiple cations in their lattice. $ACu_3Fe_4O_{12}$ perovskites with A = Ca, Sr, or Ba exhibit ferromagnetic ordering. Below the Curie temperature ~210 K, $CaCu_3Fe_4O_{12}$ exhibits ferrimagnetic behavior

TABLE 5.2
A List of Double Perovskites That Exhibit Ferrimagnetic Ordering (Vasala and Karppinen 2015)

Formula	Tolerance Factor, *t*	A Cation	B′ Cation	B″ Cation	Space Group	T_C/T_N (K)
Ba_2FeMoO_6	1.020811	Ba^{2+}	Fe^{2+}	Mo^{6+}	$Fm\bar{3}m$	308
Ba_2FeReO_6	1.057586	Ba^{2+}	Fe^{3+}	Re^{5+}	$Fm\bar{3}m$	334
Ba_2FeUO_6	1.012315	Ba^{2+}	Fe^{3+}	U^{5+}	$Fm\bar{3}m$	120
Ba_2MnReO_6	1.018369	Ba^{2+}	Mn^{2+}	Re^{6+}	$Fm\bar{3}m$	120
Ba_2MnUO_6	0.976326	Ba^{2+}	Mn^{2+}	U^{6+}	$Fm\bar{3}m$	55
Ba_2NiReO_6	1.053659	Ba^{2+}	Ni^{2+}	Re^{6+}	$Fm\bar{3}m$	18
Ca_2CoOsO_6	0.947419	Ca^{2+}	Co^{2+}	Os^{6+}	*P*21/*n*	145
Ca_2CrReO_6	0.969949	Ca^{2+}	Cr^{3+}	Re^{5+}	*P*21/*n*	360
Ca_2CrWO_6	0.960333	Ca^{2+}	Cr^{3+}	W^{5+}	*P*21/*n*	161
Ca_2FeMoO_6	0.929243	Ca^{2+}	Fe^{2+}	Mo^{6+}	*P*21/*n*	380
Ca_2FeOsO_6	0.963917	Ca^{2+}	Fe^{3+}	Os^{5+}	*P*21/*n*	320
Ca_2FeReO_6	0.962719	Ca^{2+}	Fe^{3+}	Re^{5+}	*P*21/*n*	540
Ca_2MnRuO_6	0.980999	Ca^{2+}	Mn^{4+}	Ru^{4+}	*Pnma*	230
La_2CuRuO_6	0.940537	La^{3+}	Cu^{2+}	Ru^{4+}	*P*21/*n*	16
La_2MnVO_6	0.927133	La^{3+}	Mn^{2+}	V^{4+}	*Pnma*	20/80
Mn_2FeReO_6		Mn^{2+}	Fe^{3+}	Re^{5+}	*P*21/*n*	520
Pb_2CoReO_6	0.998065	Pb^{2+}	Co^{2+}	Re^{6+}	-	<77
Pb_2FeMoO_6	1.007911	Pb^{2+}	Fe^{3+}	Mo^{5+}	$Fm\bar{3}m$	272
Pb_2FeReO_6	1.015423	Pb^{2+}	Fe^{3+}	Re^{5+}	*I*4/*m*	420
Pb_2FeTaO_6	1.000508	Pb^{2+}	Fe^{3+}	Ta^{5+}	-	133
Pb_2MnReO_6	0.97777	Pb^{2+}	Mn^{2+}	Re^{6+}	*C*2/*m*	90
Pb_2NiReO_6	1.011653	Pb^{2+}	Ni^{2+}	Re^{6+}	-	<77
Sr_2CrMoO_6	0.997855	Sr^{2+}	Cr^{3+}	Mo^{5+}	$Fm\bar{3}m$	450
Sr_2CrOsO_6	1.006608	Sr^{2+}	Cr^{3+}	Os^{5+}	$R\bar{3}$	725
Sr_2FeMoO_6	0.990473	Sr^{2+}	Fe^{3+}	Mo^{5+}	*I*4/*m*	450
Sr_2FeReO_6	0.997855	Sr^{2+}	Fe^{3+}	Re^{5+}	*I*4/*m*	445
Sr_2MnReO_6	0.960853	Sr^{2+}	Mn^{2+}	Re^{6+}	$Fm\bar{3}m$	120
Sr_2NiReO_6	0.99415	Sr^{2+}	Ni^{2+}	Re^{6+}	-	17
Y_2CrMnO_6	0.48766	Y^{3+}	Cr^{3+}	Mn^{3+}	*Pbnm*	75

because of the disproportion of Fe^{4+} to Fe^{3+}/Fe^{5+} at low temperatures with the spin arrangement of [Cu_3^{2+} (↓) Fe_2^{3+} (↑) Fe_2^{5+} (↑)]. This disproportion disappears at a high temperature with a nominal valence of Fe^{4+}. $CaCu_3Mn_4O_{12}$, $YCu_3Fe_4O_{12}$, $LnCu_3Fe_4O_{12}$, and $CaCu_3Fe_2Nb_2O_{12}$ are a few other examples that exhibit ferromagnetic behavior following the above principle. In general, in all the above compounds, it is assumed that the ferrimagnetic ordering takes place through antiferromagnetic ordering, because of the intersite charge transfer. The representative governing equations for $ACu_3Fe_4O_{12}$ are presented in Eqs. 5.12 and 5.13.

$$3Cu^{2+} + 4Fe^{3.75+} \rightarrow 3Cu^{3+} + 4Fe^{3+} \quad (5.12)$$

$$8Fe^{3.75+} \rightarrow 5Fe^{3+} + 3Fe^{5+} \quad (5.13)$$

5.4 FERROMAGNETIC PEROVSKITES

Ferromagnetic perovskites are characterized by the magnetic moments aligned to one direction; therefore, the atoms or ions in ferromagnets have permanent magnetic moments. The magnetic moment of an atom comes from its electrons, and the interatomic forces keep the magnetic moments of many atoms parallel to each other. In many perovskites, the ferromagnetic ordering is disordered by thermal agitation, during which the moments of neighboring atoms are canceled each other. Several perovskites with ferromagnetic cations in the B-site, such as Co, Ni, Fe, and Ru, display ferromagnetism below the Curie temperature. $SrCoO_3$ is one such example that shows metallic and ferromagnetic behaviors at T_c ~305 K, and above this temperature, it displays paramagnetic property.

The ferromagnetic properties can be induced by introducing the low-valent cations to the lattice site. $LaMnO_3$ is an antiferromagnetic perovskite at all the temperatures which transforms to ferromagnets by the introduction of Ca or Sr elements in the lattice site of La, as presented in $La_{1-x}A_xMnO_3$ (A = Ca or Sr). In $LaMnO_3$, the B-site Mn^{3+} is in high valence state, also a Jahn–Teller distorted ion. The divalent substitution of La^{3+} with Ca^{2+} or Sr^{2+} results in the formation of Mn^{4+} ions to neutralize the subsequent charge imbalance. In such a perovskite structure, the electron from high-spin state Mn^{3+} is transferred to low-spin Mn^{4+} ions through the intervening oxygen atoms called as double-exchange interaction, resulting in a ferromagnetic ordering. The schematic double-exchange interaction is shown in Figure 5.14. Here, one Mn^{3+} ion transfers electron to neighboring O^{2-} ion, due to the completely filled *p*-orbitals, and the O^{2-} ion concurrently transfers it to the adjacent Mn^{4+} ion. During the charge transfer process, the spin-up electron from Mn^{3+} displaces a spin up electron from O^{2-} ion on to Mn^{4+} ion. The double exchange is only possible between two cations with parallel spins, which ultimately leads to ferromagnetic alignment, also metallic behavior.

Let us consider a series of $La_{1-x}A_xBO_3$ (A = Ca, Sr, and Ba; and B = Mn, Co, Fe, and Ni) perovskite systems. The presence of B^{3+} and B^{4+} ions in the lattice permits double exchange, and thereby the accompanying ferromagnetism and metallic behavior are expected from these perovskites. Among the above systems, Co has the most prominent ferromagnetism (Jonker and Van Santen 1953), but the conductivity

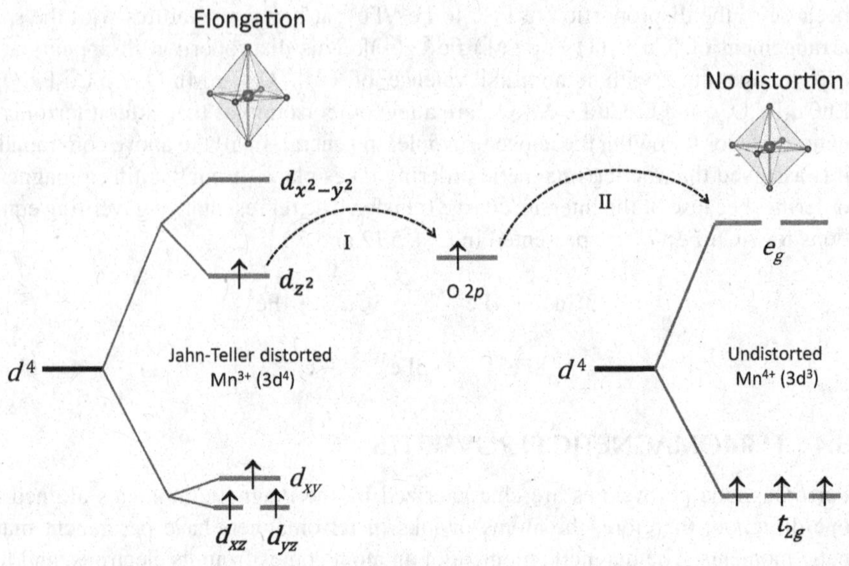

FIGURE 5.14 Schematic representation of double-exchange mechanism.

is poorer than the other systems. The ferromagnetic behavior is more likely due to the Co^{3+}–Co^{4+} double-exchange interaction, as observed in the case of Mn^{3+}/Mn^{4+} ions. Such double-exchange interactions occur at a sufficiently high A^{2+} doping.

The double exchange is prominent in perovskites with excess anions also, as in the case of oxygen-rich representative phase $LaMnO_{3+\delta}$. The presence of additional oxygen ions in the above system requires charge balance, which is achieved through the formation of Mn^{4+} ions, resulting in the formula $LaMn^{3+}_{1-x}Mn^{4+}_{x}O_{3+x}$. This composition favors double-exchange charge transfer and the accompanying ferromagnetic ordering and metallic behavior.

The double exchange is common among perovskites with cations of two different valence states. However, the double exchange is not favorable when the potential double exchangeable cations are present in two different coordination environments, such as tetrahedral and octahedral coordination. The crystal field energy level splitting of *d*-orbitals are characteristic to the coordination environment, and the *d*-orbital energy of cation present in octahedra is different from cation present in tetrahedra, and this difference results in failure of double exchange between those cations. The ferromagnetic and antiferromagnetic phases are also characteristics of A-site ions, especially in lanthanide perovskites, as we have seen in the previous sections. In general, the antiferromagnetic behavior is easily disordered to a resultant ferromagnetic state by large A-site ions, cationic replacements, excess anions, anionic replacement, etc.

The ferromagnetic ordering in double perovskites is generally due to superexchange interactions between the two B-site cations, and not by itinerant electrons. However, in some double perovskites, the superexchange mechanism is failed to explain this phenomenon. For example, Sr_2FeMoO_6 perovskite shows ferromagnetic metallic behavior due to electron transfer between Fe^{3+} ($3d^5$ HS)–O–Mo^{5+} ($4d^1$).

A similar complex perovskite $Sr_4M_3ReO_{12}$ (M = Fe or Co) also shows ferromagnetic behavior at T_c ~250 K. Therefore, in perovskite structures, by tuning the interactions of 3*d* transition cations, one can achieve paramagnetic, antiferromagnetic, or ferromagnetic behavior at a required temperature.

5.5 SPIN-GLASS BEHAVIOR

Spin glasses are a type of magnets in which the magnetic moments are ordered randomly, without a periodicity. Spin glasses are considered as magnetic materials with both ferromagnetic and antiferromagnetic interactions that coexist and compete with each other due to some frozen-in structural disorder, leading to a net magnetic moment of zero, identical to a paramagnetic phase. In spin glasses, the spins are frozen in a spatially random manner irrespective of time, whereas in paramagnets, spatially random spin pattern randomly fluctuates in time. Figure 5.15 compares the magnetic spins of spin glass with ferro- and antiferromagnetic ordering. However, the spin glasses possess a certain periodic long-range order of spins characterized by a certain wave vector (Kawamura and Taniguchi 2015). The term spin glass is originated from its analogy to structural glasses, where the molecules are arranged in random and frozen without any periodicity since glass is a supercooled liquid. Theoretically speaking, the spin-glass ordering is a characteristic of structurally disordered materials.

The spin-glass behavior is identical to a quenched randomness below a certain freezing temperature called as glass forming (T_g) or frozen temperature (T_f) or spin-glass temperature (T_{SG}). The spin-glass behavior is often exhibited by paramagnetic perovskites, below a certain freezing temperature. At T_{SG}, the spins are subject to different types of interactions, such as ferromagnetic and antiferromagnetic interactions. If a particular spin has a conflicting interaction than its neighboring spins, it is not possible for the system to arrange in a certain spin ordering, such a phenomenon is called as frustration (Anderson 1978). A schematic of frustration in a regular lattice with antiferromagnetic ordering and structural distortion is shown in Figure 5.16. Figure 5.16a corresponds to a lattice without frustration, since all the positive and negative spins are balanced. In the case of structural distortion, as in Figure 5.16b, the frustration arises from unbalanced negative and positive spins as

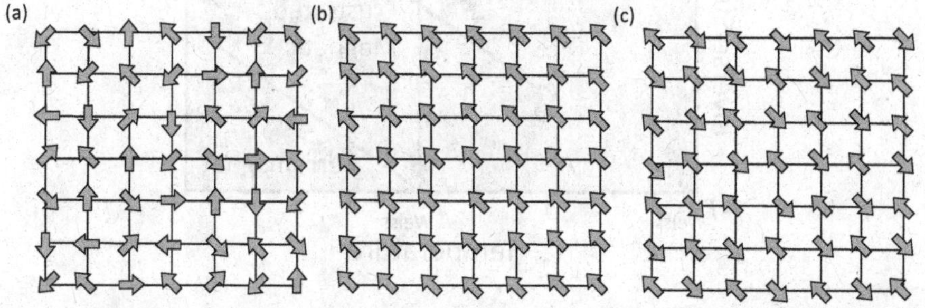

FIGURE 5.15 A comparison of the spin ordering in (a) spin glass, (b) ferromagnetic state, and (c) antiferromagnetic states.

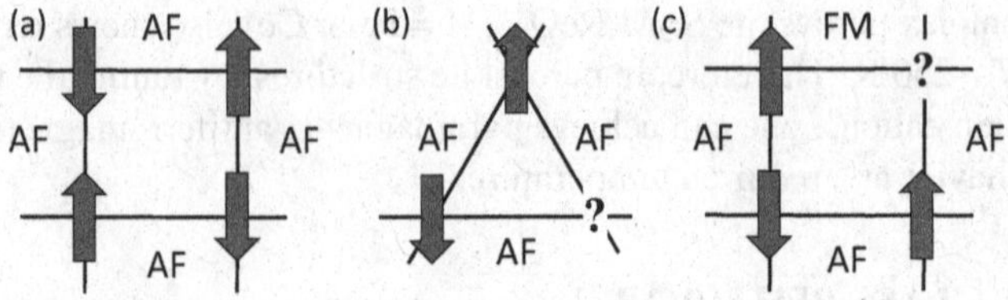

FIGURE 5.16 (a) Antiferromagnetic ordering in a regular lattice, (b) geometrical frustration, and (c) frustration during a spin-glass transition, reproduced with permission of the Cambridge University Press (Ramirez 2005).

there is no even number of positive and negative spins. In Figure 5.16c, the frustration originates from the disorder of the interactions, since there is no even number of positive and negative spins.

The spin-glass transition is distinguished from other magnetic ordering by the Curie–Weiss plot, as shown in Figure 5.17, which compares the inverse susceptibility curves of an antiferromagnet, a frustrated magnet, and a ferromagnet. From Figure 5.17, one can ascertain that the frustrated magnets with spin-glass behavior $\chi(T)$ continue to exhibit the Curie–Weiss behavior down to very low temperatures, because their long-range magnetic order is suppressed by the inability of the system to find a unique ground state (Ramirez 2005). If the magnetic phase is exhibiting small islands of spin-glass-like magnetic moments with a net moment zero is called as spin cluster glass or cluster glass. A schematic comparing spin glass and cluster glass is shown in Figure 5.18.

Many perovskite structures with B-site occupy two equal amounts of cations and exhibit a spin-glass transition at a very low temperature. In B-site-disordered $SrTi_{0.5}Mn_{0.5}O_3$ perovskite, the transition takes place at ~14 K (Sharma et al. 2019).

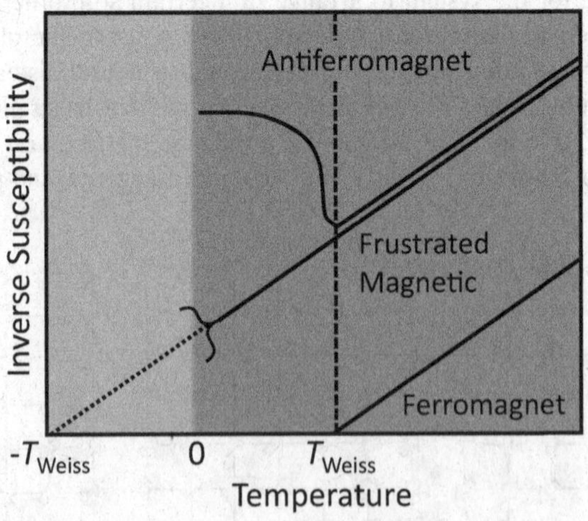

FIGURE 5.17 Inverse susceptibility versus temperature for an antiferromagnet, a frustrated magnet, and a ferromagnet, reproduced with permission of the Cambridge University Press (Ramirez 2005).

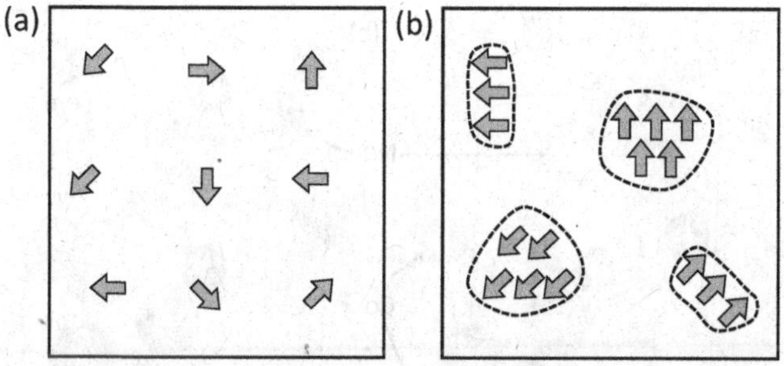

FIGURE 5.18 Comparison of (a) spin-glass and (b) cluster-glass behaviors.

Such transitions have appeared as cusps in the magnetic susceptibility *vs.* temperature plots. $BiCr_{0.5}Ni_{0.5}O_3$ is another perovskite structure that exhibits a spin-cluster behavior. In $BiCr_{0.5}Ni_{0.5}O_3$, Cr–O–Cr and Ni–O–Ni couplings are expected to be antiferromagnetic because of dominant π- and σ-superexchange interactions, respectively; however, Cr–O–Ni interactions are ferromagnetic ordered. Since the number of B-site cations is equal, the antiferromagnetic and ferromagnetic interactions are also assumed to be equal, resulting in a spin-glass configuration (Arévalo-López et al. 2015). $SrFe_{0.9}Ti_{0.1}O_3$ perovskite presents spin-glass nature at a temperature of ~48 K, but the parent compound $SrFeO_3$ shows helicoidal spin structure. The substitution of Fe^{4+} cations of this perovskite with Ti^{4+} cations results in low-temperature cluster glass, instead of antiferromagnetic ordering by disturbing the magnetic ordering interaction among Fe^{4+} cations.

Similarly, La_2CoMnO_6 double perovskite exhibits spin-glass behavior at 50 K, due to the interaction between the ferromagnetic and antiferromagnetic ordering of Co and Mn ions. Sr_2MgOsO_6 double perovskite displays antiferromagnetic behavior at a Néel temperature of ~110 K. Interestingly, near-related double perovskite Ca_2MgOsO_6 does not show any magnetic ordering and exhibits spin-glass behavior at a temperature of ~19 K. If only two valence states present among three valence states of A′-site, antiferromagnetic or ferromagnetic orderings are possible through superexchange or double-exchange mechanisms, resulting in a net-zero moment.

Spin-glass behavior is most frequently observed in perovskites with both paramagnetic cations Mn and Ru. For instance, at all temperatures, hexagonal 6*H*-$BaRuO_3$ exhibits paramagnetic behavior, but solid solution containing both Mn and Ru hexagonal 6H-$BaRu_{1-x}Mn_xO_3$ having $x = 0.1$–0.5 displays spin-glass ordering with glass-forming temperature T_f raising to maximum ~50 K at $x = 0.5$. In $LaMn_3Ti_4O_{12}$ perovskites, the spin-glass behavior is characteristic of the Mn ions. In $LaMn_3Ti_4O_{12}$, Mn cations exist in three valence states (Mn^{3+}, Mn^{2+}, and Mn^{+}), and the interaction between them results in spin-glass behavior.

5.6 SPIN CANTING

Spin canting is a phenomenon associated with magnetic dipoles in antiferromagnets at very low temperatures, during which the antiferromagnetic moments slightly deviate from their linear order, leading to a minute misalignment in spins called canted

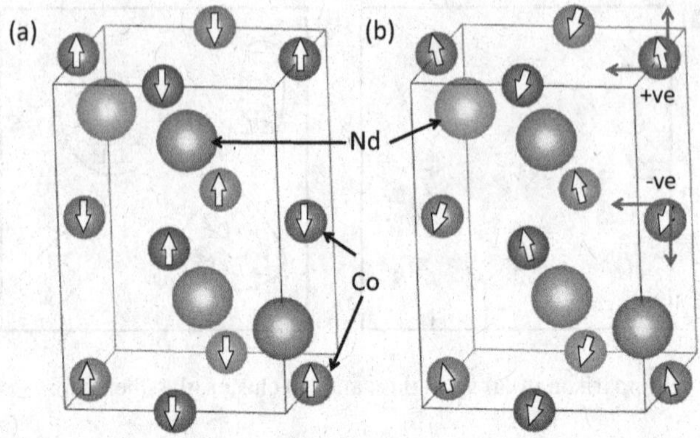

FIGURE 5.19 Canted spin arrangement of Co^{3+} in Nd_2CoTiO_6 at 3 K.

ordering, and such an arrangement is called as spin canting. Single perovskites such as $DyCoO_3$ and $TbCoO_3$ display antiferromagnetic ordering owing to the magnetic moments of Dy^{3+} ($4f^9$) and Tb^{3+} ($4f^8$) ions, and they have $GdFeO_3$-like structure at all temperatures. The magnetic moments of both these perovskites are ordered into canted antiferromagnetic phase at the lowest temperatures T_N ~3.6 and 3.3 K for $DyCoO_3$ and $TbCoO_3$, respectively.

For double perovskite with antiferromagnetic ordering, Sr_2CoOsO_6, for example, displays two T_N values at low temperatures: one at ~67 K because of freezing of magnetic moments of Co^{2+} into an antiferromagnetic canted state, whereas the second one at ~108 K is due to average effective magnetic moments ordered into antiferromagnetic phase. Likewise, Nd_2CoTiO_6 displays spin canting at ~3 K, and the schematic representation Co and Nd ions in the normal and canted states is shown in Figure 5.19. In Nd_2CoTiO_6, all the spins of Nd and Co sublattices are arranged antiferromagnetically. Usually, low-temperature spin canted phases display weak ferromagnetism owing to the misalignment since a canted spin has a vertical and horizontal component, as shown in Figure 5.19. Let us consider a pair of positive and negative spins of an antiferromagnetic ordering, the positive spin has the vertical component pointing upwards and the horizontal component towards the left, and negative spin has the vertical components pointing downwards and horizontal component towards left. Now, the vertical components cancel each other. Still, the horizontal components of both the spins towards the left are added up, resulting in a net ferromagnetic moment in the horizontal direction.

5.7 MULTIFERROIC PEROVSKITES

A material is said to be multiferroic when they exhibit two or more ferroic phases (magnetic, electric, or piezo-elastic) simultaneously. These materials switch their internal structure from one phase to another under the influence of an external specific motive force in a certain direction. This possible cross-coupling between electric polarization–electric field–strain, magnetization–magnetic field–strain, and stress-strain–electric polarization–magnetization is graphically illustrated in Figure 5.20. In multiferroics, the electric and magnetic phases represent any type of ferroic ordering, including ferromagnetic, antiferromagnetic, ferrimagnetic, paramagnetic, superparamagnetic, and their electrical equivalents. Multiferroics, in general, are promising materials for magnetic field sensors, data storage, photovoltaic, thermal energy harvesting, and solid-state cooling (Vopson 2015).

Perovskites such as $TbMnO_3$, $HoMnO_3$, $BiMnO_3$, $YMnO_3$, and $BiFeO_3$ are some representative single-phase multiferroic materials that exhibit magnetism and ferroelectricity simultaneously. Single-phase multiferroic materials are defined as homogenous compounds that are chemically isotropic, and the electric- and magnetic-order states coexist at any point or given location within the material. Only 13 point groups out of a total of 122 crystallographic point groups exhibit multiferroic state (Hill 2000); therefore, the number of multiferroic materials are less.

The magneto-electric effect facilitates the modification of electric polarization under the external magnetic field; conversely, the net magnetization is altered by the application of an external electric field. In the case of the above materials, the magnetic field is switched by an electric field and the electric field switched by the magnetic field spontaneously. The magneto-electric coupling coefficient for the applied electric field, α_E, can be mathematically described as follows (Vopson 2015):

$$\alpha_E = \left(\frac{\partial M}{\partial E}\right) \tag{5.14}$$

where M is the change in the magnetization due to the application of an electric field, E. Similarly, the magneto-electric coefficient for applied magnetic field is α_H which can be defined as

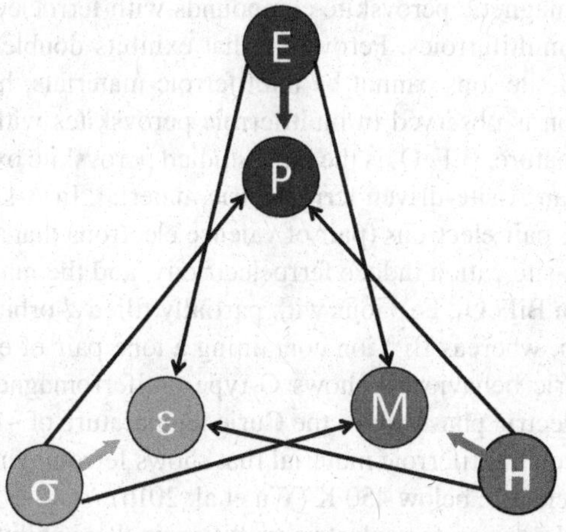

FIGURE 5.20 Possible cross-couplings in multiferroics. E – electric field; P – electric polarization; σ – applied mechanical stress; ε – strain; H – magnetic field; M – magnetization, reproduced with permission of the Taylor & Francis (Vopson 2015).

$$\alpha_{\mathrm{H}} = \left(\frac{\partial P}{\partial H}\right) \cong \varepsilon_0 \varepsilon_{\mathrm{r}} \left(\frac{\partial E}{\partial H}\right) \tag{5.15}$$

where ε_0 and ε_{r} are the dielectric permittivity of the vacuum and the relative permittivity of the medium, respectively.

$$P = \varepsilon_0 \chi E = \varepsilon_0 (\varepsilon_r - 1) E \approx \varepsilon_0 \varepsilon_{\mathrm{r}} E \text{ for } \varepsilon_{\mathrm{r}} \gg 1 \tag{5.16}$$

Since $E = V/t$, where V is the voltage and t is the thickness of the dielectric structure, the coefficient of magnetically induced magneto-electric effect can be written as

$$\alpha_{\mathrm{H}} = \left(\frac{\partial P}{\partial H}\right) \cong \varepsilon_0 \varepsilon_{\mathrm{r}} \left(\frac{\partial E}{\partial H}\right) = \frac{\varepsilon_0 \varepsilon_{\mathrm{r}}}{t} \left(\frac{\partial V}{\partial H}\right) = \alpha_{\mathrm{H}}^{\mathrm{V}} \varepsilon_0 \varepsilon_{\mathrm{r}} \tag{5.17}$$

where $\alpha_{\mathrm{H}}^{\mathrm{V}}$ is the magnetically induced voltage magneto-electric coefficient defined as

$$\alpha_{\mathrm{H}}^{\mathrm{V}} = \left(\frac{\partial E}{\partial H}\right) = \frac{1}{t} \left(\frac{\partial V}{\partial H}\right) \tag{5.18}$$

The SI unit of $\alpha_{\mathrm{H}}^{\mathrm{V}}$ is VA and the CGS units V/Cm·Oe.

Mostly, ferroelectric properties are manifested by the perovskite oxides with transition metal ions at B-sites having empty d orbitals (d^0). However, perovskite with Pb^{2+} or Bi^{3+} in the A-site displays ferroelectric behavior driven by these cations. In the perovskite structures, either A- or B-site-driven ferroelectrics, the motive is the long-range Coulombic interaction, due to which one or several B–O or A–O bond lengths are shortened (Vopson 2015).

In general, the magnetic perovskite compounds with ferroelectric properties are mostly studied as multiferroics. Perovskite that exhibits double-exchange interaction between the B-site ions cannot be multiferroic materials; however, the super-exchange interaction is observed in multiferroic perovskites with ferromagnetic or antiferromagnetic nature. $BiFeO_3$ is the most studied perovskite oxide as multiferroic material, which is an A-site-driven ferroelectric material. In A-site driven multiferroic materials, lone-pair electrons (pair of valence electrons that are not shared with another atom) of A-site cation induce ferroelectricity, and the magnetism is induced by the B-site ion. In $BiFeO_3$, Fe^{3+} ions with partially filled *d*-orbitals are responsible for ferromagnetism, whereas Bi^{3+} ion containing a lone pair of electrons is responsible for ferroelectric behavior. It shows G-type antiferromagnetic behavior at T_N ~640 K and ferroelectric phase above the Curie temperature of ~1110 K. $BaMnO_3$ is another A-site-driven multiferroic material that shows ferromagnetism below 150 K and becomes ferroelectric below 450 K (Wu et al. 2016).

$EuTiO_3$ is a B-site-driven ferroelectric multiferroic that exhibits a G-type antiferromagnetic behavior below 5.3 K under a compressive strain. $EuTiO_3$ crystallizes to an ideal cubic structure with center symmetry, which prohibits the possibility of ferroelectricity under normal conditions. Many rare earth perovskites with the general formula $LnBO_3$ (Ln = lanthanides, and B = Fe, Cr, and Mn) show multiferroic properties with weak ferromagnetic behavior accompanied by the room temperature ferroelectric behavior. In $LnFeO_3$ structures, Fe^{3+} ions exhibit superexchange coupling with weak ferromagnetic interactions. Since the $LnFeO_3$ compounds are included in the family of centrosymmetric ferrites, the ferroelectric property can be induced by the application of stress. $LnCrO_3$ are promising multiferroic materials due to the weak ferromagnetic properties of Cr^{3+} ions and the electric polarization arising from the B-site displacement (Lone et al. 2019). Similarly, $LnMnO_3$ (Ln = Dy, Ho, Tb) with magnetic rare earth in the A-site is also identified as promising multiferroics.

Double-perovskite structures with magnetic and ferroelectric cations can exhibit magnetically ordered ferroelectrics nature. Examples for double-perovskite multiferroics include Bi_2NiMnO_6, Bi_2FeCrO_6, Lu_2MnCoO_6, and Y_2NiMnO_6. All the above compounds have a rock-salt-type arrangement of B-site cations in a double-perovskite unit cell, even though the polar and magnetic phase transition temperatures are very low. Among them, Bi_2FeCrO_6 is a room temperature multiferroic in thin-film form. Interestingly, among $A_2B'B''O_6$ perovskites, if A^{2+}-site is occupied by Pb^{2+} ions, then those compounds exhibit ferroelectric transitions above room temperature, although most are antiferroelectric. The $6s^2$ electron pair in Pb^{2+} increases the magnetic ordering temperature. Pb_2FeTiO_6 is the only perovskite in this category that has been reported to be both ferromagnetic and ferroelectric at room temperature. The perovskites with A^{3+} cations and $3d$ transition metals, such as Mn on the B-sites, exhibit multiferroic properties when the A-site is occupied by Bi^{3+} ions. Bi_2NiMnO_6 belongs to this class of multiferroic double perovskites with ferroelectric behavior below ~460–480 K and ferromagnetic below 140 K (Vasala and Karppinen 2015).

The perovskites with the layered structure also promise multiferroic properties. In these materials, the A-site cation containing separating layer can be ferroelectric, and the transition metal containing BO_6 slabs can be magnetic. For instance, Aurivillius-structured $Bi_6Ti_{2.8}Fe_{1.52}Mn_{0.68}O_{18}$ is a layered ferroelectric perovskite that also exhibits multiferroic properties. The corresponding multiferroic behavior is distinguished by the ferroelectric domain switching under a magnetic field at room temperature. In this phase, the magnetic characteristics are originated from Mn^{4+} cations than the Fe^{3+} cations. $Sr_{0.25}Bi_{6.75}Fe_{1.5}Co_{1.5}Ti_3O_{12}$ is another example of the Aurivillius phase, which displays ferroelectric and ferromagnetic behaviors at room temperature.

5.8 NANOMATERIALS AND THIN FILMS

In general, nanostructured thin films and particles differ from their bulk counterparts in many aspects. The high surface-to-volume ratio of nanostructured magnetic perovskites makes them to behave differently from their bulk counterparts. Due to the high surface area of the nanostructures than volume, atoms on the surface have a less number of nearest neighboring magnetic ions as compared to the core of the particles; therefore, the surface cations always have an uncompensated spin. A magnetic frustration related to the structure can also arise as shown in Figure 5.16b. The thinning of lattice parameters at the surface lowers the local symmetry at the surface, which ultimately leads to spin canting, which is prominent if the nanostructures are in the thickness of few atomic layers. The asymmetry near the surface of the nanoparticles, therefore, leads to magnetic anisotropy in the same particle, within the nanoscale domain.

Many perovskite materials behave differently at the nanoscale, as compared to their bulk counterparts. For instance, $BiFeO_3$ nanowires exhibit a weak ferromagnetic nature, in contrast to the antiferromagnetic behavior of typical bulk $BiFeO_3$ samples at room temperature. This is related to the spin canting (Figure 5.21a) due to the asymmetry of local structure on the surface, leading to an uncompensated spin on the surface of the nanowires, as shown in Figure 5.21b (Patel et al. 2018). In nanostructures, the characteristics of the surface atoms contribute more than the atoms in the volume of the structures.

Superparamagnetism is another characteristic of a nanosized magnetic structure which is applicable to perovskite oxides as well. Superparamagnetism is observed in small ferromagnetic or ferrimagnetic nanoparticles. If the nanoparticles are sufficiently small, their magnetic moments are randomly oriented under the influence of temperature, and the time between two subsequent flips is called the Néel relaxation time. Under normal conditions, in the absence of an external magnetic field, the average value of magnetization of nanoparticles appears to be zero, and the nanoparticles are said to be in the superparamagnetic state. A magnetic perovskite nanostructure with one of its dimensions falls between 10 and 100 nm sizes can exhibit superparamagnetism. Many ferromagnetic perovskites at nanoscale exhibit a superparamagnetic state. Interestingly, $LaCoO_3$ perovskite shows an increase in the magnetic moment with a decrease in the particle size (Figure 5.22) (Zhou et al. 2016). As the particle size decreases, the paramagnetic nature of the perovskite increases. For instance, ferromagnetic $LaFeO_{3+\delta}$, $YFe_{1-x}Co_xO_3$ (Pomiro et al. 2017), $La_{0.5}Sr_{0.5}Ti_{0.5}Fe_{0.5}O_3$ (Choudhary et al. 2018), etc. are few examples for superparamagnetic perovskite nanostructures. Most importantly, at the nanoscale level, the

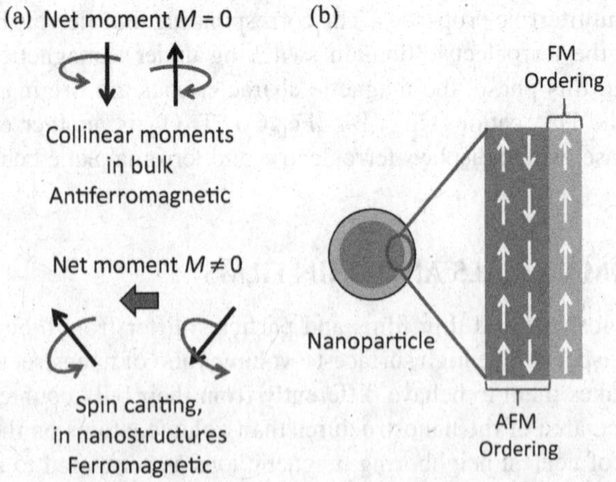

FIGURE 5.21 (a) Canted spin configuration in nanoparticles and (b) uncompensated magnetic moment causing net ferromagnetic properties, reproduced with permission of the Royal Society of Chemistry (Patel et al. 2018).

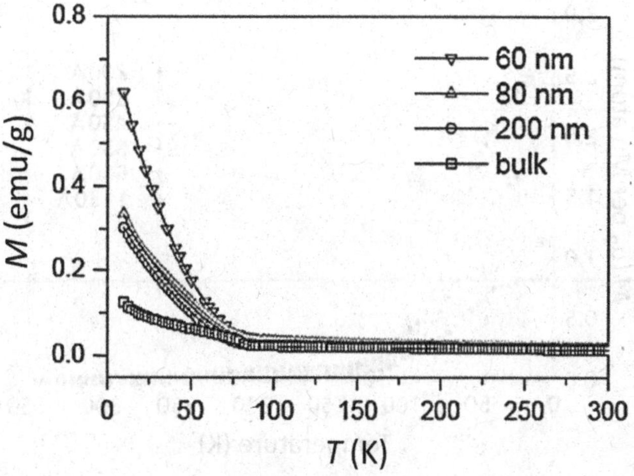

FIGURE 5.22 Temperature-dependent magnetization curves of $LaCoO_3$ nanoparticles as compared to the bulk under $H = 1$ k Oe (Zhou et al. 2016).

exchange interaction between surface atoms of the nearest particles themselves can influence the magnetic properties that ultimately lead to superparamagnetic relaxation and the spin structure (Issa et al. 2013). In addition, nanomaterials show lower Curie temperature, higher coercivity, and lower field-saturated magnetization as compared to bulk materials.

In thin films, the magnetic alignment depends upon the various factors such as the thickness of the film, the number and spacing of layers in multi-layer films, the deposited substrate, and the interfacial region between substrate and film. For example, a thin film of $BiMnO_3$ shows a ferromagnetic transition of ~105 K like bulk, but it shows an increase in the magnetic moment along with the increase in thickness (Figure 5.23) (Ohshima et al. 2000).

The controlled optimization of the interface will result in thin films with potential applications or unexpected properties due to the formation of sharp edges at grain boundaries. Especially, the properties of epitaxial-grown, single-crystal thin film of perovskite can be changed by optimizing both substrate and film thickness due to lattice mismatch. For instance, $LuMnO_3$, with an orthorhombic $GdFeO_3$ structure, exhibits E-type antiferromagnetic ordering. If the thin film of $LuMnO_3$ is grown on the $YAlO_3$ nonmagnetic substrate, this thin film on the $YAlO_3$ substrate displays combined antiferromagnetic and ferromagnetic regions due to lattice strain at thin film and substrate interface. When $LuMnO_3$ films are grown on (001) planes of the $YAlO_3$ substrate, the films show compression at the interface with substrate owing to the lattice parameters of film larger than the substrate. Hence, the strain from compression results in monoclinic distortion in the unit cell of $LuMnO_3$; this induces variations in the orbital ordering of Mn, Mn-O bonding, and octahedral tilting at the interface of the substrate. Therefore, ferromagnetic ordering takes place at the interface, and this will transform into antiferromagnetic ordering on moving away from the interface region.

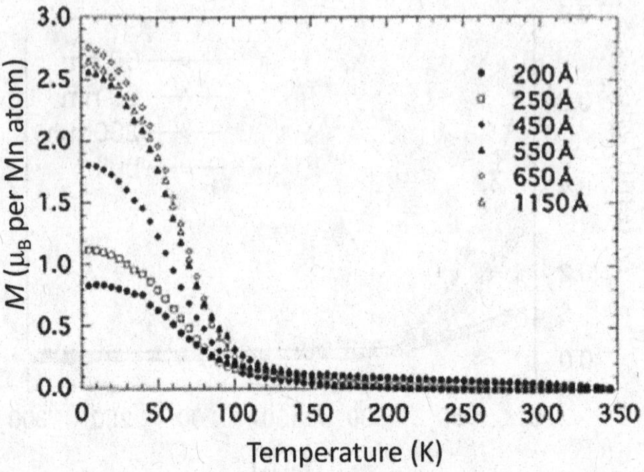

FIGURE 5.23 Temperature variation of the magnetization of the $BiMnO_3$ thin films with various thickness grown on the (100) $SrTiO_3$ substrate measured at 1 T, reproduced with permission of the Elsevier (Ohshima et al. 2000).

Similarly, $La_{0.67}Ca_{0.33}MnO_3$ in bulk form exhibits ferromagnetic ordering with T_c ~370 K. The thin films show the change in magnetic ordering with respect to substrates. This perovskite becomes CO insulator at low temperatures when it is grown on $LaAlO_3$ substrate owing to the compressive strain. The ferromagnetic metallic character reverts when it is subjected to several Tesla of the magnetic field. This transformation is due to the formation of islands of metals within the matrix of CO insulator; thus, the regions experiencing low stress are ferromagnetic, whereas regions of high stress remain insulating. Films of $La_{0.67}Ca_{0.33}MnO_3$ with a thickness less than 4 nm on a $SrTiO_3$ substrate display spin canting with a T_c ~540 K. If the same perovskite is grown on $BaTiO_3$ substrate, it shows raise in T_c ~650 K owing to a high degree of strain at the interface. Likewise, $CaRuO_3$ is a metallic perovskite with paramagnetic behavior; when it is grown on the (100) planes of $SrTiO_3$ substrate, it exhibits ferromagnetic behavior due to tensile strain at the epitaxial layers.

The magnetic moment of thin films increases with increasing strain at the interface. For example, the films of $TbMnO_3$ (001) grown on $SrTiO_3$ (001) substrate show interfacial strain, and this strain is released by the formation of ferromagnetic ordering at the interface. Further, the $TbMnO_3$ films show a decrease in the magnetic moment with a decrease in film thickness due to a decrease in ferromagnetic ordering at the interface.

The antiferromagnetic perovskites ($LaFeO_3$ and $LaMnO_3$) deposited on $LaCrO_3$, paramagnetic metallic $CaRuO_3$ grown on antiferromagnetic $CaMnO_3$, and antiferromagnetic $LaMnO_3$ deposited on diamagnetic $SrTiO_3$ substrate display ferromagnetic regions near the interface of the substrate and thin film. In all the above cases, the characteristic properties are originated due to the charge transfer across the boundary or superexchange between cations at the boundary instead of interfacial strain. Interestingly, the interface between nonmagnetic $LaAlO_3$ grown on $SrTiO_3$ displays magnetic behavior. This is due to the charge imbalance between charged layers

(LaO^+, AlO_2^-) of $LaAlO_3$ film and uncharged layers (SrO, TiO_2) of the $SrTiO_3$ substrate. This imbalance induces charge transfer and leads to an electron transfer of 0.5 e^- per perovskite unicell from thin-film Ti 3*d* levels to substrate and makes the interface both ferromagnetic and electronically conductive. Oxygen deficiency in the surface layers of the $SrTiO_3$ insulator causes a similar effect. In TiO_2-terminated thin films, vacancies are created due to loss of oxygen, which have adjacent Ti^{3+} cations. The spins are aligned in an antiferromagnetic ordering owing to the superexchange coupling between oxygen anions and anion vacancies. This anion vacancy creation, along with layer thickness, produces different magnetic orderings.

The magnetic properties of perovskite materials are accurately determined by the magnetic measurements and neutron diffraction data.

REFERENCES

Abakumov, A. M., M. G. Rozova, A. M. Aleekseva, et al. 2003. Synthesis and structure of $Sr_2MnGaO_{5+\delta}$ brownmillerites with variable oxygen content. *Solid Stat. Sci.* 5:871–882.

Anderson, P. W. 1978. The Concept of Frustration in Spin Glasses. *J. Common Met.* 62:291–294.

Arévalo-López, Á. M., A. J. Dos santos-García, J. R. Levin, et al. 2015. Spin-Glass Behavior and Incommensurate Modulation in High-Pressure Perovskite $BiCr_{0.5}N_{i0.5}O_3$. *Inorg. Chem.* 54:832–836.

Bao, X., Y. Maeno, F. Nakamura, et al. 1995. Effect of Tb Substitution on the Superconductivity of $La_{2-x}Ba_xCuO_4$. In *Advances in Superconductivity. VII*, ed. K. Yamafuji and T. Morishita, 241–244. Tokyo: Springer Japan.

Battle, P. D., J. B. Goodenough, and R. Price. 1983. The Crystal Structures and Magnetic Properties of Ba_2LaRuO_6 and Ca_2LaRuO_6. *J. Solid State Chem.* 46:234–244.

Choudhary, N., M. K. Verma, N. D. Sharma, et al. 2018. Superparamagnetic Nanosized Perovskite Oxide $La_{0.5}Sr_{0.5}Ti_{0.5}Fe_{0.5}O_3$ Synthesized by Modified Polymeric Precursor Method: Effect of Calcination Temperature on Structural and Magnetic Properties. *J. Sol–Gel Sci. Technol.* 86:73–82.

Coey, M. 2004. Charge-Ordering in Oxides. *Nature* 430:155–157.

Deng, Z., M. Retuerto, S. Liu, et al. 2018. Dynamic Ferrimagnetic Order in a Highly Distorted Double Perovskite Y_2CoRuO_6. *Chem. Mater.* 30:7047–7054.

Ganguli, A. K., and J. Gopalakrishnan. 1986. Magnetic Properties of $Ca_2F_{2-x}Mn_xO_5$. *Proc. Indian Acad. Sci. – Chem. Sci.* 97:627–630.

Greaves, C., A. J. Jacobson, B. C. Tofield, et al. 1975. A Powder Neutron Diffraction Investigation of the Nuclear and Magnetic Structure of $Sr_2Fe_2O_5$. *Acta Crystallogr. B* 31:641–646.

Hill, N. A. 2000. Why Are There so Few Magnetic Ferroelectrics? *J. Phys. Chem. B* 104:6694–6709.

Issa, B., I. M. Obaidat, B. A. Albiss, et al. 2013. Magnetic Nanoparticles: Surface Effects and Properties Related to Biomedicine Applications. *Int. J. Mol. Sci.* 14:21266–21305.

Izumiyama, Y., Y. Doi, M. Wakeshima, et al. 2001. Magnetic properties of the antiferromagnetic double perovskite Ba_2PrRuO_6. *J. Phys. Condens. Matter.* 13:1303.

Jonker, G. H., and J. H. Van Santen. 1953. Magnetic Compounds Wtth Perovskite Structure III. Ferromagnetic Compounds of Cobalt. *Physica* 19:120–130.

Kawamura, H., and T. Taniguchi. 2015. Chapter 1- Spin Glasses. In *Handbook of Magnetic Materials*, eds. K. H. J. Buschow, Vol. 24:1–137. Amsterdam: Elsevier.

Kundu, A. K. 2016. *Magnetic Perovskites: Synthesis, Structure and Physical Properties.* Engineering Materials. Jabalpur: Springer India.

Lone, I. H., J. Aslam, N. R. E. Radwan, et al. 2019. Multiferroic ABO_3 Transition Metal Oxides: A Rare Interaction of Ferroelectricity and Magnetism. *Nanoscale Res. Lett.* 14:142.

MacChesney, J. B., H. J. Williams, J. F. Potter, et al. 1967. Magnetic Study of the Manganate Phases: $CaMnO_3$, $Ca_4Mn_3O_{10}$, $Ca_3Mn_2O_7$, Ca_2MnO_4. *Phys. Rev.* 164:779–785.

Medarde, M. L. 1997. Structural, magnetic and electronic properties of perovskites (R = rare earth). *J. Phys. Condens. Matter* 9:1679–1707.

Ohshima, E., Y. Saya, M. Nantoh, et al. 2000. Synthesis and magnetic property of the perovskite $Bi_{1-x}Sr_xMnO_3$ thin film. *Solid Stat. Commun. C* 116:73–76.

Patel, S. K. S., J.-H. Lee, M.-K. Kim, et al. 2018. Single-Crystalline Gd-Doped BiFeO3 Nanowires: *R*3*c*-to-*Pn*2_1*a* Phase Transition and Enhancement in High-Coercivity Ferromagnetism. *J. Mater. Chem. C* 6:526–534.

Philipp, J. B., P. Majewski, L. Alff, et al. 2003. Structural and Doping Effects in the Half-Metallic Double Perovskite A_2CrWO_6 (A = Sr, Ba, and Ca). *Phys. Rev. B* 68:144431.

Pomiro, F., D. M. Gil, V. Nassif, et al. 2017. Weak Ferromagnetism and Superparamagnetic Clusters Coexistence in $YFe_{1-x}Co_xO_3$ ($0 \leq x \leq 1$) Perovskites. *Mater. Res. Bull.* 94:472–482.

Ramirez, A. P. 2005. Geometrically Frustrated Matter–Magnets to Molecules. *MRS Bull.* 30:447–451.

Rykov, A. I., K. Nomura, Y. Ueda, et al. 2008. An Ising Ferrimagnet with Layered and Chained Magnetic Sublattices: Ca_2FeMnO_5. *J. Magn. Magn. Mater.* 320:950–956.

Sakai, C., Y. Doi, Y. Hinatsu, et al. 2005. Magnetic Properties and Neutron Diffraction Study of Double Perovskites Ca_2LnRuO_6 (Ln = Y, La–Lu). *J. Phys. Condens. Matter* 17:7383–7394.

Sharma, S., P. Yadav, T. Sau, et al. 2019. Evidence of a Cluster Spin-Glass State in B-Site Disordered Perovskite $SrTi_{0.5}Mn_{0.5}O_3$. *J. Magn. Magn. Mater.* 492:165671.

Takeda, T., Y. Yamaguchi, S. Tomiyoshi, et al. 1968. Magnetic Structure of $Ca_2Fe_2O_5$. *J. Phys. Soc. Jpn.* 24:446–452.

Vasala, S., and M. Karppinen. 2015. $A_2B'B''O_6$ Perovskites: A Review. *Prog. Solid State Chem.* 43:1–36.

Vopson, M. M. 2015. Fundamentals of Multiferroic Materials and Their Possible Applications. *Crit. Rev. Solid State Mater. Sci.* 40:223–250.

Wu, J., Z. Fan, D. Xiao, et al. 2016. Multiferroic Bismuth Ferrite-Based Materials for Multifunctional Applications: Ceramic Bulks, Thin Films and Nanostructures. *Prog. Mater. Sci.* 84:335–402.

Zhao, Y. G., W. Cai, J. Zhao, et al. 2002. Electrical Transport and Magnetic Properties of $La_{0.5}Ca_{0.5}MnO_{3-\delta}$ with Varying Oxygen Content. *Phys. Rev. B* 65:144406.

Zhou, S., X. Miao, X. Zhao, et al. 2016. Engineering Electrocatalytic Activity in Nanosized Perovskite Cobaltite Through Surface Spin-State Transition. *Nat. Commun.* 7:11510.

6 Electronic Properties of Perovskite Oxides

The properties of perovskite oxides are primarily governed by the B-site cations owing to their *d*-electron configuration. The electronic properties of perovskite oxides can be divided as metallic, semiconducting, superconducting, insulating, dielectric, piezoelectric, thermoelectric, pyroelectric, electrocaloric, etc. Many of these properties are temperature-dependent, and they are altered by the crystal structure, phase transformation, presence of oxygen vacancies, etc. This chapter addresses the electronic and related properties of the perovskite oxides concerning the possible applications.

The electrons in perovskites exhibit several means of electron–electron interactions; therefore, the conduction of electrons in perovskites occurs through the collective action of several electron interactions. As a result, perovskite oxides display diverse electronic properties. Insulating perovskites, in general, show dielectric, ferroelectric, and piezoelectric properties, and the conducting perovskites show thermoelectric properties. These properties of perovskite-structured materials are highly reliant on the electronic interaction between *d*/*f* orbitals of B-site cations (Tomioka et al. 2000).

6.1 BAND STRUCTURE IN PEROVSKITE OXIDES

Like several other oxides, perovskite oxides are ionic with a certain degree of covalent characteristics. The ionic model is a simplified model for explaining the electronic properties of perovskites. According to the ionic model, A- and B-site cations lose electrons to the oxygen anions for charge balance, thus forming O^{2-} ions. If the ions are at the noble gas configuration, then they are in a closed-shell configuration. If all the ions in the perovskite lattice are in a closed-shell configuration, then the perovskite is an insulator. For example, in $BaTiO_3$, the ion Ba^{2+} has Xe configuration, O^{2-} has Ne configuration, and Ti^{4+} has Ar configuration; since all the ions are in the closed-shell structure, $BaTiO_3$ is an insulator at room temperature. Even though the ionic model is handy for determining the electronic nature of simple oxides, such models cannot be extended for doped perovskite structures, and to explain other properties. Moreover, the ions are more likely to interact with each other in different ways, e.g. electrostatic interactions.

Since the cations in the perovskite oxides are surrounded by negatively charged oxygen ions, the electrons orbiting in these ions experience a repulsive force and destabilization of their energy, and the repulsive electrostatic effect is called as repulsive Madelung potential. Similarly, the electrons in the negatively charged oxygen experience an attractive Madelung potential, which stabilizes the positively charged cations. The Madelung potential is calculated by the sum of electrostatic

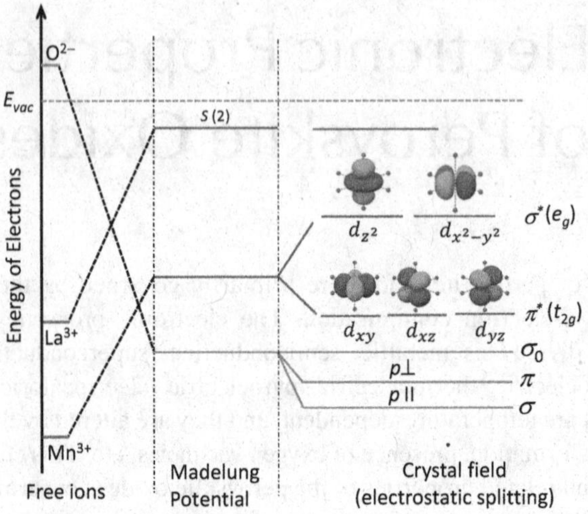

FIGURE 6.1 The electronic band structure for the insulating perovskite oxide corresponding to octahedral oxygen coordination around transition metal ions.

interactions, q^2/r. The interatomic distance r varies with the crystal structure, while the charge, q, is a constant; therefore, the Madelung potential varies with the crystal structure. The Madelung potential determines the stability of the perovskite structure. For instance, in $SrTiO_3$, the attractive potential at the oxygen sites allows the oxygen ions to bind a pair of electrons. Since the electron affinity of O^{1-} for the second electron is positive, the second electron would not bind O^{2-} ion in the absence of the Madelung potential. Therefore, O^{2-} is stable in the lattice because of the attractive site Madelung potential. Conversely, in the absence of the repulsive site Madelung potential, the donation of an electron from the Ti^{3+} to an O^{1-} ion in $SrTiO_3$ would be energetically unfavorable. The electronic energy of the individual ions is higher in free space than the ions placed in the perovskite crystal lattice due to the Madelung potential. The representative ionic energy in free space with respect to them in the lattice of $LaMnO_3$ is shown in Figure 6.1.

In most perovskites, the energy bands of the A cation correspond to the *s* state, which is much higher than the primary valence and conduction bands of perovskite, and these bands are unoccupied. As a result, the *s* state of the A cation usually does not play any significant role in determining the electronic properties (Wolfram and Ellialtioglu 2006). However, the electrostatic potentials of the A ions have a strong influence on the energy of the *p-d* valence and conduction bands. Even though the A ion is omitted from the calculations of electronic structures, the size of A-site ion determines whether the crystal structure is distorted from the ideal cubic form, which in turn modifies the electronic properties (Wolfram and Ellialtioglu 2006). Moreover, omitting A-site ions simplifies the electronic structure calculations tremendously. Therefore, it is assumed that the electronic properties of perovskites are solely arising from the BO_3 part of the ABO_3 structure, which implies that perovskites with the same B-site ions should behave the same, unless there are doping or oxygen vacancies, e.g. $BaTiO_3$ and $SrTiO_3$ have the same insulating properties.

The nature of B-site transition metal is one of the major factors that determines the electronic properties in perovskite materials. The simple electronic structure of perovskites can be deduced by considering electrons as gas (being free and non-interacting) surrounding the B-site ion. In such a model, the electronic structure is derived by assuming the splitting of atomic orbitals as upper and lower energy bands depending on the phase of the crystal structure. The upper energy band is filled with non-interacting electrons. In perovskite oxides, the electron filling of upper energy band governs the properties of perovskite oxide. Many perovskites have *d*-group elements as B-site cations; hence, the splitting of *d*-orbitals mainly decides the property of perovskite, as in Figure 6.1.

The energy band structure of perovskite oxides is also affected by the covalent bond between the metal and oxygen ions. This covalent bonding leads to the hybridization of *p* and *d* orbitals of metal and oxygen ions, respectively. Often the band structures obtained from DFT calculations are complex, but it is possible to interpret the conductivity by simplifying the energy band model (Hamada et al. 1997). When the transition metal occupies the B-site of the perovskite lattice, it will be in an octahedral environment surrounded by oxygen atoms. Therefore, the d_{z^2} and $d_{x^2-y^2}$ atomic orbitals of transition metals will have a strong spatial overlap with the oxygen *p*-orbitals, forming σ bonding molecular orbital (MO) and antibonding MO* (σ^*) molecular orbitals with the destabilized energy represented as e_g. The orbitals that are not directly overlapped form π bonding MO (π) and antibonding MO* (π*) states with less destabilized energy than the e_g and are represented as t_{2g} states. The bonding and antibonding MO and MO* states have the characters of both metallic and oxygen, based on the difference in electronegativity of the cationic and oxygen states.

Since there are three oxygen (each having three 2*p*-orbitals) per metal (having five *d*-orbitals) in the AMO_3 perovskite, some oxygen O 2*p* states may not hybridize with metal *d* states, depending on the symmetry of the structure, hence forming non-bonding (σ_0) states. These states can self-hybridize (i.e. 2*sp* hybridization) and bond with the *sp*-orbitals of other nearby O^{2-} ions. These states are called ligand states, which alone do not contribute to the diverse physical properties in oxides. In general, M *d*-band O 2*p* bands can be designated as "metal" and "oxygen" bands, respectively, describing the character of the band. Even though the metal *d*-bands in oxides are formed by the hybridization of metal and oxygen orbitals, MO* states are cationic metallic in nature.

The metal–oxygen hybridization is controlled by the oxidation state of transition metal occupying the B-site, which in turn determines the number of *d*-electrons and the electronegativity of the metal ion. In metals, the electronegativity increases with the decrease in the oxidation state, and the electronegativity determines the energy of the transition metal *d* states. The degree of metal–oxygen hybridization is increased as the electronegativity increases as the metal *d* states move closer to the O 2*p* states with an increase in electronegativity. Altogether, the electronic conductivity in perovskite occurs due to the metal–oxygen hybridization and the ligand–field splitting. Therefore, the perovskite materials can be characterized as conductors, half metals, and insulators based on the hybridization and ligand–field splitting of the B-site elements (Kundu 2016).

In insulating perovskites, B-site elements are occupied by less electronegative atoms; therefore, the M *d*-band is higher in energy than the O 2*p* band, resulting in antibonding M *d*-bands and bonding O 2*p* bands. In such a case, the highest energy electrons are predominantly M *d*-electrons, which are more localized than the O 2*p* electrons. If the M *d* states are below O 2*p* states, the resulting hybridization is bonding M *d*-bands and antibonding O 2*p* bands. Hence, the highest energy levels are comprised of *p*-orbital electrons; as a result, the overlapping between O 2*p* and M *d* states governs the semiconducting/metallic nature of the perovskite material. In perovskite oxides, the electronic structure is a cumulative effect of the cooperative ordering of cation orbitals, cation displacement ordering, octahedral tilting/distortion, Jahn–Teller (J–T) distortion, and spin polarization (Rondinelli and Fennie 2012).

6.2 PEROVSKITE INSULATORS

The electronic behavior of oxides can be related to the relative occupancy of the e_g and t_{2g} states (also known as the spin state). Assume that the electrons of opposite spin occupy the same orbital; due to Coulombic repulsion, they experience competition between electron pairing and filling of the higher energy e_g states. If the electron splitting energy is greater than the pairing energy, electrons completely occupy the lower energy t_{2g} states before filling the higher energy e_g states – typically known as a low-spin configuration. If the splitting is less than the pairing energy, the electrons occupy the e_g states before pairing in the t_{2g} states – known as a high-spin configuration. The configuration discussed above does not apply to all the transition metal-containing perovskites. For instance, cubic $SrTiO_3$ perovskite is an insulator because Ti^{4+} does not have any electrons in *d*-orbital ($4s^0 3d^0$); therefore, t_{2g} and e_g conduction bands are empty (Piskunov et al. 2004). Hence, the filled valence band and empty conduction band of $SrTiO_3$ lead to an insulating phase. The electronic configuration of Ti^{4+} in $SrTiO_3$ can be extended to other similar compounds such as $PbTiO_3$ and $BaTiO_3$ (Piskunov et al. 2004). The same is applicable for the perovskite structures containing Zr^{4+} and Hf^{4+} on the B-site.

The occupancy of electrons in the e_g states alone does not guarantee the metallic behavior in many perovskite structures at room temperature. For example, perovskite manganite $LaMnO_3$ having Mn^{3+} in the B-site, with electronic configuration $3d^4\left(t_{2g}^3 e_g^1 \text{ high spin(HS)}\right)$ exhibits J–T distortion in BO_6 octahedra with elongated apical oxygen coordination (Alonso et al. 2000), and these distorted perovskite phases are antiferromagnetic insulators. At room temperature, the MnO_6 octahedra of $LaMnO_3$ have two short, two long, and two medium Mn–O bond distances; as a result, the energy of the twofold degenerate e_g orbital (singly occupied) is lifted as compared to undistorted MnO_6. Additionally, Mn–O–Mn bond angle is lower than 180°; therefore, an orbital ordered type insulating state is reached due to the localization of electrons to each MnO_6 octahedra.

In addition to J–T distortion, octahedral tilting and B-cations displacement affect the electronic band structure, even though the BO_6 octahedron remains undistorted in the tilted perovskite structures. Tilting stabilizes the perovskite structure by lowering the total energy when accommodating cations with different sizes. Though the energies involved are low, the phase transition by titling can significantly affect the electronic structure of perovskite, leading to subsequent changes in resistivity, ferroelectricity, magnetic order, superconductivity, etc. The energy band structure of the tilted perovskite structure differs from the ideal cubic structure due to the following reasons (Kundu 2016):

1. the strong interaction of the A-cation orbitals with oxygen;
2. the changes of the ionic site potentials;
3. the changes in B–O–B and O–O interaction angles; and
4. splitting of degeneracies due to lower symmetry.

Like manganites with J–T tilting, lanthanide nickelites with perovskite structure also show an increase in resistivity on moving from La to Lu. The decrease in the size of lanthanide ion changes the crystal structure from rhombohedral to orthorhombic, *via* octahedral tilting (Sarma et al. 1994). Because of that, the bond angle Ni–O–Ni alters from 180° to lower or higher angles. The above changes fail the conducting *d*-band formation due to the localization of *d*-electrons of Ni^{3+} cations. However, the rise in temperature leads to the transformation of the insulating phase to metallic in perovskites with larger lanthanide cations (La and Pr) with smaller octahedral tilting. However, the lanthanides with larger octahedral tilting, particularly with smaller lanthanide cations, do not undergo insulator to metal transitions even at high temperatures.

The insulating nature of perovskites with B-cations having partially filled *d*-orbital occurs due to the splitting of partially filled 3*d*-orbitals into a filled and empty sub-band, originated from the Coulomb repulsion among electrons, as shown in Figure 6.2. The gap between filled and empty bands is named as Hubbard or Mott–Hubbard gap (U) (Katsufuji et al. 1997), and the compounds that depict this phenomenon are termed as Mott insulators (Huijben et al. 2006). In transition metal oxides, the oxygen *p*-bands remain unchanged when interaction (U) occurs, and the *d*-band splits into two sub-bands (upper and lower Hubbard bands). The half-filled state becomes an insulator with the opening of a charge gap (Biswas et al. 2016).

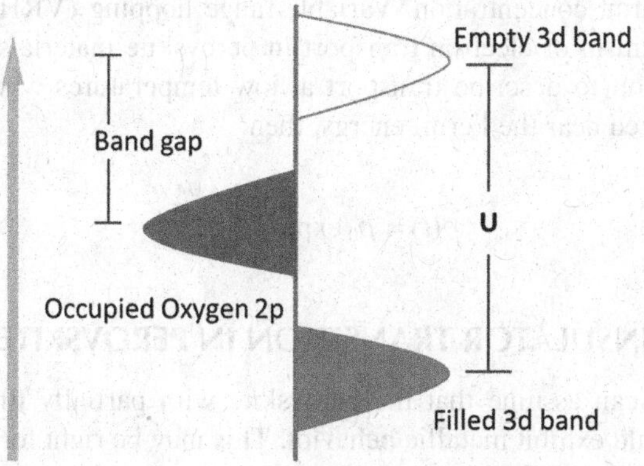

FIGURE 6.2 Schematic representation of the Hubbard or Mott–Hubbard gap (U).

The Hubbard model is widely used to describe two opposing tendencies: the first one is to explain the metallic behavior due to electron hopping, which delocalizes the electrons into itinerant states, and latter is to explain the Mott insulators, which is the localization of electrons (electrons adherent to individual atoms). The inter-atomic Coulombic repulsion is determined by the degrees of freedom surrounding the *d*-electrons, i.e. charge, orbital, spin, lattice, etc. (Tsuda et al. 2000). Additionally, the transition metals occupying the B-site in a perovskite lattice have 3*d* (e.g. Fe, Co, Ni), 4*d* (e.g. Mo, Ru, Rh), or 5*d* (e.g. W, Re, Ir) electronic configuration. 3*d*-Orbitals are well localized and thus form a narrow band (W) with a large on-site Coulomb interaction (U). 5*d*-Orbitals are spatially more extended than 4*d* and 3*d* counterparts; as a result, nearest-neighbor orbitals overlap significantly, and therefore W is wider in 5*d*-orbitals than in 3*d* or 4*d* orbitals, i.e. W3*d* < W4*d* < W5*d* (Biswas et al. 2016). Therefore, perovskite structures with 5*d* and 4*d* transition elements behave like insulators at all the temperatures.

In insulating perovskites, temperature causes the electrons to excite from the valence to conduction band, and the resistivity is therefore a temperature-controlled process. For such a process, log ρ versus $1/T$ is linear, and it follows the modified Arrhenius equation as follows (Majumder and Karan 2013):

$$\rho(t) = A\exp\left(\frac{E_0}{k_B T}\right) \tag{6.1}$$

where E_0 is the activation energy, k_B is Boltzmann's constant, and A is the constant of proportionality.

The electrical transport mechanism in the insulating region can be due to the formation of polarons (a strong coupling between an electron and phonons). For which, Eq. 6.1 can be modified as follows:

$$\rho(t) = BT\exp\left(\frac{E_0}{k_B T}\right) \tag{6.2}$$

where B is a measure of ideal conductivity at elevated temperatures, and it is a function of polaron concentration. Variable-range hopping (VRH) is also considered as a mechanism of electron transport in perovskite materials. VRH has been suggested by Mott to describe transport at low temperatures when the electronic states are localized near the Fermi energy, then:

$$\rho(t) = \rho_0\exp\left(\frac{E_0}{k_B T}\right)^{1/4} \tag{6.3}$$

6.3 METAL-INSULATOR TRANSITION IN PEROVSKITES

In general, one can assume that the perovskite with partially filled *d*-orbitals of B-site cation could exhibit metallic behavior. This may be right for few perovskites, such as $CaVO_3$, $SrVO_3$, $LaCuO_3$, and $LaNiO_3$, which are conducting at room temperature. The electronic conductivity of perovskites at normal temperatures is generally explained in terms of small polarons which are thermally activated. During the process, the electrons hop from one B-site to the other via B–O–B bridge which is mostly increased by the overlap of anion and B-site cation, which is, in turn, dependent on the B–O distance and B–O–B superexchange angle (Torrance et al. 1992).

At the molecular scale, the structure of the perovskite includes the valence state of the A- and B-site cations, spin state of the B-site cations, and the oxygen deficiency (Erat et al. 2010). The high resistivity of the perovskite materials at and below room temperature can be due to the localization of electrons by several factors. The localized cluster states can be delocalized by overlapping the wavefunctions of the adjacent clusters, and the formation of conducting *d*-band is possible by the transfer of electrons between cations *via* intervening oxygen ions. Therefore, oxygen in ABO_3 perovskites plays a decisive role in the electronic transport properties of these materials (Harrell et al. 2017). For example, the oxygen mediates the electron hopping or transfer between neighboring cations, Mn–Mn, Fe–Fe, or Fe–Ni, through exchange interactions. However, the oxidation state and the spin state of the metal ions determine whether hopping across oxygen ions can take place or not.

The common electron exchange interactions between the cations through anions in perovskite materials are superexchange and double exchange interactions. Superexchange is the spin-related antiferromagnetic coupling between two next-to-nearest neighbor positive ions through a non-magnetic oxygen anion. It strongly depends on the electronic and crystallographic structures such as electron occupancy, orbital configuration, and geometry. In the case of double exchange, the electrons move between positive ions having different *d*-shell occupancy via a non-magnetic oxygen anion (Erat et al. 2010).

6.3.1 Metal Insulator Transition by Doping and Temperature

The desired metallic or semiconducting properties from insulating perovskites can be achieved by introducing A-, B-, and X-site substitutions in simple perovskites. These substitutions directly affect the spin polarization of band structure, thus leading to the desired electronic properties in perovskites. For instance, if J–T distortion is the factor that prevents the formation of metallic perovskite, then suppressing this distortion by optimizing reaction conditions (high temperature or pressure) or doping with an aliovalent element can help to achieve metallic properties (Dai et al. 1996).

Electronic band structure of perovskite oxides is sensitive to doping of A-, B-, or X-site with aliovalent ions, due to the different modes of charge balance in the system, and the associated occupancy of electrons in the conduction band. One can easily tailor the electronic properties by doping either A- or B- or X-site to generate electrons or holes as charge carriers. If electrons are the charge carriers, then the perovskite exhibits *n*-type behavior, or *p*-type if holes are the charge carriers. As a general rule of thumb followed in silicon and germanium semiconductors, doping ions with a higher valence than the parent lattice site creates more electrons, and lower valence creates holes and the subsequent change in the conductivity.

Titanates are the most widely studied model materials to understand the mechanism of the metal–insulator transition in perovskite oxides. For instance, the alkaline earth titanate perovskites, such as $CaTiO_3$, $SrTiO_3$, and $BaTiO_3$, are insulators because of $3d^0$ configuration of B-site Ti^{4+} ions (Lee et al. 2008). These materials can be tuned to be metallic by introducing electrons into the empty $3dt_{2g}$ band by introducing oxygen vacancies, or doping B-site with aliovalent elements. The oxygen-deficient form of $BaTiO_3$, $BaTiO_{2.75}$, is metallic with low resistivity. In $BaTiO_{2.75}$, the vacant oxygen sites compensate for the structural distortions that lead to insulating ferroelectric behavior in a fully stoichiometric $SrTiO_3$ phase, in such a way that the Ti–O bonds are equal and Ti–O–Ti bond angle is 180°, which allows the formation of conduction band. Moreover, the charge balance due to the presence of oxygen vacancies is achieved by the formation of Ti^{3+} ($3d^1$) cations, which add two electrons in the conduction band, and ultimately resulting in a metallic perovskite (Jeong et al. 2011). Similarly, doping H– with O^{2-}, by forming oxyhydride, allows the charge balancing by the formation of Ti^{3+} ($3d^1$), which populates the e_g level of *d*-band (Yamamoto et al. 2015). Moreover, the doping in X-sites urges to attain a cubic phase with the random distribution of H^- and O^{2-}. Such insulator to metal transitions are also observed in $CaTiO_{3-\delta}H_\delta$, $SrTiO_{3-\delta}H_\delta$ (Sakaguchi et al. 2012), and $BaTiO_{3-\delta}H_\delta$ (Kobayashi et al. 2012).

Doping pentavalent Nb^{5+} in $SrTiO_3$ and $BaTiO_3$ by partially substituting B-site Ti^{4+} ions adds electrons to the conduction band, therefore leading to a metallic nature (Horikiri et al. 2007). Here, the excess charge of Nb^{5+} is balanced by the formation Ti^{3+} ($3d^1$) ions. However, $BaNb_{0.01}Ti_{0.99}O_3$ remains insulator, even though Ti^{3+} ($3d^1$) is formed; the structural study reveals that Ti–O bonds are irregular and Ti–O–Ti bond remains distorted, on the contrary to the expected Ti–O–Ti bond angle of 180° (Nowotny and Rekas 1994). The increase in the conductivity by the addition of electrons as charge carriers to the conduction band, as we have seen above, is a typical example of *n*-type behavior.

The insulator to metal transition is also observed in perovskites with partial replacement of A-site with an aliovalent ion, which can bring a similar effect discussed previously by the introduction of holes. They are called *p*-type conductors. One such example is $LaTiO_3$, which is an insulator. $LaTiO_3$ can be transitioned from insulator to metallic behavior by doping La-site with Sr ions. In contrast to Nb^{5+} substation of Ti^{4+}, each substation of a trivalent La-site with divalent Sr atom creates one hole per Ti-site. Therefore, Sr doping leads to a decrease in the electronic band filling and enhancement of hole mobility, creating a *p*-type conductor.

The insulator–metal behavior in magnetite depends on the size of the lanthanide ion present on the A-site, which determines the degree of octahedral tilt, and as the tilting increases with a decrease in the size of lanthanide ion, at the same time bond angle Mn–O–Mn drops from 180° (Autret et al. 2004). The reduction in Mn–O–Mn bond angle from 180° causes an increase in resistivity. Therefore, the metallic nature of lanthanide manganite decreases from left to right in the periodic table. The effect of J–T distortions in the manganites can be diluted by doping A-site of lanthanide manganites with alkaline earth metals such as Ca, Sr, and Ba. $LaMnO_3$ shows various magnetic phases when A-site doped with Sr or Ca having an empirical formula $La_{1-x}Sr_xMnO_3$ or $La_{1-x}Ca_xMnO_3$, as presented in Figure 6.3a and b, respectively. Each magnetic transition is accompanied by a change in electronic conductivity. In both above materials, the antiferromagnetic insulating phase is observed at either ends of the phase diagram ($x < 0.1$ and $x > 0.5$), and it exhibits ferromagnetic

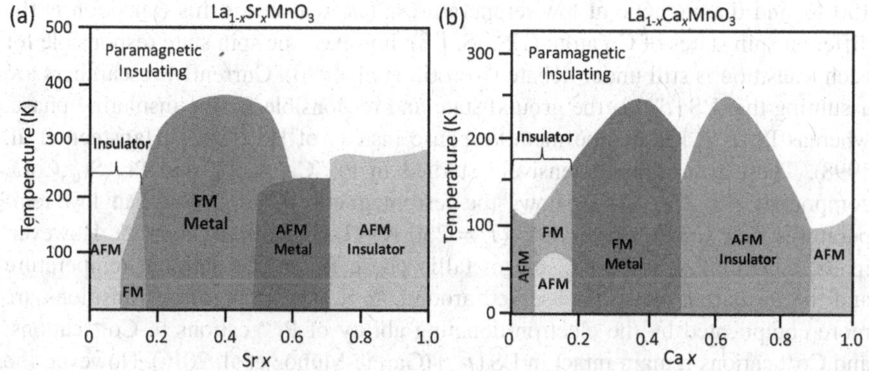

FIGURE 6.3 Phase diagram of (a) $La_{1-x}Sr_xMnO_3$ and (b) $La_{1-x}Ca_xMnO_3$ (Rao 2000). The labels of phases: AFM – antiferromagnetic and FM – ferromagnetic, reproduced with permission of the American Chemical Society.

metal behavior in-between states ($0.2 < x < 0.5$), which is the conducting phase. The presence of divalent ion in the trivalent A-site creates a mixed valence of Mn^{3+} and Mn^{4+}. Doping these ions facilitates the transfer of electrons from Mn^{3+} to Mn^{4+} ions through a double-exchange mechanism (Shah et al. 2012). If the Mn–O–Mn bond angle is 180°, then the Mn e_g orbitals can interact directly with O 2*p*-orbitals, which are almost at the same energy level. Since Mn^{3+} has more electrons than Mn^{4+}, an electron from O 2*p* tends to move to Mn^{4+} if there is no localization of electrons. The vacant O 2*p*-orbital is later filled by the electron from Mn^{3+}, and such an exchange is called as double exchange, as discussed in the previous chapter. The double-exchange mechanism can be observed in several perovskite structures with transition metal with mixed M^{3+}/M^{4+} or M^{2+}/M^{3+} (M = Co, Ni, Fe, etc.) valence states, which are formed when the A-site with higher valence is doped with ions of lower valence. The double exchange also favors the ferromagnetic coupling between the transition metal ions.

The optimized composition for Ca doping for double exchange is $x = (0.2–0.5)$ for $La_{1-x}Ca_xMnO_3$ (Meneghini et al. 2002) and Sr doping $x = (0.16–0.5)$ in $La_{1-x}Sr_xMnO_3$ (Branković et al. 2010). These compositions provide enough Mn^{4+} cations to promote conductivity through double exchange leading to metallic conductivity. The double-exchange mechanism of electrons is sensitive to high temperatures, as the ferromagnetic phase transforms to paramagnetic phase due to the tampering of electron delocalization at high temperatures. Generally, $LaMnO_3$ is insulating at high temperatures, and the transition temperature of the metal to insulator phase is governed by the doped alkaline earth metal. The transition temperatures for $La_{0.7}Sr_{0.3}MnO_3$, $La_{0.7}Ca_{0.3}MnO_3$, and $La_{0.7}Ba_{0.3}MnO_3$ are 350, 250, and 320 K, respectively.

The superexchange is the antiferromagnetic coupling of the nearest transition metal ions with antiparallel spins through a non-magnetic anion with paired spins. In magnetically ordered perovskites, such as $LaCrO_3$, $PbCrO_3$, $CaMnO_3$, and $LaFeO_3$, the electronic conductivity is through superexchange. On the other hand, $LaCoO_3$ behaves completely opposite; it is metallic at temperature 500 K, paramagnetic at

100 K, and diamagnetic at low temperatures. The reason for this transition is the different spin states of Co atom (HS, IS, LS); however, the spin state responsible for each transition is still under debate (Korotin et al. 1996). Currently, researchers are assuming that LS $\left(t_{2g}^{6}\right)$ is the ground state and responsible for the insulating phase, whereas IS $\left(t_{2g}^{5}e_{g}^{1}\right)$ is for the metallic phase instead of HS $\left(t_{2g}^{4}e_{g}^{2}\right)$ (Hansteen et al. 1998). These transitions extensively studied in $Pr_{0.5}Ca_{0.5}CoO_3$ and $Pr_{0.5}Sr_{0.5}CoO_3$ compounds. $Pr_{0.5}Ca_{0.5}CoO_3$ shows the ferromagnetic metallic phase at low temperatures near Curie temperature (T_c = 230 K) (Tsubouchi et al. 2002). However, in the case of $Pr_{0.5}Sr_{0.5}CoO_3$, the metallic phase is noticed at room temperature and the insulating phase is observed around 75 K. The changes in transitions are more complicated by the electron-donating ability of Pr^{3+} cations to Co^{3+} cations, and Co^{4+} cations remain intact in LS $\left(t_{2g}^{5}\right)$ (García-Muñoz et al. 2016). However, the $Pr_{0.5}Ca_{0.5}CoO_3$ and $Pr_{0.5}Sr_{0.5}CoO_3$ compounds have different phases at room temperature and low temperatures, suggesting that Ca^{2+} and Sr^{2+} cations play their part in governing metal–insulator transition.

B-site substitution of perovskites also influences the conductivity due to carriers scattering. This phenomenon is well explained by considering conducting $La_{0.67}Sr_{0.33}Mn_{1-x}Co_xO_3$ perovskite; the corresponding phase diagram is shown in Figure 6.4. At low temperatures <100 K, as the Mn-site is doped with Co (0%–0.2%) as in $La_{0.67}Sr_{0.33}MnO_3$, the resistivity is slightly increased, but there is a dramatic increase in the resistivity as the percent of Co is increased further (0.3%–0.7%). The resistivity is increased due to the weakening of double-exchange interaction of Mn ions in the presence of Co, and *vice versa* in heavy Co doping. When Co % is increased further (0.7%–1%), the resistivity is decreased again. Similar to the double exchange between Mn^{3+}/Mn^{4+} ions, the double exchange is possible between Co^{3+}/Co^{4+}; however, the double-exchange pathways are disrupted in the presence of intermediate doping of ions. At these doping levels, the interaction between Mn and Co ions results in localization of electrons at a specific area, forming the insulating-conducting character as well as the superparamagnetic like free spins and reentrant spin-glass behavior. The superparamagnetic-like free spins and reentrant spin

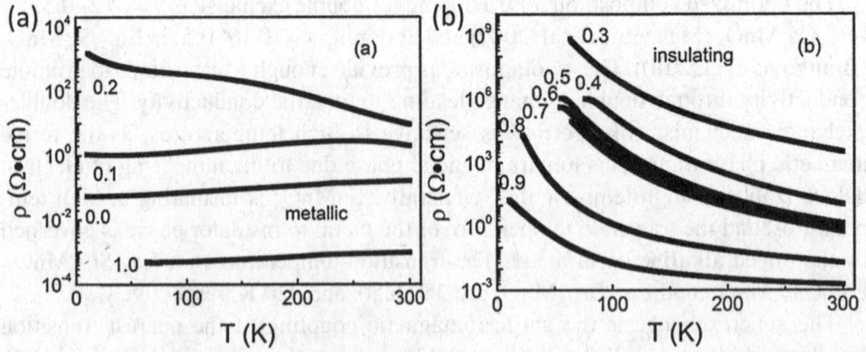

FIGURE 6.4 The temperature dependence of resistivity for $La_{0.7}Sr_{0.3}Mn_{1-x}Co_xO_3$ samples (semilog scale): (a) $0.0 \le x \le 0.2$ and $x = 1.0$, and (b) $0.3 \le x \le 0.9$, reproduced with permission of the American Chemical Society (Chen et al. 2014).

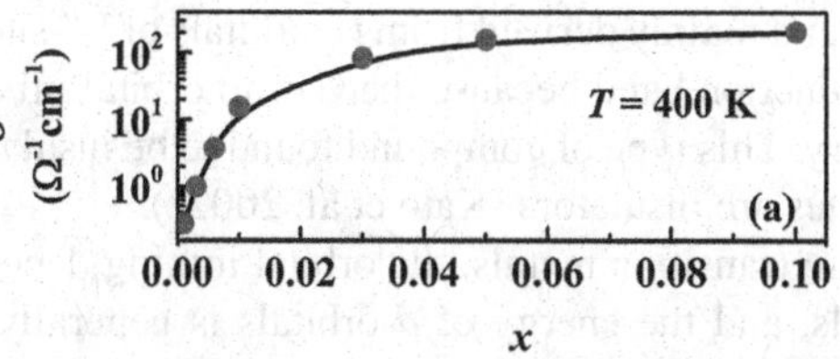

FIGURE 6.5 The electrical conductivity as a function of Nb contents at T ~ 400 K of $Ba_{0.7}Eu_{0.3}Ti_{1-x}Nb_xO_3$, reproduced with permission of the Elsevier (Rubi and Mahendiran 2020).

glass originate from the competition between antiferromagnetic superexchange (Mn^{3+}–Mn^{3+} and Co^{3+}–Co^{3+}, etc.) and ferromagnetic double-exchange interactions (Mn^{3+}–Mn^{4+}, Co^{3+}–Co^{4+}, etc.) (Chen et al. 2014).

The insulator–metallic transition temperatures for Sm and Eu nickelites are 460 and 400 K, respectively. In perovskite systems with more than one lanthanide ion, the insulator–metal transition temperature is governed by the larger lanthanide cation. For example, in $Eu_xLa_{1-x}NiO_3$, the phase transition temperature (T) increases with an increase in Eu content ($x = 0.4$, $T = 125$ K; $x = 0.5$, $T = 190$ K; $x = 0.6$, $T = 260$ K and $x = 0.8$, $T = 380$ K). Doping of B-site significantly enhances the electronic properties, considering $Ba_{0.7}Eu_{0.3}Ti_{1-x}Nb_xO_3$ ($x = 0$ to 0.1); the electronic properties will vary along with an increase in dopant Nb concentration as depicted in Figure 6.5 (Rubi and Mahendiran 2020).

$NaOsO_3$ is an example of metal–insulator transition other than 3*d* transition element in the B-site. At room temperature, $NaOsO_3$ is an antiferromagnetic insulator, and it will undergo a transition to metallic phase at 410 K. It displays orthorhombic crystal structure in both insulating and metallic phases; hence, the structural transitions or distortions are not the reason for the metal–insulator transition (Du et al. 2012). This is the first real example of the Slater transition, in which magnetic spin ordering is the driving force for the metal–insulator transition. This transition occurs due to the splitting of half-filled 5*d*-orbital $\left(5d^3e_{2g}^3\right)$ of Os^{5+} cation into empty and filled bands by antiferromagnetic spin ordering of Os^{5+} cation spins.

Double perovskites are another class of perovskites with a formula of $A_2B'B''O_6$. The bandwidth of double perovskites mostly depends on the spatial orbital overlap between the different B-site elements as well as the crystal structure (Borges et al. 1999). The bandwidth of these compounds can be estimated using two major parameters: bond length and bond angle. The double perovskites with 3*d* group elements on the B-sites usually show a narrow bandwidth, owing to the contracted nature of 3*d*-orbitals and corresponding reluctant overlapping with 2*p*-orbitals. In contrast, 4*d* and 5*d* group elements have extended orbitals, which results in the wider band in perovskites having 4*d*/5*d* B-site cations (Sánchez et al. 2002). Hence, spin delocalization is more favorable in compounds with 4*d*/5*d* B-site cations. Most frequently, this type of perovskites are reported as metallic or near to the metallic phase (Kato et al. 2002a). In $A_2B'B''O_6$ perovskites, the electron delocalization depends on two B-site cations, unlike simple perovskites (ABO_3). In ordered $A_2B'B''O_6$ perovskites, B′ cation might inhibit the electron depolarization of B″ cation by distortion of $B''O_6$ octahedra. If the B′ cation is from the main group and B″ cation is a transition metal, the π* conduction band mainly derived from t_{2g} orbitals of B″-site. B′-site has no contribution to the conduction band because there is no orbital mixing with π* orbitals because of symmetry. This type of compound found to be insulators, and most of the $A_2B'B''O_6$ perovskites are insulators (Kato et al. 2002b).

If both B-sites are transition metals, the orbital mixing depends on the energetic overlap of *d*-orbitals, and the energy of *d*-orbitals is generally governed by redox potentials (Ramesha et al. 2001). If the band energies of *d*-orbitals are different, then the mixing of orbitals is poor. For instance, Sr_2FeWO_6 is an insulator because of the high band energy of tungsten *d*-orbitals than iron (Matteo et al. 2003). There are another two reasons for insulating phase: one is partial electronic isolation of two B cations results in narrow bandgap, and the other is no orbital mixing of t_{2g} and e_g orbitals due to different symmetries of B-site cations, even though the structural distortion of octahedra might allow some σ to π interactions. Matching of orbital energies of two B-site cations is a rare scenario because it depends on several parameters. In the case of transition metals, the orbital energy of the elements decreases from right to left of the periodic table, because of the decrease in atomic size. Likewise, when moving from left to right in periodic table, the ionic radius decreases, which leads to the corresponding decrease in crystal field splitting and bond covalency (Zhao et al. 2014).

Further, there are some notable factors that govern the crystal field splitting energies in transition elements. 3*d* Elements tend to have high spin states because their small crystal field splitting energies correspond to the electron–electron repulsions (Harvey et al. 2003). In contrast, 4*d*/5*d* elements preferably found in low spin states because of their high crystal field splitting energies and negligible inter electronic repulsions. Another reason is that high cation charge of 4*d*/5*d* elements results in high crystal field splitting energies (Lee et al. 2003). The higher cation charge means that 4*d*/5*d* cation sites tend to have fewer electrons, and often they found in t_{2g} orbitals, while in 3*d* the electrons often filled in e_g orbitals owing to their lower oxidation states, and this occurrence may hinder electron transfer because of the symmetry. The low spin state is also possible with 3*d* elements, especially with cobalt, by the weakening of crystal field splitting of B′ cation through the decrease of electron density around it due to the strong covalency of B″–O bond (Taguchi 1996). For example, $Ba_2CoB''O_6$ with B″ = Nb or Ta and La_2CoFeO_6 are proposed to have low or intermediate spin Co^{3+} cation (Yoshii 2000). Comparing unit cell volume of a series of $R_2B'PtO_6$ (B′ = Mg, Co, Ni or Zn) compounds reveals that Co yields a low unit cell volume compared to other compounds, which confirms the low spin or intermediate spin state of Co (Vasala and Karppinen 2015). Spin orbital coupling (SOC) is another factor that regulates the electronic structure 4*d*/5*d* B-site cations in double perovskites (Chen et al. 2010). In the case of 3*d* B-site cations, SOC effect is negligible because it is much weaker compared to other energy parameters (Svoboda et al. 2017). For instance, in the case of Ir^{4+} and Ir^{5+} B-site cations, the SOC found to split the degenerated t_{2g} orbitals (Narayanan 2010). Nevertheless, the crystal structure has prime importance in governing electronic structure by lifting the degeneracy of *d*-orbitals and orbital moment quenching through crystal structure distortions (Varignon et al. 2019).

Various modes of conductivities are observed in semiconducting $A_2B'B''O_6$ perovskites. The common reason for these properties is thermally excited small-polaron hopping (SPH) (Saxena et al. 2017). Some perovskites exhibit semiconducting nature due to Mott VRH (Deshpande et al. 2011), as observed in disordered compounds such as Sr_2CoSbO_6 (Wei et al. 2010) and Sr_2RuMnO_6 (Woodward et al. 2008). Some compounds display both VRH and SPH at different temperatures; for instance, $Sr_2B'MoO_6$ when B′ = Cr or Mn (Poddar et al. 2004) and A_2CoIrO_6 when A = La or Sr are found to display VRH at low temperatures and SRH at high temperatures (Narayanan et al. 2010). The VRH conduction is due to the formation of an impurity band near the Fermi level because of structural distortions, oxygen vacancies, and cation disordering. SRH will prevail at high temperatures owing to the conductivity because of thermal excitation. There are some semiconducting compounds that display very low resistivity in the order of magnitude 10^{-3}–10^{-4} Ω. Sr_2CoIrO_6 (Narayanan 2010) and $Sr_2B'MoO_6$ (B′ = Ti or Zr) exhibit the resistivity close to metal–insulator transition (Saxena et al. 2017). Some of the insulating compounds reported with low resistivity values could be due to the presence of polycrystalline grain boundaries.

A_2FeMoO_6 perovskite, when A = Ca, Sr, or Ba, shows changes in conductivity with respect to temperature in H_2 atmosphere (Figure 6.6) (Zhang et al. 2010). Ca_2FeMoO_6 shows metallic behavior in this temperature range, while Sr_2FeMoO_6 behaves differently. Sr_2FeMoO_6 shows metallic nature up to 150°C; in the temperature range of 150°C–550°C, it behaves as a semiconductor and reverts to metallic nature above 550°C. Ba_2FeMoO_6 also shows different conduction behaviors. It is metallic below 150°C, and above 150°C, it displays only semiconducting behavior. The reports on metallic double perovskites are very scarce, and most of them belong to half-metallic ferrimagnets, and others are near metallic with metal–insulator transitions.

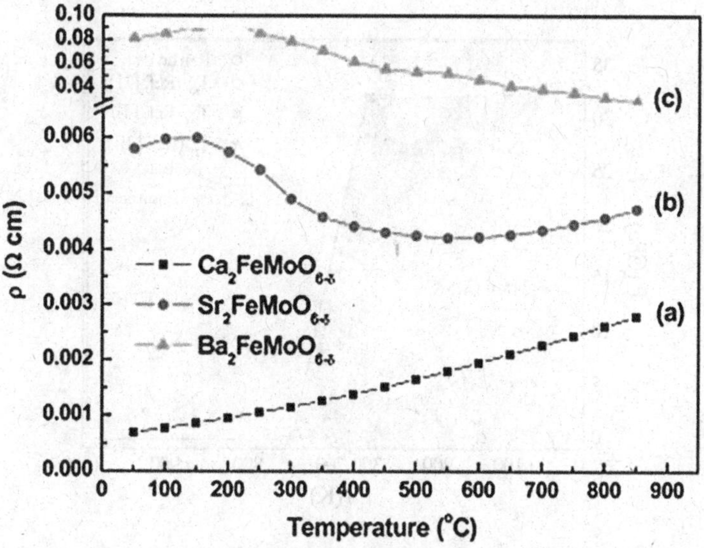

FIGURE 6.6 Electrical resistivities as functions of temperatures of A_2FeMoO_6 double perovskites in H_2: (a) Ca_2FeMoO_6, (b) Sr_2FeMoO_6, and (c) Ba_2FeMoO_6, reproduced with permission of the Elsevier (Zhang et al. 2010).

For example, A_2VMoO_6 compounds, when A = Ca or Sr, are metallic in nature with Pauli paramagnetic magnetic behavior, and the reason for their metallic phase is valence mixing of $V^{3/4+}/Mo^{4/5+}$. Sr_2MnRuO_6 and La_2CuRhO_6 are insulators, although they have a similar type of valence mixing like A_2VMoO_6. Metal–insulator transitions with respect to temperature are noticed in compounds such as Ca_2MnMoO_6, Sr_2CoTiO_6, and Y_2NiNiO_6. Ca_2MnMoO_6 display metal–insulator transition at 209 K, while Sr_2CoTiO_6 at 700 K and Y_2NiNiO_6 at 580 K.

6.3.2 Metal–Insulator Transition under Pressure and Temperature

Electronic structure significantly varies with temperature and pressure due to the changes in the BO_6 octahedra. The changes in the octahedra change the crystal field and the associated electronic states of B-site ions. For instance, $LaMnO_3$ is a paramagnetic insulator at ambient conditions and an example for cooperative J–T and an orbitally ordered system. With reference to Figure 6.7, one can observe that $LaMnO_3$ is insulating up to ~750°C. As the pressure is increased, the material remains insulated for a wide temperature range. The insulating behavior of $LaMnO_3$ is arising from the localization of electrons by the J–T distortions of MnO_6, and the J–T distortions of MnO_6 units are observed up to 32 GPa, without considering the temperature. On the further increase of the pressure to 34 GPa, the material turns to be metallic, which is due to the suppression of J–T distortion in the lattice (Baldini et al. 2011). However, it is interesting to note that the metal–insulator transition temperature is reduced to room temperature under a pressure of 35 GPa.

The pressure dependence of metal–insulator transition is observed in $LnNiO_3$ (Ln = Nd, Pr, Eu, Y, Lu) as well. The pressure–temperature relationship on the

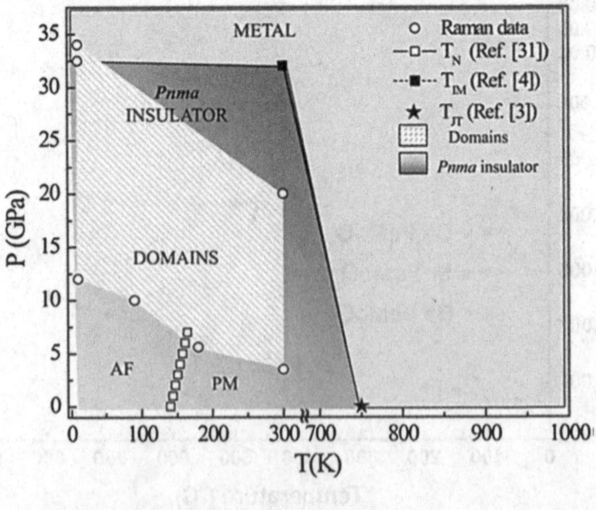

FIGURE 6.7 P-T phase diagram of $LaMnO_3$. PM – paramagnetic phase and AFM – antiferromagnetic phase. White squares represent TN. The star and black squares indicate the JT transition and the IMT temperatures, respectively, reproduced with permission of the American Physical Society (Baldini et al. 2011).

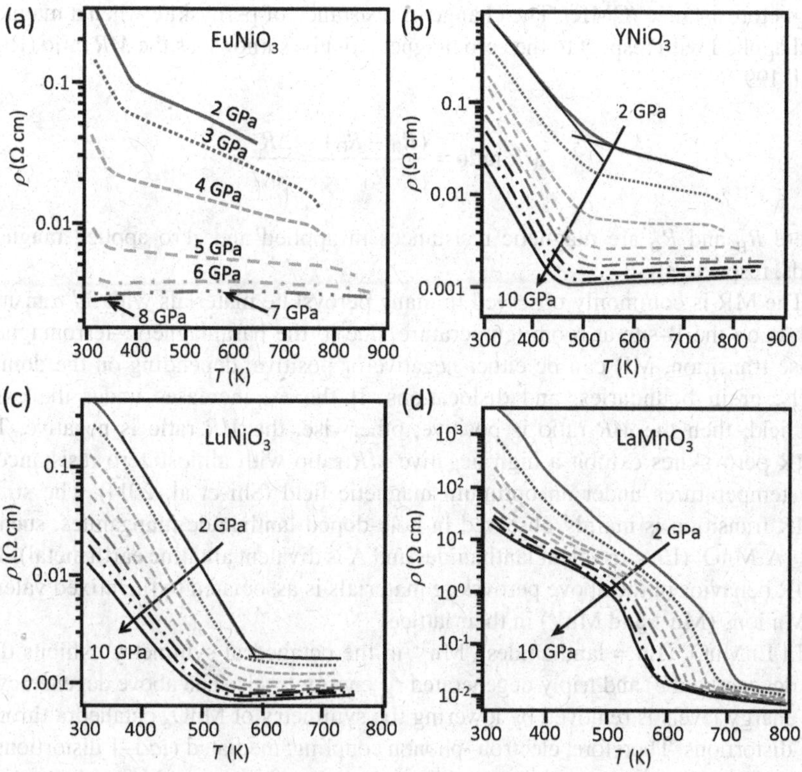

FIGURE 6.8 Temperature–pressure dependence of resistivity for $LnNiO_3$ (Ln = Eu, Y, Lu) and $LaMnO_3$, reproduced with permission of the American Physical Society (Cheng et al. 2010).

resistivity for representative perovskites is shown in Figure 6.8, with a comparison to $LaMnO_3$. The $LnNiO_3$ structures with a smaller Ln^{3+} (e.g. Lu) radius exhibit J–T distortions, which is identical to $LaMnO_3$. The J–T distortions are less sensitive to the pressure-induced metal–insulator transition. Therefore, one cannot expect a significant change in the metal to insulator transition temperature at low pressures, as shown in Figure 6.8c and d. In contrast, $LnNiO_3$ with a large ionic radius exhibits a substantial change in the metal–insulator transition temperature with pressure (Cheng et al. 2010; Obradors et al. 1993).

6.3.3 Metal–Insulator Transition under Magnetic Field

6.3.3.1 Colossal Magnetoresistance

Magnetoresistance (MR) is the property of some materials, due to which its resistance changes when an external magnetic field is applied. The MR is often observed in many materials, where the resistance drops by 5%, but in the case of most perovskite manganites, the resistance drop is by several orders in magnitude, called colossal magnetoresistance (CMR). The change in resistance of perovskite when a magnetic field applied with respect to the zero magnetic fields is known as the *MR* ratio (Righi et al. 1997):

$$MR\text{ ratio} = \frac{(R_H - R_O)}{R_O} = \frac{\Delta R}{R_O} \tag{6.4}$$

where R_H and R_O are magnetic resistances in applied and zero applied magnetic fields, respectively.

The MR is commonly observed in many perovskite materials with 3*d* transition metals on the B-site at room temperature, due to the paramagnetic–ferromagnetic phase transition. MR can be either negative or positive, depending on the domain walls, grain boundaries, and dislocations. If the R_H increases under the external field, then the *MR* ratio is positive; otherwise, the *MR* ratio is negative. The CMR perovskites exhibit a high negative *MR* ratio with almost zero resistance at low temperatures under an optimum magnetic field (Shi et al. 2014). The strong CMR transition is mainly observed in hole-doped lanthanide manganites, such as $Ln_{1-x}A_xMnO_3$ (Ln = trivalent lanthanides and A is divalent alkaline earth metal). The CMR behavior in the above perovskite materials is associated to the mixed valence of Mn ions (Mn^{3+} and Mn^{4+}) in their lattice.

In $LnMnO_3$ (Ln = lanthanides), Mn^{3+} in the octahedral symmetry exhibits doubly degenerated e_g and triply degenerated t_{2g} energy levels. The above degeneracy of the energy levels is removed by lowering the symmetry of MnO_6 octahedra through J–T distortions. Therefore, electron–phonon coupling; mediated *via* J–T distortions of the Mn^{3+} ion, is prominent in perovskite manganites than perovskite with any other transition metal on the B-site. The strong J–T coupling of the manganites leads to interesting properties such as charge and orbital ordering, ferrimagnetism, metal–insulator transition, CMR, and giant MR (Ramirez 1997). These materials are always paramagnetic and adopt different antiferromagnetic ordering with the magnetic moments perpendicular to *c*-axis at lower temperatures as discussed in the previous chapter. In all the antiferromagnetic ordering, the magnetic moments flip in the subsequent planes; therefore, it can be called as the canted antiferromagnetic state.

The CMR property in $LnMnO_3$ is observed only for a certain level of doping of A^{2+} ions in the trivalent Ln-sites. When the Ln-sites are doped with A = Ca, Sr, Ba or Pb to form $Ln_{1-x}A_xMnO_3$ (Shah et al. 2012), two different oxidation states of Mn ions (Mn^{3+} and Mn^{4+}) are formed (Luo et al. 2005, 2009). Let us consider the phase diagrams of $La_{1-x}Sr_xMnO_3$ and $Pr_{1-x}Sr_xMnO_3$ perovskites. As shown in Figure 6.9, one can notice the change in the magnetic ordering with respect to the level of A^{2+} dopants. However, the level of Sr^{2+} doping to accomplish a certain magnetic ordering is different in both $La_{1-x}Sr_xMnO_3$ and $Pr_{1-x}Sr_xMnO_3$ perovskite structures. With very low Sr^{2+} doping, both the above structures adopt a canted insulating phase, with $x \approx 0.1$. As the Sr doping is between 0.1% and 0.17% in $La_{1-x}Sr_xMnO_3$ and 0.1%–0.3% in $Pr_{1-x}Sr_xMnO_3$, they adopt a ferromagnetic insulating phase. The further increase of Sr^{2+} doping in the respective lattices leads to the ferromagnetic metallic phase at the ground state. At these compositions, the ferromagnetic metallic behavior is associated to double-exchange mechanism as discussed in Chapter 5, and

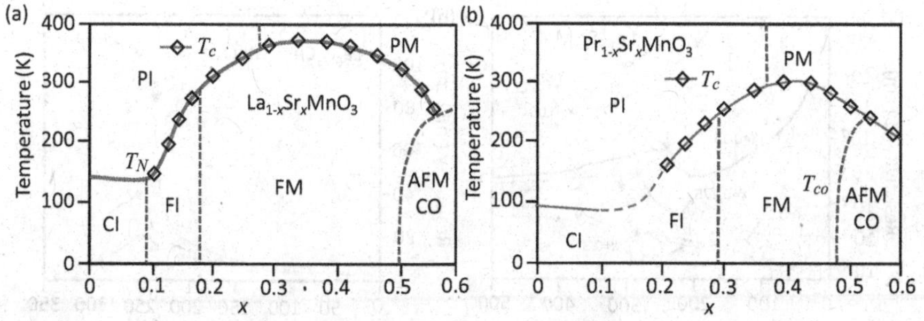

FIGURE 6.9 The magnetic–electronic phase diagrams for (a) $La_{1-x}Sr_xMnO_3$ (Urushibara et al. 1995) and (b) $Pr_{1-x}Sr_xMnO_3$ as function of Sr-doping (Tokura et al. 1996). PI, PM, and CI denote the paramagnetic insulating, paramagnetic metallic, and spin-canted insulating states, respectively. FI and FM denote the ferromagnetic insulating and ferromagnetic metallic states, respectively, reproduced with permission of the American Physical Society.

the ferromagnetic metallic phase is transformed to paramagnetic insulator at Curie temperature (T_c): $T_c \approx 375$ K for $La_{1-x}Sr_xMnO_3$ and $T_c \approx 250$ K for $Pr_{1-x}Sr_xMnO_3$.

CMR is observed in the perovskite compositions, where the materials exhibit a ferromagnetic metallic behavior at the ground state. For the ferromagnetic-metallic compositions, if the temperature is raised above the Curie temperature, i.e. above the ferromagnetic to paramagnetic transition, the materials loose their ferromagnetic-metallic behavior. On the contrary, in CMR materials, the ferromagnetic-metallic behavior can be retained by applying a sufficiently high magnetic field, meaning that the magnetic field reestablishes the ferromagnetic metallic property, resulting in a metal–insulator transition. Thus, the applied magnetic field aligns the t_{2g} spins and reduces scattering of the carriers by the local spins. Curie temperature is the boundary above which the materials exhibit CMR behavior. The effect of the Sr-doping level on the resistivity in $La_{1-x}Sr_xMnO_3$ perovskites is shown in Figure 6.10a. The change in resistance with respect to temperature under different magnetic fields of representative CMR materials is shown in Figure 6.10b–d.

It is interesting to note that perovskite-structured compounds exhibiting CMR are usually orthorhombic, but the symmetry can be modified near the metal–insulator boundary by the application of a magnetic field. The above structural transformation can also occur by changing the temperature. For instance, $Pr_{1-x}Ca_xMnO_3$ exhibits a phase transition between a high-temperature pseudo-cubic phase and an orthorhombic phase above 900 K for $x = 0$, which decreases to room temperature at around $x = 0.3$, which is again the doping level where the ferromagnetic state appears, suggesting a strong magneto-elastic coupling. The CMR plots of $Pr_{0.7}Ca_{0.3}MnO_3$ are shown in Figure 6.10d.

The tolerance factor of $Ln_{1-x}A_xMnO_3$ compounds varies depending on the size of A^{2+} dopants such as Ca, Sr, Ba, and Pb, with the resultant perovskite structures with a stable tolerance factor, t, between 0.85 and 0.91. However, the charge balance and the associated metallic behavior are achieved only at a certain fraction x, as observed in Figure 6.9a and b, creating sufficient tetravalent Mn^{4+} in a random fashion throughout the crystal, with the remainder in the Mn^{3+} state. In all the above cases, the divalent

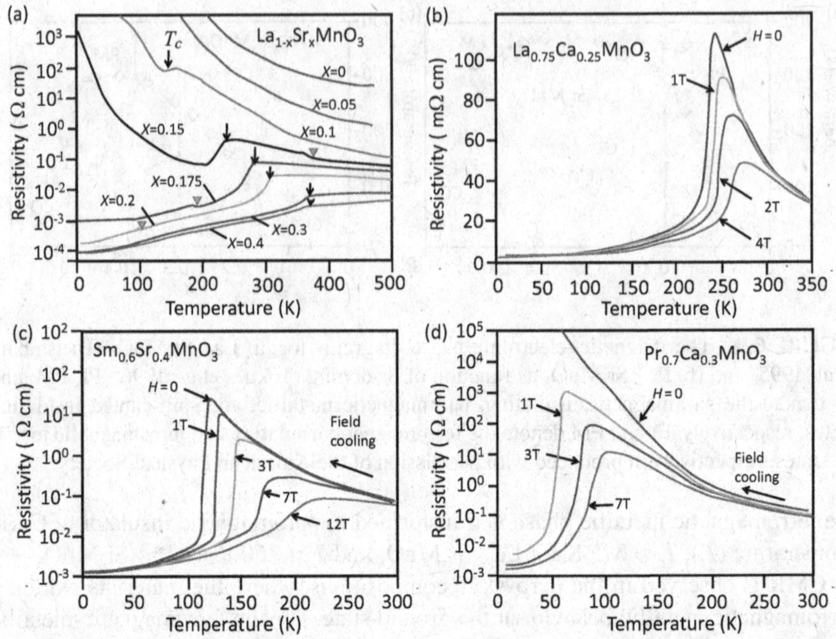

FIGURE 6.10 (a) Temperature dependence of resistivity for $La_{1-x}Sr_xMnO_3$ (arrows indicate the critical temperature for the ferromagnetic phase transition, and the triangles are anomalies due to the phase transition from rhombohedral (*R*3*c*) to orthorhombic (*Pbnm*)) (Urushibara et al. 1995), and MR as a function of temperature at various magnetic fields (b) $La_{075}Ca_{0.25}MnO_3$ (Schiffer et al. 1995), (c) $Sm_{0.6}Sr_{0.4}MnO_3$ (Tokura et al. 1996), and (d) $Pr_{0.7}Ca_{0.3}MnO_3$. (Tomioka et al. 1996), reproduced with permission of the American Physical Society.

substitution is equivalent to hole doping, which is similar to increasing the oxygen content (Hansteen et al. 2004). Additionally, above the ferromagnetic Curie point, T_c, the resistivity behaves like a semiconductor, $\frac{d\rho}{dT} < 0$, but below T_c, not only there is a sharp reduction in resistivity, but also a transition to metallic behavior, $\frac{d\rho}{dT} < 0$. Figure 6.11a and b shows the phase diagram and the change in Curie temperature for $Ln_{1-x}A_xMnO_3$ perovskite-type manganites as a function of the tolerance factor.

The smaller the ionic radius of the A-site, the smaller the electronic bandwidth, *W*. The double-exchange interaction in $Ln_{1-x}A_xMnO_3$ is the resultant of itinerancy of the doped holes, which favors a reduction in *W*, so as the charge-ordered state. At the same time, the ferromagnetic state is destabilized. When the charge-ordering occurs, the resistivity increases abruptly by several orders of magnitude, depending on the crystallographic and/or magnetic structures. In accordance with the electronic change, the ferromagnetic to antiferromagnetic transition occurs simultaneously. The ferromagnetic transition temperature T_c changes steeply as a function of the tolerance factor, and the competition between the ferromagnetic double-exchange and antiferromagnetic charge order interactions is sensitive to *W* (Kuwahara et al. 1997).

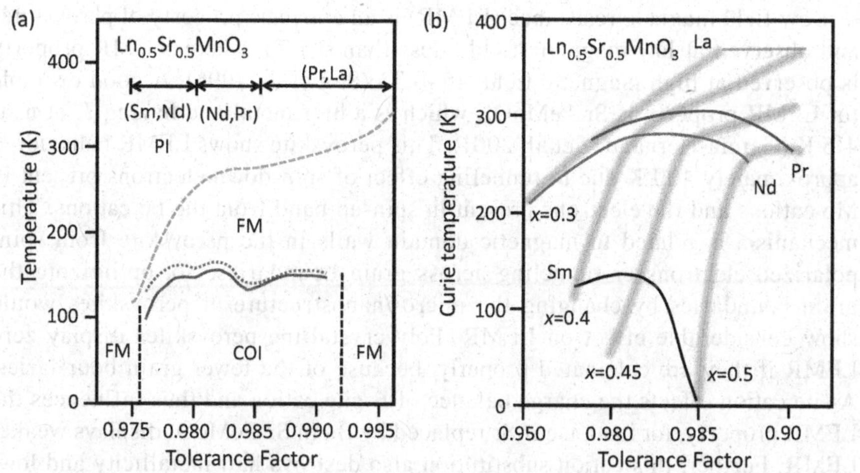

FIGURE 6.11 Effect of tolerance factor on phase transition (a) temperature and (b) Curie temperature of representative $Ln_{1-x}Sr_xMnO_3$ perovskites.

In $Ln_{1-x}A_xMnO_3$ perovskite materials, the strong CMR phenomenon is arising from the strong mutual coupling among spin, charge, and orbital degrees of freedom of Mn *d*-electrons. In the presence of mixed-valent Mn ions, Mn^{4+} and Mn^{3+}, the electronic configurations are t_{2g}^3 and $t_{2g}^3e_g^1$, respectively. In divalent ion-doped perovskite manganites, the mixed electronic configuration results in a competition between the antiferromagnetic superexchange interaction (Mn^{3+}–O–Mn^{3+}) and the ferromagnetic double-exchange interaction (Mn^{3+}–O–Mn^{4+}). Additionally, as the e_g electrons are localized, the J–T interaction lifts the orbital degeneracy and favors the occupation of either the $d_{x^2-y^2}$ or d_{z^2} orbitals of the e_g states. Therefore, the orbital degree of freedom increases, which ultimately controls the competition between the superexchange and the double-exchange interactions. As a result, one can expect a connection between the charge and spin dynamics in the doped manganites. Layered perovskites such as $La_{1-x}Sr_{1+x}MnO_4$, $La_{2-2x}Sr_{1+2x}Mn_2O_7$, and $La_{2-2x}Ca_{1+2x}Mn_2O_7$ exhibit CMR characteristics analogous to the parent $LaMnO_3$. The MR for $La_{2-2x}Sr_{1+2x}Mn_2O_7$ is much larger as compared to 3D ($n = \infty$) system with the general trend of increasing MR with decreasing T_c. However, $n = 2$ layered systems exhibit anisotropy in their transport properties: $d\rho_{ab}/dT > 0$ near ferromagnetic T_c, while $d\rho_c/dT < 0$ in the same temperature region (Ramirez 1997).

Further, the doping of Mn with magnetic or non-magnetic cations such as Ni^{2+}, Mg^{2+}, In^{3+}, Ga^{3+}, Al^{3+}, Cr^{3+}, Fe^{3+}, Sn^{4+} and Ti^{4+} influences the CMR property considerably. For example, $Pr_{0.7}Ca_{0.2}Sr_{0.1}MnO_3$ shows the R_O/R_H value of 230, and this value increases to $4{\times}10^5$ if it is modestly doped with Mg^{2+} ($Pr_{0.7}Ca_{0.2}Sr_{0.1}Mn_{0.98}Mg_{0.02}O_3$) (Budhani et al. 2001). The actual physical reasons or factors are not explored for CMR property. Initially, researchers believed that it is associated with double exchange or superexchange. However, after studying Mn-related perovskites, researchers are now assuming ferro metallic half-metallicity as a major factor to influence CMR (Jakob et al. 2001).

Low-field magneto resistance (LFMR) is an extrinsic property of perovskites and observed at low magnetic fields (less than 0.5 T), whereas CMR property is observed at high magnetic fields (5–6 T) (Sun et al. 1996). A good example for LFMR property is Sr_2FeMoO_6, which is a half metal and having T_c at near 415 K (García-Hernández et al. 2001). This perovskite shows LFMR behavior at approximately 400 K due to tunneling effect of spin-down electrons present in Mo cations and the electrons present in spin-up band from the Fe cations. This mechanism is related to magnetic domain walls in the perovskite from spin-polarized electrons or tunneling across grain boundaries. So, optimizing the grain boundaries by changing the micro-/nanostructure of perovskites would show considerable effect on LFMR. Poly crystalline perovskites display zero LFMR if they are calcinated properly, because of the fewer grain boundaries. A-site cation effects the charge balance of B-site cation and thus influences the LFMR property, for instance if Sr replaced by Ba in Sr_2FeMoO_6 displays weaker LFMR. Further, this cation substitution also destroys half metallicity and lowers the LFMR, but a small amount of doping improves the LFMR property of perovskite (Maignan et al. 1999).

Giant magnetoresistance (GMR) is yet another property associated to MR mostly exhibited by perovskite thin films. The difference between CMR and GMR is that CMR deals with intrinsic properties of a material, and the magnetic fields required to induce CMR are very large as compared to those needed for real-time applications. On the other hand, the GMR is a characteristic of layered structures that by themselves are weak MR materials, but that in combination, produce a large MR effect (Dagotto 2003). For instance, Ruddlesden–Popper(RP)-layered perovskite-type structure $La_{2-2x}Sr_{1+2x}Mn_2O_7$ ($n = 2$), formed by a rock-salt-type layer $(La, Sr)_2O_2$ separating every MnO_6 octahedral layers, exhibits a large drop in resistivity as compared to $La_{1-x}Sr_{1+x}MnO_4$ ($n = 1$) or $La_{1-x}Sr_xMnO_3$ ($n = \infty$) under a low magnetic field. $La_{2-2x}Sr_{1+2x}Mn_2O_7$ is a ferromagnetic metal below $T_c = 126$ K and exhibits a rapid reduction in resistance without any hysteresis just above T_c. The improved MR is attained at the cost of decreased T_c, owing to the reduction of the kinetic energy of the carriers as compared to cubic $La_{1-x}Sr_xMnO_3$. The enhanced MR property in the layered manganite is due to the anisotropic exchange interaction. The intra-layer exchange interaction is much stronger than the inter-layer double-exchange interaction (Moritomo et al. 1996). $La_{2-2x}Ca_{1+2x}Mn_2O_7$ is another layered manganite that exhibits GMR. Similarly, one can expect GMR behavior in double or complex perovskites with layered ordering.

So far, the ongoing discussions are on negative CMR and GMR characteristics of perovskites, where the resistivity decreases with the applied magnetic field. There are several perovskite oxides that exhibit positive CMR and GMR characteristics, where the resistance increases with the applied magnetic field. $SrIrO_3$ is an orthorhombic perovskite that exhibits positive MR near 170 K (Zhao et al. 2008). A positive CMR is observed at the *p-n* junction of $La_{0.9}Sr_{0.1}MnO_3$ and $SrNb_{0.01}Ti_{0.99}O_3$ (Lu et al. 2004).

Double perovskite structured Sr_2CrWO_6 and Mn_2FeReO_6 are examples of positive GMR materials. The negative MR characteristics are very common in perovskite structures, which is associated to the ferromagnetic metallic behavior. But the positive GMR is observed in the above double perovskites are associated to ferromagnetic nature of the materials. In Sr_2CrWO_6 films deposited on $SrTiO_3$ substrate, the antiferromagnetic coupling between Cr^{3+} ($3d^3$, $S = 3/2$) and W^{5+} ($5d^1$, $S = 1/2$) leads to a macroscopic metallic ferrimagnetic nature, and the resistivity increases as the temperature increases. In the above system, the applied magnetic field suppresses the long-range antiferromagnetic ordering, leading to a state with short-range antiferromagnetic correlations and strong electronic scattering, which induces the positive GMR with 17,200% (at 2 K and 7 T) increase (Zhang et al. 2017). Mn_2FeReO_6 is ferrimagnetically ordered up to 500 K due to the coupling between Mn and Re sublattices, and the temperature-dependent positive GMR results an increase in the resistivity up to 220% (Li et al. 2015).

6.4 HALF METALS

Half metal is a material that shows the conductance of electrons of specific spin orientation and exhibits semiconductive or insulating nature for opposite spin-oriented electrons (Figure 6.12). All half metals are ferromagnets, but not all ferromagnets are half metals. The simple electronic band structure of perovskites becomes more complex when considering the spin of electrons present in the orbitals. For example, calculations show that t_{2g} and e_g orbitals further split into sub-bands called spin up (↑) and spin down (↓) in $CaMnO_3$. According to theoretical prediction, $CaMnO_3$ should be metallic with $Mn^{4+}\left(d^3t_{2g}^3\right)$ cation as B-site, but it is an antiferromagnetic semiconductor. Because of electron correlation, according to Hund's principle, both t_{2g} and e_g orbitals further split into "spin up (↑)" and "spin down (↓)" sub-bands (Filippetti and Pickett 1999). In $CaMnO_3$, the t_{2g} band split into lower energy spin-up (↑) and higher energy spin-down (↓) bands; here, e_g band is not taken into consideration because it is empty. The lowest energy spin-up band of t_{2g} is located just above the O $2p$ band, and it is filled with three electrons $\left(Mn^{4+} \cdot d^3t_{2g}^3\right)$, and most of the time, high-energy spin-up band of t_{2g} remains unfilled, as a result, antiferromagnetic semiconducting nature is observed in $CaMnO_3$.

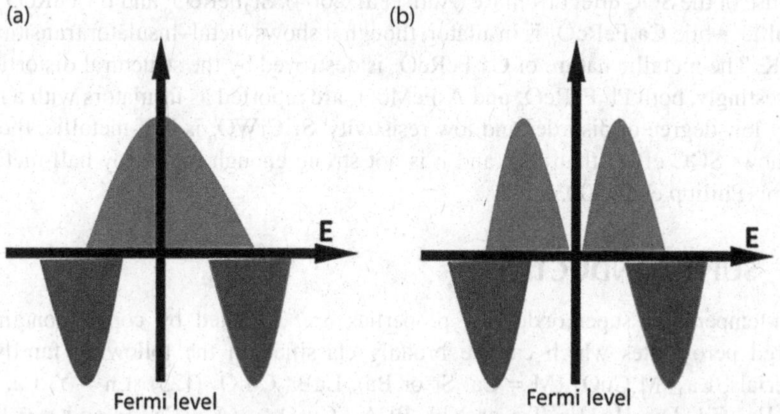

FIGURE 6.12 (See online for color version.) Schematic representation of (a) a half-metal and (b) a half-semiconductor. Green and sky blue present the spin-down and spin-up states, respectively.

The above example gives us information about how the spin state of electrons affects the conductivity of perovskites. If the electrons present in both spin-up (↑) and spin-down (↓) sub-bands of t_{2g}, it will lead to mixed phase called half-metallic phase, and the electrons in the spin-up band is responsible for metallic phase. In contrast, electrons in the spin down band result in the insulating phase. Half-metallic conductivity is noticed in A-site doped lanthanum manganite perovskites, and the doping elements are alkaline earth metals (Ca and Sr). The half-metallic nature is found in $La_{1-x}Ca_xMnO_3$ in the range of $x = 0.2–0.4$ and particularly in $La_{0.7}Sr_{0.3}MnO_3$ (Geng and Zhang 2006; Pickett and Singh 1996). The half-metallic nature of perovskites depends on temperature; for instance, the half-metallic ferromagnetic phase of $La_{0.7}Sr_{0.3}MnO_3$ changes above the Curie temperature, and at this point, electron spin disordering happens by losing single electron spin current.

A_2FeMoO_6 (A = Ca, Sr and Ba) compounds are the most studied double perovskites for their half-metallic nature (Borges et al. 1999). The band structure of these compounds is made of hybridization of Fe^{3+} $\left(3d^5t_{2g}^5\right)$, Mo^{5+} $\left(4d^1t_{2g}^1\right)$, and orbitals from O 2*p*. The spin-up band does not intersect the Fermi level, and these compounds are insulators. The spin-down band is mostly made up of electrons from Fe^{3+} t_{2g}^5 and Mo^{5+} t_{2g}^1, which intersects the Fermi level, and these compounds are metallic. The conductivity mainly originates from the *d*-orbital electrons of Mo^{5+} cations, and some extent of half-metallic nature depends on the structural disorder of Fe- and Mo-sites, especially the antisite disorder in which Mo occupies the sites of Fe and *vice versa*. This antisite disorder plays a key role in optimizing half-metallic nature of double perovskites. The half-metallic properties of perovskites are applicable in spintronic devices, and careful synthetic methods are essential to make half-metallic perovskites by optimizing antisite disorder (Hirohata and Takanashi 2014).

However, finding half-metallic property is very difficult, and sometimes polycrystalline grain boundaries mask the electrical conductivity. Other factors such as SOC and cation disorder may alter the conductive nature or destroy half-metallic nature. For instance, a series of A_2FeReO_6 (A = Ca, Sr and Ba) compounds have an electronic structure similar to A_2FeMoO_6 but show a different conductive nature because of the SOC effect from Re (Auth et al. 2004). Sr_2FeReO_6 and Ba_2FeReO_6 are metallic, while Ca_2FeReO_6 is insulator, though it shows metal–insulator transition at 150 K. The metallic nature of Ca_2FeReO_6 is destroyed by the structural distortions. Interestingly, both Pb_2FeReO_6 and A_2FeMoO_6 are reported as insulators with a relatively low degree of disorder and low resistivity. Sr_2CrWO_6 is half-metallic, though W shows SOC effect than Mo; and it is not strong enough to destroy half-metallic nature (Philipp et al. 2003).

6.5 SUPERCONDUCTIVITY

High-temperature superconducting properties are exhibited by copper-containing layered perovskites which can be broadly classified in the following family of materials, $La_{2-x}M_xCuO_4$ (M = Ca, Sr or Ba), $LnBa_2Cu_3O_7$ (123) (Ln = Y, La, Nd, Sm, Eu, Gd, Dy, Ho, Er, Tm or Yb), $Bi_2A_{n+1}Cu_nO_{2n+4}$ (A = Ca, Sr and n = 1–4), $Tl_2Ca_{n-1}Ba_2Cu_nO_{2n+4}$ (n = 1–4), $TlCa_{n-1}Ba_2Cu_nO_{2n+3}$ (n = 1–4), and $Pb_2Sr_2LnCu_3O_8$ (Ln = lanthanides). Among these perovskite structures, all the Bi- and Tl-containing cuprates adopt Aurivillius-type layered perovskite structure and most others are RP phases. In general, all the cuprate superconductors are composed of anion-deficient perovskite layers of a square planar-coordinated CuO_2 sheets, and the 123 compounds have one-dimensional CuO chains in addition. The square planar coordination has a Cu–O bond distance around 1.98 Å, which is an indication of the high covalency of these compounds.

$La_{2-x}M_xCuO_4$ perovskite family is an intermediate nonstoichiometric phase in between simple and double perovskites, adopt a tetragonal structure at room temperature and orthorhombic at low temperature around 180 K. The superconducting transition temperature (T_{SC}) of these compounds is in the range of 25–40 K depending on the percentage of M dopants. Substitution of La in the above compounds with smaller lanthanide ions or Cu by other transition metals reduces the superconducting transition temperature. Sr_2RuO_4 is another well-known non-cuprate superconductor that has a similar structure to (Sr, La)$_2$CuO$_4$ and (Ba, La)$_2$CuO$_4$. The metallic behavior of Sr_2RuO_4 is owing to the presence of *d*-electrons (Ru^{4+}) in partly occupied t_{2g} band (Bergemann et al. 2003). However, isostructural Ca_2RuO_4 is a Mott insulator because Ru–O–Ru bond angle is disturbed from 180° due to RuO_6 octahedral tilt. This tilting effect can be minimized at higher temperatures, and Ca_2RuO_4 becomes metallic near 350 K (Mizokawa et al. 2001). The doping of Ru with Cr also shows a similar effect. A small amount of Cr^{3+} lessens the octahedral tilting and angle of Ru–O–Ru bond. Hence, after doping, metal–insulator transition temperature decreases in $Ca_2Ru_{0.968}Cr_{0.032}O_4$ (284 K) and $Ca_2Ru_{0.908}Cr_{0.092}O_4$ (81 K) (Shen et al. 2017).

In the case of $LnBa_2Cu_3O_7$ compounds, replacing the Ln ion does not affect the T_{SC} significantly; however, T_{SC} strongly depends on oxygen stoichiometry (δ), as shown in Figure 6.13. Generally, the T_c increases and falls according to the oxygen composition changes in cuprates. For instance, in the case of $YBa_2Cu_3O_{7-\delta}$, T_{SC} remains constant as 90 K up to $\delta = 0.2$, but drops to 55 K, when $\delta = 0.2$ and 0.4. Further increase in δ lowers the T_{SC}, and $YBa_2Cu_3O_{7-\delta}$ becomes non-superconducting when $\delta \approx 0.6$.

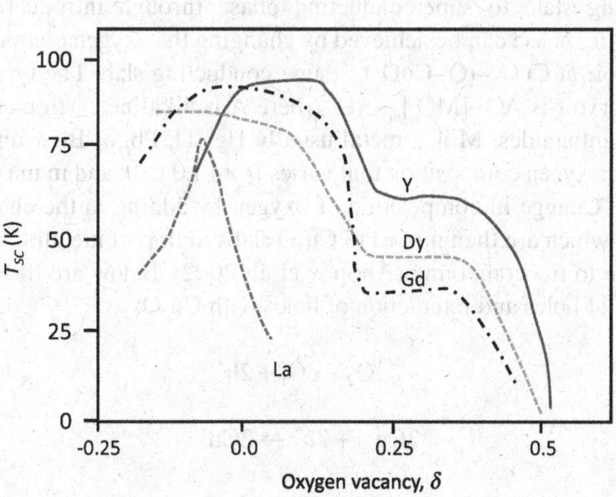

FIGURE 6.13 A representative image of T_{SC} *vs.* δ for $LnBa_2Cu_3O_7$.

The structure is transformed simultaneously from orthorhombic (0.0–0.60) to tetragonal as the $\delta \geq 0.6$, and one can observe similar transformation if Y is replaced by Gd or Dy as shown in Figure 6.13 (Rao and Raveau 1989). On the other hand, the excess of oxygen is not favorable for superconductivity, as $YBa_2Cu_3O_8$ shows a T_{SC} of only 80 K. Additionally, in $Bi_2Sr_2CaCu_2O_{8+\delta}$ superconductor, the T_{SC} increases to a maximum of 95 K at oxygen content near to $Bi_2Sr_2CaCu_2O_{8.19}$ (Krishana et al. 1997). The T_{SC} also depends on the number of Cu_2O layers; three Cu_2O layers will cause a rise in T_c. The slab of Cu_2O layers derived from CuO_6 octahedra by the elimination of apical oxygen ions is the key for the superconducting nature of all the above cuprates.

There are some metallic cuprates that do not show superconductivity, such as $La_4BaCu_5O_{13}$, in which the absence of two-dimensional Cu_2O layers is the reason for not displaying superconductivity. $La_4BaCu_5O_{13}$ comprises apex-linked CuO_5 pyramids and CuO_6 octahedra of CuO_6 (Guo et al. 1988). The transition temperature (T_{SC}) of superconducting cuprates is dependent on various factors such as oxygen stoichiometry and external factors including crystal elastic strain and pressure (Hanaguri et al. 2009).

It is assumed that the mixed valence of Cu is considered to be essential for the superconductivity in layered cuprates (Alvarez et al. 2005). However, the superconductivity of $YBa_2Cu_3O_{6.5}$ (T_c – 45 K), with Cu^{2+} ions only, can be associated with the intergrowth of $YBa_2Cu_3O_7$ and $YBa_2Cu_3O_6$ phases. In stoichiometric Ln_2CuO_4, Cu is present in the 2+ state, but in the case of $La_{2-x}M_xCuO_4$ or $YBa_2Cu_3O_7$, it has a mixed valence of Cu^{2+} and Cu^{3+} states, which is vital to explain the superconducting properties. But there is no experimental evidence for the presence of Cu^{3+} in the above lattices; instead, Cu^{1+} state and holes on oxygen, O^{1-}, and the dimerization of oxygen holes result in peroxo-type species, O^{2-}. These mobile oxygen holes are responsible for the superconductivity in the above layered cuprates. The formation of holes is more favored by the d^{10} state Cu^{1+} ions than by Cu^{2+} (d^9) ions.

Oxygen composition is the main factor that plays a vital role to induce the change from insulating state to superconducting phase through introduction of holes. Superconducting phase can be achieved by changing the oxygen composition within charge reservoir at CuO_2–$(Q–CuO_2)_{n-1}$ superconducting slab. The typical structure of charge reservoir is AO–$[MO_x]_m$–AO, where A is alkaline earth metal (generally Sr or Ba) or lanthanides; M is a metal usually Hg, Tl, Pb, or Bi; x represents non-stoichiometric oxygen composition that varies from 1.0 to 0; and m makes the values like 0, 1, 2.... Change in composition of oxygen by adding to the charge reservoir creates holes, which are then moved to Cu_2O slabs to impart metallic and superconducting nature to the compound (Antipov et al. 2002). Below are the equations for the formation of holes and interaction of holes with Cu_2O:

$$^{1/2}O_2 \rightarrow O_i'' + 2h^{\bullet} \tag{6.5}$$

$$2Cu_{cu} + 2h^{\bullet} \rightarrow 2Cu^{\bullet} \tag{6.6}$$

For example, considering $La_{1-x}Sr_xCuO_4$ compound, a small amount of doping has a minute effect on changing the insulating property when the Sr ratio crosses the concentration of 0.07, the formation of hole concentration becomes significant, and at low temperature, the compound displays superconducting nature. The Cu^{3+} number reaches a maximum when the concentration of Sr reaches 0.16, which results in a T_{SC} of 30 K. Continuing the replacement of La with Sr results in the formation of oxygen vacancies, and the oxygen composition falls below 4.0. At ~0.27 Sr concentration, the compound is no longer superconductor because the number of Cu^{3+} ions decreases owing to the increase of oxygen vacancies. Superconductivity observed within a composition-temperature range called a superconducting dome. But at high oxygen pressures, the compound shows the superconductivity until the Sr ratio reaches ~0.32. Recently, decoupling of the hole and Sr concentration induced superconductivity in pristine La_2CuO_4 by avoiding conventional doping. The superconductivity in La_2CuO_4 by Sr doping is an example of *p*-type superconductivity. A phase diagram of *p*- and *n*-type superconductors is shown in Figure 6.14.

Nd_2CuO_4 has a similar structure to La_2CuO_4, but the Nd_2CuO_4 compound does not show a transition from the insulating phase to the superconductor when doped with alkaline earth metals (Ca, Sr, Ba). However, Nd_2CuO_4 becomes superconductive when Nd-site doped with Ce^{4+}, for example, considering $Nd_{2-x}Ce_xCuO_4$ in which there is the formation of vacancy at sublattice of Nd^{3+} for every three Ce^{4+} ions. For every single substitution of Nd^{3+} with Ce^{4+} results in generation Cu^+ in place of Cu^{2+} and every two substitutions with Ce^{4+}, there is the inclusion of one oxygen in interstitial ($Nd_{2-x}Ce_xCu^{2+}{}_{1-x}Cu^+O_4$ or $Nd_{2-x}Ce_xCu^{2+}O_{4+x}$) (Spinolo et al. 1995). This compound becomes superconductive at the concentration of ~0.12–0.18, and the maximum T_{SC} of 24 K achieved with the $Nd_{1.85}Ce_{0.15}CuO_4$ compound. Such superconducting behavior is called *n*-type superconductors.

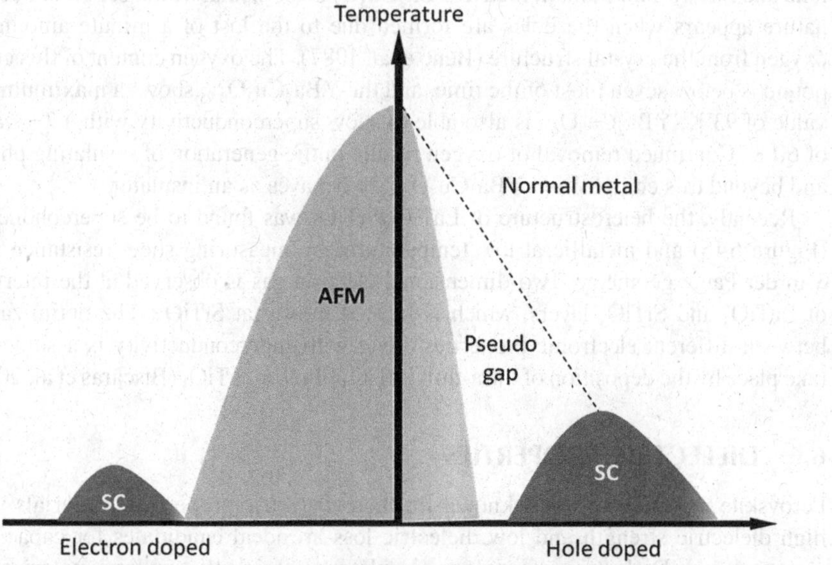

FIGURE 6.14 Phase diagram of representative high-temperature electron-doped and hole-doped Cuprate perovskites (SC: superconductor; AFM: antiferromagnetic insulator).

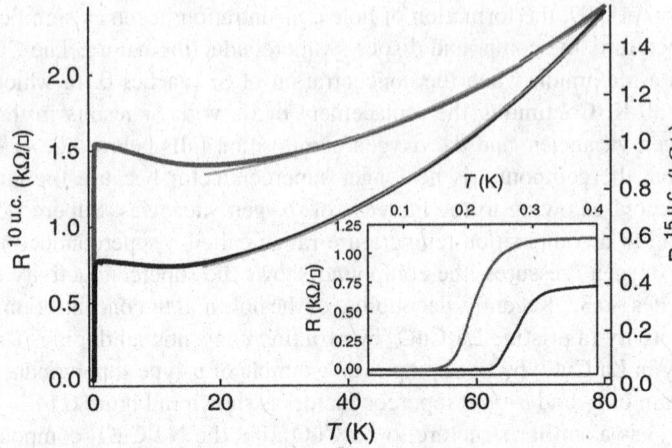

FIGURE 6.15 Sheet resistance of $LaTiO_3/SrTiO_3$ samples in an intermediate range of temperature. (Inset) The superconducting transitions at T_c^{onset} = 310 mK for the 10 u.c. (black dots) and T_c^{onset} = 260 mK for the 15 u.c. (blue dots), where T_c^{onset} is defined by a 10% drop of the resistance (Biscaras et al. 2010).

$YBa_2Cu_3O_7$ is widely studied by researchers because of its ability to show superconductive behavior above the boiling point of liquid nitrogen temperature. The crystal structure comprises a stacking of three perovskite units, and if all atoms (Y, Ba, and O) take their respective oxidation states, the Cu must possess the average charge of 2.33, which can be considered the unit cell containing Cu with two +2 ions and one +3 ion. Stoichiometric $YBa_2Cu_3O_7$ is an insulator; the superconductive nature appears when the holes are formed due to the lost of a minute amount of oxygen from the crystal structure (Beno et al. 1987). The oxygen content of this compound is below seven most of the time, and the $YBa_2Cu_3O_{6.95}$ shows a maximum T_{SC} value of 93 K. $YBa_2Cu_3O_{6.5}$ is also able to show superconductivity with a T_{SC} value of 60 K. Continued removal of oxygen results in the generation of insulating phase, and beyond this composition $YBa_2Cu_3O_{6.35}$, it behaves as an insulator.

Recently, the heterostructure of $LaTiO_3/SrTiO_3$ was found to be superconductive (Figure 6.15) and metallic at low temperatures by measuring sheet resistance in a Van der Pauw geometry. Two-dimensional electron gas is observed at the interface of $LaTiO_3$ and $SrTiO_3$ layers, which is located mostly at $SrTiO_3$. The optimization between different electronic properties along with superconductivity is assumed to take place by the deposition of ultra-thin $LaTiO_3$ films on $SrTiO_3$ (Biscaras et al. 2010).

6.6 DIELECTRIC PROPERTIES

Perovskite materials are best known for their dielectric properties. Materials with high dielectric strength and low dielectric loss are ideal candidates for capacitors and insulators. Dielectric materials are insulator materials that undergo polarization in the presence of an electric field (Figure 6.16). Even though dielectric materials are insulators, the term dielectric represents the ability of a material to store energy by

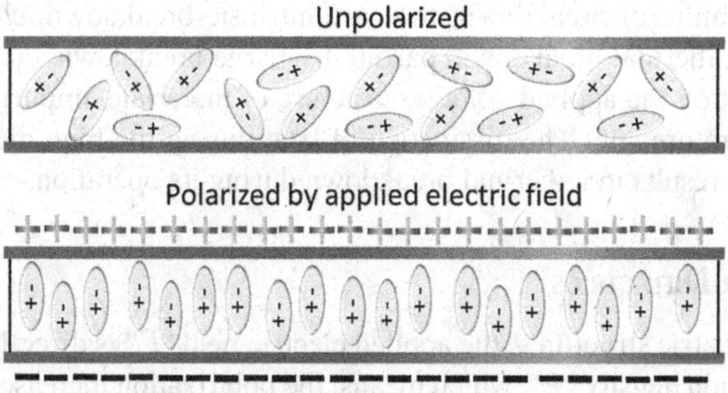

FIGURE 6.16 Schematic representation of a dielectric material.

means of polarization. Under the application of an external field, the positive charges of the dielectric materials are aligned towards the direction of the field and negative charges away from the field. Many perovskites are identified as excellent dielectric materials, and often the dielectric materials exhibit piezoelectric characteristics and the associated properties like pyroelectric properties.

The dielectric characteristics of a material can be considered as a response to an external electric field, and the response here is electric polarization. The dielectric properties of the materials can be used for storing energy since the polarization reverses when the applied field is removed. The ability of a material to store energy is called capacitance, *C*:

$$C = \frac{\varepsilon_0 \varepsilon_r A}{d} \tag{6.7}$$

where ε_0 is the dielectric permittivity in a vacuum (~8.85×10^{-12} F m^{-1}), ε_r is the dielectric constant (or relative dielectric permittivity, $\varepsilon_r = \frac{\varepsilon}{\varepsilon_0}$, here ε is the permittivity of the material) of the dielectric layer, *A* is the overlapping area of the two electrodes, and *d* is the thickness of the dielectric layer.

In a capacitor, during the charging process, charges with opposite sign and equal magnitude are accumulated at the electrodes, creating an internal electric field with a direction opposite to the applied electric field. The strength of the internal electric field increases as the accumulated charges increase, which is a function of the applied field. A capacitor is fully charged when the above internal electric field is equal to the external field. The movement of charges occurs by the external electric field, and electrostatic energy is stored in the dielectric layer. Since the created internal potential is opposite to the external potential, the dielectric materials are excellent candidates for insulators as well.

The polarization characteristic of dielectric material is determined by dielectric loss and dielectric breakdown in addition to its permittivity. Dielectric loss is the energy due to the movement of charge in an alternating applied field. The materials with high permittivity feature high dielectric loss and *vice versa*. Dielectric breakdown represents the upper limit of the applied voltage across the dielectric material. There are several proposed breakdown mechanisms for dielectric materials based on the nature of breakdown; they are intrinsic breakdown, electromechanical breakdown, thermal breakdown, partial discharge breakdown, etc. These are, in turn, dependent on the applied voltage, structure of materials, impurities, humidity, ambient temperature, etc. The dissipation of heat during the high dielectric loss of materials often results in a thermal breakdown during its operation.

6.6.1 Linear Dielectrics

In highly symmetric structures, the applied electric field "*E*" is directly proportional to the polarization density "*P*", which means, the polarization increases linearly with increasing electric field, the so-called as linear dielectrics. In other words, for an ideal linear dielectric, the permittivity is independent of the electric field:

$$P = \varepsilon_0 \chi_e E \tag{6.8}$$

χ_e is known as the dielectric susceptibility, which is a measure of the polarizability of a material under an applied electric field. ε_0 is the permittivity of free space.

In linear dielectrics, there is no hysteresis in the polarization *vs.* applied electric field curve, as shown in Figure 6.17. Therefore, all the polarized charges are reversed back to the original state upon the removal of the electric potential. These dielectric materials are ideal for analog to digital (A/D) conversion and filter applications, which require high thermal stability, property reliability, tolerance, and predictability with respect to frequency (Ulrich and Schaper 2000). Perovskites with the general formulas of $AZrO_3$ and $AHfO_3$ (A = Ba, Ca, Sr) exhibit linear dielectric properties with low permittivity and dielectric loss (Feteira et al. 2008). $CaZrO_3$ and their solid solutions exhibit stable dielectric properties over the temperature range from −55°C to 150°C (Motoki et al. 2000). For instance, $SrTiO_3$-doped $CaZrO_3$ exhibits extremely low dielectric loss over the temperature range from −55°C to 200°C, with a linear P-E hysteresis loop even at 170°C, which may be associated with the large bandgap of $CaZrO_3$ (~4.0 eV).

6.6.2 Non-linear Dielectrics

Non-linear dielectrics are materials whose permittivity changes with the applied electric field. The change in permittivity variation is usually associated with the realignment of

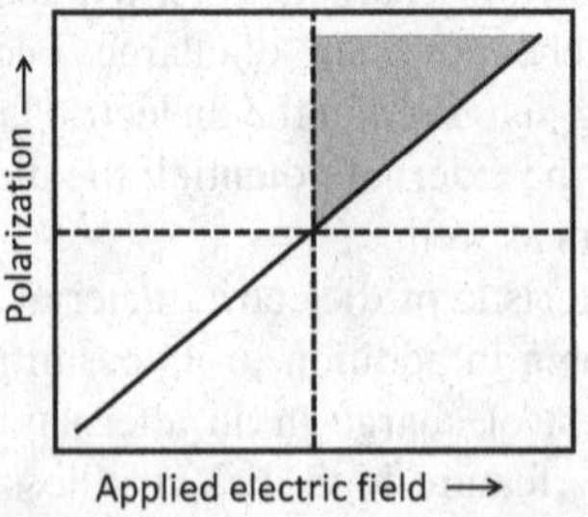

FIGURE 6.17 Schematic representation of hysteresis loops for linear dielectrics.

dipoles, domain switching, domain wall motion, etc. The typical non-linear dielectrics are paraelectrics (PEs), ferroelectrics (FEs), and antiferroelectrics (AFEs).

6.6.2.1 Paraelectrics

Paraelectric materials exhibit dielectric polarizations under an electric field, and the polarization disappears as the electric field is removed. A typical hysteresis loop for a paraelectric material is shown in Figure 6.18. Polarization in parametric crystallites take place by two different means: (1) electronic polarization and (2) ionic polarization. In electronic polarization, the electrons and nucleus of the individual atoms are oriented in response to the applied field, and the ionic polarization takes place by the orientation of positive and negative ions in an applied field. Many paraelectric materials display a small permittivity and a low dielectric loss. If the material remains as a paraelectric at all the temperatures, then it is called as quantum paraelectric. Often the paraelectric materials refer to the materials that exhibit paraelectric state at a high temperature, and they mostly adopt a cubic perovskite structure. The dielectric constant of a paraelectric is always lower than the typical ferroelectric state, but higher than that of a linear dielectric.

$SrTiO_3$ with an ideal cubic perovskite structure exhibits paraelectric property from room temperature to 50 K. In the above temperature range, the inverse dielectric *vs.* temperature curve follows the Curie–Weiss law, as shown in Figure 6.19. $SrTiO_3$ possesses relatively high permittivity (~300) and low dielectric loss (<1%) at room temperature. At low temperature, it approaches a non-linear behavior with a very high dielectric constant, $\varepsilon_r = \sim 18{,}000$ at 1.4 K (Weaver 1959). However, the transition to a ferroelectric phase is not achieved even at 30 mK, and the dielectric constant ε_r stabilizes below 3 K to a constant value of 10^4 (Müller et al. 1991). The paraelectric phase at low temperature is stabilized by quantum fluctuations (temporary change in energy) at cryogenic temperatures, and thus, it is also called a quantum paraelectric or incipient ferroelectric. Room-temperature paraelectric materials with high dielectric constant and breakdown strength are ideal for energy storage. Doping trivalent ions in the Sr-site as in the case of $Sr_{1-1.5x}Bi_xTiO_3$ can increase the permittivity with a breakdown strength of 217.6 kV cm^{-1} (Zhang et al. 2013). The single crystal $SrTiO_3$ exhibits a breakdown strength of ~10^4 kV cm^{-1}, but the sintered structures show a breakdown strength between 80 and 200 kV cm^{-1} (Shende et al. 2001), due to the presence of defects.

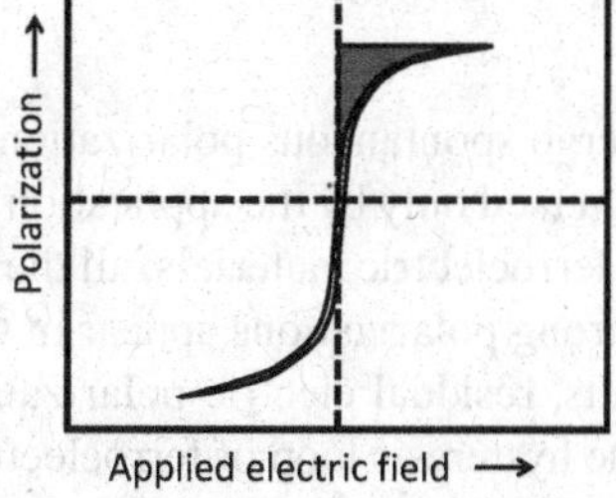

FIGURE 6.18 Schematic representation of hysteresis loops for paraelectric material.

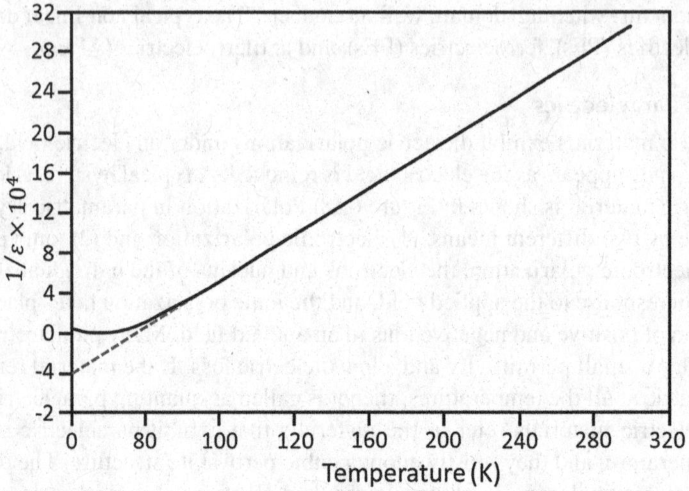

FIGURE 6.19 Inverse dielectric *vs.* temperature curve for $SrTiO_3$. The dotted line represents the Curie–Weiss law.

$BaTiO_3$ is a ferroelectric material with a ferroelectric–paraelectric phase transition temperature at about 393 K, but the addition of Ba to $SrTiO_3$ lattice as in $Ba_{0.4}Sr_{0.6}TiO_3$ results in a ferroelectric–paraelectric transition temperature of about 208 K, with a dielectric constant of ~1,000, and a breakdown strength of ~150 kV cm^{-1} (Song et al. 2014). The breakdown strength is a function of grain size, and in general for ceramic materials, the breakdown strength increases as the grain size decreases due to the increased grain boundary density (Song et al. 2014). Similarly, Ca-doped $SrTiO_3$ also exhibits high permittivity, low loss, and higher breakdown strength (Zhang et al. 2015). Another effective way to improve the dielectric strength is by substituting the B-site Ti ions. For instance, small quantities of Zr doping in $SrTi_{1-x}Zr_xO_3$ ($x = 0$, 0.002, 0.006, 0.01, and 0.014) obtained by solid-state reaction followed by sintering in N_2 atmosphere at 1500°C, resulted in a giant permittivity (>10^4) and a very low dielectric loss (<0.01) (Wang et al. 2015). $Ln_{0.5}Na_{0.5}TiO_3$ (Ln = La, Pr, Nd, Sm, Eu, Gd and Tb) perovskite materials also exhibit paraelectric properties at room temperature. The dielectric constants are decreased as moving from Pr to Tb due to the increase in the volume of TiO_6 octahedron. Unlike $SrTiO_3$, these materials exhibit a high-temperature quantum paraelectricity (~30–50 K) (Sun et al. 1997).

6.6.2.2 Ferroelectrics

Ferroelectric materials undergo spontaneous polarization under an external electric field, and the same can be reversed only by the application of an external electric field in the opposite direction. In ferroelectric materials, all the domains are aligned in the same direction; therefore, strong polarizations appear in the direction of the electric field. Similar to ferromagnets, residual electric polarizations remain even after the electric field is removed. The hysteresis loop of ferroelectric polarization against the external field is shown in Figure 6.20. Many perovskite materials exhibit ferroelectricity below a phase transformation temperature, called the Curie temperature T_c,

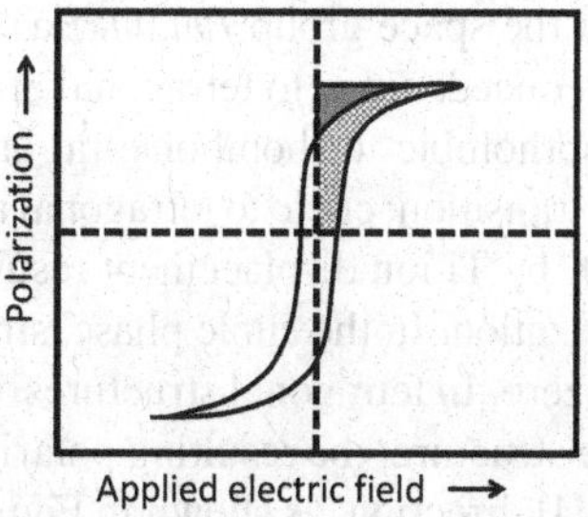

FIGURE 6.20 Schematic representation of hysteresis loops for a ferroelectric material.

and are paraelectric above this temperature. A representative transition of paraelectric to ferroelectric behavior in perovskite materials is shown in Figure 6.21.

Unlike paraelectric materials, the ferroelectric materials do not lose their polarization after removing the electric field. In ferroelectrics, the lattice restricts the relaxation of the ionic configurations formed under the electric field to their initial states. In ferroelectrics, the dipoles in one domain are orientated in the same direction, and the orientation of individual domains is aligned towards the direction of the applied electric field. As a result, ferroelectrics possess high permittivity and dielectric loss, along with a large remnant polarization and low breakdown strength, and ultimately the recoverable energy is very small. In general, ferroelectric to paraelectric phase occurs at Curie temperature (T_c); below T_c, the ferroelectric materials are polar, while above T_c, they are in the nonpolar paraelectric phase. However, every ferroelectric has a temperature point above which the material becomes substantially non-electric, i.e. dielectric, known as its Curie temperature (Nikulin 1988).

$BaTiO_3$ and $PbTiO_3$ are two well-known ferroelectric perovskite materials. Both $BaTiO_3$ and $PbTiO_3$ exhibit identical unit cell volume, but their ferroelectric properties are different. Both the materials are paraelectric at high temperatures with a

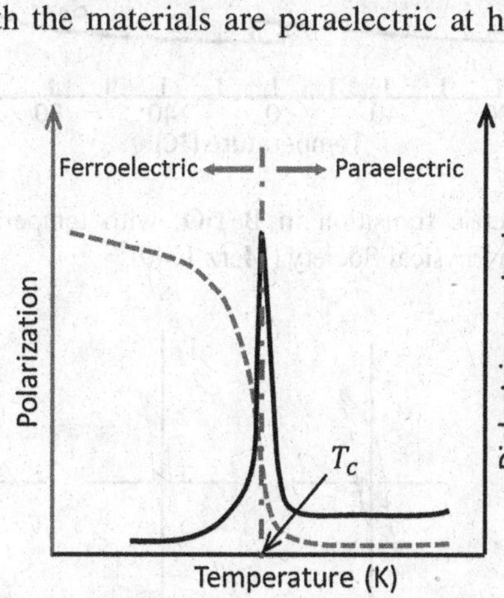

FIGURE 6.21 Illustration of the changes in a ferroelectric material that transforms from a paraelectric cubic into a ferroelectric tetragonal phase with temperature observed in $PbTiO_3$ and $BaTiO_3$.

cubic structure belonging to the space group *Pm3m*. $BaTiO_3$ undergoes three transitions as the temperature is reduced: cubic to tetragonal at 393 K, tetragonal to orthorhombic at 278 K, and orthorhombic to rhombohedral at 1,843 K. On the contrary, PbTiO3 undergoes only one transition: cubic to tetragonal at 766 K. In $BaTiO_3$, the off-centering distortions of TiO_6 by Ti ion displacement result in the structural transformations as well as the polarization. In the cubic phase, since there are no distortions, the resulting polarization is zero. In tetragonal structures, the resultant polarization is along [110]; in orthorhombic structure, the resulting polarization is along [110]; and in rhombohedral, it is along [111] direction, as shown in Figure 6.22. The changes in the P-E curves at and above room temperature are shown in Figure 6.23.

The distinguished perovskites, $BaTiO_3$ and $PbTiO_3$, do not have structures with octahedral rotations, whereas perovskites with octahedral rotations tend not to be polar or ferroelectric. In $BaTiO_3$, the maximum alignment of domains is observed at 380 K where the cubic-to-tetragonal transition takes place, near the Curie

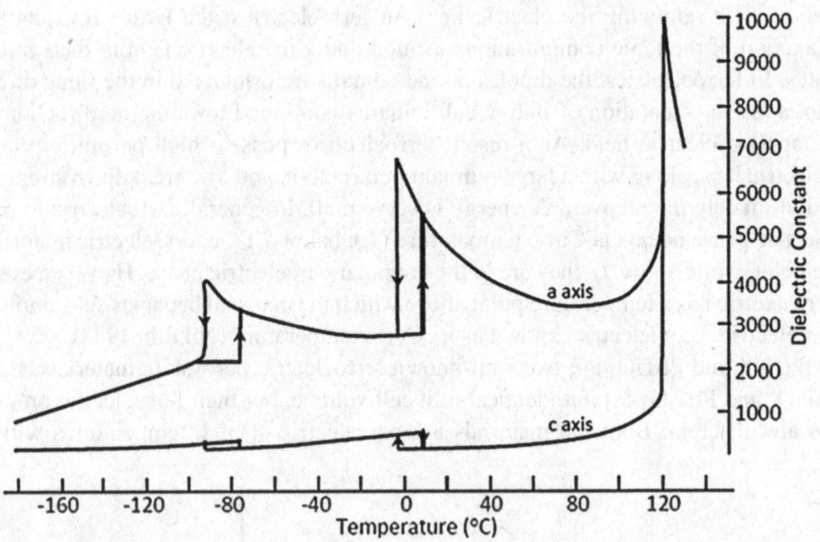

FIGURE 6.22 The dielectric transition in $BaTiO_3$ with temperature, reproduced with permission of the American Physical Society (Merz 1949).

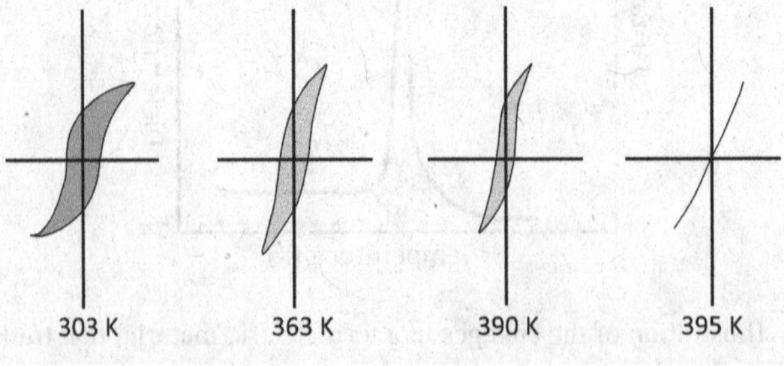

FIGURE 6.23 Change in the hysteresis loop of $BaTiO_3$ under different temperatures.

temperature, under an external electric field. The shape of the ferroelectric hysteresis loop of $BaTiO_3$ changes with temperature. Below the Curie temperature, it becomes wider, and above the Curie temperature, $BaTiO_3$ becomes paraelectric. In $BaTiO_3$, the octahedral interstitial sites occupied by Ti^{4+} ions are quite large compared to the size of the Ti^{4+} ions; therefore, Ti ions are not stable in this octahedral position. In the cubic structure, >380 K, with the applied electric field, the Ti ions tend to move in the octahedral site to a minimum energy position, resulting in polarization, but it is reversed as the electric field is removed. But below 380 K, the off-centering of Ti ions with respect to the surrounding oxygen takes place, and the structure transforms into tetragonal. The off-centering towards one of the apex oxygen results in a spontaneous increase in positive charge in this direction. The application of an electric field opposite to the polarity of this original dipole will cause the Ti^{4+} ion to move through the center of the octahedral site and to an equivalent off-center position. This results in a reversal polarization, hysteresis in the *E vs. P* curve, and ferroelectricity. Since each Ti ion has a +4 charge, the degree of polarization can be very high (Richerson and Lee 2018).

One can observe such displacement in perovskite oxides with highly charged cations such as Ti^{4+}, Zr^{4+}, Nb^{5+}, Ta^{5+}, and W^{6+} displaying a similar ferroelectric mechanism to that of $BaTiO_3$. Others contain asymmetrical ions with a "lone-pair" electron configuration; examples are Pb^{2+}, Bi^{3+}, Sn^{2+}, and Te^{4+}. Each of these ions has two electrons outside a closed *d* shell. These electrons form a lone-pair orbital on one side of the ion and promote a directional bonding. The resulting structure has dipoles that result in spontaneous polarization when the dipoles do not cancel each other (Richerson and Lee 2018).

In general, the dielectric properties of perovskites are dependent on the grain size and temperature. For instance, for $BaTiO_3$, a high dielectric constant is recorded for large grains at Curie point. It is because of the formation of multiple domains within a single grain, and the motion of those walls increases the dielectric constant at Curie point. If the grain size is <1 μm, only a single domain is observed, and the grain boundaries arrest the movement of domain walls; as a result, dielectric constant is low at the Curie point as compared to coarse-grained $BaTiO_3$. Since the internal stresses are higher in fine-grained materials, they exhibit a high permittivity than the coarse-grained materials at room temperature.

In thin films, the B-site displacements are governed by the orientation and fitting between overlying ferroelectric film and substrate. For instance, tetragonal ferroelectric films of the $PbZrO_3$–$PbTiO_3$ system grown on (100) plane have domains parallel and perpendicular to the surface of the substrate, whereas the same films oriented in [111]-direction have domains that meet the surface of the substrate with 45° angle. The non-lead perovskites such as $NaNbO_3$, $KNbO_3$ (Dungan and Golding 1965), and $K_{0.5}Na_{0.5}NbO_3$ (Sen et al. 2011) are also studied widely in order to replace lead-based ferroelectric perovskites.

In double-perovskite compounds, the lead-based ($Pb_2B'B''O_6$) materials are mostly explored by the research community, due to the lone $6s^2$ polarization effect (Barbur and Ardelean 2001). The B-site ordering in these compounds governs the ferroelectric properties if they are highly ordered, which results in ferro- or antiferroelectrics. B-site with disordering or with short-range ordering display relaxor ferroelectric nature with diffuse ferroelectric anomaly and the same also depends on changes in frequency. Interestingly, some compounds display both relaxor and normal ferroelectric behaviors, which is due to heterogeneous cation ordering. As discussed earlier, sintering conditions also play a key role in governing ferroelectric behavior. For example, $Pb_2ScB''O_6$ with B″ = Nb and Ta (Baba-Kishi et al. 2001) shows a change in cation ordering with respect to synthesis parameters; sintering these compounds at high temperatures results in relaxor ferroelectrics, while sintering at low temperatures leads to the formation of ferroelectrics.

6.6.2.3 Relaxor Ferroelectrics

The phase transition and long-rage order of ferroelectric materials are disrupted when the dopant is present in the lattice. As a result, the temperature of maximum permittivity (T_m) is sometimes shifted to a higher temperature as the measurement frequency increases; such materials are called relaxor ferroelectrics (Hilton et al. 1989). The long-range macroscopic polarization observed in ferroelectric materials also becomes short-range ordered with nanosized domains in relaxors, and they are highly sensitive to the applied electric field (Li et al. 2016). Below the freezing temperature (T_f) of the ferroelectric moments, if the relaxors exhibit ferroelectric properties, then they are called as nonergodic relaxors. At normal temperatures, if the dipoles are aligned under the external electric field and return to their initial state after removing the electric field, resulting in a narrow hysteresis loop, then they are called as ergodic relaxors (Bokov and Ye 2006).

The solid solutions such as $Pb(Mg_{1/3}Nb_{2/3})O_3$, $Pb(Zn_{1/3}Nb_{2/3})O_3$, and (Pb, La) (Zr, Ti) O_3 are well-known relaxor ferroelectrics with excellent piezoelectric and electro-optic properties. In $Pb(Mg_{1/3}Nb_{2/3})O_3$ or $Pb(Zn_{1/3}Nb_{2/3})O_3$, if Nb^{5+} ions, and Mg^{2+} or Zn^{2+} ions are in 2:1 ratio, then it retains a local stoichiometry. However, in the lattice, each B^{2+} ions occupy a site alternate with two B^{5+} ions along the [100] directions. This implies the existence of the negatively charged compositionally ordered non-stoichiometric nanoregions in the positively charged disordered non-stoichiometric matrix, ultimately leading to chemical nanoregions and resulting relaxor behavior.

6.6.2.4 Piezoelectric Properties

Piezoelectric materials generate an electric charge by the application of mechanical stress (Figure 6.24). Ferroelectrics can be considered as a subset of piezoelectrics, so all ferroelectrics are piezoelectrics, and most of the commercially available piezoelectrics are ferroelectrics. The basic reason for ferroelectricity in perovskite oxides is the change in off-centered position of B-cation in BO_6 octahedra in the presence of applied electric field (Cohen 1992) as observed in $BaTiO_3$ (Choi et al. 2004).

This change in the crystal structure of $BaTiO_3$ due to the off-centered movement of Ti^{4+} ions along the *c*-axis slightly changes the dimensions of octahedron so that the oxygen atoms at equatorial position move parallel to *c*-axis. Because of this, off-centered movement of Ti^{4+} ions makes *c*-axis as polar axis, and in the applied electric field, the octahedral deformation undergoes changes, which results in the ferroelectric phase of $BaTiO_3$.

Further, the tetragonal structure undergoes a transition resulting in an orthorhombic structure *via* the elongation of a face diagonal of the unit cell in the temperature

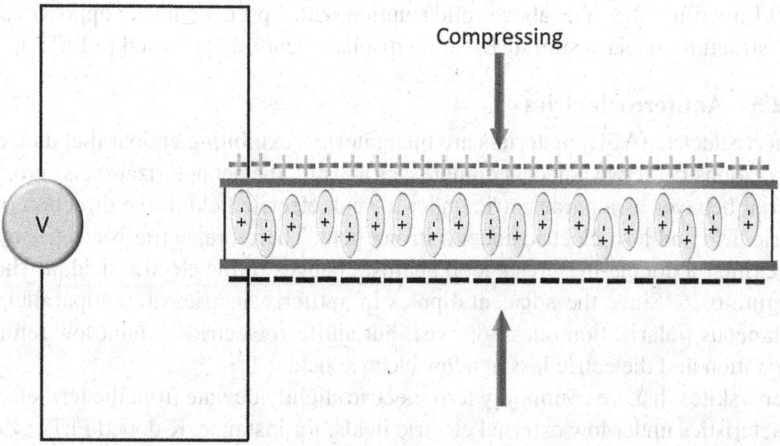

FIGURE 6.24 Schematic representation of piezoelectric behavior.

range between 278 and 183 K. There is another structural change below 183 K, during which, the orthorhombic becomes rhombohedral. Along with B-cation displacement, there are some other important parameters that influence the ferroelectric phase of perovskites. They are A-cation displacements, especially cations like Bi^{3+} and Pb^{2+}, irregularities in oxygen positions present in BO_6 octahedra, and octahedral distortions.

As discussed previously, the grain size affects the ferroelectric properties, so as the piezoelectric properties. For instance, $BiScO_3$–$PbTiO_3$ perovskite solid solution exhibits a temperature-dependent polarization, as shown in Figure 6.25a. The above solid solution also exhibits a change in piezoelectric coefficient (d_{33}) and remnant polarization with respect to size, as shown in Figure 6.25b (Amorín et al. 2015). d_{33} drops slowly as the grain size is decreased from 2 to 0.5 μm, but it drops suddenly in

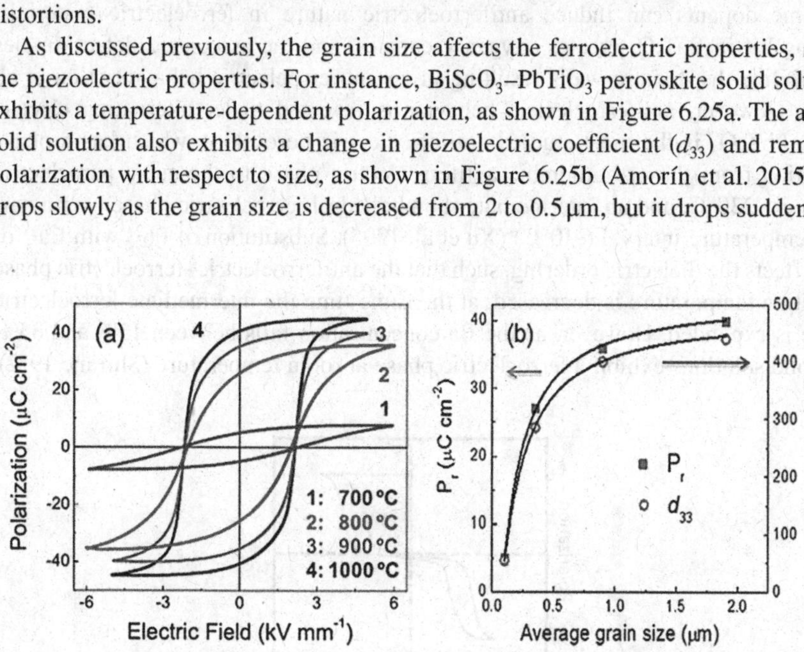

FIGURE 6.25 (a) Ferroelectric hysteresis loops for $BiScO_3$–$PbTiO_3$ perovskites obtained at different temperatures, and (b) evolution of the remnant polarization (P_r) and d_{33} piezoelectric coefficient with increasing grain size (Amorín et al. 2015).

0.3–0.1 micron range. The above solid solution with a pseudocubic or approximately cubic structure displays similar B-cation displacements, as observed in $BaTiO_3$.

6.6.2.5 Antiferroelectrics

Antiferroelectric (AFE) materials are the materials exhibiting antiparallel dielectric polarizations from two adjacent domains; as a result, the net polarization is zero. The material behaves as a paraelectric under a weak electric field in the direction of an electric field and ferroelectric under a strong field. Therefore, antiferroelectric materials exhibit a double-hysteresis loop against changes in the electric field, as shown in Figure 6.26. Since the adjacent dipoles in antiferroelectrics are antiparallel, the spontaneous polarization does not exist, but antiferroelectrics exhibit low remnant polarization and dielectric loss at a low electric field.

Perovskites that are commonly ferroelectric slightly deviate from the ferroelectric characteristics under low external electric fields, for instance, K-doped $PbTi_{1-x}Zr_xO_3$ (Tan et al. 1999). $BaTiO_3$ also shows a pinched and antiferroelectric-like hysteresis loop at a temperature slightly higher than its phase transition temperature T_c (Merz 1953). To identify the antiferroelectric behavior, one must add both paraelectric and ferroelectric loops. Interestingly, some perovskite structures that exhibit metastable ferroelectric phase under the applied external field can be transformed to the antiferroelectric phase under a shock force. This sudden transformation from a polarized to a completely depolarized state results in the release of a large amount of energy in a short time (Jiang et al. 2012).

Some dopants can induce antiferroelectric nature in ferroelectric materials. Doping Li^+ in $BaTiO_3$ creates oxygen vacancies, and the associated defect dipoles immobilize the domain walls resulting in an antiferroelectric behavior (Lou et al. 2018). However, antiferroelectric behavior is reversed under alternating electric fields. $PbZrO_3$ is the most studied antiferroelectric perovskite, which has a phase transition from the orthorhombic antiferroelectric phase to the cubic paraelectric phase at ~230°C, and an intermediate rhombohedral ferroelectric phase over a narrow temperature interval (~10°C) (Xu et al. 1995). Substitution of Pb^{2+} with Ba^{2+} or Sr^{2+} affects the dielectric ordering, such that the antiferroelectric–ferroelectric phase transition temperature is decreased, at the same time the intermediate ferroelectric phase is expanded. However, as the Ba concentration falls between 15% and 35%, the solid solutions exhibit a ferroelectric phase at room temperature (Shirane 1952).

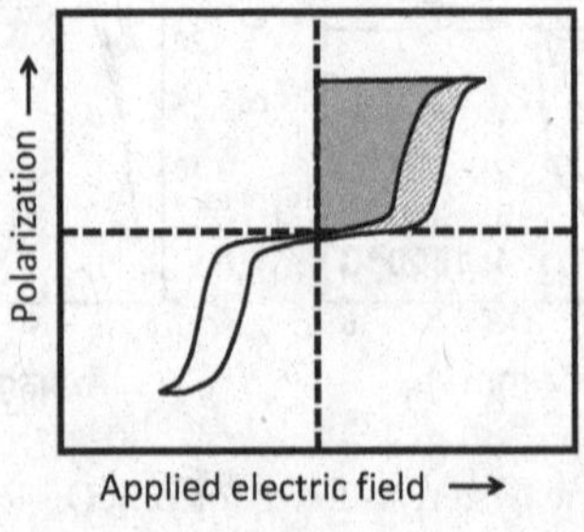

FIGURE 6.26 Schematic representation of hysteresis loops for an antiferroelectric material.

In the above cases, it is assumed that the antiferroelectric phase is stabilized due to the change in the tolerance factor due to the difference in the ionic radius of Ba^{2+} and Pb^{2+}. While doping trivalent Ln^{3+} ions, the charge ordering due to the charge difference between Pb^{2+} and La^{3+} cations prevents the long-range ordering and destabilization of the ferroelectric phase. In general, doping the B-site cation with smaller ions destabilizes the antiferroelectric phase. Perovskite-structured $NaNbO_3$ is a promising lead-free dielectric material that exhibits ferroelectric–antiferroelectric transition.

In double perovskites ($A_2B'B''O_6$), many different cations with varying dielectric properties can be incorporated (Kim and Woodward 2007). For many applications, dielectrics should have a high dielectric constant (ε_r) and strong temperature dependency, especially for communication applications. Double perovskites make it possible to achieve both high ε_r and strong temperature dependency by mixing desired cations. For example, a series of $A_2B'B''O_6$, A = Ca, Sr and Ba; B′ = La, Nd, Sm and Yb; B″ = Ta and Nb compounds having $\varepsilon_r > 20$ and with positive or negative temperature coefficients (Kagata and Kato 1994). Likewise, $A_2B'WO_6$ series compounds with A = Sr and Ba; B′ = Co, Ni and Zn display high ε_r values and negative temperature coefficients (Zhao et al. 2005). Microwave applications also depend on the cation ordering, and high ordering of cations in double perovskites is preferred to reduce the dielectric loss. Few double perovskites, particularly the ones with transition metals at B-sites, show high ε_r values as equal as colossal dielectric solids, but the reason for this behavior is unclear. For instance, compounds like $La_2B'MnO_6$ with B′ = Co, Ni and Mg (Lin et al. 2009; Lin and Chen 2010; Lin and Chen 2011), Ca_2CrTiO_6 (Yan-Qing et al. 2012), La_2NiRuO_6 (Yoshii et al. 2006), and La_2CuTiO_6 (Yang et al. 2010) have ε_r values in the order of 10^4.

Colossal dielectric constant materials are usually RP oxides with relative permittivity or dielectric constant in the order of 10^5, nearly independent of frequency, at a temperature range of 100–500 K. These compounds also show very low energy loss tangent around 0.017. $CaCu_3Ti_4O_{12}$ (Bender and Pan 2005) and closely related materials such as $La_{2/3}Cu_3Ti_4O_{12}$ (Jin et al. 2009), $SrCu_3Ti_4O_{12}$ (Li et al. 2004), $Na_{0.5}Bi_{0.5}Cu_3Ti_4O_{12}$ (Ren et al. 2010), and $Y_{2/3}Cu_3Ti_4O_{12}$ (Deng et al. 2017) are the best-known ceramic solids for these properties.

6.7 THERMOELECTRIC PROPERTIES

Thermoelectric materials convert thermal energy into electrical energy. Thermoelectric materials are sought to convert waste heat from industries and automobiles to useful electricity. Thermoelectric energy converters contain no moving mechanical parts in addition to relatively small size, long lifetime, and high reliability. Thermal conduction and electrical conductivity are related by three coefficients, namely, the Seebeck coefficient, Peltier coefficient, and Thomson coefficient. The Seebeck coefficient is the primary factor that governs the thermoelectric property of a compound. In perovskites, this coefficient is proportional to the density states either electron transport in the conduction band or hole transport in the valence band. The perovskites are ionic, which comprises charge carriers moving from one cation to another, and the charge carriers can be either electrons or holes. The Seebeck coefficient can be generally expressed as follows:

$$S = -\left(\frac{k_B}{e}\right)\left[\frac{1}{k_B T}\left(\frac{J_q}{J}\right)_{\nabla T=0} - \frac{\mu}{k_B T}\right] \quad (6.9)$$

where J_q is the isothermal energy current, J is the particle current that develops in response to a weak electrical field, and k_B is Boltzmann's constant, e is the electronic charge, T is the absolute temperature, and μ is the electronic chemical potential. $k_B/e = 86.17\ \mu V\ K^{-1}$. The value of k/e is negative for n-type materials and positive for p-type materials. The Eq. 6.8 can be deduced for p-type perovskite materials as follows (Marsh and Parris 1996):

$$S = \frac{k}{e}\ln\left(\frac{1-\rho_h}{\rho_h}\right) \quad (6.10)$$

where ρ_h is the hole concentration, and this equation is known as the Hikes equation.

For n-type,

$$S = \frac{k}{e}\ln\left(\frac{2-\rho_e}{\rho_e - 1}\right) \quad (6.11)$$

where ρ_e is the electron concentration.

For n-type thermoelectric materials, the A-site is doped with ions of higher valence. For instance, $CaMnO_3$ doped with lanthanides as in $Ca_{1-x}Ln_xMnO_3$. The doping at Ca-site enhances the electrical conductivity dramatically and exhibits good thermoelectric performance at high temperatures. A similar improvement is observed in Ln-doped $SrTiO_3$ as well. However, the ionic radius of Ln^{3+} ions impacts the thermoelectric properties of these materials. The thermoelectric properties of $SrTiO_3$ are also affected by doping the B-site with Ta^{5+} and Nb^{5+}. The thermoelectric energy conversion efficiency is estimated as a dimensionless term called the figure of merit, zT, which is a function of electrical and thermal conductivities (Snyder and Toberer 2008):

$$zT = \frac{S^2 \sigma T}{k} \quad (6.12)$$

where S is the Seebeck coefficient, σ is the electrical conductivity, k is the total thermal conductivity, and T is the absolute temperature.

The figure of merit of $SrTiO_3$ doped with different trivalent elements on the A-site and pentavalent elements on B-site is shown in Figure 6.27. The thermoelectric figure of merit is not significantly improved by B-site dopants in $SrTiO_3$ as compared to trivalent A-site doping because the doping on the B-sites sensitively affects the structure of TiO_6 octahedral unit and therefore affects the electrical conductivity. However, a reduction in thermal conductivity can be achieved by doping B-site with heavy elements. Increasing the electrical conductivity or decreasing the thermal conductivity can improve the figure of merit. When $SrTiO_3$ is prepared in a

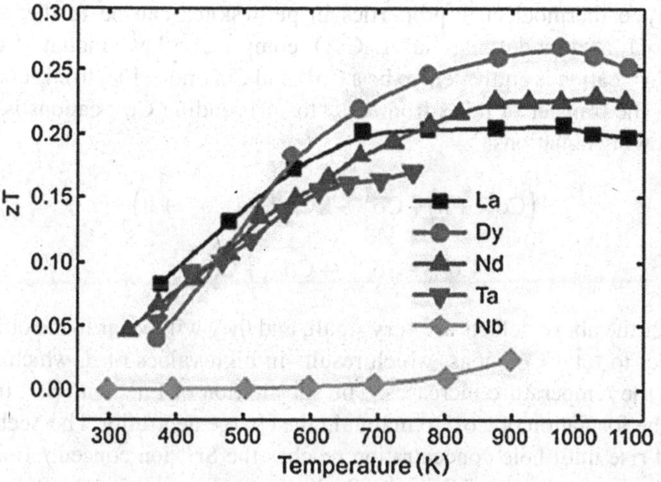

FIGURE 6.27 Comparison of the figure of merit of $SrTiO_3$ doped with different elements.

reducing atmosphere, oxygen vacancies are introduced to the system, which affects the electrical and thermal conductivities. The A-site doping of $SrTiO_3$ with Pr^{3+}, as in $Sr_{1-x}Pr_xTiO_{3\pm\delta}$ and $Sr_{1-1.5x}Pr_xTiO_{3\pm\delta}$ perovskite series ($0.20 \leq x \leq 0.30$), shows improved electrical conductivity with a change from metallic to semiconductor behavior with increasing temperature, as shown in Figure 6.28a. Enhancement in the Seebeck coefficients of A-substituted $SrTiO_3$ can be observed in Figure 6.28b (Kovalevsky et al. 2014).

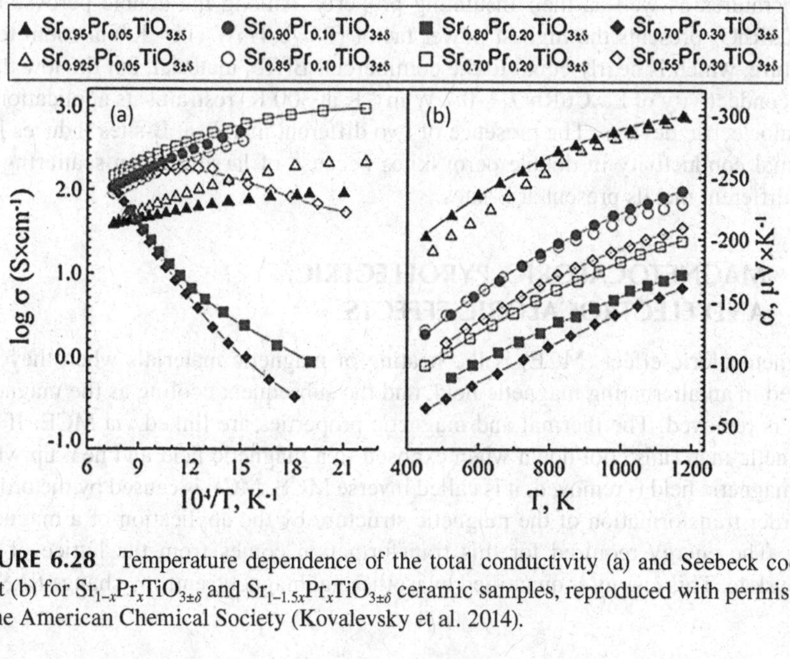

FIGURE 6.28 Temperature dependence of the total conductivity (a) and Seebeck coefficient (b) for $Sr_{1-x}Pr_xTiO_{3\pm\delta}$ and $Sr_{1-1.5x}Pr_xTiO_{3\pm\delta}$ ceramic samples, reproduced with permission of the American Chemical Society (Kovalevsky et al. 2014).

The *p*-type thermoelectric properties in perovskites can be further elucidated using $LaCoO_3$ as a model material. $LaCoO_3$ comprises a low amount of Co^{4+} ions, and each Co^{4+} cation is equivalent to be a Co^{4+} and $Co^{\bullet}_{Co}$ hole. Electronic conductivity assisted by the transfer of holes from Co^{4+} to surrounding Co^{3+} cations is governed by the following equations:

$$\left(Co^{3+}+h\right)+Co^{3+}\rightarrow Co^{3+}+\left(Co^{3+}+h\right) \tag{6.13}$$

$$Co^{\bullet}_{Co}+Co_{Co}\rightarrow Co_{Co}+Co^{\bullet}_{Co} \tag{6.14}$$

The number the above defects are very small, and they will occur in the order of 10^{-4} with respect to total Co^{3+} ions, which results in high values of *S*, which decreases slightly as the temperature increases. The substitution of La^{3+} with Sr^{2+} in $LaCoO_3$ results in the formation of Co^{4+} to maintain the charge neutrality. The Seebeck coefficient will rise until hole concentration reaches the Sr^{2+} ion concentration. Once it reaches the concentration of Sr^{2+}, the Seebeck coefficient starts to fall. A similar phenomenon could occur if the B-site is doped with an appropriate substituent. For instance, if Co^{3+} is doped with Ti^{4+} ion in $LaCoO_3$, the structure forces the Co^{3+} to transform to Co^{2+} cations to maintain charge balance in the doped perovskite ($LaTi_xCo_{1-x}O_3$). In the above structure, Co^{2+} cation and their electrons contribute to electronic conductivity.

The thermoelectric properties of double perovskites exhibit both *n*- and *p*-types of conduction with high Seebeck coefficients. For example, A_2FeMoO_6 (A = Ba or Sr) shows a relatively higher thermoelectric performance compared to other double perovskites with figure of merit, ZT = 0.3, at 1100 K. However, $Sr_2B'RuO_6$ (B′ = Y or Eu) shows low thermoelectric performance due to low thermal conductivity at high temperatures as well as their insulating property. Among the double perovskites, La_2CuRhO_6 presents the highest power factor (P = 0.4 mW cm^{-1} K^2) at room temperature, which is nearly equal to the commercial Bi_2Te_3 material, but the low thermal conductivity of La_2CuRhO_6 (~0.2 W m^{-1} K at 300 K) restrains its application in thermoelectric devices. The presence of two different metals at B-sites induces low thermal conductivity in double perovskites because of large phonon scattering by two different metals present at B-sites.

6.8 MAGNETOCALORIC, PYROELECTRIC, AND ELECTROCALORIC EFFECTS

Magnetocaloric effect (MCE) is the heating of magnetic materials when they are placed in an alternating magnetic field, and the subsequent cooling as the magnetic field is removed. The thermal and magnetic properties are linked via MCE. If the magnetic materials cool down when exposed to a magnetic field and heat up when the magnetic field is removed, it is called inverse MCE. MCE is caused by the order–disorder transformation of the magnetic structure by the application of a magnetic field. The energy required for this transformation comes from the lattice of the perovskite. This energy is quantified by isothermal magnetic entropy change ($|\Delta S_M|$), a large value of $|\Delta S_M|$ associated with a change in spin magnetization. The entropy change by varying the applied magnetic field is (Kuhn et al. 2011)

$$\Delta S_M = \mu_0 \int_{H_1}^{H_2} \left(\frac{\partial M}{\partial T}\right)_H dH \tag{6.15}$$

Applying the definition of the heat capacity at constant pressure C_p and the second law of thermodynamics leads to the expression for the related adiabatic temperature change ΔT_{ad} as follows:

$$\Delta T_{ad} = \mu_0 \int_{H_1}^{H_2} \frac{T}{C_p} \left(\frac{\partial M}{\partial T}\right)_H dH \tag{6.16}$$

Several perovskites exhibit MCE, for example, $La_{0.7}Ca_{0.3}MnO_3$ and $La_{0.7}Sr_{0.3}MnO_3$ are ferromagnetic below T_c of approximately 220 and 370 K, respectively. These two perovskites show poor MCE, and MCE of these compounds can be improved by doping A-, B-, and X-sites. A-site substituted phase like $La_{0.67}Ca_{0.33-x}Sr_xMn_{1.05}O_3$ and $La_{0.8}Ca_{0.2-x}Na_xMnO_3$, B-site substituted phase like $La_{0.8}Ca_{0.2}Mn_{1-x}Cr_xO_3$, and both A- and B-sites substituted phases like $La_{0.6}Pr_{0.1}Sr_{0.3}Mn_{0.9}Fe_{0.1}O_3$, $La_{0.67}Sr_{0.33}Mn_{1-x}V_xO_3$, and $La_{0.67}Ba_{0.33}Mn_{1-x}Fe_xO_3$ display increase in their $|\Delta S_M|$ value near their respective T_c. The representative MCE in $La_{0.67}Ca_{0.33-x}Sr_xMn_{1.05}O_3$ for increasing x is shown in Figure 6.29.

The magnetic caloric effect can be applied to refrigeration. The measure of heat transfer between the cold sink and hot volume during a single ideal refrigeration cycle is given by refrigerant capacity or relative cooling power (RCP). RCP is calculated from $|\Delta S_M|$ *vs.* T curves shown in Figure 6.29.

$$CP = |\Delta S_M| \times \delta T \tag{6.17}$$

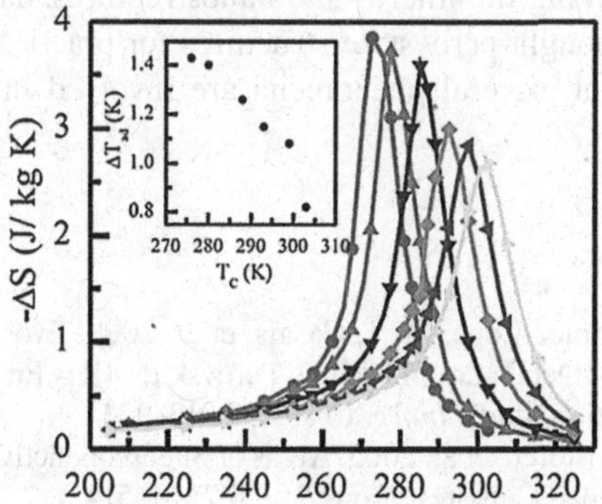

FIGURE 6.29 The MCE in $La_{0.67}Ca_{0.33-x}Sr_xMn_{1.05}O_3$ for increasing x (from left to right: $x = 0.0375$ to $x = 0.09$). The inset shows the corresponding adiabatic temperature change. ΔS has been determined indirectly by magnetization measurements, and ΔT_{ad} has been measured directly; both are measured under 1 T applied field change (Kuhn et al. 2011).

where δT is the full-width half-maximum of $|S_M|$ *vs.* T plots.

The pyroelectric effect is the generation of the electric field due to spontaneous spin polarization because of a change in temperature. The pyroelectric nature is shown by the ferroelectric perovskites, which contain spontaneous electric polarization (P_s) and unique polar axis. This effect comprises two components. One is measured in perovskites having constant crystal structure, and the other is in unconstrained crystal. Among the several potential ferroelectric materials for pyroelectrics, lead-based perovskites such as Pb(Zr, Ti)O_3 and (PbLa)(Zr, Ti)O_3 have a high value of pyroelectric coefficient and dielectric constant. $PbZrO_3$ doped with Fe, Nb, Ti, and U, such as $Pb_{1.02}(Zr_{0.58}Fe_{0.20}Nb_{0.20}Ti_{0.02})_{0.994}U_{0.006}O_3$, gives a higher pyroelectric coefficient and a lower dielectric constant.

The electrocaloric effect is opposite to the pyroelectric effect and is applicable in refrigeration. In the electrocaloric effect, the applied electric field along the polar axis induces changes in temperature of the crystal structure of perovskites. $LiTaO_3$ with $T_c \sim 618°C$, $PbTiO_3$ with $T_c \sim 490°C$, and the relaxor ferroelectrics such as $Pb(Sc_{0.5}Ti_{0.5})O_3$, $Pb(Mg_{0.33}Nb_{0.67})O_3$, and $Pb(Sc_{0.5}Sb_{0.5})O_3$ are good examples of electrocaloric and pyroelectric materials. However, the theoretical study of these two phenomena is still under debate, but few researchers have suggested the relationship between fundamental properties and the estimated change in temperature in electrocaloric perovskite as follows (Pirc et al. 2011):

$$\Delta T_s = (T \ln \Omega) P_s^2 / (3\varepsilon_0 \theta C) \tag{6.18}$$

where Ω is the number of polar states accessible in the system, T is the temperature, P_s is the saturation polarization, θ is the Curie constant of the phase, C is the specific heat capacity, and ε_0 is the dielectric permittivity of free space. The value of Ω is most varied in the relaxor ferroelectrics, in which the thermal and chemical methods are used in optimizing the micro- and nanostructures. Therefore, ferroelectric relaxors are the most sought perovskite structures for practical applications. At the relevant transition point, several phenomena are involved in magnetic or electric transitions.

REFERENCES

Alonso, J. A., M. J. Martínez-Lope, M. T. Casais, et al. 2000. Evolution of the Jahn–Teller Distortion of MnO_6 Octahedra in $RMnO_3$ Perovskites (R = Pr, Nd, Dy, Tb, Ho, Er, Y): A Neutron Diffraction Study. *Inorg. Chem.* 39:917–923.

Alvarez, G., M. Mayr, A. Moreo, et al. 2005. Areas of Superconductivity and Giant Proximity Effects in Underdoped Cuprates. *Phys. Rev. B* 71:014514.

Amorín, H., M. Algueró, R. D. Campo, et al. 2015. High-Sensitivity Piezoelectric Perovskites for Magnetoelectric Composites. *Sci. Technol. Adv. Mater.* 16:016001.

Antipov, E. V., A. M. Abakumov, and S. N. Putilin. 2002. Chemistry and Structure of Hg-Based Superconducting Cu Mixed Oxides. *Supercond. Sci. Technol.* 15:R31–R49.

Auth, N., G. Jakob, W. Westerburg, et al. 2004. Crystal Structure and Magnetism of the Double Perovskites A_2FeReO_6 (A = Ca, Sr, Ba). *J. Magn. Magn. Mater.* 272–276:E607–E608, Proceedings of the International Conference on Magnetism (ICM 2003).

Autret, C., M. Gervais, F. Gervais, et al. 2004. Signature of Ferromagnetism, Antiferromagnetism, Charge Ordering and Phase Separation by Electron Paramagnetic Resonance Study in Rare Earth Manganites, $Ln_{1-x}A_xMnO_3$ (Ln = rare Earth, A = Ca, Sr). *Solid State Sci.* 6:815–824.

Baba-Kishi, K. Z., P. M. Woodward, and K. Knight. 2001. The Crystal Structures of Pb_2ScTaO_6 and Pb_2ScNbO_6 in the Paraelectric and Ferroelectric States. *Ferroelectrics* 261:21–26. Taylor & Francis.

Baldini, M., V. V. Struzhkin, A. F. Goncharov, et al. 2011. Persistence of Jahn–Teller Distortion up to the Insulator to Metal Transition in $LaMnO_3$. *Phys. Rev. Lett.* 106:066402.

Barbur, I., and I. Ardelean. 2001. Structures and Phase Transitions in Some $Pb_2B'B''O_6$ Complex Perovskite-Type Compounds. *Phase Transit.* 74:367–373. Taylor & Francis.

Bender, B. A., and M.-J. Pan. 2005. The Effect of Processing on the Giant Dielectric Properties of $CaCu_3Ti_4O_{12}$. *Mater. Sci. Eng. B* 117:339–347.

Beno, M. A., L. Soderholm, D. W. Capone, et al. 1987. Structure of the Single-phase High-temperature Superconductor $YBa_2Cu_3O_{7-\delta}$. *Appl. Phys. Lett.* 51:57–59.

Bergemann, C., A. P. Mackenzie, S. R. Julian, et al. 2003. Quasi-Two-Dimensional Fermi Liquid Properties of the Unconventional Superconductor Sr_2RuO_4. *Adv. Phys.* 52:639–725.

Biscaras, J., N. Bergeal, A. Kushwaha, et al. 2010. Two-Dimensional Superconductivity at a Mott Insulator/Band Insulator Interface $LaTiO_3/SrTiO_3$. *Nat. Commun.* 1:89.

Biswas, A., K.-S. Kim, Y. H. Jeong. 2016. Metal–Insulator Transitions and Non-Fermi Liquid Behaviors in 5*d* Perovskite Iridates. In *Perovskite Materials – Synthesis, Characterisation, Properties, and Applications*. 221–259, London, UK: InTechOpen.

Bokov, A. A., and Z.-G. Ye. 2006. Recent Progress in Relaxor Ferroelectrics with Perovskite Structure. *J. Mater. Sci.* 41:31–52.

Borges, R. P., R. M. Thomas, C. Cullinan, et al. 1999. Magnetic Properties of the Double Perovskites A_2FeMoO_6 A = Ca, Sr, Ba. *J. Phys. Condens. Matter* 11:L445–L450.

Branković, Z., K. Đuriš, A. Radojković, et al. 2010. Magnetic Properties of Doped $LaMnO_3$ Ceramics Obtained by a Polymerizable Complex Method. *J. Sol-Gel Sci. Technol.* 55:311–316.

Budhani, R. C., N. K. Pandey, P. Padhan, et al. 2001. Electric- and Magnetic-Field-Driven Nonlinear Charge Transport and Magnetic Ordering in Epitaxial Films of $Pr_{0.7}Ca_{0.3-x}Sr_xMnO_3$. *Phys. Rev. B* 65:014429.

Chen, G., R. Pereira, and L. Balents. 2010. Exotic Phases Induced by Strong Spin-Orbit Coupling in Ordered Double Perovskites. *Phys. Rev. B* 82:174440.

Chen, X. G., J. B. Fu, C. Yun, et al. 2014. Magnetic and Transport Properties of Cobalt Doped $La_{0.7}Sr_{0.3}MnO_3$. *J. Appl. Phys.* 116:103907.

Cheng, J.-G., J.-S. Zhou, J. B. Goodenough, et al. 2010. Pressure Dependence of Metal-Insulator Transition in Perovskites $RNiO_3$(R = Eu, Y, Lu). *Phys. Rev. B* 82:085107.

Choi, K. J., M. Biegalski, Y. L. Li, et al. 2004. Enhancement of Ferroelectricity in Strained $BaTiO_3$ Thin Films. *Science* 306:1005–1009.

Cohen, R. E. 1992. Origin of Ferroelectricity in Perovskite Oxides. *Nature* 358:136–138.

Dagotto, E. 2003. Brief Introduction to Giant Magnetoresistance (GMR). In *Nanoscale Phase Separation and Colossal Magnetoresistance: The Physics of Manganites and Related Compounds*, ed. E. Dagotto, 395–405. Springer Series in Solid-State Sciences. Berlin, Heidelberg: Springer.

Dai, P., J. Zhang, H. A. Mook, et al. 1996. Experimental Evidence for the Dynamic Jahn–Teller Effect in $La_{0.65}Ca_{0.35}MnO_3$. *Phys. Rev. B* 54: R3694–R3697.

Deng, J., L. Liu, X. Sun, et al. 2017. Dielectric Relaxation Behavior and Mechanism of $Y_{2/3}Cu_3Ti_4O_{12}$ Ceramic. *Mater. Res. Bull.* 88:320–329.

Deshpande, S. K., S. N. Achary, R. Mani, et al. 2011. Low-Temperature Polaronic Relaxations with Variable Range Hopping Conductivity in $FeTiMO_6$ (M = Ta, Nb, Sb). *Phys. Rev. B* 84:064301.

Du, Y., X. Wan, L. Sheng, et al. 2012. Electronic Structure and Magnetic Properties of $NaOsO_3$. *Phys. Rev. B* 85:174424.

Dungan, R. H., and R. D. Golding. 1965. Polarization of $NaNbO_3$-$KNbO_3$ Ceramic Solid Solutions. *J. Am. Ceram. Soc.* 48:601–601.

Erat, S., A. Braun, C. Piamonteze, et al. 2010. Entanglement of Charge Transfer, Hole Doping, Exchange Interaction and Octahedron Tilting Angle and Their Influence on the Conductivity of $La_{1-x}Sr_xFe_{0.75}Ni_{0.25}O_{3-\delta}$: A Combination of X-Ray Spectroscopy and Diffraction. *J. Appl. Phys.* 108:124906.

Feteira, A., D. C. Sinclair, K. Z. Rajab, et al. 2008. Crystal Structure and Microwave Dielectric Properties of Alkaline-Earth Hafnates, $AHfO_3$ (A = Ba, Sr, Ca). *J. Am. Ceram. Soc.* 91:893–901.

Filippetti, A., and W. E. Pickett. 1999. Magnetic Reconstruction at the (001) $CaMnO_3$ Surface. *Phys. Rev. Lett.* 83:4184–4187.

García-Hernández, M., J. L. Martínez, M. J. Martínez-Lope, et al. 2001. Finding Universal Correlations between Cationic Disorder and Low Field Magnetoresistance in FeMo Double Perovskite Series. *Phys. Rev. Lett.* 86:2443–2446.

García-Muñoz, J. L., J. Padilla-Pantoja, X. Torrelles, et al. 2016. Magnetostructural Coupling, Magnetic Ordering, and Cobalt Spin Reorientation in Metallic $Pr_{0.5}Sr_{0.5}CoO_3$ Cobaltite. *Phys. Rev. B* 94:014411.

Geng, T., and N. Zhang. 2006. Electronic Structure of the Perovskite Oxides $La_{1-x}Sr_xMnO_3$. *Phys. Lett. A* 351:314–318.

Guo, Y., J.-M. Langlois, and W. A. Goddard. 1988. Electronic Structure and Valence-Bond Band Structure of Cuprate Superconducting Materials. *Science* 239:896–899.

Hamada, N., H. Sawada, I. Solovyev, et al. 1997. Electronic Band Structure and Lattice Distortion in Perovskite Transition-Metal Oxides. *Phys. B Condens. Matter.* 237–238:11–13, Proceedings of the Yamada Conference XLV, the International Conference on the Physics of Transition Metals.

Hanaguri, T., Y. Kohsaka, M. Ono, et al. 2009. Coherence Factors in a High-T_c Cuprate Probed by Quasi-Particle Scattering Off Vortices. *Science* 323:923–926.

Hansteen, O. H., Y. Bréard, H. Fjellvåg, et al. 2004. Divalent Manganese in Reduced $LaMnO_{3-\delta}$ – Effect of Oxygen Nonstoichiometry on Structural and Magnetic Properties. *Solid State Sci.* 6:279–285.

Hansteen, O. H., H. Fjellvåg, and B. C. Hauback. 1998. Crystal Structure, Thermal and Magnetic Properties of $La_3Co_3O_8$. Phase Relations for $LaCoO_{3-\delta}$ ($0.00 \leq \delta \leq 0.50$) at 673 K. *J. Mater. Chem.* 8:2081–2088.

Harrell, Z., E. Enriquez, A. Chen, et al. 2017. Oxygen Content Tailored Magnetic and Electronic Properties in Cobaltite Double Perovskite Thin Films. *Appl. Phys. Lett.* 110:093102.

Harvey, J. N., R. Poli, and K. M. Smith. 2003. Understanding the Reactivity of Transition Metal Complexes Involving Multiple Spin States. *Coord. Chem. Rev.* 238–239:347–361, Theoretical and Computational Chemistry.

Hilton, A. D., C. A. Randall, D. J. Barber, et al. 1989. TEM Studies of $Pb(Mg_{1/3}Nb_{2/3})O_3$-$PbTiO_3$ Ferroelectric Relaxors. *Ferroelectrics* 93:379–386. Taylor & Francis.

Hirohata, A., and K. Takanashi. 2014. Future Perspectives for Spintronic Devices. *J. Phys. Appl. Phys.* 47:193001.

Horikiri, F., L. Han, N. Iizawa, et al. 2007. Electrical Properties of Nb-Doped $SrTiO_3$ Ceramics with Excess TiO_2 for SOFC Anodes and Interconnects. *J. Electrochem. Soc.* 155: B16.

Huijben, M., G. Rijnders, D. H. A. Blank, et al. 2006. Electronically Coupled Complementary Interfaces between Perovskite Band Insulators. *Nat. Mater.* 5:556–560.

Jakob, G., W. Westerburg, F. Martin, et al. 2001. Magnetotransport Properties of Thin Films of Magnetic Perovskites. In *Advances in Solid State Physics*, ed. B. Kramer, 589–600. Vol. 41. Berlin, Heidelberg: Springer.

Jeong, I.-K., S. Lee, S.-Y. Jeong, et al. 2011. Structural Evolution across the Insulator-Metal Transition in Oxygen-Deficient $BaTiO_{3-\delta}$ Studied Using Neutron Total Scattering and Rietveld Analysis. *Phys. Rev. B* 84:064125.

Jiang, D., J. Du, Y. Gu, et al. 2012. Self-Generated Electric Field Suppressing the Ferroelectric to Antiferroelectric Phase Transition in Ferroelectric Ceramics under Shock Wave Compression. *J. Appl. Phys.* 111:024103.

Jin, S., H. Xia, and Y. Zhang. 2009. Effect of La-Doping on the Properties of $CaCu_3Ti_4O_{12}$ Dielectric Ceramics. *Ceram. Int.* 35:309–313.

Kagata, H., and J. Kato. 1994. Dielectric Properties of Ca-Based Complex Perovskite at Microwave Frequencies. *Jpn. J. Appl. Phys.* 33:5463.

Kato, H., T. Okuda, Y. Okimoto, Y. Tomioka, K. Oikawa, et al. 2002a. Metal-Insulator Transition of Ferromagnetic Ordered Double Perovskites: Ca_2FeReO_6. *Phys. Rev. B* 65:144404.

Kato, H., T. Okuda, Y. Okimoto, Y. Tomioka, Y. Takenoya, et al. 2002b. Metallic Ordered Double-Perovskite Sr_2CrReO_6 with Maximal Curie Temperature of 635 K. *Appl. Phys. Lett.* 81:328–330.

Katsufuji, T., Y. Taguchi, and Y. Tokura. 1997. Transport and Magnetic Properties of a Mott-Hubbard System Whose Bandwidth and Band Filling Are Both Controllable: $R_{1-x}Ca_xTiO_{3+y/2}$. *Phys. Rev. B* 56:10145–10153.

Kim, Y.-I., and P. M. Woodward. 2007. Crystal Structures and Dielectric Properties of Ordered Double Perovskites Containing Mg^{2+} and Ta^{5+}. *J. Solid State Chem.* 180:2798–2807.

Kobayashi, Y., O. J. Hernandez, T. Sakaguchi, et al. 2012. An Oxyhydride of $BaTiO_3$ Exhibiting Hydride Exchange and Electronic Conductivity. *Nat. Mater.* 11:507–511.

Korotin, M. A., S. Y. Ezhov, I. V. Solovyev, et al. 1996. Intermediate-Spin State and Properties of $LaCoO_3$. *Phys. Rev. B* 54:5309–5316.

Kovalevsky, A. V., A. A. Yaremchenko, S. Populoh, et al. 2014. Effect of A-Site Cation Deficiency on the Thermoelectric Performance of Donor-Substituted Strontium Titanate. *J. Phy. Chem. C.* 76:1221–1225.

Krishana, K., N. P. Ong, Q. Li, et al. 1997. Plateaus Observed in the Field Profile of Thermal Conductivity in the Superconductor $Bi_2Sr_2CaCu_2O_8$. *Science* 277:83–85.

Kuhn, L. T., N. Pryds, C. R. H. Bahl, et al. 2011. Magnetic Refrigeration at Room Temperature – from Magnetocaloric Materials to a Prototype. *J. Phys. Conf. Ser.* 303:012082.

Kundu, A. K. 2016. *Magnetic Perovskites: Synthesis, Structure and Physical Properties.* Engineering Materials. Jabalpur: Springer India.

Kuwahara, H., Y. Moritomo, Y. Tomioka, et al. 1997. Low-Field Colossal Magnetoresistance in Bandwidth-Controlled Manganites. *J. Appl. Phys.* 81:4954–4956.

Lee, S., W. H. Woodford, and C. A. Randall. 2008. Crystal and Defect Chemistry Influences on Band Gap Trends in Alkaline Earth Perovskites. *Appl. Phys. Lett.* 92:201909.

Lee, Y. S., J. S. Lee, T. W. Noh, et al. 2003. Systematic Trends in the Electronic Structure Parameters of the 4*d* Transition-Metal Oxides $SrMO_3$(M = Zr, Mo, Ru, and Rh). *Phys. Rev. B* 67:113101.

Li, F., S. Zhang, T. Yang, et al. 2016. The Origin of Ultrahigh Piezoelectricity in Relaxor-Ferroelectric Solid Solution Crystals. *Nat. Commun.* 7:1–9.

Li, J., M. A. Subramanian, H. D. Rosenfeld, et al. 2004. Clues to the Giant Dielectric Constant of $CaCu_3Ti_4O_{12}$ in the Defect Structure of "$SrCu_3Ti_4O_{12}$". *Chem. Mater.* 16:5223–5225.

Li, M.-R., M. Retuerto, Z. Deng, et al. 2015. Giant Magnetoresistance in the Half-Metallic Double-Perovskite Ferrimagnet Mn_2FeReO_6. *Angew. Chem. Int. Ed.* 54:12069–12073.

Lin, Y. Q., and X. M. Chen. 2010. Dielectric Relaxation and Polaronic Conduction in Double Perovskite La_2MgMnO_6. *Appl. Phys. Lett.* 96:142902.

Lin, Y. Q., and X. M. Chen. 2011. Dielectric, Ferromagnetic Characteristics, and Room-Temperature Magnetodielectric Effects in Double Perovskite La_2CoMnO_6 Ceramics. *J. Am. Ceram. Soc.* 94:782–787.

Lin, Y. Q., X. M. Chen, and X. Q. Liu. 2009. Relaxor-like Dielectric Behavior in La_2NiMnO_6 Double Perovskite Ceramics. *Solid State Commun.* 149:784–787.

Lou, Q., X. Shi, X. Ruan, et al. 2018. Ferroelectric Properties of Li-Doped $BaTiO_3$ Ceramics. *J. Am. Ceram. Soc.* 101:3597–3604.

Lu, H. B., G. Z. Yang, Z. H. Chen, et al. 2004. Positive Colossal Magnetoresistance in a Multilayer p–n Heterostructure of Sr-Doped $LaMnO_3$ and Nb-Doped $SrTiO_3$. *Appl. Phys. Lett.* 84:5007–5009.

Luo, Z. P., D. J. Miller, J. F. Mitchell. 2005. Electron microscopic evidence of charge-ordered bi-stripe structures in the bilayered colossal magnetoresistive manganite $La_{2-2x}Sr_{1+2x}Mn_2O_7$. *Phys. Rev. B* 71:014418.

Luo, Z. P., D. J. Miller, J. F. Mitchell. 2009. Structure and charge ordering behavior of the colossal magnetoresistive manganite $Nd_{0.5}Sr_{0.5}MnO_3$. *J. Appl. Phys.* 105:07D528.

Maignan, A., B. Raveau, C. Martin, et al. 1999. Large Intragrain Magnetoresistance above Room Temperature in the Double Perovskite Ba_2FeMoO_6. *J. Solid State Chem.* 144:224–227.

Majumder, D. D., and S. Karan. 2013. Ceramic Nanocomposites. In *Magnetic Properties of Ceramic Nanocomposites*, 51–91. Cambridge, UK: Woodhead Publishing.

Marsh, D. B., and P. E. Parris. 1996. High-Temperature Thermopower of $LaMnO_3$ and Related Systems. *Phys. Rev. B Condens. Matter* 54:16602–16607.

Matteo, S. D., G. Jackeli, and N. B. Perkins. 2003. Effective Spin-Orbital Hamiltonian for the Double Perovskite Sr_2FeWO_6 Derivation of the Phase Diagram. *Phys. Rev. B* 67:184427.

Meneghini, C., C. Castellano, S. Mobilio, et al. 2002. Local Structure of Hole-Doped Manganites: Influence of Temperature and Applied Magnetic Field. *J. Phys. Condens. Matter* 14:1967–1974.

Merz, W. J. 1949. The Electric and Optical Behavior of $BaTiO_3$ Single-Domain Crystals. *Phys. Rev.* 76:1221–1225.

Merz, W. J. 1953. Double Hysteresis Loop of $BaTiO_3$ at the Curie Point. *Phys. Rev.* 91:513–517.

Mizokawa, T., L. H. Tjeng, G. A. Sawatzky, et al. 2001. Spin-Orbit Coupling in the Mott Insulator Ca_2RuO_4. *Phys. Rev. Lett.* 87:077202.

Moritomo, Y., A. Asamitsu, H. Kuwahara, et al. 1996. Giant Magnetoresistance of Manganese Oxides with a Layered Perovskite Structure. *Nature* 380:141–144.

Motoki, T., M. Naito, H. Sano, et al. 2000. Effect of Microstructure on Dielectric Properties of $CaZrO_3$-Based Ceramics. In *Key Engineering Materials*. 7–10. Kapellweg, Switzerland: Trans Tech Publications.

Müller, K. A., W. Berlinger, and E. Tosatti. 1991. Indication for a Novel Phase in the Quantum Paraelectric Regime of $SrTiO_3$. *Z. Für Phys. B Condens. Matter* 84:277–283.

Narayanan, N. 2010. Physical Properties of Double Perovskites $La_{2-x}Sr_xCoIrO_6$. Phd, Darmstadt: Technische Universität. https://tuprints.ulb.tu-darmstadt.de/2328/.

Narayanan, N., D. Mikhailova, A. Senyshyn, et al. 2010. Temperature and Composition Dependence of Crystal Structures and Magnetic and Electronic Properties of the Double Perovskites $La_{2-x}Sr_xCoIrO_6$. *Phys. Rev. B* 82:024403.

Nikulin, N. 1988. *Fundamentals of Electrical Materials*. Moscow: Mir Publishers.

Nowotny, J., and M. Rekas. 1994. Defect Structure, Electrical Properties and Transport in Barium Titanate. VII. Chemical Diffusion in Nb-Doped $BaTiO_3$. *Ceram. Int.* 20:265–275.

Obradors, X., L. M. Paulius, M. B. Maple, et al. 1993. Pressure Dependence of the Metal-Insulator Transition in the Charge-Transfer Oxides $RNiO_3$ (R = Pr, Nd, $Nd_{0.7}La_{0.3}$. *Phys. Rev. B* 47:12353–12356.

Philipp, J. B., P. Majewski, L. Alff, et al. 2003. Structural and Doping Effects in the Half-Metallic Double Perovskite A_2CrWO_6 (A = Sr, Ba, and Ca). *Phys. Rev. B* 68:144431.

Pickett, W. E., and D. J. Singh. 1996. Electronic Structure and Half-Metallic Transport in the $La_{1-x}Ca_xMnO_3$ System. *Phys. Rev. B* 53:1146–1160.

Pirc, R., Z. Kutnjak, R. Blinc, et al. 2011. Upper Bounds on the Electrocaloric Effect in Polar Solids. *Appl. Phys. Lett.* 98:021909.

Piskunov, S., E. Heifets, R. I. Eglitis, et al. 2004. Bulk Properties and Electronic Structure of $SrTiO_3$, $BaTiO_3$, $PbTiO_3$ Perovskites: An Ab Initio HF/DFT Study. *Comput. Mater. Sci.* 29:165–178.

Poddar, A., S. Das, and B. Chattopadhyay. 2004. Effect of Alkaline-Earth and Transition Metals on the Electrical Transport of Double Perovskites. *J. Appl. Phys.* 95:6261–6267.

Ramesha, K., J. Gopalakrishnan, V. Smolyaninova, et al. 2001. $ALaFeVO_6$ (A = Ca, Sr): New Double-Perovskite Oxides. *J. Solid State Chem.* 162:250–253.

Ramirez, A. P. 1997. Colossal Magnetoresistance. *J. Phys. Condens. Matter* 9:8171–8199.

Rao, C. N. R. 2000. Charge, Spin, and Orbital Ordering in the Perovskite Manganates, $Ln_{1-x}A_xMnO_3$ (Ln = Rare Earth, A = Ca or Sr). *J. Phys. Chem. B* 104:5877–5889.

Rao, C. N. R., and B. Raveau. 1989. Structural Aspects of High-Temperature Cuprate Superconductors. *Acc. Chem. Res.* 22:106–113.

Ren, H., P. Liang, and Z. Yang. 2010. Processing, Dielectric Properties and Impedance Characteristics of $Na_{0.5}Bi_{0.5}Cu_3Ti_4O_{12}$ Ceramics. *Mater. Res. Bull.* 45:1608–1613.

Richerson, D. W., and W. E. Lee. 2018. *Modern Ceramic Engineering: Properties, Processing, and Use in Design*, 1–836. Boca Raton, FL: CRC Press, Taylor & Francis Group.

Righi, L., P. Gorria, M. Insausti, et al. 1997. Influence of Fe in Giant Magnetoresistance Ratio and Magnetic Properties of $La_{0.7}Ca_{0.3}Mn_{1-x}Fe_xO_3$ Perovskite Type Compounds. *J. Appl. Phys.* 81:5767–5769.

Rondinelli, J. M., and C. J. Fennie. 2012. Ferroelectricity: Octahedral Rotation-Induced Ferroelectricity in Cation Ordered Perovskites. *Adv. Mater.* 24:1918–1918.

Rubi, K., and R. Mahendiran. 2020. Large Thermoelectric Response of B-Site Doped Ferroelectrics: $Ba_{0.7}Eu_{0.3}Ti_{1-x}NbO_3$ ($x = 0$ to 0.1). *J. Solid State Chem.* 281:121050.

Sakaguchi, T., Y. Kobayashi, T. Yajima, et al. 2012. Oxyhydrides of (Ca, Sr, Ba)TiO_3 Perovskite Solid Solutions. *Inorg. Chem.* 51:11371–11376.

Sánchez, D., J. A. Alonso, M. García-Hernández, et al. 2002. Origin of Neutron Magnetic Scattering in Antisite-Disordered Sr_2FeMoO_6 Double Perovskites. *Phys. Rev. B* 65:104426.

Sarma, D. D., N. Shanthi, and P. Mahadevan. 1994. Electronic Structure and the Metal-Insulator Transition in $LnNiO_3$(Ln = La, Pr, Nd, Sm and Ho): Bandstructure Results. *J. Phys. Condens. Matter* 6:10467–10474.

Saxena, M., K. Tanwar, and T. Maiti. 2017. Environmental Friendly Sr_2TiMoO_6 Double Perovskite for High Temperature Thermoelectric Applications. *Scr. Mater.* 130:205–209.

Schiffer, P., A. P. Ramirez, W. Bao, et al. 1995. Low Temperature Magnetoresistance and the Magnetic Phase Diagram of $La_{1-x}Ca_xMnO_3$. *Phys. Rev. Lett.* 75:3336–3339.

Sen, C., B. Alkan, I. Akin, et al. 2011. Microstructure and Ferroelectric Properties of Spark Plasma Sintered Li Substituted $K_{0.5}Na_{0.5}NbO_3$ Ceramics. *J. Ceram. Soc. Jpn.* 119:355–361.

Shah, W. H., K. Safeen, and G. Rehman. 2012. Effects of Divalent Alkaline Earth Ions on the Magnetic and Transport Features of $La_{0.65}A_{0.35}Mn_{0.95}Fe_{0.05}O_3$ (A = Ca, Sr, Pb, Ba) Compounds. *Curr. Appl. Phys.* 12:742–747.

Shen, S., M. Williamson, G. Cao, et al. 2017. Non-Destructive Reversible Resistive Switching in Cr Doped Mott Insulator Ca_2RuO_4: Interface vs Bulk Effects. *J. Appl. Phys.* 122:245108.

Shende, R. V., D. S. Krueger, G. A. Rossetti, et al. 2001. Strontium Zirconate and Strontium Titanate Ceramics for High-Voltage Applications: Synthesis, Processing, and Dielectric Properties. *J. Am. Ceram. Soc.* 84:1648–1650.

Shi, J., Y. Zhou, and S. Ramanathan. 2014. Colossal Resistance Switching and Band Gap Modulation in a Perovskite Nickelate by Electron Doping. *Nat. Commun.* 5:1–9.

Shirane, G. 1952. Ferroelectricity and Antiferroelectricity in Ceramic $PbZrO_3$ Containing Ba or Sr. *Phys. Rev.* 86:219–227.

Snyder, G. J., and E. S. Toberer. 2008. Complex Thermoelectric Materials. *Nat. Mater.* 7:105–114.

Song, Z., H. Liu, S. Zhang, et al. 2014. Effect of Grain Size on the Energy Storage Properties of $(Ba_{0.4}Sr_{0.6})TiO_3$ Paraelectric Ceramics. *J. Eur. Ceram. Soc.* 34:1209–1217.

Spinolo, G., M. Scavini, P. Ghigna, et al. 1995. Nature and Amount of Carriers in Ce Doped Nd_2CuO_4 I. High-Temperature Characterization. *Phys. C Supercond.* 254:359–369.

Sun, J. Z., W. J. Gallagher, P. R. Duncombe, et al. 1996. Observation of Large Low-field Magnetoresistance in Trilayer Perpendicular Transport Devices Made Using Doped Manganate Perovskites. *Appl. Phys. Lett.* 69:3266–3268.

Sun, P.-H., T. Nakamura, Y. J. Shan, et al. 1997. High Temperature Quantum Paraelectricity in Perovskite-Type Titanates $Ln_{1/2}Na_{1/2}TiO_3$ (Ln = La, Pr, Nd, Sm, Eu, Gd and Tb). *Ferroelectrics* 200:93–107. Taylor & Francis.

Svoboda, C., M. Randeria, and N. Trivedi. 2017. Effective Magnetic Interactions in Spin-Orbit Coupled d^4 Mott Insulators. *Phys. Rev. B* 95:014409.

Taguchi, H. 1996. Spin State of Cobalt Ion in $Nd(Cr_{1-x}Co_x)O_3$. *J. Solid State Chem.* 122:297–302.

Tan, Q., J. Li, and D. Viehland. 1999. Role of Lower Valent Substituent-Oxygen Vacancy Complexes in Polarization Pinning in Potassium-Modified Lead Zirconate Titanate. *Appl. Phys. Lett.* 75:418–420.

Tokura, Y., Y. Tomioka, H. Kuwahara, et al. 1996. Origins of Colossal Magnetoresistance in Perovskite-type Manganese Oxides (Invited). *J. Appl. Phys.* 79:5288–5291.

Tomioka, Y., A. Asamitsu, H. Kuwahara, et al. 1996. Magnetic-Field-Induced Metal-Insulator Phenomena in $Pr_{1-x}Ca_xMnO_3$ with Controlled Charge-Ordering Instability. *Phys. Rev. B* 53:R1689–R1692.

Tomioka, Y., T. Okuda, Y. Okimoto, et al. 2000. Magnetic and Electronic Properties of a Single Crystal of Ordered Double Perovskite Sr_2FeMoO_6. *Phys. Rev. B* 61:422–427.

Torrance, J. B., P. Lacorre, A. I. Nazzal, et al. 1992. Systematic Study of Insulator-Metal Transitions in Perovskites $RNiO_3$ (R = Pr, Nd, Sm, Eu) Due to Closing of Charge-Transfer Gap. *Phys. Rev. B* 45:8209–8212.

Tsubouchi, S., T. Kyômen, M. Itoh, et al. 2002. Simultaneous Metal-Insulator and Spin-State Transitions in $Pr_{0.5}Ca_{0.5}CoO_3$. *Phys. Rev. B* 66:052418.

Tsuda, N., K. Nasu, A. Fujimori, et al. 2000. Electron–Electron Interaction and Electron Correlation. In *Electronic Conduction in Oxides*, eds. Nobuo Tsuda, Keiichiro Nasu, Atsushi Fujimori, and Kiiti Siratori, 119–155. Springer Series in Solid-State Sciences. Berlin, Heidelberg: Springer.

Ulrich, R., L. Schaper. 2000. Materials Options for Dielectrics in Integrated Capacitors. In *Proceedings International Symposium on Advanced Packaging Materials Processes, Properties and Interfaces (Cat. No.00TH8507)*, Braselton, GA, USA, 38–43, doi: 10.1109/ISAPM.2000.869240.

Urushibara, A., Y. Moritomo, T. Arima, et al. 1995. Insulator-Metal Transition and Giant Magnetoresistance in $La_{1-x}Sr_xMnO_3$. *Phys. Rev. B* 51:14103–14109.

Varignon, J., M. Bibes, and A. Zunger. 2019. Origin of Band Gaps in 3*d* Perovskite Oxides. *Nat. Commun.* 10:1–11.

Vasala, S., and M. Karppinen. 2015. $A_2B'B''O_6$ Perovskites: A Review. *Prog. Solid State Chem.* 43:1–36.

Wang, Z., M. Cao, Q. Zhang, et al. 2015. Dielectric Relaxation in Zr-Doped $SrTiO_3$ Ceramics Sintered in N_2 with Giant Permittivity and Low Dielectric Loss. *J. Am. Ceram. Soc.* 98:476–482.

Weaver, H. E. 1959. Dielectric Properties of Single Crystals of $SrTiO_3$ at Low Temperatures. *J. Phys. Chem. Solids* 11:274–277.

Wei, H., Y. Chen, G. Huo, et al. 2010. Crystal Structure, Infrared Spectroscopic Characterization and Electrical Property of Double Perovskite Sr_2CoSbO_6. *Phys. B Condens. Matter* 405:1369–1373.

Wolfram, T., and S. Ellialtioglu. 2006. *Electronic and Optical Properties of D-Band Perovskites*. Cambridge, UK: Cambridge University Press.

Woodward, P. M., J. Goldberger, M. W. Stoltzfus, et al. 2008. Electronic, Magnetic, and Structural Properties of Sr_2MnRuO_6 and $LaSrMnRuO_6$ Double Perovskites. *J. Am. Ceram. Soc.* 91:1796–1806.

Xu, Z., X. Dai, D. Viehland, et al. 1995. Ferroelectric Domains and Incommensuration in the Intermediate Phase Region of Lead Zirconate. *J. Am. Ceram. Soc.* 78:2220–2224.

Yamamoto, T., R. Yoshii, G. Bouilly, et al. 2015. An Antiferro-to-Ferromagnetic Transition in $EuTiO_{3-x}H_x$ Induced by Hydride Substitution. *Inorg. Chem.* 54:1501–1507.

Yang, W. Z., M. M. Mao, X. Q. Liu, et al. 2010. Structure and Dielectric Relaxation of Double-Perovskite La_2CuTiO_6 Ceramics. *J. Appl. Phys.* 107:124102.

Yan-Qing, T., Y. Meng, and H. Yong-Mei. 2012. Structure and Colossal Dielectric Permittivity of Ca_2TiCrO_6 ceramics. *J. Phys. Appl. Phys.* 46:015303.

Yoshii, K. 2000. Magnetic Transition in the Perovskite Ba_2CoNbO_6. *J. Solid State Chem.* 151:294–297.

Yoshii, K., N. Ikeda, and M. Mizumaki. 2006. Magnetic and Dielectric Properties of the Ruthenium Double Perovskites La_2MRuO_6 (M = Mg, Co, Ni, and Zn). *Phys. Status Solidi A* 203:2812–2817.

Zhang, G.-F., M. Cao, H. Hao, et al. 2013. Energy Storage Characteristics in $Sr_{(1-1.5x)BixTiO3\ Ceramics}$. *Ferroelectrics* 447:86–94. Taylor & Francis.

Zhang, G.-F., H. Liu, Z. Yao, et al. 2015. Effects of Ca Doping on the Energy Storage Properties of (Sr, Ca)TiO_3 Paraelectric Ceramics. *J. Mater. Sci. Mater. Electron.* 26:2726–2732.

Zhang, J., W.-J. Ji, J. Xu, et al. 2017. Giant Positive Magnetoresistance in Half-Metallic Double-Perovskite Sr_2CrWO_6 Thin Films. *Sci. Adv.* 3: e1701473.

Zhang, L., Q. Zhou, Q. He, et al. 2010. Double-Perovskites $A_2FeMoO_{6-\delta}$ (A = Ca, Sr, Ba) as Anodes for Solid Oxide Fuel Cells. *J. Power Sources* 195:6356–6366.

Zhao, F., Z. Yue, Z. Gui, et al. 2005. Preparation, Characterization and Microwave Dielectric Properties of A_2BWO_6 (A = Sr, Ba; B = Co, Ni, Zn) Double Perovskite Ceramics. *Jpn. J. Appl. Phys.* 44:8066.

Zhao, H. J., X. Q. Liu, X. M. Chen, et al. 2014. Effects of Chemical and Hydrostatic Pressures on Structural, Magnetic, and Electronic Properties of R_2NiMnO_6 (R = rare earth ion) Double Perovskites. *Phys. Rev. B* 90:195147.

Zhao, J. G., L. X. Yang, Y. Yu, et al. 2008. High-Pressure Synthesis of Orthorhombic $SrIrO_3$ Perovskite and Its Positive Magnetoresistance. *J. Appl. Phys.* 103:103706.

7 Diffusion, Thermal, and Optical Properties of Perovskites

Due to their unique structural features, perovskite oxides offer various distinguished properties through structural design. As described in the previous chapters, their magnetic and electronic properties are closely related to their electronic structures controlled by materials composition, crystal structure, and processing. This chapter deals with other properties, including diffusion, thermal, and optical properties.

7.1 DIFFUSION

Diffusion is generally described as the movement of atoms or ions or molecules from a high concentration region to a lower concentration region, as shown in Figure 7.1. In addition to the concentration gradient, diffusion can take place due to an electrical potential gradient called drift; or a temperature gradient called thermodiffusion. In perovskite oxides, the transport of oxygen or cation or electrolytic ions is the major focus of interest.

7.1.1 Laws of Diffusion

The problems based on diffusion can be tackled macroscopically and microscopically. One can use the macroscopic approach to find the diffusion coefficients. Even though the macroscopic approach does not consider the atomic

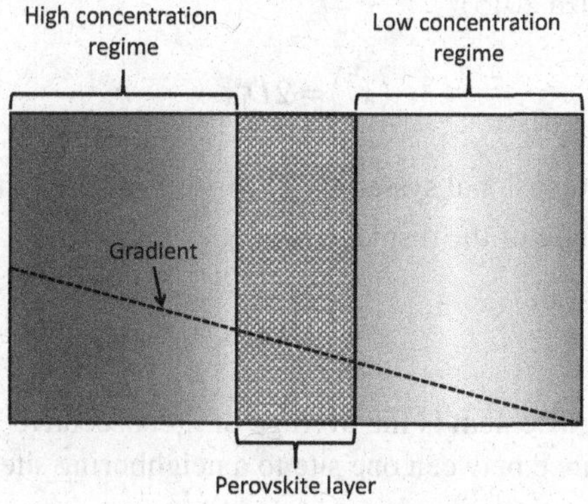

FIGURE 7.1 Schematic of a diffusion process.

nature of the diffusing elements. The diffusion in solids ultimately results from the individual jumps of the diffusing particles, through point defects, such as vacancies and interstitials. The macroscopic approach is based on Fick's law, which is the basis for determining the diffusion coefficients at an atomic scale. The Fick's laws of diffusion are the major governing equations at a macroscopic level. Fick's first law is

$$J_i = -D_i \frac{\partial C_i}{\partial x} \tag{7.1}$$

where J_i is the diffusion flux, $\frac{\partial C_i}{\partial x}$ is the concentration gradient, and D_i is the constant of proportionality called diffusion coefficient in a certain system. The above equation is applicable for isotropic materials.

For crystals that have two mutually perpendicular axes of symmetry, e.g., tetragonal, hexagonal, and orthorhombic, (Eq. 7.1) changes to

$$J_i = -D_{xx} \frac{\partial C_i}{\partial x} - D_{yy} \frac{\partial C_i}{\partial x} \tag{7.2}$$

In many cases, diffusion is a non-steady process; therefore, the change in concentration with time in any direction is proportional to the change in concentration gradient at that instance. This is known as Fick's second law of diffusion, and the corresponding governing equation is called the diffusion equation:

$$\frac{\partial C_i}{\partial t} = D_i \frac{\partial^2 C_i}{\partial x^2} \tag{7.3}$$

Microscopically, in perovskite structures, the diffusion of anions and cations essentially requires the presence of vacancies or open regions in the lattice, as in layered perovskites. The mean square displacement $\langle x^2 \rangle$ of the diffusing ions in the x-direction is related to the x-component of the diffusion coefficient, D_i^x, in a certain time interval t as (Souza 2015)

$$\langle x^2 \rangle = 2D_i^x t \tag{7.4}$$

For isotropic three-dimensional systems, $\langle x^2 \rangle = \langle y^2 \rangle = \langle z^2 \rangle = \langle R_i^2 \rangle / 3$

Then the mean square of the displacement is

$$\langle R_i^2 \rangle = 6D_i t \tag{7.5}$$

The mean square displacement is the average of n consecutive jumps for a distance a, where a is the distance between one site to a neighboring site.

$$\langle R_i^2 \rangle = na^2 \tag{7.6}$$

The diffusion in perovskites occurs by swapping of ions and vacancies, which can be defined by jump rate, Γ.

Then,

$$D_{\mathrm{i}} = \frac{1}{6}a^2 Z\Gamma_{\mathrm{i}} \tag{7.7}$$

Since the jump rate of the particle to one of its Z neighbors is, $Z\Gamma_{\mathrm{i}} = n/t$

To enhance the diffusivity in perovskites, an appropriate doping is required via generating vacancies in the sublattice.

Then the diffusion coefficient of the vacancies is

$$D_{\mathrm{V}} = \frac{1}{6}a^2 Z\Gamma_{\mathrm{V}} \tag{7.8}$$

Similarly, the diffusion coefficient of ions is

$$D_{\mathrm{O}} = \frac{1}{6}a^2 Z\Gamma_{\mathrm{O}} \tag{7.9}$$

Each time as the vacancy moves, an ion has to move, and the two species swap places. Thus, the total number of displacements of the ions and the vacancies have to be equal, i.e., $\Gamma_{\mathrm{O}}[\mathrm{O}] = \Gamma_{\mathrm{V}}[\mathrm{V}]$, where [O] and [V] represent the concentration of ions and vacancies, respectively.

Therefore,

$$D_{\mathrm{O}} = D_{\mathrm{V}}\frac{[\mathrm{V}]}{[\mathrm{O}]} \tag{7.10}$$

Eq. 7.10 reveals that the diffusion of ions is a function of vacancy diffusion.

Often the diffusion of ions in perovskites is a temperature-assisted phenomena as in the solid oxide fuel cells (SOFC), then the diffusion of ions can be written as

$$D = D_0 \exp\left(-\frac{\Delta H_{\mathrm{D}}}{k_{\mathrm{B}}T}\right) \tag{7.11}$$

where D is diffusivity, D_0 is the pre-exponential factor, T is the temperature, k_{B} is the Boltzmann's constant, and ΔH_{D} is activation enthalpy for diffusion.

7.1.2 Ionic Conductivity

In ionic conductors, the transport of cation/anions across the medium takes place by applying an external electric field. However, the ionic conductivity requires vacancies or open regions in the sublattice. The substitution of A-site ions with lower valent ions is a widely used technique for the tailoring of vacancies. Moreover, the absence of mixed valence is necessary to ensure the ionic conductivity by suppressing electronic conductivity. The ionic conductivity of perovskite oxides decreases rapidly with decreasing temperature since the activation energy for the ionic conductivity is very high, and the ionic conductivity drop is dominant than the temperature drop. In ionic conductors, the temperature-dependent ionic transport is governed by the Arrhenius equation, a modification of Eq. 7.11.

$$\sigma_T = \sigma_0 \exp\left(-\frac{E_a}{RT}\right) \quad (7.12)$$

where σ is the ionic conductivity, T the temperature in K, σ_0 is the pre-exponential ionic conductivity factor, and E_a is the activation energy for conduction. The values of E_a can be calculated from the slope of $\ln(\sigma_T)$ *vs.* $1/T$.

The transport mechanism in perovskite materials is termed as lattice diffusion. In general, lattice diffusion takes place through point defects such as vacancies and interstitial atoms. The following are the possible diffusion mechanisms in crystals (Liu et al. 2006):

1. *Vacancy mechanism.* When an atom or ion on a normal site jumps into an adjacent unoccupied lattice site, the vacancy moves into the site left by the atom. i.e., the movement of vacancies is opposite to the movement of atoms, as shown in Figure 7.2.
2. *Interstitial mechanism.* Here atoms or ions move through the interstitial site by jumping from one site to another. Such jumps through interstitial sites create a considerable distortion of the crystal lattice; therefore, this mechanism is possible for the transport of ions, which is considerably smaller than the lattice site. A schematic of the interstitial diffusion of ions is shown in Figure 7.3. Carburizing, nitriding, etc. of steels are typical examples of interstitial diffusion.

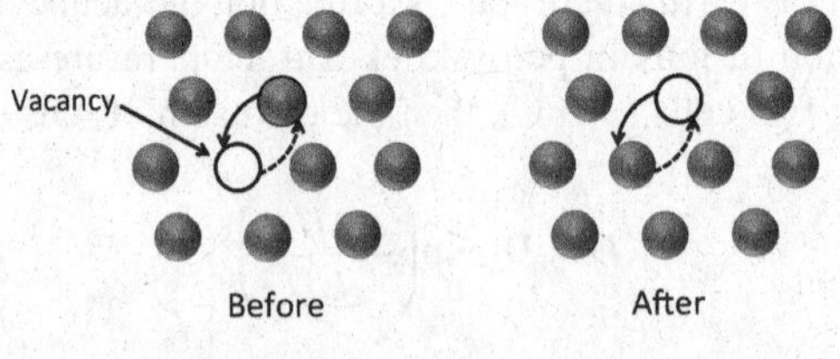

FIGURE 7.2 Vacancy-assisted diffusion process.

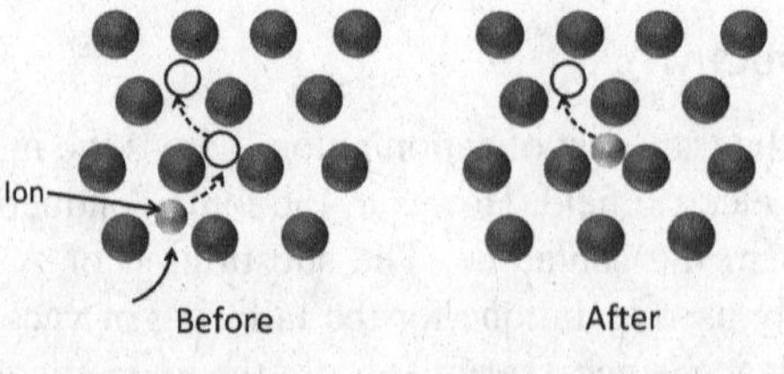

FIGURE 7.3 Diffusion process through interstitial sites.

3. *Interstitialcy mechanism.* In interstitialcy, atoms in the regular lattice are replaced by ions, and the regular atom is moved to the interstitial position, as shown in Figure 7.4, and the process repeats for the migration of ions. Interstitialcy takes place if the distortion of the regular lattice is too large.
4. *Direct exchange.* In direct exchange, diffusion occurs by the direct exchange of atoms in the lattice with the ions. The direct exchange is also possible through a ring-type movement between the three neighboring lattice atoms. Both the above processes are non-vacancy diffusion process, and the schematic representations are shown in Figure 7.5.
5. *Crowd-ion mechanism.* In this case, an extra atom has crowded into a line of atoms, which displaces several atoms along the line, as shown in Figure 7.6, from their equilibrium positions. The energy to move such a defect is small, but it only happens along an equivalent direction, as shown in Figure 7.6.

7.1.2.1 Oxygen Conductivity

The oxygen conductivity is guided by the vacancies in the regular perovskite lattice or empty regions in the layered lattices. A large number of oxygen vacancies are created when the perovskites are doped with aliovalent ions. The oxygen vacancies

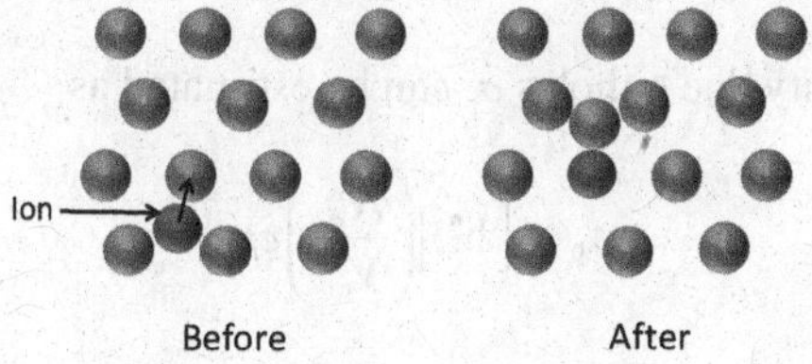

FIGURE 7.4 Diffusion process through interstitialcy.

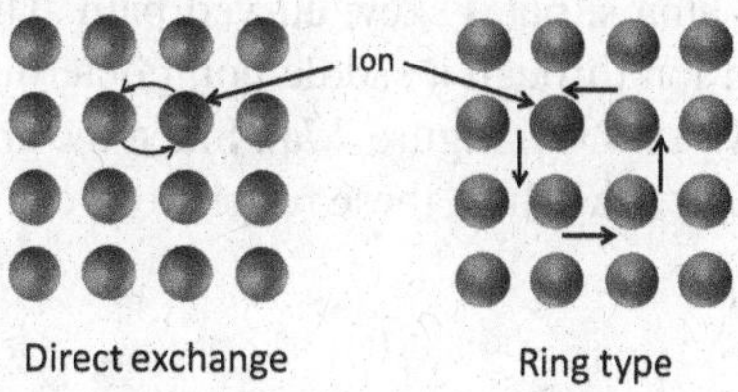

FIGURE 7.5 Diffusion through direct exchange.

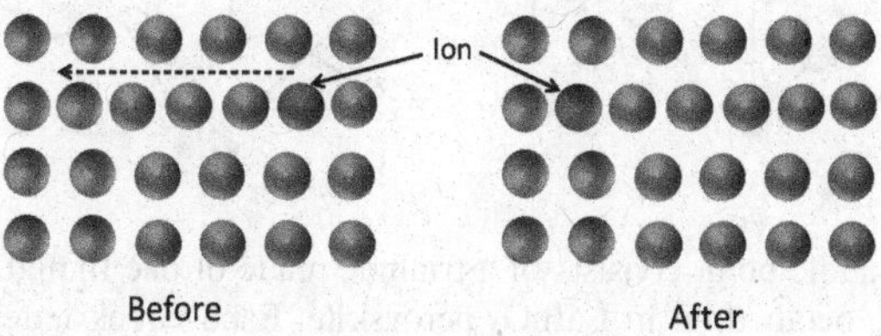

FIGURE 7.6 Diffusion through the Crowd-ion mechanism.

are filled with atmospheric oxygen atoms by forming two-electron holes at high temperatures to accommodate charge compensation as (Lybye 2000)

$$\frac{1}{2}O_2 + V_O^{\bullet\bullet} \leftrightarrow O_O^x + 2h^\bullet \tag{7.13}$$

where $V_O^{\bullet\bullet}$ is the oxygen vacancy, O_O^x represents the oxygen in the lattice, and 2 is the electron-hole.

The equilibrium constant, K_{ox}, for the above reaction is

$$\frac{1}{2}O_2 + V_O^{\bullet\bullet} \leftrightarrow O_O^x + 2h^\bullet K_{ox} = \frac{\left[O_O^x\right]\left[h^\bullet\right]^2}{P_{O_2}{}^{1/2}\left[V_O^{\bullet\bullet}\right]} \tag{7.14}$$

where P_{O_2} is the partial pressure of oxygen.

Since the oxygen non-stoichiometry is controlled by the defect induced by the dopants, not just by the partial pressure of oxygen (Ishihara et al. 1997), the concentration of *p*-type defects or holes can be calculated as

$$\left[h^\bullet\right] = K_{ox}^{1/2} P_{O_2}{}^{1/4}\left[V_O^{\bullet\bullet}\right]\left[O_O^x\right]^{-1/2} \tag{7.15}$$

Further the conductivity due to holes σ_p can be estimated as

$$\sigma_p = \left[h^\bullet\right]\left(\frac{N_A}{V_m}\right)e\mu_p \tag{7.16}$$

where N_A is the Avogadro constant, V_m is the molar volume, e is the charge of an electron, and μ_p is the mobility of holes.

The oxygen ion migration does not take place in a straight trajectory between two neighboring oxygen atoms, but a skew curved path. This process involves the crossing of the migrating ion through a saddle point determined by one In and two La atoms in representative $LaInO_3$ (Figure 7.7a). Since the anions squeeze through a triangular face made of La^{3+}-La^{3+}-In^{3+}, these restrictions can be overcome by doping

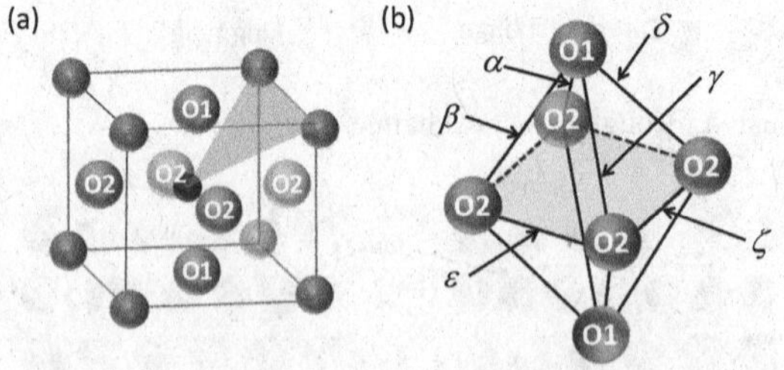

FIGURE 7.7 (a) Saddle point consists of a triangle made of one In and two La atoms and (b) a distorted InO_6 octahedron in $LaInO_3$ perovskite. Each Greek letter indicates a jump marked by arrows. All the paths are curved, reproduced with permission of the Elsevier (Ruiz-Trejo et al. 2003).

A-site with alkaline earth metals such as Ca, Sr, and Ba. The charge imbalance created due to doping in $LaInO_3$ or $LaGaO_3$ is not balanced by valence change, but it is balanced by the oxygen vacancy creation (Ruiz-Trejo et al. 2003). These high concentrations of vacancies lead to an enhancement in oxide ion conductivity, and it obeys Eq. 7.12.

The hopping of oxygen ion from one vacancy to another is determined by the energy barrier (activation energy) of the trajectory. In cubic perovskites, this energy barrier is almost the same due to equal bond lengths. In perovskite lattices with lower symmetry than cubic symmetry, for instance, $LaInO_3$ with orthorhombic symmetry has different O–O and metal–O distances; therefore, the activation energy is different for each trajectory (Ruiz-Trejo et al. 2003). The schematic of the trajectories is shown in Figure 7.7b. The corresponding activation energy for each trajectory is presented in Table. 7.1. From Table 7.1, one can understand that the most favorable jump is between O1 to O2 through β trajectory. It is also possible to have a jump from O1 to O1 (η) or O2 to O2 (θ) of the neighboring octahedra.

The changes in the energies of oxygen vacancy sites are also determined by the MO_6 octahedral tilt. For instance, the theoretical predictions on $LnNiO_3$ perovskite reveal that the migration barrier decreases when the O–Ni–O bond angle deviates from 180°. Especially, the ionic conductivity of epitaxial thin films which are used in devices mostly depends upon the tensile strain of octahedral tilt, and it is observed that an increase in tensile strength lowers the migration barrier.

Among the different dopants in the A-site as in $La_{0.9}Mg_{0.1}InO_{2.95}$, $La_{0.9}Ca_{0.1}InO_{2.95}$, $La_{0.9}Sr_{0.1}InO_{2.95}$, and $La_{0.9}Ba_{0.1}InO_{2.95}$ compounds display different point defect energies. The defects created at apical O1 sites are not favorable for oxygen hopping through vacancies; rather, at O2 sites are favorable. Among the different divalent alkaline earth metals listed above, Sr doping displays the lowest energy and results in oxygen transport. From the theoretical studies, Sr doping is more favorable for defects associated with O2 site, which is more frequent than O1 in the lattice

TABLE 7.1

The Activation Energy for Some Oxygen Migration in InO_6 Octahedra in $LaInO_3$ and the Corresponding Migrations Are Shown in Figure 7.7

Path	Jump	Distance (Å)	Activation Energy (eV)
α	O1–O2	2.94	0.45
β	O1–O2	2.93	0.25
Γ	O1–O2	3.10	1.17
δ	O1–O2	3.11	0.88
ε	O2–O2	2.98	0.38
ζ	O2–O2	3.05	0.62
η	O1–O1	3.18	0.97
θ	O2–O2	3.36	1.00

Reproduced with Permission of the Elsevier (Ruiz-Trejo et al. 2003).

(Ruiz-Trejo et al. 2003). However, the oxygen conductivities increased irrespective of the divalent alkaline earth metal dopants when compared with pure $LaInO_3$.

Similarly, doping of higher valent Zr^{4+} in small quantities ($x = 0.05$) in the lattice site of either La or In in $LaInO_3$ results in an increase in oxygen conductivity, but inferior to the Sr-doped counterparts. Additionally, the excess of Zr doping over the above critical level reduces the oxygen conductivity, due to the formation of a secondary phase. Such doping creates cationic vacancies in the lattice resulting in excess oxygen, and the conduction takes place due to the oxygen in excess (Takao and Binti 2010). In addition to $LaInO_3$, perovskite structures based on $LaYO_3$, $LaGeO_3$, $LaCrO_3$, $BaZrO_3$, and $SrCeO_3$ are promising oxygen-conducting materials.

Various layered perovskites with Aurivillius phase, Ruddlesden–Popper phase, double perovskites, brownmillerites, hexagonal perovskites, and the $LnInO_4$-based oxides were reported to exhibit high oxide-ion conductivities (Zhang et al. 2020). Often, in layered perovskites, the separating layer between each perovskite slabs also takes part in the conductivity in addition to oxygen hopping in the vacancies of octahedral slabs. For instance, $La_2CuO_{4+\delta}$ and $Ln_2NiO_{4+\delta}$ are high oxygen-conducting materials due to the presence of additional oxygen in the interstitial sites in Ln-O comprising layers between perovskite slabs. This oxide conductivity follows the interstitialcy mechanism, in which the interstitial ion occupies the normal site by displacing the lattice ion into the adjacent interstitial site, as in Figure 7.3 (Bochkov et al. 1999).

7.1.2.2 Proton Conductivity

Proton-conducting perovskites are useful for the separation of hydrogen fuel cells, electrochromic displays, etc. In general, perovskite materials containing Ce or Zr or Ti in the B-site exhibit good proton-conducting behavior. When the perovskite oxides with oxygen vacancies (by doping) are exposed to humid atmospheres at moderate temperatures, the vacancies dissociate the water molecules (undergo hydration) to hydroxide ion and a proton, leading to the formation of protonic defects at the interstitials (Kreuer 2003). The hydroxide ion sits on the oxide ion vacancy, and the protons form a covalent bond with lattice oxygen. The dissociation of water molecules in the presence of oxygen defect can be expressed as

$$H_2O + V_O^{\bullet\bullet} + O_O^x \leftrightarrow 2OH_O^{\bullet} \tag{7.17}$$

As two hydroxide ions substitute for oxide ions, two positively charged protonic defects $\left(OH_O^{\bullet}\right)$ are formed. $\left(V_O^{\bullet\bullet}\right)$ is the oxygen vacancy. The charged protonic defects can diffuse into the bulk through the oxygen vacancy-assisted diffusion process (Kreuer 1999). These materials exhibit oxide ion conductivity in the dry state and have an affinity for chemical diffusion for water (Kreuer et al. 1994). Assuming the ideal behavior of all species involved in the dissociation reaction (Eq. 7.15), which leads to the formation of two protonic defects, the equilibrium constant for the reaction, K, can be written as (Kreuer et al. 1994)

$$K = \frac{\left[OH_O^{\bullet}\right]^2}{\left[V_O^{\bullet\bullet}\right]\left[O_O^x\right]P_{H_2O}} \tag{7.18}$$

Since the number of mobile H^+ is equal to the concentration of $OH_O^{\bullet}$ species, the number of mobile H^+ available in the structure for conductivity increases in proportion to the partial pressure of water vapor, i.e.,

$$\left[OH_O^{\bullet}\right]^2 = K\left[V_O^{\bullet\bullet}\right]\left[O_O^x\right]P_{H_2O} \tag{7.19}$$

Since the water molecule is eventually split into a hydroxide ion and a proton, the formation reaction for protonic defects may be considered an exothermic amphoteric reaction, where the oxide acts as an acid (absorption of hydroxide ions by oxide ion vacancies) as well as a base (protonation of lattice oxide ions) (Kreuer 2003). The enthalpy of the hydration reaction tends to become more exothermic, with decreasing electronegativity of the cations interacting with the lattice oxygen. The variation of the electronegativity for the B-site element is large among the proton-conducting perovskites, and the equilibrium constant of the hydration reaction decreases in the order cerate→zirconate→stannate→niobate→titanate, i.e., with increasing electronegativity of the B-site cation (Kreuer 1999). The hydration reaction is temperature-sensitive; as a result, higher proton conductivity is observed at lower temperatures due to higher water uptake.

Estimation of the true stoichiometry of the perovskites and positions of protons in the lattice using conventional analytical, diffraction, and spectroscopic techniques is quite difficult; therefore, a precise conduction mechanism is not predicted. However, the principal features of the transport mechanism are rotational diffusion of the protonic defect and proton transfer toward a neighboring oxide ion, i.e., only the proton shows long-range diffusion. In contrast, the oxygen remains in their crystallographic positions. A detailed mechanism of proton diffusion in perovskite lattice is shown in Figure 7.8. In Figure 7.8, the top left and bottom left boxes (a and e) show the proton diffusion mechanism in 3D structures: hydroxide ion rotation in AO plane and proton transfer in BO_2 plane. Four figures in both rows (b and C, and f and g) describe the elementary steps underlying hydroxide ion rotation: outward O–B–O bending motion and hydroxide ion reorientation take place in the AO plane, and

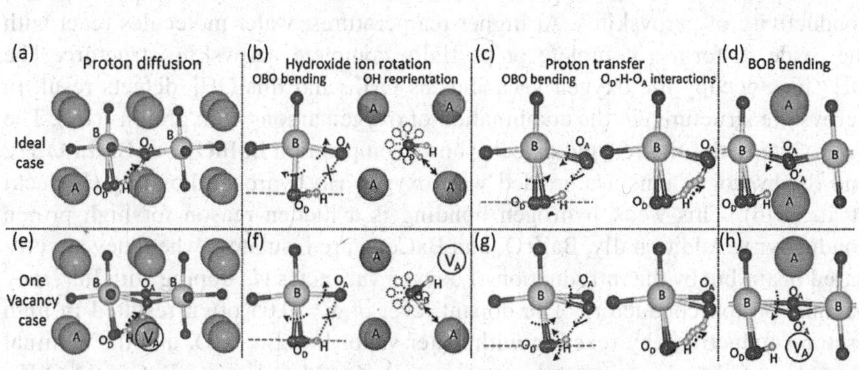

FIGURE 7.8 A schematic illustration of the proton diffusion mechanism in an ABO_3 perovskite oxide without (top row) and with (bottom row) one A cation vacancy.

proton transfer: inward O–B–O bending motion, hydroxide ion rotation out of AO plane, and OD–H–OA interactions take place in the BO_2 plane. Top right (d) and bottom right (h) figures show additional B–O–B bending motion involved in both hydroxide ion rotation and proton transfer. In all the figures, the red balls denote the initial positions, and the brown balls show the final positions.

Since the oxygen vacancies are the primary components necessary for proton conductivity, a large number of perovskites and their derivatives are extensively studied for proton conductivity. $BaZrO_3$, $BaCeO_3$, $BaTiO_3$, and $Ba_2In_2O_5$, are the major classes of proton-conducting perovskites. Perovskites with Ce at B-site are excellent proton conductors, but the stability of cerium is poor under acidic CO_2 and SO_2 gases. The Zr-based perovskites are more stable, but their proton conductivity is poor than Ce-perovskites.

Y-doped $BaZrO_3$ is a cubic perovskite with a high lattice constant, and the high crystal symmetry of these materials allows an upper solubility limit of protonic defects for the high isotropic proton mobility. The high lattice constant and the covalency of the Zr/O bond reduce the Zr/H-repulsive interaction; therefore, the activation enthalpy for the mobility of protonic defects reduces (Kreuer et al. 2001). The deviations from the ideal cubic perovskite structure result in an increase of the activation enthalpy, i.e., energy for the migration of protons. On comparing the Y:$BaCeO_3$ and Y:$SrCeO_3$, the large orthorhombic distortion of Y:$SrCeO_3$ has tremendous effects on the energies of the lattice oxygen, as discussed in the previous section (Figure 7.7a), due to different chemical interactions with the cations. When Sr is occupying the A-site, the oxygen atoms at O1 and O2 have distinct characteristics, and one can expect different binding energies for the proton. In $SrCeO_3$, O1 has the lowest basicity, but for $BaCeO_3$ it is O2. If the photons are binding to low basic sites, then the long-range proton transport is through the most frequent O2 sites in $BaCeO_3$, whereas long-range proton transport in $SrCeO_3$ is through O1. As a result, $SrCeO_3$ exhibits a lower proton conductivity when compared with $BaCeO_3$ (Münch et al. 1999). $Ba_2In_2O_5$ shows brownmillerite structure at low temperatures, and it starts to disorder at ~900°C to form a cubic structure with a large number of oxygen vacancies, favorable for proton conduction.

Incorporation of water vapor to the lattice is another way to improve proton conductivity of perovskites. At higher temperatures, water molecules react with the oxide to form a complete or partially complete perovskite structure. The OH^- ions occupy the oxygen vacancies as $OH_O^\bullet$, and this $OH_O^\bullet$ defects result in perovskite structure *via* the combination of oxygen anions with protons (H^+). The perovskite with fully reacted defects shows composition $BaInO_3H$ or $Ba_2In_2O_6H_2$, and the hydrogen ions are bonded with oxygen *via* hydrogen bonding (Bielecki et al. 2016). This weak hydrogen bonding is a hidden reason for high proton conductivity. Additionally, $BaZrO_3$ and $BaCeO_3$ are insulators when they are prepared in air, but by the introduction of oxygen vacancies *via* doping with In^{3+}, they become proton conductors. The dopant level of $x = 0.05$ often resulted in high proton conductivity by reacting with water vapor like $Ba_2In_2O_5$ and the nominal formulae of the phases after hydration is $BaZr_{1-x}M_xO_3H_x$ and $BaCe_{1-x}M_xO_3H_x$, respectively (Mazzei et al. 2019).

7.1.2.3 Li-ion Conductivity

Li-ion conductivity of perovskites is promising to use them as a separation membrane in Li-ion batteries. Perovskite structured A-site deficient titanium, niobium, hafnium, and tantalum bronzes (the term bronzes correspond to $A_{1-x}WO_3$ structure as discussed in chapter 3). For example, $La_{2/3}TiO_3$ and its derivatives show Li-ion diffusion through A-site vacancies. The above perovskite structured titanium bronze comprises of 1/3 vacant A sites and has Ti^{4+} as B-site. Li^+ ion diffusion is highly favored in $Li_3La_{2/3}TiO_3$ phase, which is the prime parameter in enhancing the efficiency of Li-ion batteries. Li^+ ion diffusion increases if a significant amount of vacancies presents in the derivative phase (Wu et al. 2017). In the $Li_{0.5}La_{0.5}TiO_3$ phase, the ionic conductivity of Li^+ ions is low because of the absence of vacancies. Maximum conductivity (10^{-3} S cm^{-1}) is noticed when the vacancies are half-filled in the $Li_{0.33}La_{0.56}TiO_3$ phase (Hu et al. 2018). In reality, perovskites have partial segregation of vacancies, and this segregation apparently leads to the formation of microdomain, which may interfere with ionic conductivity. In thin films, microdomains have an influence on the conductivity of ions, especially in the thin films of $Li_3La_{2/3}TiO_3$ compound.

7.1.2.4 Mixed Conductors

The large class of perovskite oxides exhibits both ionic and electronic conduction, and they are called mixed conductors. In mixed conductors, the oxygen or hydrogen is transported through a dissociated or ionized form rather than the conventional molecular diffusion. In mixed conductors, the ionic diffusion takes place by chemical potential gradient, but the overall charge neutrality is maintained by a counterbalancing flux of electrons. Such systems do not require an external circuit. The mixed conductivity is observed in perovskite materials with a large specific free volume and a tolerance factor of around 0.96 (Hayashi et al. 1999).

One can obtain the free volume by subtracting the unit cell volume by the volume of the constituent ions.

$$\text{Specific free volume} = \text{free volume/unit cell volume}$$
$$= \left(V - \text{total volume of the constituent ions}\right)/V$$

where V is the volume of the unit cell.

The tolerance factor for perovskite materials is calculated as

$$t = \frac{r_A + r_O}{\sqrt{2}\,(r_B + r_O)} \tag{7.20}$$

where r_A and r_B are the ionic radii for A- and B-site cations based on the coordination numbers (Shannon 1976), and r_O is the radius of oxygen, respectively.

From Figure 7.9, one can observe that, if the tolerance factor falls between 0.95 and 1.0, the conductivity is highest as the tolerance factors fall below or above 0.95–1.0, and the structure deviates from the ideal cubic structure. The deviation

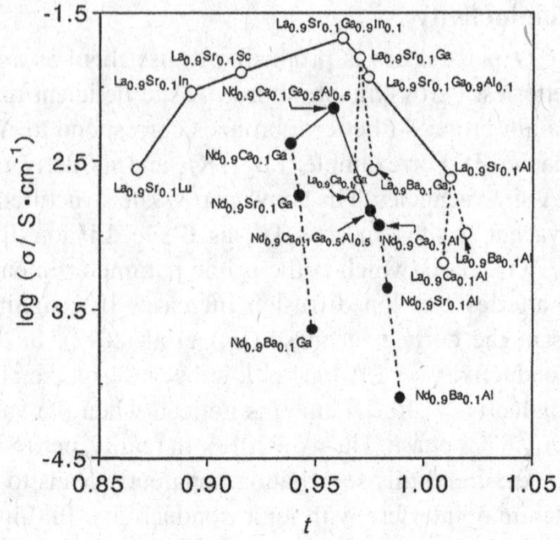

FIGURE 7.9 Comparison of the electrical conductivity between $LaAMO_3$ (open symbols) and $NdAMO_3$ (A = Sr, Ca, Ba, and M = In, Al, Ga, Sc) (filled symbols) at 1,000 K, reproduced with permission of Elsevier (Hayashi et al. 1999).

from the cubic geometry leads to a high anisotropy in oxygen sites; as a result, the electric conductivity decreases. The specific volume and tolerance factor have a linear relationship when the specific free volume becomes larger; the tolerance factor becomes smaller and *vice versa*. Such perovskites have larger B-site cations as compared to the transition metals, since both the specific volume and the tolerance factor are functions of the ionic radius. When considering the specific volume, a larger specific free volume is favored to obtain higher electrical conductivity at the same tolerance factor (Hayashi et al. 1999).

In order to achieve a high unit cell volume, large B-site ions must be incorporated into the lattice, so as to the A-site ions without affecting the tolerance factor. In the case of $LaGaO_3$, the size of the trivalent La^{3+} and Ga^{3+} is optimal for the largest free volume and a tolerance factor of 0.96, which has the highest electrical conductivity. Additionally, the substitution of a B-site ion by the same valent cation changes the specific free volume and the tolerance factor without the change of charge neutrality in perovskites. Doping divalent alkaline earth metal in the A-site increases the electrical conductivity of the doped oxide as compared to the parent oxide. La is substituted by Sr, and Nd is substituted by Ca in $AGaO_3$, or $AAlO_3$ (A = La or Nd) increases the electrical conductivity. Moreover, a small amount of transition metal doping in the B-site increases electrical conductivity (Liu et al. 2006).

The conducting property of mixed-conducting materials is usually expressed by conductivity for both ions and electrons. The total conductivity of a mixed-conducting material can be given as the sum of the ionic conductivity and electronic conductivity:

$$\sigma_{\text{total}} = \sigma_{\text{ion}} + \sigma_{\text{e}} \tag{7.21}$$

Electronic conductivity can be obtained by subtracting the ionic conductivity from the total conductivity.

A useful concept of transport (or transference) number t_k is more frequently used to describe the property of the mixed-conducting materials in practice. For any carrier k, t_k is defined by

$$t_k = \sigma_k / \sigma_{total} \tag{7.22}$$

In the case of mixed conduction

$$\sigma_{total} = \sigma_{ion} + \sigma_e = \sigma_{total}\left(t_{ion} + t_e\right) \tag{7.23}$$

The quantification of the ionic component in mixed perovskites is highly difficult because its magnitude is lower than the electrical component in several orders. For example, $Sr_3Fe_2O_{6+\delta}$ Ruddesden-Popper phase nearer to the $Sr_3Fe_2O_7$. $Sr_3Fe_2O_7$ phase comprising only Fe^{4+} and preparation of this compound in air results in the composition of $Sr_3Fe_2O_{6.73}$, which shows a mixed valence state ($Sr_3Fe_{0.54}{}^{3+}Fe_{1.46}{}^{4+}O_{6.73}$). This composition is stable up to 400°C; after that, it loses oxygen and shows a composition $Sr_2Fe_2O_6$ above 1,000°C, which displays only Fe^{3+}. In this compound, the oxygen vacancies are distributed randomly at the equatorial sites present around FeO_6 octahedra (Mogni et al. 2005). The Ruddlesden–Popper oxide undergoes changes at high oxygen partial pressures due to the presence of open regions at the interlayers of perovskite blocks. Hence, the total conductivity of this material is majorly contributed from interstitialcy migration of oxygen ions (*n*-type) and hole (*p*-type) conduction. These phenomena are well explained in terms of the defect chemistry of the crystals under high and low oxygen partial pressures. The oxygen conductivity is shown as the following equation if the bulk diffusion is a rate-limiting step (Itoh et al. 2006),

$$J_{O_2} = \frac{RT\sigma_e\sigma_{ion}}{16F^2\left(\sigma_e + \sigma_{ion}\right)t}\ln\left(\frac{P_h}{P_l}\right) \tag{7.24}$$

where J_{O_2} is the oxygen flux, σ_e is the electron conductivity, σ_{ion} is the oxide ion conductivity, F is Faraday constant, R is gas constant, t is the membrane thickness, T is temperature, and p_h and p_l are high and low oxygen partial pressures, respectively.

The electronic conductivity of perovskites at elevated temperatures is influenced by the partial pressure of atmospheric oxygen, P_{O_2}, *n*-type electronic conductivity σ_n increases with decreasing P_{O_2}, whereas *p*-type electronic conductivity σ_p increases with increasing P_{O_2}. Therefore, the respective conductivities can be written as a function of oxygen partial pressure (Iwahara 2009).

$$\sigma_n = \sigma_n^O \exp P_{O_2}^{-1/n} \tag{7.25}$$

and

$$\sigma_p = \sigma_p^O \exp P_{O_2}^{1/n} \tag{7.26}$$

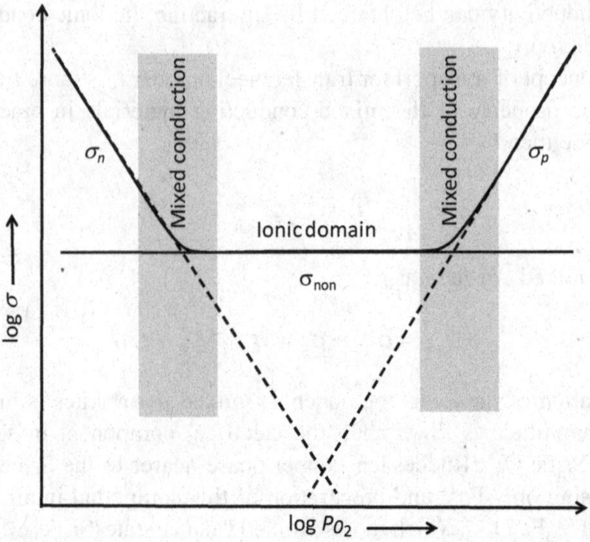

FIGURE 7.10 Dependence of conductivity on the partial pressure of oxygen, reproduced with permission of the Springer Nature (Iwahara 2009).

where σ_n^O and σ_p^O are constants independent of the partial pressure P_{O_2}.

The ionic conductivity is independent of P_{O_2}.

Therefore, the total conductivity is

$$\sigma = \sigma_i + \sigma_n^O \exp P_{O_2}^{-1/n} + \sigma_p^O \exp P_{O_2}^{1/n} \tag{7.27}$$

The relationship between the logarithm of each conductivity and P_{O_2} is shown in Figure 7.10. The shaded regions indicate the mixed conduction domains, and between them, there exists an ionic conduction domain where electronic conductivity is negligibly low. The outer sides of the mixed conduction domains are electronic conduction regions (Iwahara 2009).

The p-type conduction arises from the defect equilibrium with oxygen in the gas phase:

$$V_O^{\bullet\bullet} + 1/2O_2 \leftrightarrow O_O^x + 2h^\bullet \tag{7.28}$$

7.2 THERMAL PROPERTIES

Thermal properties of perovskite materials are important since they are considered as an ideal candidate for several high-temperature applications, such as SOFCs, thermal barrier coating, oxygen separation membranes, and thermoelectrics. In the following section, we will be discussing the major factors affecting the performance of perovskite materials in the above applications.

7.2.1 Thermal Conductivity

Thermal conductivity can be defined as the transport of heat across a temperature gradient. Low thermal conductivity perovskites are sought for high-temperature solid oxide fuels, thermal barrier coatings, and thermoelectrics. According to Fourier's law of heat conduction, the heat transfer is proportional to the negative of the temperature gradient.

$$Q_x = -k\frac{dT}{dx} \tag{7.29}$$

where Q_x is the heat flow along a certain direction x, k is the constant of proportionality known as conductivity and dT/dx is the temperature gradient.

In general, thermal conductivity in a material takes place by electrons and phonons. Therefore,

$$k_{tot} = k_e + k_{ph} \tag{7.30}$$

Experimentally, the thermal conductivity can be calculated using the thermal diffusivity, α, a measure of the rate of change of temperature with respect to time (Hofmeister 2010).

Then,

$$k = \rho C_p \alpha \tag{7.31}$$

where C_p is the heat capacity at constant pressure and ρ is the density.

Many perovskite oxides exhibit very low thermal conductivities, especially the ferroelectric perovskites. Perovskites composed of elements of high atomic mass exhibit low thermal conductivities, for example, k = 2.1 W mK^{-1} for $SrZrO_3$ and k = 3.4 W mK^{-1} for $BaZrO_{3.18}$ at 1,000°C on account of the structural rigidity of corner-shared octahedra (Pan et al. 2012). Ba ions are bulkier than the Sr ions; therefore, it is assumed that the phonon transport is affected by the ionic sizes, since phonons are nothing but lattice vibrations. Similarly, the bulk thermal conductivity of this $BaTiO_3$ is 2.65 W mK^{-1} (He 2004). Doping in the perovskite lattice sites can result in unique thermal characteristics. For instance, $Ba(Mg_{1/3}Ta_{2/3})O_3$ has a melting temperature of ~3,000°C with a thermal conductivity of 2.5 W mK^{-1} at 1,100°C and it is a promising refractory material (Vaßen et al. 2010). Likewise, the complex $La(Al_{1/4}Mg_{1/2}Ta_{1/4})O_3$ perovskite has a thermal conductivity of ~2 W mK^{-1} and a high coefficient of thermal expansion ~11.9×10^{-6} K^{-1} (Jarligo et al. 2010).

The reduction in thermal conductivity is favorable for high thermoelectric properties. To reduce the thermal conductivity of perovskites, there are several approaches. For instance, the thermal conductivity of $SrTiO_3$ is sensitive to non-stoichiometry, with Sr-excess and -deficiency resulting in 65% and 35% reduction in thermal conductivity, respectively (Breckenfeld et al. 2012). Doping A-site with higher valence ions such as La, or doping B-site with a higher valence like Nb in $SrTiO_3$ reduced the thermal conductivities significantly. However, as the grain size

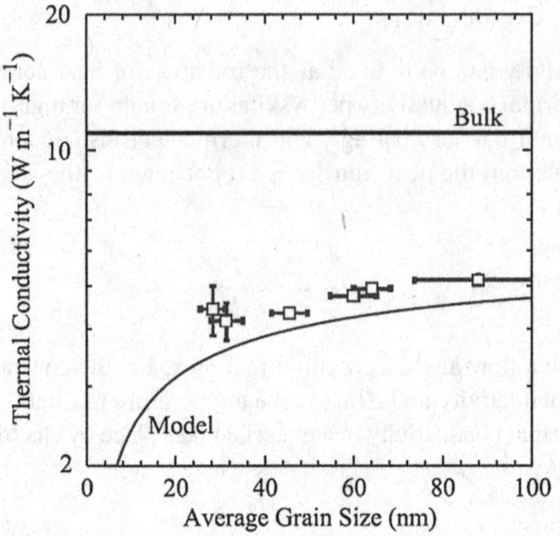

FIGURE 7.11 Thermal conductivity of ng-$SrTiO_3$ as a function of average grain size (hollow squares) and bulk $SrTiO_3$ (solid line), reproduced with permission of the American Institute of Physics (Foley et al. 2012).

of $SrTiO_3$ decreases, the thermal conductivity decreases considerably, as shown in Figure 7.11, which is due to the different modes of phonon scattering at the grain boundary. Materials with a nanograin structure have more grain boundaries than the ones with large grains due to the size effects (Foley et al. 2012). Similarly, single crystals exhibit a smaller thermal conductivity than polycrystalline materials (Suemune 1965).

The perovskite oxides exhibit a higher thermal conductivities at high temperatures. Some perovskites undergo a phase transformation with temperature. The thermal conductivity changes with the subsequent phase transformation as well. For instance, $BaTiO_3$ experiences phase transformation from rhombohedral to orthorhombic to tetragonal with increasing temperatures. The thermal conductivity changes in accordance with the phase change are shown in Figure 7.12, which indicates the anharmonicity (the springs that tie atoms together in a lattice model do not precisely obey Hooke's law) of phonon transfer with the change in the crystal structure (Suemune 1965).

When considering the layered perovskite, the thermal conductivity is anisotropic along parallel and perpendicular to the layers. For example, La-doped $Sr_2Nb_2O_7$ has a thermal conductivity of 1.0 W mK^{-1}. The thermal diffusivity α is an indirect measure of thermal conductivity according to Eq. 7.31. One can observe the anisotropy in thermal diffusivity along parallel and perpendicular to the layered structure of La-doped $Sr_2Nb_2O_7$ up to 1,000°C in Figure 7.13. The in-plane thermal conductivity is always larger than that is along the *c*-axis, parallel to the layer stacking direction (Sparks et al. 2010). A similar observation is made among $Ln_2SrAl_2O_7$ (Ln = La, Nd, Sm, Eu, Gd, or Dy) series with anisotropy in the thermal conductivity based on density functional theory (DFT) calculations (Feng et al. 2012).

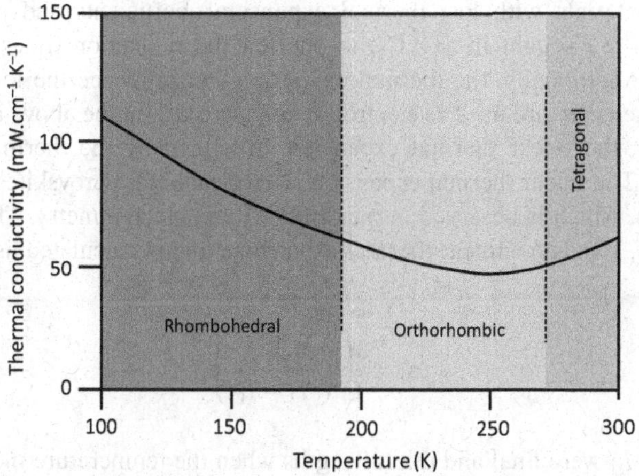

FIGURE 7.12 The thermal conductivity of $BaTiO_3$ single crystals, where dotted lines indicate the transition temperature (Suemune 1965).

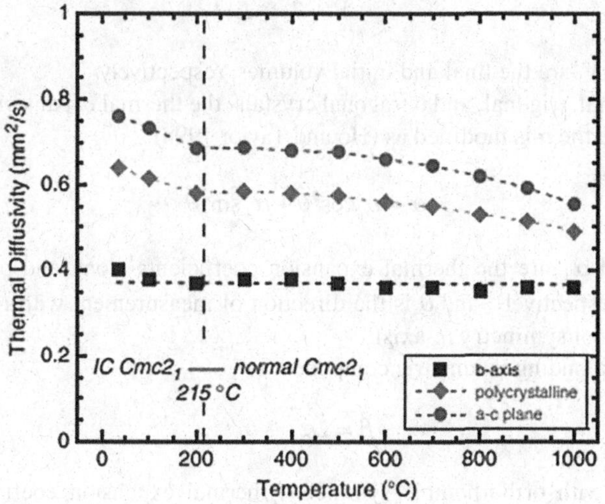

FIGURE 7.13 Thermal diffusivity of La-doped $Sr_2Nb_2O_7$ as a function of temperature along the *b*-axis, *a-c* plane, and for randomly oriented polycrystalline material, reproduced with permission of the John Wiley & Sons (Sparks et al. 2010).

7.2.2 Thermal Expansion

Many perovskite materials are promising candidates for high-temperature applications. At high temperatures, the mismatch of the thermal expansion of components can lead to thermal stresses and, ultimately, failure of the systems. For instance, in SOFCs, the cells are made up of all solid components so that thermal stresses may result from thermal expansion coefficient mismatch among the cell components.

Therefore, materials with low thermal expansion coefficients and high thermal conductivities are sought in SOFCs, to shorten the relaxation times for thermal fluctuations. Additionally, the thermal expansion can influence the ionic transport properties when they are used as electrolyte or electrodes in the above applications.

In general, the linear thermal expansion is defined by the thermal expansion coefficient α. The linear thermal expansion is applicable for perovskites with isotropic expansion, which is observed in materials with cubic symmetry. The expansion coefficient α_L, stands for linear thermal expansion, and is calculated as (Mori et al. 2000)

$$\alpha_L = \frac{1}{L_1}\left(\frac{L_2 - L_1}{T_2 - T_1}\right) \tag{7.32}$$

where L_2 and L_1 were final and initial lengths when the temperature increased from a low-temperature T_1 to a high-temperature T_2.

Similarly the volumetric thermal expansion

$$\beta = \frac{1}{V_1}\left(\frac{V_2 - V_1}{T_2 - T_1}\right) \tag{7.33}$$

where V_2 and V_1 are the final and initial volumes, respectively.

In hexagonal, trigonal, and tetragonal crystals, the thermal expansion is anisotropic. Therefore the α is modified as (Ho and Taylor 1998)

$$\alpha = \alpha_{\parallel}\cos^2\hat{\theta} + \alpha_{\perp}\sin^2\hat{\theta} \tag{7.34}$$

where $\alpha_{\parallel}$ and $\alpha_{\perp}$ are the thermal expansion coefficients along and perpendicular to the axis, respectively, and $\hat{\theta}$ is the direction of measurement with respect to the principal axis of symmetry (c-axis).

The corresponding volumetric component is

$$\beta = 2\alpha_{\perp} + \alpha_{\parallel} \tag{7.35}$$

For crystals with orthorhombic symmetry, thermal expansion coefficient in any arbitrary direction is

$$\alpha = \alpha_a l^2 + \alpha_b m^2 + \alpha_c n^2 \tag{7.36}$$

where l, m, and n are cosines of angles between the direction of measurement and crystallographic axis, and α_a, α_b, and α_c are thermal expansion coefficients along the crystallographic axes.

At the microscopic scale, thermal expansion is calculated as the change in the lattice parameter with temperature using the following equation (Krishnan et al. 2013):

$$\alpha_t = \frac{\Delta a}{\Delta T}\frac{1}{a_0} \tag{7.37}$$

where $\frac{\Delta a}{\Delta T}$ is the change in lattice parameter with temperature and a_0 is the original lattice parameter.

For non-linear thermal expansions, α is empirically related to temperature as

$$\alpha_t = A + Bt + Ct^2 \tag{7.38}$$

where A, B, and C are constants.

The thermal expansion coefficient is sensitive to the structure, and it changes with any transitions in the crystal structure. The thermal expansion of perovskites is often due to the phase change associated with BO_6 octahedral tilts, B-cation displacement, and the change in B-O bond length at high temperatures. Generally, all the perovskite materials exhibit a positive thermal expansion coefficient as the temperature increases. Still, it depends on several factors such as structural phase transition, ferroelectric transitions, magnetic transitions, vacancies, dopants, grain growth, oxidation or reduction, etc. For instance, ferroelectric $BaTiO_3$ exhibits positive and negative thermal expansion coefficients as phase transformation occurs at different temperatures, as discussed in the previous chapters. In Figure 7.14a, one can observe the change in strain with the phase transformation as the temperature increases. The slope of the curves indicates the thermal expansion coefficient. For cubic structured $BaTiO_3$ at a temperature higher than 393 K, the thermal strain is the highest. The thermal expansion coefficient depends on the polarization; one can note the drastic change in the thermal expansion coefficient at each boundary of phase transformation shown in Figure 7.14b. However, the thermal expansion coefficient is recovered from the sudden drop on the further slight increase in the temperature, which indicates the involvement of saturation polarization (Rao et al. 1997).

The thermal expansion coefficient is higher for doped samples than their pure counterparts. In the case of Sr-doped $LaMnO_{3+\delta}$, the isotropic thermal expansion increases slightly with increasing Sr content. However, the anisotropic thermal expansion along the c-axis is doubled as compared to the same along a-axis, since

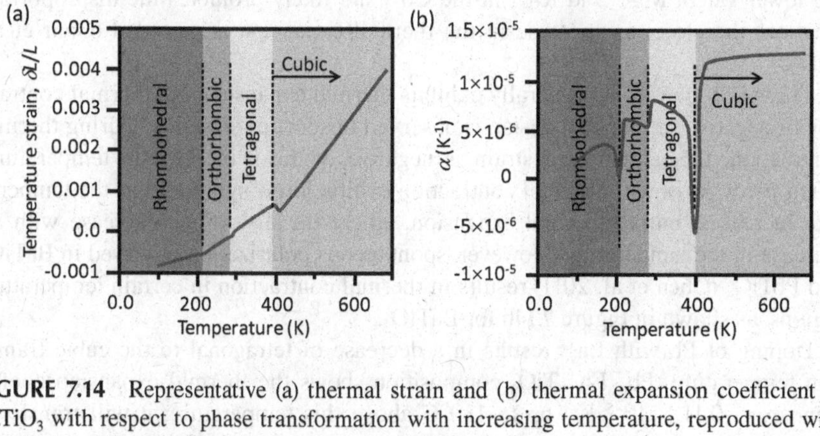

FIGURE 7.14 Representative (a) thermal strain and (b) thermal expansion coefficient of $BaTiO_3$ with respect to phase transformation with increasing temperature, reproduced with permission of the Springer Nature (Rao et al. 1997).

the crystal is compressed along the c-axis due to the tilting of the MnO_6 octahedra. Such anisotropic thermal expansion is also observed in the case of $LaAlO_3$ (Howard et al. 1999), $La_{1-x}A_xCoO_{3-\delta}$ (A = Sr, Ca) (Mastin et al. 2006), and for $La_{1-x}Sr_xFeO_{3-\delta}$ (Fossdal et al. 2004).

Unlike doping, the non-stoichiometry results in a lattice contraction. The lattice expansion is likely to decrease as the amount of oxygen increases, as in the case of $LaMnO_{3+\delta}$, lattices with excess oxygen has an expansion lower than the one with composition near stoichiometry. An oxygen excess lattice is identical to non-stoichiometry due to the cation vacancies. In the above case, the reduction in thermal expansion with increasing oxygen content is due to the reduced rhombohedral distortion with the increase in the temperature and decreased partial pressure of oxygen (Miyoshi et al. 2004). Additionally, the excess oxygen in the lattice requires charge compensation, which is achieved by the formation of electron holes at oxygen sublevels (O^-). O^- is smaller than O^{2-}, and during isothermal expansion, the average ionic radius, therefore, decreases (Miyoshi et al. 2003), resulting in a reduction in thermal expansion with oxygen content. On the other hand, in the case of oxygen-deficient perovskite structure, $LaCoO_{3-\delta}$, the thermal expansion increases with the decrease in the oxygen content. In the case of oxygen deficiency, the disproportion of B-site occurs as per Eq. 7.39, at normal temperatures. During such a reaction, the average ionic radius decreases as the size of Co^{3+} is larger than Co^{4+}. The disproportion reaction reverses at high temperature as per Eq. 7.40, resulting in an increase in the ionic size, ultimately leading to a large thermal expansion at high temperatures (Zuev et al. 2008).

$$2Co^{3+} \rightarrow Co^{4+} + Co^{2+} \tag{7.39}$$

$$Co^{4+} + Co^{2+} \rightarrow 2Co^{3+} \tag{7.40}$$

Further, doping of cobalt with a higher valent Nb^{5+} or Ti^{4+} suppresses the above disproportion reactions of Co^{3+} due to the charge compensation by the dopants; therefore, a reduction in the thermal expansion is observed. On the other hand, doping lower valent Mg^{2+} and Rh^{2+} in the Co^{3+} site likely promotes the disproportion reaction; therefore, an increase in the thermal expansion is recorded (Gaur et al. 2017).

The ABO_3 perovskite generally exhibits thermal expansion, but thermal contraction or negative thermal expansion is observed in some perovskites. During thermal contraction, the temperature strain is negative, with an increase in temperature. Often perovskites with thermal contraction exhibit lower symmetry as the temperature increases, unlike thermal expansion, where the distortions decrease with an increase in the temperature. However, spontaneous polarization observed in $BaTiO_3$ and $PbTiO_3$ (Chen et al. 2011) results in thermal contraction in certain temperature regions, as shown in Figure 7.14b for $BaTiO_3$.

Doping of Pb with La^{3+} results in a decrease of tetragonal to the cubic transition temperature. $Pb_{0.8}La_{0.2}TiO_3$ composition shows the thermal expansion coefficient $\alpha = -0.11 \times 10^{-5}\ K^{-1}$ up to 130°C above this temperature; it will transform into cubic phase and show the same thermal expansion coefficient like the undoped one (3.72×10^{-5} K^{-1}) (Chen et al. 2005). Similarly, in $BiNiO_3$, the strain $\Delta L/L$ increases with increasing temperature up to 270 K, indicating the normal positive thermal expansion, but decreases above 270 K. This volume change results from charge transfer from Ni to Bi as the temperature increases from $Bi^{3+}_{0.5}Bi^{5+}_{0.5}Ni^{2+}O_3$ to $Bi^{3+}Ni^{3+}O_3$. The substitutions for Bi may suppress the charge disproportionation in the $Bi^{3+}_{0.5}Bi^{5+}_{0.5}Ni^{2+}O_3$ phase and thereby shift the charge-transfer transition to near ambient conditions. But $Bi_{0.95}La_{0.05}NiO_3$ undergoes a first-order phase transition from the larger-volume low-*T* phase to the smaller-volume high-*T* phase (Azuma et al. 2011), but the above intersite transitions take place under high pressure.

A similar charge disproportion reaction is observed in $ACu_3Fe_4O_{12}$ (A = alkaline earth or rare earth metals). In $LaCu_3Fe_4O_{12}$, low-temperature $LaCu_3^{3+}Fe_4^{3+}O_{12}$ is altered to $LaCu_3^{2+}Fe_4^{3.75+}O_{12}$ structure at high temperatures. This charge disproportion reaction leads to thermal contraction (Long et al. 2009). The isostructural $SrCu_3Fe_4O_{12}$ double perovskite also exhibits a high thermal contraction between 170 and 270 K with a thermal expansion coefficient of $\alpha = -2.26 \times 10^{-5}$ K^{-1}, but outside of this temperature range, both above and below, it shows a normal thermal expansion. The reason for this thermal contraction is charge transfer between Fe on B-sites and Cu on A′-sites. The charge distribution is $Sr^{2+}Cu_3{}^{`2.8+}$ $Fe_4{}^{`3.4+}$ O_{12} at low temperature, Fe charges are ($Fe_{2.4}{}^{3+}$ $Fe_{1.6}{}^{4+}$), and the Cu charges are ($Cu_{0.6}{}^{2+}$ $Cu_{2.4}{}^{3+}$). At extreme temperature the charge distribution is $Sr^{2+}Cu_3{}^{`2.4+}$ $Fe_4{}^{`3.7+}$ O_{12} with Cu charges ($Cu_{1.8}{}^{2+}$ $Cu_{1.2}{}^{3+}$) and Fe charges ($Fe_{1.2}{}^{3+}$ $Fe_{2.8}{}^{4+}$). The coordination polyhedra around the cations changes continuously due to changes occurring in charge transfer, especially in the temperature range of 170–270 K. In this temperature range, Cu–O bond distance increases while Fe–O and Sr–O bond distances slowly decrease, and the Fe–O–Fe bond angle (ψ) in octahedral tilt raises. The relation between the lattice constant, Fe–O bond length d_{Fe-O}, and the Fe–O–Fe bond angle (ψ) is depicted below (Yamada et al. 2011).

$$a = 4d_{Fe\text{-}O}\sin\left(\Psi/2\right) \tag{7.41}$$

The expansion or contraction of the above perovskite structures largely depends on the Fe–O bond length compared to the Fe–O–Fe bond angle, and it plays a key role in anomalous thermal contraction (Yamada et al. 2011).

Materials that do not expand along with the change in temperature are known as zero thermal expansion (ZTE) materials, and they are found to be essential in miniature electronic devices. In the past, ZTE materials are fabricated by making composition of two materials, one with thermal expansion and another with thermal contraction. But these ZTE materials are often accompanied by several drawbacks. A ZTE is found in perovskite structured superlattice of $PbTiO_3$–$Bi(Zn_{1/2}Ti_{1/2})O_3$ and related solid solutions over a wide temperature range, starting from room temperature to 500°C. By changing the ratio of individual components in the superlattice, we can control the ZTE temperature range. For example, $0.7PbTiO_3$-$0.3Bi(Zn_{0.5}Ti_{0.5})O_3$ (0.7PT-0.3BZT) displays ZTE in the temperature range of 25 and 400°C and $0.6PbTiO_3$-$0.3Bi(Zn_{0.5}Ti_{0.5})O_3$-$0.1BiFeO_3$ (0.7PT-0.3BZT-0.1BF) shows ZTE up to ~700°C from room temperature (Figure 7.15) (Chen et al. 2008).

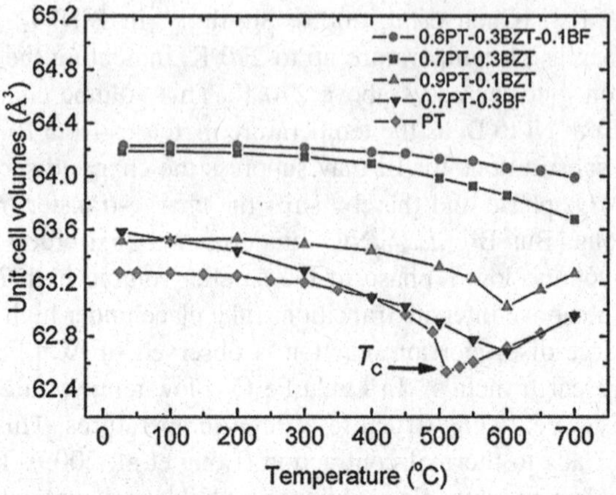

FIGURE 7.15 Temperature evolution of unit cell volume in $(1-x-y)PbTiO_3$-$xBi(Zn_{1/2}Ti_{1/2})O_3$-$yBiFeO_3$ solid solutions. The data of 0.7PT-0.3BF is shown for comparison, reproduced with permission of the American Chemical Society (Chen et al. 2008).

A ZTE near the room temperature range is also observed in cubic perovskite oxide $SrCu_3Fe_{4-x}Mn_xO_{12}$ for $x = 1.5$. Partial substitution of Mn for Fe in $SrCu_3Fe_4O_{12}$ systematically changes the linear thermal expansion coefficient from negative [−6.4, −5.7, and −1.88 ppm K^{-1} for $x = 0.5$, 1, and 1.25, respectively] to almost zero [0.67 ppm K^{-1} for $x = 1.5$] to positive [2.97 ppm K^{-1} for $x = 1.75$]. This change in thermal expansion and contraction behavior is due to the intermetallic charge disproportion between Cu and (Fe, Mn) ions, as discussed previously (Yamada et al. 2015).

7.3 OPTICAL PROPERTIES

The optical nature of perovskites can be defined in terms of refraction, reflectance, absorption, luminescence, transmittance, etc. when they interact with light. In general, polycrystalline perovskites show better optical behavior as compared to single crystalline counterparts. Many perovskites are identified as promising candidates for optical and related applications such as photovoltaics, photocatalysis, displays, scintillation, etc. In this section, the interaction of perovskites with light and other electromagnetic radiations is discussed.

7.3.1 Refraction

The term refraction represents the change in the direction of light when it interacts with a medium or passes from one medium to another. The schematic representation of refraction is shown in Figure 7.16a. Refraction of a material is specified in terms of its refractive index (n), which is the ratio of the velocity of light ($c = 3 \times 10^8$ m s^{-1}) in a vacuum and the velocity of the light wave in the respective material or medium (v):

$$n = c/v \tag{7.42}$$

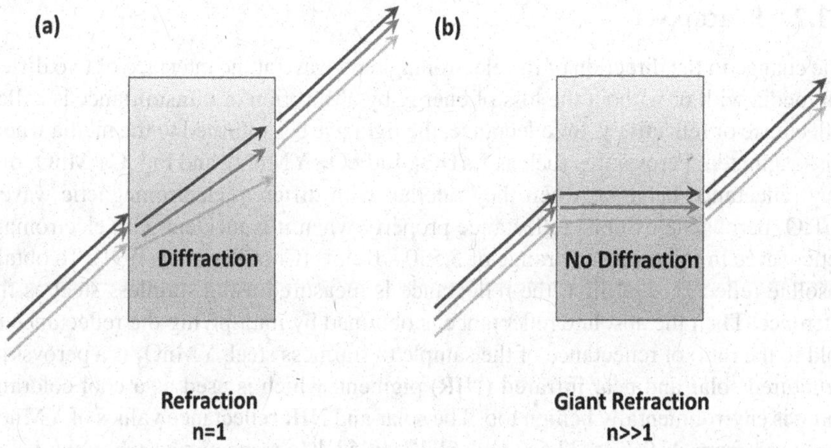

FIGURE 7.16 Schematic representations of (a) refraction and (b) giant refraction.

The refractive behavior of perovskite oxides and their crystals in the photon energy range below 3 eV is most promising. Cubic structured $SrTiO_3$ thin films and crystals display the refractive index (*n*) in the range of ~2.3–2.8. The films are shown smaller refractive index values compared to crystals, which is attributed to the unstressed paraelectric nature of $SrTiO_3$ films at room temperature. $SrTiO_3$ films display ferroelectric behavior when they grow epitaxially over $KTaO_3$ substrate due to induced strain; in this case, the refractive index decreased as compared to unstressed film. Likewise, $NaNbO_3$ exhibits a decrease in the refractive index; however, the reduction is associated with the strain-induced antiferroelectric property. But, at room temperatures ferroelectric perovskites such as $BaTiO_3$ and $K_{0.5}Na_{0.5}NbO_3$, display a large decrease in refractive index with induced lattice strain. In these perovskites, the refractive index decreases with a further increase in lattice strain, unlike $SrTiO_3$ and $NaNbO_3$ perovskites. This observed difference refractive index is attributed to the interaction of light with the polar phonons (Dejneka and Tyunina 2018).

Some perovskites show another unusual refractive behavior, which is called as giant refraction ($n \gg 1$). In giant refraction, the light wave propagates through perovskite without suffering diffraction and chromatic dispersion, as shown in Figure 7.16b. $K_{0.997}Li_{0.003}Ta_{0.64}Nb_{0.36}O_3$ (KLTN) shows giant refraction, which allows the propagation of the light beam without significant diffraction and chromatic dispersion at room temperature irrespective of the angle of incidence, intensity, and beam size (Mei et al. 2018). The giant refractive property of KTLN perovskite paves a new way to solve the chromatic dispersion in transmission imaging of nano metasurfaces (Aieta et al. 2015). This property offers much flexibility in the fabrication of transformation optics (Pendry et al. 2006), optical components, and lithography (Mansfield and Kino 1990). The giant refraction of visible light irrespective of its incidence can play a critical role in reducing the cost of photovoltaic devices considerably (Stranks and Snaith 2015).

7.3.2 Reflectance

The change in the direction of the electromagnetic wave at the interface of two different media with or without the loss of energy by absorption or transmittance is called reflectance or reflectivity. In reflectance, the light wave is returned to the media where it is originated. Perovskites such as $LaTiO_3$, $LaFeO_3$, $YMnO_3$, and $Pr_{1-x}Ca_xMnO_3$ display reflectance behavior when they interact with different electromagnetic waves. $LaTiO_3$ perovskite exhibits reflectance property when it is incident with electromagnetic waves in the frequency range of 5,500–70 cm^{-1} (Crandles et al. 1991). To obtain absolute reflectance, at first, the reflectance is measured using stainless steel as the reference. Then the absolute reflectance is obtained by multiplying the reflectance of gold to the ratio of reflectance of the sample to stainless steel. $YMnO_3$ is a perovskite structured solar and near-infrared (NIR) pigment, which is used as a cool colorant, and it is environmentally benign too. The solar and NIR reflectance values of $YMnO_3$ increased from 23.9% to 32.5% and 51.4% to 53.4%, respectively, when the calcination temperature increases from 700°C to 900°C. The enhancement is due to a decrease in the bandgap from 1.89 to 1.58 eV (Han et al. 2013).

The thin films of perovskites also exhibit reflectance behavior; for example, epitaxially grown $LaFeO_3$ perovskite thin film on $(LaAlO_3)_{0.3}(Sr_2AlTaO_6)_{0.7}$ substrate displays an ultrafast transient reflectance. The transient reflectance spectra of $LaFeO_3$ portrays two negative transients, which are having two local maxima at ~3.5 and ~2.5 eV. These are belonging to the optical transitions of $LaFeO_3$ thin film measured using the ellipsometry technique. The transients are assigned to the recombination of photoexcited carriers, and this spectroscopy provides quantitative information that is highly useful in fabrication photovoltaic cells from perovskite thin films (Smolin et al. 2014). Researchers also revealed the influence of electric pulses on the infrared reflectance of $Pr_{1-x}Ca_xO_3$ thin films. In this method, initially, $Pr_{1-x}Ca_xO_3$ thin film is grown on $SrTiO_3$ substrate, and on the top film, two electrodes (Al and Au) are deposited by sputtering or thermal evaporation. The electrodes are connected to a pulse generator, and the place between two electrodes incident with light and reflected light is detected using a photodiode. A positive pulse is applied through the Au electrode, and a negative pulse is applied using the Al electrode at room temperature (Figure 7.17). A decrease in reflectance is observed by applying a positive pulse, whereas an increase in reflectance is noticed when applying negative pulses, and this change in reflectance is reversible and non-volatile. The difference in reflectance is attributed to metal to insulator transitions when a positive pulse is applied. As the positive pulse is applied to the perovskite thin film, it exhibits a metallic behavior by displaying a decrease in resistance and the opposite when a negative pulse was applied (Aoyama et al. 2004).

7.3.3 Absorption

The wavelength or energy of an electromagnetic wave can change when it passes through a medium. The difference in energy of the wave as it passes through the medium is said to be absorbed by the medium or simply absorption. The absorbed energy from electromagnetic waves will transform into the internal

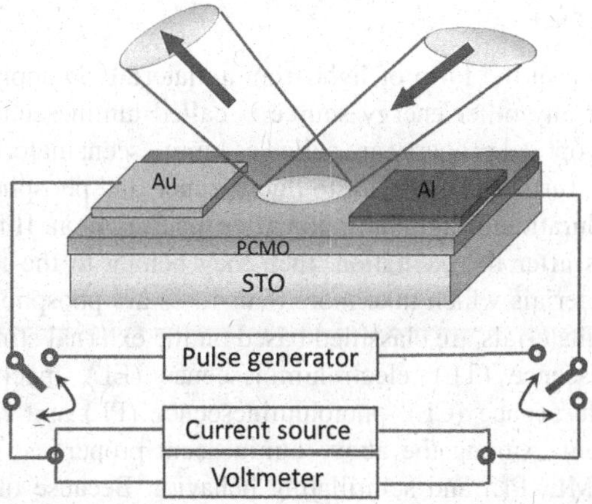

FIGURE 7.17 Schematic picture of the experimental setup for measuring the electric-pulse-induced change of reflectance and resistance, reproduced with permission of the American Institute of Physics (Aoyama et al. 2004).

energy of medium (for example, thermal energy). Optical absorption is typically exhibited by ferroelectric perovskites, and these materials are known for large bandgaps due to high electronegative difference between oxygen and cations (A–O or B–O) (Cohen 1992). The perovskite oxides such as $BaTiO_3$ and $KTaO_3$ show optical absorption characteristics in the vicinity of the inter-band absorption edge, where the photon excites the electrons from the valence to the conduction band. The data from the inter-band absorption edge displayed an Urbach tail (where the absorption increases exponentially) in both the above perovskites up to absorption coefficients ~10^{-3} cm^{-1} (DiDomenico and Wemple 1968). Likewise, the optical absorption characteristics of $LiTiO_3$ and $LiNbO_3$ also follow Urbach's equation with an exponential increase in the absorption (Kase and Ohi 1974). In the case of semiconducting $KTa_{0.65}Nb_{0.35}O_3$ the optical absorption is majorly from the NIR region, and the absorption occurs in the form of photoionization (ionization of an atom or a molecule by a photon). As a consequence, a large dichroism (splitting of visible light into two distinct beams) is observed (DiDomenico and Wemple 1968).

Recently, the solid solution of $[KNbO_3]_{1-x}[BaNi_{1/2}Nb_{1/2}O_{3-}\delta]_x$ (KBNNO) is evaluated for its light absorption properties. The Ni containing KBNNO is green in color with a bandgap in the range of 1.1–2.0 eV, which is much lower than pristine $KNbO_3$ perovskite (3.8 eV). KBNNO shows an absorption coefficient of ~2.5×10^4 cm^{-1} at 885 nm, which is comparable to the GaAs and CdTe compounds. More importantly, KBNNO shows promising activity for photovoltaic applications in photoresponse measurements (Grinberg et al. 2013). $LaFeO_3$ with the composite of carbon studied as microwave absorption materials (MAMs), particularly the perovskite with La deficiency ($La_{0.62}FeO_3$/C) shows superior microwave absorption property. This enhanced activity is attributed to the polarization loss of many dipoles due to A-site deficiency (Liu et al. 2017).

7.3.4 Luminescence

The energy emitted in the form of light from a material *via* applying temperature, pressure, light, or any other energy source is called luminescence. The materials which exhibit this optical property are called as luminescent materials. The luminescent materials are further classified into fluorescence and phosphorescence materials based on the duration of light emission after the excitation. If the materials emit light within 10^{-5}s after the excitation, then they belong to the fluorescence type, whereas those materials which take more than 10^{-5}s are phosphorescent materials. The luminescent materials are classified based on the external stimuli applied, such as thermoluminescence (TL), electroluminescence (EL), mechanoluminescence (ML), chemiluminescence (CL), photoluminescence (PL) and radiative luminescence (scintillation). Among the above luminescent properties, perovskite oxides exhibit TL, EL, ML, PL, and scintillating behavior. Because of the above luminescence properties, they are used in the fabrication of optoelectronic devices and microelectronics.

7.3.4.1 Thermoluminescence

The luminescence which results from temperature changes is known as TL. The first TL reported among perovskites is from $BaTiO_3$, under X-ray irradiation in the temperature range of 180°C–400°C (Fasasi et al. 2007). The same perovskite also shows TL by irradiating with gamma rays. Interestingly, the TL peak is enhanced with the increase in gamma dose indicating the potential of $BaTiO_3$ as a dosimetry material. When gamma dose is varied from 2.27 to 9.08 kGy, a shift in TL peak is noticed from 480 to 530 K along with an increase in the sharpness and intensity. The TL behavior of $BaTiO_3$ perovskite is mainly assigned to the traps which formed due to vacancies or irradiation (Fasasi et al. 2007). Another best example of TL is $CaTiO_3$. The TL peak for $CaTiO_3$ is observed at 160°C irrespective of the particle size, but as the particle size is reduced, an additional peak appeared at 102°C due to quantum confinement (Shivaram et al. 2013).

7.3.4.2 Mechano and Electroluminescence (ML & EL)

If the luminescence originated by mechanical stress, then it is known as ML. The luminescence which results from electric voltage as stimuli is called EL. The ML intensity of $LiNbO_3$ perovskite is improved significantly by doping with Pr and Gd. Interestingly, the ML intensity of $LiNbO_3{:}Pr^{3+}$ significantly enhanced ~177% when it is co-doped with Gd^{3+}. This augmentation in ML intensity is attributed to the regulated trap quantities due to the appropriate co-doping of Pr and Gd (Qiu et al. 2018). Layered perovskites such as $Sr_3(Sn_{2-x}Si_x)O_7$ and $Sr_3(Sn_{2-x}Ge_x)O_7$ also exhibit ML property, and the intensity of the ML is improved further by doping with Sm^{3+}. The enhancement in ML is assigned to a partial decrease in bandgap due to doping, thereby enhancing the trapping of electrons during UV pre-excitation, and the simultaneous widening of energy gap arrests the de-trapping of electrons (Li et al. 2018).

Tb-Mg doped $CaSnO_3$ thin films deposited on silica glass substrates or $SrTiO_3$ (100) single crystal substrates exhibit an EL behavior. For the above system, EL is observed at 543 nm when the applied voltage is above 500 V; below 500 V, there is no EL behavior. As the applied voltage is increased further (above and beyond 500 V), the EL intensity increases. The maximum EL intensity is observed at 1,000 V. Interestingly, the perovskite film deposited on $SrTiO_3$ required a higher voltage than the film deposited on the glass substrate (Ueda and Shimizu 2010). Pr^{3+}-doped $BaTiO_3$-$CaTiO_3$ perovskites display simultaneous ML and EL. In the above material, electron voltage induces mechanical strain and emission of light, whereas mechanical stress input stimulates electric signal and light emission. The composition $x \sim 0.25$ [$(1-x)$ $BaTiO_3$-$(x)CaTiO_3$]:Pr shows strong EL and ML emissions, and both the emissions are strongly related to the imperfections and piezoelectricity in the perovskite (Wang et al. 2005).

7.3.4.3 Photoluminescence

The luminescence which results from the absorption of electromagnetic waves is known as PL. Generally, undoped perovskites do not show any PL behavior at room temperature because of their indirect-bandgap semiconducting nature, although they display broad PL emission at low temperatures (Leonelli and Brebner 1986; Eglitis et al. 2002). Recent reports reveal that electron-doped perovskite can show PL at room temperature. For example, $SrTiO_3$ exhibits blue emission at room temperature when it is doped with electrons under continuous-wave photoexcitation (Yasuda et al. 2008; Yamada et al. 2009b). More importantly, PL behavior can be induced in perovskites by creating oxygen deficiencies on the surfaces by irradiating with Ar^+ (Reagor and Butko 2005). The results show that the Ar^+ irradiated $SrTiO_3$ display high PL intensity compared to pristine perovskites; this enhancement is attributed to the metallic conduction in the irradiated samples, which is caused by surface oxygen deficiencies (Kan et al. 2005). The PL properties of Ar^+ irradiated $SrTiO_3$ are like the Nd- and La-doped $SrTiO_3$ (Kan et al. 2005; Yamada et al. 2009a). Such higher valent doping at Sr-site also induces metallic behavior in $SrTiO_3$ (Suzuki et al. 1996). Likewise, $BaTiO_3$ and $KTaO_3$ show enhancement in the PL signal after Ar^+ irradiation (Yamada et al. 2013). However, in the case of $LiNbO_3$ and $LiTaO_3$, no change in the PL intensity is observed after irradiating with Ar^+. This could be due to the large bandgap energies of $LiNbO_3$ and $LiTaO_3$ compared to $SrTiO_3$, $BaTiO_3$, and $KTaO_3$ perovskites (Yamada et al. 2013).

The most common route to induce PL characteristics to perovskite oxides is by doping luminescent rare earth metals or transition metals like Mn^{4+} (Song et al. 2019). For example, the PL behavior of Pr^{3+}-doped $Ca_xBa_{1-x}TiO_3$ perovskite varies with the Ca composition (0.3–0.998). PL emission peak intensity of this perovskite increases along with an increase in Ca content (0.3–0.998) (He et al. 2009). Eu^{3+} doping in $MSnO_3$ (M = Ba, Sr and Ca) result in a PL spectrum with red emission at 614 nm of Eu^{3+} ions (Lu et al. 2005), and the PL spectra are red-shifted from Ba to Sr to Ca (Zhang et al. 2007). In Sr_2CaMoO_6 double perovskite, the Eu^{3+} can be doped either at A or B sites, and the B-site doping shows higher PL intensity compared with A-site doping. $Sr_2Ca_{0.80}Li_{0.10}Eu_{0.10}$ $Mo_{0.10}W_{0.90}O_6$ composition shows better PL intensity and it is comparable to commercial Y_2O_2S:Eu. The introduction of W enhances the energy trapping ability of Eu^{3+} species; thus, it enhances the PL emission (Ye et al. 2008).

Mn^{4+}-doped Sr_2LaNbO_6 emits promising red PL band suitable for white-light-emitting diodes. This phosphor can be excited in the range of 300–500 nm, and the emission is observed at 694 nm from the spin forbidden transition in Mn^{4+}. More importantly, this perovskite displays good thermal stability up to 500 K (Fu et al. 2017). Along with Mn, chromium also induces PL behavior in perovskite oxides, for example, Cr^{3+}-doped $LaAlO_3$ perovskite exhibits deep red persistent luminescence at 734 nm, due to $^2E \rightarrow {}^4A_2$ transition Cr^{3+} by ultraviolet excitation. Co-doping of Sm^{3+} further improved the PL intensity by 35-fold, and the deep red emission is longlasting too. The PL emission of Cr^{3+} and Sm^{3+} co-doped $LaAlO_3$ is comparable to the $ZnGa_2O_4$:Cr^{3+} phosphor used in the application of *in vivo* imaging (Katayama et al. 2014).

7.3.4.4 Scintillation

The luminescence as a result of absorption of ionization radiation is known as scintillation, and the materials which exhibit this property are called scintillators (Luo et al. 2017). Scintillators convert the high-energy photon (X-ray or gamma-ray) or particle (electron, neutron-proton, or alpha particle) into lower-energy photons (UV-Vis), which can be easily detectable by common photomultiplier detectors (Chen et al. 2018). Ce^{3+}-doped $MAlO_3$ (M = Y, Lu) and Pr^{3+}-doped $LuAlO_3$ perovskites show scintillating properties. The $YAlO_3$:Ce exhibits broad UV absorption bands in the range of 180–300 nm due to the presence of Ce (Takeda et al. 1980). The major emission band is observed in near UV-Vis region 349–380 nm, which is in accordance with the splitting of 5*d* orbitals of Ce. The fast decay time of this emission band is about 17 ns, and another two slow decay components are observed in between 22 and 38 ns (Baryshevsky et al. 1991; Moszyński et al. 1998). High light yield (LY) ~18,000 photons/MeV is also reported for $YAlO_3$:Ce scintillator, especially for emitting green spectral band. The emission peaks of Ce^{3+} in $YAlO_3$ falls in the range of 340–370 nm for different synthesizing methods (Mares et al. 1991; Asatryan et al. 1997). Pr^{3+}-doped $YAlO_3$ shows shorter decay time and low LY compared to $YAlO_3$:Ce, which is attributed to the presence of emission bands in the shorter UV region (Pedrini et al. 1994).

$LuAlO_3$:Ce displays similar spectroscopic properties like $YAlO_3$:Ce scintillator due to their isostructural properties. However, they have different scintillation properties, and $LuAlO_3$:Ce displays a little longer decay time ~21 ns (Korzhik and Trower 1995). The mixed scintillator of $LuAlO_3$:Ce and $YAlO_3$:Ce ($Lu_{1-x}Y_xO_3$) also shows similar optical properties like parent materials; however, the scintillation properties are different (Kuntner et al. 2005). Pr^{3+}-doped $LuAlO_3$ presents the same mechanism of transfer, like Ce-doped ones. The Pr^{3+}-doped $LuAlO_3$ is expected to be a more efficient scintillator compared to the Ce-doped one due to the presence of 4*f* level near to the valence band. Although under gamma-ray excitation, $LuAlO_3$:Ce is able to show a photo peak, $LuAlO_3$:Pr failed to display the same behavior. This might be due to the large overlap between emission and absorption bands (Dujardin et al. 1997).

7.3.5 Transmittance

Transmittance is a measure of incident light transmitted through a medium. Transparent perovskites can be obtained by limiting them with no impurities or pores, especially at grain boundaries during the sintering process of the precursors.

Further, the transparency of perovskites can be optimized by applying an external electric field. For instance, $Pb_{1-x}La_x(Zr_{1-y}Ti_y)_{1-x/4}O_3$, which is also written as PZLT (x/1–y/y), especially, PZLT with x, y composition (8/70/30) shows this effect, and they are used in attenuators and light modulators without using adjunct polarizers (Yang et al. 2016).

Highly transparent perovskites are usually synthesized by using conventional ceramic processing techniques. The transparency of perovskites is governed by reflectivity and composition. For example, considering refractive index (n) of a perovskite as ~2.95, then surface reflectivity (R) is calculated using the equation $R = [(n - 1)/(n + 1)]^2$ as 0.184. The loss in R due to reflectivity from both sides is ~0.37. The variation in the optical nature of perovskites is due to the destruction and formation of polar nano and microscopic domains. The different polarization direction of domains results in considerable changes in refractive indexes when light crosses domain boundaries, which leads to a significant light scattering. This optical property is majorly governed by the temperature and composition, and it changes the formation of domain wall and ease the domain wall movements. Along with PZLT other perovskites such as $Pb(Zn_{1/3}Nb_{2/3})O_3$–$PbTiO_3$ (PZN-PT) $Pb(Mg_{1/3}Nb_{2/3})O_3$–$PbTiO_3$ (PMN-PT) (Yang et al. 2016), and, lead-free perovskites like $K_{0.5}Na_{0.5}NbO_3$ (Taghaddos et al. 2015), optimized compositions of all of these phases are applicable as modulators due to their high electro-optic coefficients.

The perovskites exhibit transmittance find application as transparent conductive oxides (TCO) in photovoltaics due to their large bandgap and low resistance. A schematic of the perovskite layer in a solar cell with TCO is shown in Figure 7.18. However, because of the wide bandgap values, most of the transparent perovskite oxides are non-conductive in nature; the conductivity of the perovskite can be improved by doping. For instance, $SrSnO_3$ has high transmittance with high resistance because

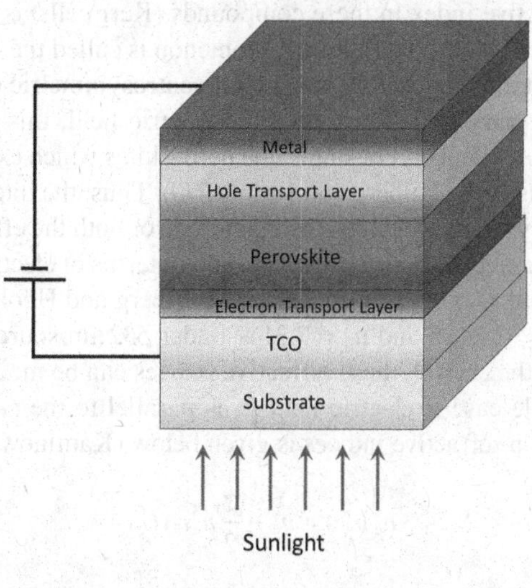

FIGURE 7.18 A schematic showing the representative position of transparent and conductive perovskite oxide layer of a perovskite in solar cell.

of the wide bandgap (~4 eV). Doping with Sb (Liu et al. 2008) and Nd significantly enhances the conductive behavior of $SrSnO_3$ perovskite. The Sb addition in 5%–13% range enhances the conductivity, and the optimized dopant level is 5% (Liu et al. 2011). Similarly, the La doping of $BaSnO_3$ improves the conductive behavior with transmittance >80% (Shan et al. 2014).

A change in the substrate influences the transmittance behavior of perovskite films. The Tb and Mg co-doped $CaSnO_3$ displays two different transmittance values; when it is deposited on a glass substrate, the transmittance value is ~85%. But when it is deposited on the $SrTiO_3$ substrate, the transmittance value is only ~75%. The transmittance of Tb, Mg-co-doped $CaSnO_3$ film on $SrTiO_3$ is retained even after annealing at 900°C, but in case of glass substrate the transmittance is decreased by 75% after annealing. This study reveals why most of the reports on perovskite TCOs are using $SrTiO_3$ as a substrate instead of glass (Ueda and Shimizu 2010).

7.4 ELECTRO-OPTICAL PROPERTIES

The electro-optic effect arises from the change in optical properties of perovskites due to the applied electric field, which includes changes in optical absorption or color, or change in refractive index and color change in electrochromic materials.

When a slowly varying or static electric field is applied to the perovskite, it causes a change in the refractive index due to the displacement of electrons or atom in the compound, as presented in Eq. 7.43:

$$n(E) = n - \frac{1}{2} rn^3 E - \frac{1}{2} Rn^3 E^2 \dots \tag{7.43}$$

where $n(E)$ is the refractive index in the electric field (E) and r and R are electro-optic coefficients. The rE is zero in centrosymmetric cubic crystal structures, and the change in refractive index in these compounds (Kerr cells) is proportional to the square of the applied electric field; this phenomenon is called the Kerr effect and R is called as Kerr constant (Thacher 1976). In non-centrosymmetric crystals, the refractive index does not vary with changes in the electric field; this effect is called the Pockels effect, and r is Pockels constant. The perovskites which exhibit Pockels effect are called as Pockels cells (Hilberg and Hook 1970). Thus, the interaction of the electric field with the crystal axis decides the magnitude of both the effects. $LiNbO_3$ is the best example for the electro-optic perovskite, and in terms of electro-optic property, it is with the hexagonal c-axis as an optical axis (Hilberg and Hook 1970). The refractive indices are $n_e = 2.2265$ and $n_o = 2.3149$ under 532 nm source, and by applying the electric field to the crystal, these refractive indices can be modified. For example, considering a simple case if electric field E_3 is parallel to the c-axis of crystal, the subsequent change in refractive indices is given below (Kaminow et al. 1973):

$$n_e(E) = n_e - \frac{1}{2} n_e^3 r_{33} E_3 \tag{7.44}$$

$$n_o(E) = n_o - \frac{1}{2} n_o^3 r_{13} E_3 \tag{7.45}$$

where r_{13} is Pockels coefficient for n_o, with a value of 9.6×10^{-12} mV^{-1} and r_{33} for n_e has a value of 30.9×10^{-12} mV^{-1}. The induced changes in refractive index are very small, for instance, applying an electric field of 10 kV on 1 cm crystal, and the change in n_e is:

$$\Delta n_e = \frac{1}{2} \times (2.2265)^3 \times 30.9 \times 10^{-12} \times 10^6 = 1.71 \times 10^{-4} \tag{7.46}$$

$LiNbO_3$ is used in several electro-optical components, and regardless of showing the small magnitude of the effect, $LiNbO_3$ is used as an electro-optic phase and intensity modulator (Cao et al. 2014). Usually, the $LiNbO_3$ phase modulator comprises of the optical axis (*c*-axis or *z*-direction) of crystal slab oriented parallel to the applied electric field, and the light beam runs parallel to the *x*- or *y*-axis, as shown in Figure 7.19. This arrangement prevents the splitting of unpolarized light into extraordinary and ordinary rays and allows it to propagate in the crystal as an extraordinary ray. The intensity modulator of $LiNbO_3$ is designed using two crystals of them in tandem. The crystals are placed in such a way that the optical axis of crystals is placed parallel to the applied electric field, and the light beam passes perpendicular to it. The light beam will split into an ordinary and extraordinary ray as it enters the first crystal (Kawanishi et al. 2006). The second crystal is arranged such that ordinary and extraordinary rays in the first crystal swap and become extraordinary and ordinary rays. After swapping, ordinary and extraordinary rays show a change in phase difference due to the applied electric field, and they will be polarized linearly perpendicular to one another.

7.4.1 Electrochromic Films

The films or materials which change their color upon applying electric field are called as electrochromic films or materials, and they are used in smart windows to control the amount of light transmitted or reflected. Color production by these materials

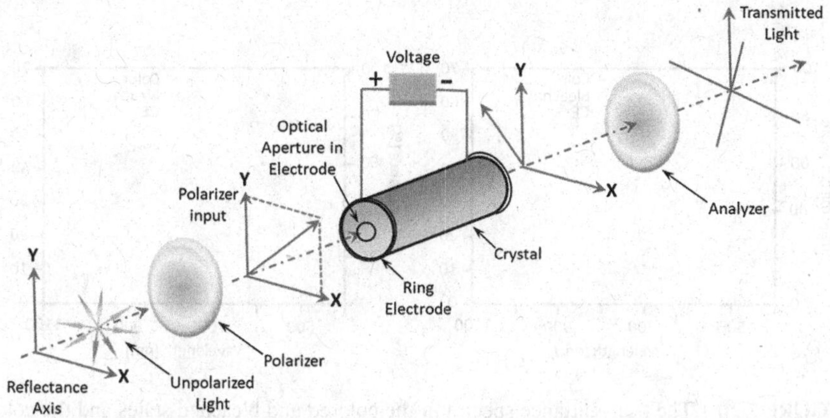

FIGURE 7.19 A schematic representation of light scattering in the electronic-optical modulator.

is related to the defect formation while applying the electric field. WO_3 is the most studied electrochromic material. The electrochromic nature of WO_3 is due to the formation of perovskite structure (A_xWO_3) when an electric field is applied (Mardare and Hassel 2019). The A-site of the above perovskite structure is occupied by cations such as H^+, Li^+, Na^+, or K^+, between corner linked WO_6 octahedra (Granqvist et al. 2018), also known as tungsten bronzes. In the presence of a low amount of the above intercalated ions, the color changes due to the charge transfer between two valence states of tungsten. They are W^{4+}–W^{6+} and W^{5+}–W^{6+} couples, as observed in the transmittance spectra of H^+ and Na^+ intercalated WO_3 (Figure 7.20) (Patel et al. 2010).

In electrochromic devices based on A_xWO_3 perovskites, a certain amount of voltage is used to drive the reduction metals (M) such as Li, Na, or K into the WO_3 film. This leads to the loss of an electron from metal atoms to form an M^+ ion. The released electrons from doped atoms reduce W^{6+} to form W^{4+} or W^{5+} ions, and they allow the charge transfer. As a result, the transparent film of WO_3 turned to blue-black perovskite if Li is used to reduce the WO_3 film. Reversal of applied voltage reverts the transparency of WO_3 film by removing intercalated metal ion, and this process is called bleaching. The mechanism of electrochromism in the WO_3 film reaction is presented in Eq. 7.47 (Niklasson and Granqvist 2006):

$$WO_3 \text{ film } (\text{Transparent}) + x\text{Li} \leftrightarrow \text{Li}_x WO_3 (\text{reduced, blue-black}) \quad (7.47)$$

In device applications, a series of thin films of WO_3 constructed on the transparent conductive substrate, usually on indium tin oxide (ITO), ion-conducting electrolyte, and a source for metal ions is used. In practice, there are many available designs; for instance, the car mirrors to cut down the bright light dazzling effects. In the above application, the hydrogen from atmospheric water vapor is used to produce hydrogen tungstate perovskite (H_xWO_3) (Figure 7.21). The reaction which happens on the outer side of the ITO glass is shown below:

$$2H_2O \rightarrow O_2(g) + 4H^+ + 4e^- \quad (7.48)$$

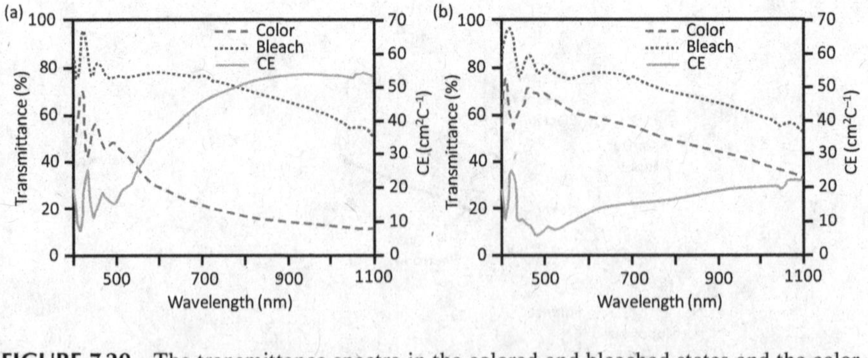

FIGURE 7.20 The transmittance spectra in the colored and bleached states and the coloration efficiency (CE) of WO_3 films grown at $T_S = 300°C$ for (a) H^+ and (b) Na^+ intercalation, reproduced with permission of the Elsevier (Patel et al. 2010).

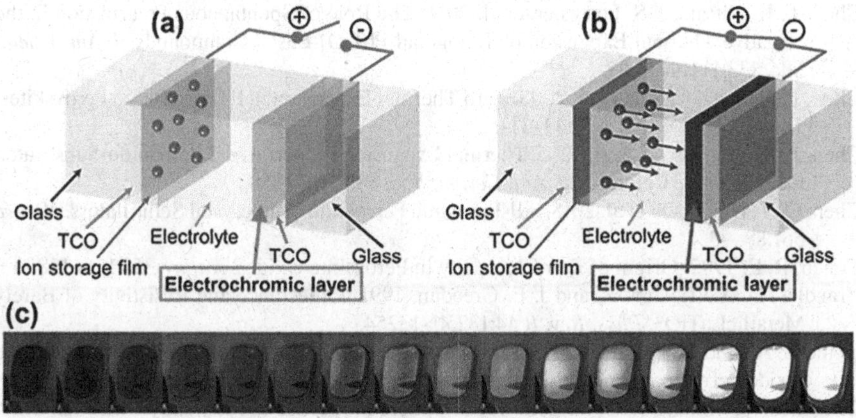

FIGURE 7.21 A schematic drawing of a smart window based on electrochromic layers: (a) transparent and (b) colored. (c) Photographs of a dimmable electrochromic window from Boeing 787 Dreamliner airplane (Mardare and Hassel 2019).

This reaction needs ~1 V at the surface of the electrode, and the electrolyte used in this application is hydrogen uranyl phosphate (HUO_2PO_4 $4H_2O$) (Howe et al. 1980). When the troublesome reflections occur, the device is switched on, and the transparent window is darkened by the formation of H_xWO_3, which curtails the dazzling reflections from bright light. If the device turns off, the voltage is reversed, and the film becomes colorless by losing H^+ ion.

REFERENCES

Aieta, F., M. A. Kats, P. Genevet, et al. 2015. Multiwavelength Achromatic Metasurfaces by Dispersive Phase Compensation. *Science* 347:1342–1345.

Aoyama, K., K. Waku, A. Asanuma, et al. 2004. Electric-Pulse-Induced Reflectance Change in the Thin Film of Perovskite Manganite. *Appl. Phys. Lett.* 85:1208–1210.

Asatryan, H. R., J. Rosa, and J. A. Mareš. 1997. EPR Studies of Er^{3+}, Nd^{3+} and Ce^{3+} in $YAlO_3$ Single Crystals. *Solid State Commun.* 104:5–9.

Azuma, M., W. Chen, H. Seki, et al. 2011. Colossal Negative Thermal Expansion in $BiNiO_3$ Induced by Intermetallic Charge Transfer. *Nat. Commun.* 2:1–5.

Baryshevsky, V. G., M. V. Korzhik, V. I. Moroz, et al. 1991. $YAlO_3$: Ce-Fast-Acting Scintillators for Detection of Ionizing Radiation. *Nucl. Instrum. Methods Phys. Res. Sect. B* 58:291–293.

Bielecki, J., S. F. Parker, L. Mazzei, et al. 2016. Structure and Dehydration Mechanism of the Proton Conducting Oxide $Ba_2In_2O_5(H_2O)_x$. *J. Mater. Chem. A* 4:1224–1232.

Bochkov, D. M., V. V. Kharton, A. V. Kovalevsky, et al. 1999. Oxygen Permeability of $La_2Cu(Co)O_{4+\delta}$ Solid Solutions. *Solid State Ion.* 120:281–288.

Breckenfeld, E., R. Wilson, J. Karthik, et al. 2012. Effect of Growth Induced (Non) Stoichiometry on the Structure, Dielectric Response, and Thermal Conductivity of $SrTiO_3$ Thin Films. *Chem. Mater.* 24:331–337.

Cao, L., A. Aboketaf, Z. Wang, et al. 2014. Hybrid Amorphous Silicon (a-Si:H)–$LiNbO_3$ Electro-Optic Modulator. *Optics Commun.* 330:40–44.

Chen, J., K. Nittala, J. S. Forrester, et al. 2011. The Role of Spontaneous Polarization in the Negative Thermal Expansion of Tetragonal $PbTiO_3$-Based Compounds. *J. Am. Chem. Soc.* 133:11114–11117.

Chen, J., X. Xing, C. Sun, et al. 2008. Zero Thermal Expansion in $PbTiO_3$-Based Perovskites. *J. Am. Chem. Soc.* 130:1144–1145.

Chen, J., X. Xing, R. Yu, et al. 2005. Thermal Expansion Properties of Lanthanum-Substituted Lead Titanate Ceramics. *J. Am. Ceram. Soc.* 88:1356–1358.

Chen, Q., J. Wu, X. Ou et al. 2018. All-Inorganic Perovskite Nanocrystal Scintillators. *Nature* 561:88–93.

Cohen, R. E. 1992. Origin of Ferroelectricity in Perovskite Oxides. *Nature* 358:136–138.

Crandles, D. A., T. Timusk, and J. E. Greedan. 1991. Reflectance and Resistivity of Barely MetallicLaTiO3. *Phys. Rev. B* 44:13250–13254.

Dejneka, A., and M. Tyunina. 2018. Elasto-Optic Behaviour in Epitaxial Films of Perovskite Oxide Ferroelectrics. *Adv. Appl. Ceram.* 117:62–65.

DiDomenico, M., and S. H. Wemple. 1968. Optical Properties of Perovskite Oxides in Their Paraelectric and Ferroelectric Phases. *Phys. Rev.* 166:565–576.

Dujardin, C., C. Pedrini, J. C. Gâcon, et al. 1997. Luminescence Properties and Scintillation Mechanisms of Cerium- and Praseodymium-Doped Lutetium Orthoaluminate. *J. Phys. Condens. Matter* 9:5229–5243.

Eglitis, R. I., E. A. Kotomin, and G. Borstel. 2002. Quantum Chemical Modelling of "Green" Luminescence in ABO Perovskites. *Eur. Phys. J. B* 27:483–486.

Fasasi, A. Y., F. A. Balogun, M. K. Fasasi, et al. 2007. Thermoluminescence Properties of Barium Titanate Prepared by Solid-State Reaction. *Sens. Actuators A* 135:598–604.

Feng, J., B. Xiao, R. Zhou, et al. 2012. Anisotropic Elastic and Thermal Properties of the Double Perovskite Slab–Rock Salt Layer $Ln_2SrAl_2O_7$ (Ln = La, Nd, Sm, Eu, Gd or Dy) Natural Superlattice Structure. *Acta Mater.* 60:3380–3392.

Foley, B. M., H. J. Brown-Shaklee, J. C. Duda, et al. 2012. Thermal Conductivity of Nano-Grained $SrTiO_3$ Thin Films. *Appl. Phys. Lett.* 101:231908.

Fossdal, A., M. Menon, I. Wærnhus, et al. 2004. Crystal Structure and Thermal Expansion of $La_{1-x}Sr_xFeO_{3-\delta}$ Materials. *J. Am. Ceram. Soc.* 87:1952–1958.

Fu, A., A. Guan, D. Yu, et al. 2017. Synthesis, Structure, and Luminescence Properties of a Novel Double-Perovskite Sr_2LaNbO_6:Mn^{4+} Phosphor. *Mater. Res. Bull.* 88:258–265.

Gaur, N. K., R. Thakur, and R. K. Thakur. 2017. Specific Heat and Thermal Expansion Effects in $LaCo_{1-x}A_xCoO_3$ (A = Mg, Rh, Ti and Nb). *J. Alloys Compd.* 691:866–872.

Granqvist, C. G., M. A. Arvizu, İ. Bayrak Pehlivan, et al. 2018. Electrochromic Materials and Devices for Energy Efficiency and Human Comfort in Buildings: A Critical Review. *Electrochim. Acta* 259:1170–1182.

Grinberg, I., D. V. West, M. Torres, et al. 2013. Perovskite Oxides for Visible-Light-Absorbing Ferroelectric and Photovoltaic Materials. *Nature* 503:509–512.

Han, A., M. Zhao, M. Ye, et al. 2013. Crystal Structure and Optical Properties of $YMnO_3$ Compound with High Near-Infrared Reflectance. *Solar Energy* 91:32–36.

Hayashi, H., H. Inaba, M. Matsuyama, et al. 1999. Structural Consideration on the Ionic Conductivity of Perovskite-Type Oxides. *Solid State Ion.* 122:1–15.

He, S., Y. Liu, and Y. Imai. 2009. Synthesis and Luminescence Properties of Pr^{3+} Doped $CaxBa_{1-x}TiO_3$ ($0.3 \leq x < 1$) Fine Particles. *J. Wuhan Univ. Technol.-Mat. Sci. Edit.* 24:689–693.

He, Y. 2004. Heat Capacity, Thermal Conductivity, and Thermal Expansion of Barium Titanate-Based Ceramics. *Thermochim. Acta* 419:135–141.

Hilberg, R. P., and W. R. Hook. 1970. Transient Elastooptic Effects and Q-Switching Performance in Lithium Niobate and KD*P Pockels Cells. *Appl. Opt.* 9:1939–1940.

Ho, C. Y., and R. E. Taylor. 1998. *Thermal Expansion of Solids*. Cleveland, OH: ASM International.

Hofmeister, A. M. 2010. Thermal Diffusivity of Oxide Perovskite Compounds at Elevated Temperature. *J. Appl. Phys.* 107:103532.

Howard, C. J., B. J. Kennedy, and B. C. Chakoumakos. 1999. Neutron Powder Diffraction Study of Rhombohedral Rare-Earth Aluminates and the Rhombohedral to Cubic Phase Transition. *J. Phys. Condens. Matter* 12:349–365.

Howe, A. T., S. H. Sheffield, P. E. Childs, et al. 1980. Fabrication of Films of Hydrogen Uranyl Phosphate Tetrahydrate and Their Use as Solid Electrolytes in Electrochromic Displays. *Thin Solid Films* 67:365–370.

Hu, Z., J. Sheng, J. Chen, et al. 2018. Enhanced Li Ion Conductivity in Ge-Doped $Li_{0.33}La_{0.56}TiO_3$ Perovskite Solid Electrolytes for All-Solid-State Li-Ion Batteries. *New J. Chem.* 42:9074–9079.

Ishihara, T., H. Furutani, T. Yamada, et al. 1997. Oxide Ion Conductivity of Double Doped Lanthanum Gallate Perovskite Type Oxide. *Ionics* 3:209–213.

Itoh, N., T. Kato, K. Uchida, et al. 1994. Preparation of Pore-Free Disk of $La_{1-x}Sr_xCoO_3$ Mixed Conductor and Its Oxygen Permeability. *J. Membr. Sci.* 92:239–246.

Iwahara, H. 2009. Ionic Conduction in Perovskite-Type Compounds. In *Perovskite Oxide for Solid Oxide Fuel Cells*, ed. T. Ishihara, 45–63. Boston, MA: Springer US.

Jarligo, M. O., D. E. Mack, G. Mauer, et al. 2010. Atmospheric Plasma Spraying of High Melting Temperature Complex Perovskites for TBC Application. *J. Therm. Spray Tech.* 19:303–310.

Kaminow, I. P., J. R. Carruthers, E. H. Turner, et al. 1973. Thin-Film LiNbO3 Electro-Optic Light Modulator. *Appl. Phys. Lett.* 22:540–542.

Kan, D., T. Terashima, R. Kanda, et al. 2005. Blue-Light Emission at Room Temperature from Ar^+ -Irradiated $SrTiO_3$. *Nat. Mater.* 4:816–819.

Kase, S., and K. Ohi. 1974. Optical Absorption and Interband Faraday Rotation in $LiTaO_3$ and $LiNbO_3$. *Ferroelectrics* 8:419–420.

Katayama, Y., H. Kobayashi, and S. Tanabe. 2014. Deep-Red Persistent Luminescence in Cr^{3+}-Doped $LaAlO_3$ Perovskite Phosphor for in Vivo Imaging. *Appl. Phys. Express* 8:012102.

Kawanishi, T., T. Sakamoto, M. Tsuchiya, et al. 2006. 70dB Extinction-Ratio $LiNbO_3$ Optical Intensity Modulator for Two-Tone Lightwave Generation. In Optical Fiber Communication Conference and Exposition and The National Fiber Optic Engineers Conference (2006), Paper OWC4, OWC4. Optical Society of America.

Korzhik, M. V., and W. P. Trower. 1995. Origin of Scintillation in Cerium-doped Oxide Crystals. *Appl. Phys. Lett.* 66:2327–2328.

Kreuer, K. D. 1999. Aspects of the Formation and Mobility of Protonic Charge Carriers and the Stability of Perovskite-Type Oxides. *Solid State Ion.* 125:285–302.

Kreuer, K. D. 2003. Proton-Conducting Oxides. *Annu. Rev. Mater. Res.* 33:333–359.

Kreuer, K. D., E. Schönherr, and J. Maier. 1994. Proton and Oxygen Diffusion in $BaCeO_3$ Based Compounds: A Combined Thermal Gravimetric Analysis and Conductivity Study. *Solid State Ion.* 70–71:278–284.

Kreuer, K. D., St. Adams, W. Münch, et al. 2001. Proton Conducting Alkaline Earth Zirconates and Titanates for High Drain Electrochemical Applications. *Solid State Ion.* 145:295–306.

Krishnan, R. S., R. Srinivasan, and S. Devanarayanan. 2013. *Thermal Expansion of Crystals: International Series in the Science of the Solid State*. New York: Pergamon Press.

Kuntner, C., E. Auffray, D. Bellotto, et al. 2005. Advances in the Scintillation Performance of LuYAP:Ce Single Crystals. *Nucl. Instrum. Methods Phys. Res. Sect. A* 537:295–301.

Leonelli, R., and J. L. Brebner. 1986. Time-Resolved Spectroscopy of the Visible Emission Band in Strontium Titanate. *Phys. Rev. B* 33:8649–8656.

Li, J., C.-N. Xu, D. Tu, et al. 2018. Tailoring Bandgap and Trap Distribution via Si or Ge Substitution for Sn to Improve Mechanoluminescence in $Sr_3Sn_2O_7$:Sm^{3+} Layered Perovskite Oxide. *Acta Mater.* 145:462–469.

Liu, Q., J. Dai, X. Zhang, et al. 2011. Perovskite-Type Transparent and Conductive Oxide Films: Sb- and Nd-Doped $SrSnO_3$. *Thin Solid Films* 519:6059–6063.

Liu, Q. Z., H. F. Wang, F. Chen, et al. 2008. Single-Crystalline Transparent and Conductive Oxide Films with the Perovskite Structure: Sb-Doped $SrSnO_3$. *J. Appl. Phys.* 103:093709.

Liu, X., L.-S. Wang, Y. Ma, et al. 2017. Enhanced Microwave Absorption Properties by Tuning Cation Deficiency of Perovskite Oxides of Two-Dimensional $LaFeO_3$/C Composite in X-Band. *ACS Appl. Mater. Interfaces* 9:7601–7610.

Liu, Y., X. Tan, and K. Li. 2006. Mixed Conducting Ceramics for Catalytic Membrane Processing. *Catal. Rev.* 48:145–198.

Long, Y. W., N. Hayashi, T. Saito, et al. 2009. Temperature-Induced A–B Intersite Charge Transfer in an A-Site-Ordered $LaCu_3Fe4O_{12}$ Perovskite. *Nature* 458:60–63.

Lu, Z., L. Chen, Y. Tang, et al. 2005. Preparation and Luminescence Properties of Eu^{3+}-Doped $MSnO_3$ (M = Ca, Sr and Ba) Perovskite Materials. *J. Alloys Compd.* 387:L1–L4.

Luo, Z., J. G. Moch, S. S. Johnson, et al. 2017. A Review on X-ray Detection Using Nanomaterials. *Curr. Nanosci.* 13:364–372.

Lybye, D. 2000. Conductivity of A- and B-Site Doped $LaAlO_3$, $LaGaO_3$, $LaScO_3$ and $LaInO_3$ Perovskites. *Solid State Ion.* 128:91–103.

Mansfield, S. M., and G. S. Kino. 1990. Solid Immersion Microscope. *Appl. Phys. Lett.* 57:2615–2616.

Mardare, C. C., and A. W. Hassel. 2019. Review on the Versatility of Tungsten Oxide Coatings. *Phys. Status Solidi A* 216:1900047.

Mares, J. A., M. Nikl, and K. Blazek. 1991. Green Emission Band in Ce^{3+}-Doped Yttrium Aluminium Perovskite. *Phys. Status Solidi A* 127:K65–K68.

Mastin, J., M.-A. Einarsrud, and T. Grande. 2006. Structural and Thermal Properties of $La_{1-x}Sr_xCoO_{3-\delta}$. *Chem. Mater.* 18:6047–6053.

Mazzei, L., A. Perrichon, A. Mancini, et al. 2019. Local Coordination of Protons in In- and Sc-Doped $BaZrO_3$. *J. Phys. Chem. C* 123:26065–26072.

Mei, F. D., L. Falsi, M. Flammini, et al. 2018. Giant Broadband Refraction in the Visible in a Ferroelectric Perovskite. *Nat. Photonics* 12:734–738.

Miyoshi, S., A. Kaimai, H. Matsumoto, et al. 2004. In Situ XRD Study on Oxygen-Excess $LaMnO_3$. *Solid State Ion.* 175:383–386.

Miyoshi, S., J.-O. Hong, K. Yashiro, et al. 2003. Lattice Expansion upon Reduction of Perovskite-Type $LaMnO_3$ with Oxygen-Deficit Nonstoichiometry. *Solid State Ion.* 161:209–217.

Mogni, L., J. Fouletier, F. Prado, et al. 2005. High-Temperature Thermodynamic and Transport Properties of the $Sr_3Fe_2O_{6+\delta}$ Mixed Conductor. *J. Solid State Chem.* 178:2715–2723.

Mori, M., Y. Hiei, N. M. Sammes, et al. 2000. Thermal-Expansion Behaviors and Mechanisms for Ca- or Sr-Doped Lanthanum Manganite Perovskites under Oxidizing Atmospheres. *J. Electrochem. Soc.* 147:1295.

Moszyński, M., M. Kapusta, D. Wolski, et al. 1998. Properties of the YAP : Ce Scintillator. *Nucl. Instrum. Methods Phys. Res. Sect. A* 404:157–165.

Münch, W., K. D. Kreuer, S. Adams, et al. 1999. The Relation between Crystal Structure and the Formation and Mobility of Protonic Charge Carriers in Perovskite-Type Oxides: A Case Study of Y-Doped $BaCeO_3$ and $SrCeO_3$. *Phase Transitions* 68:567–586.

Niklasson, G. A., and C. G. Granqvist. 2006. Electrochromics for Smart Windows: Thin Films of Tungsten Oxide and Nickel Oxide, and Devices Based on These. *J. Mater. Chem.* 17:127–156.

Pan, W., S. R. Phillpot, C. Wan, et al. 2012. Low Thermal Conductivity Oxides. *MRS Bull.* 37:917–922.

Patel, K. J., C. J. Panchal, M. S. Desai, et al. 2010. An Investigation of the Insertion of the Cations H^+, Na^+, K^+ on the Electrochromic Properties of the Thermally Evaporated WO_3 Thin Films Grown at Different Substrate Temperatures. *Mater. Chem. Phys.* 124:884–890.

Pedrini, C., D. Bouttet, C. Dujardin, et al. 1994. Fast Fluorescence and Scintillation of Pr-Doped Yttrium Aluminum Perovskite. *Opt. Mater.* 3:81–88.

Pendry, J. B., D. Schurig, and D. R. Smith. 2006. Controlling Electromagnetic Fields. *Science* 312:1780–1782.

Qiu, G., H. Fang, X. Wang, et al. 2018. Largely Enhanced Mechanoluminescence Properties in Pr^{3+}/Gd^{3+} Co-Doped $LiNbO_3$ Phosphors. *Ceram. Int.* 44:15411–15417.

Rao, M. V. R., and A. M. Umarji. 1997. Thermal Expansion Studies on Ferroelectric Materials. *Bull. Mater. Sci.* 20:1023–1028.

Reagor, D. W., and V. Y. Butko. 2005. Highly Conductive Nanolayers on Strontium Titanate Produced by Preferential Ion-Beam Etching. *Nat. Mater.* 4:593–596.

Ruiz-Trejo, E., G. Tavizón, and A. Arroyo-Landeros. 2003. Structure, Point Defects and Ion Migration in $LaInO_3$. *J. Phys. Chem. Solids* 64:515–521.

Shan, C., T. Huang, J. Zhang, et al. 2014. Optical and Electrical Properties of Sol–Gel Derived $Ba_{1-x}La_xSnO_3$ Transparent Conducting Films for Potential Optoelectronic Applications. *J. Phys. Chem. C* 118:6994–7001.

Shannon, R. D. 1976. Revised Effective Ionic Radii and Systematic Studies of Interatomic Distances in Halides and Chalcogenides. *Acta Cryst. A* 32:751–767.

Shivaram, M., R. H. Krishna, H. Nagabhushana, et al. 2013. Synthesis, Characterization, EPR and Thermoluminescence Properties of $CaTiO_3$ Nanophosphor. *Mater. Res. Bull.* 48:1490–1498.

Smolin, S. Y., M. D. Scafetta, G. W. Guglietta, et al. 2014. Ultrafast Transient Reflectance of Epitaxial Semiconducting Perovskite Thin Films. *Appl. Phys. Lett.* 105:022103.

Song, Z., J. Zhao, and Q. Liu. 2019. Luminescent Perovskites: Recent Advances in Theory and Experiments. *Inorg. Chem. Front.* 6:2969–3011.

Souza, R. A. D. 2015. Oxygen Diffusion in $SrTiO_3$ and Related Perovskite Oxides. *Adv. Funct. Mater.* 25:6326–6342.

Sparks, T. D., P. A. Fuierer, and D. R. Clarke. 2010. Anisotropic Thermal Diffusivity and Conductivity of La-Doped Strontium Niobate $Sr_2Nb_2O_7$. *J. Am. Ceram. Soc.* 93:1136–1141.

Stranks, S. D., and H. J. Snaith. 2015. Metal-Halide Perovskites for Photovoltaic and Light-Emitting Devices. *Nat. Nanotechnol.* 10:391–402.

Suemune, Y. 1965. Thermal Conductivity of $BaTiO_3$ and $SrTiO_3$ from 4.5° to 300°K. *J. Phys. Soc. Jpn.* 20:174–175.

Suzuki, H., H. Bando, Y. Ootuka, et al. 1996. Superconductivity in Single-Crystalline $Sr_{1-x}La_xTiO_3$. *J. Phys. Soc. Jpn.* 65:1529–1532.

Taghaddos, E., M. Hejazi, and A. Safari. 2015. Lead-Free Piezoelectric Materials and Ultrasonic Transducers for Medical Imaging. *J. Adv. Dielect.* 05:1530002.

Takao, E., and S. S. B. C. Abdullah. 2010. Oxide Ion Conduction in the Perovskite-Type $LaYO_3$ Doped with ZrO_2. *Electrochemistry* 78:907–911.

Takeda, T., T. Miyata, F. Muramatsu, et al. 1980. Fast Decay U.V. Phosphor – $YAlO_3$: Ce. *J. Electrochem. Soc.* 127:438.

Thacher, P. D. 1976. Optical Effects of Fringing Fields in Kerr Cells. *IEEE Trans. Electr. Insul.* EI-11:40–50.

Ueda, K., and Y. Shimizu. 2010. Fabrication of Tb–Mg Codoped $CaSnO_3$ Perovskite Thin Films and Electroluminescence Devices. *Thin Solid Films* 518:3063–3066.

Vaßen, R., M. O. Jarligo, T. Steinke, et al. 2010. Overview on Advanced Thermal Barrier Coatings. *Surf. Coat. Technol.* 205:938–942.

Wang, X., C.-N. Xu, H. Yamada, et al. 2005. Electro-Mechano-Optical Conversions in Pr^{3+}-Doped $BaTiO_3$–$CaTiO_3$ Ceramics. *Adv. Mater.* 17:1254–1258.

Wu, J., L. Chen, T. Song, et al. 2017. A Review on Structural Characteristics, Lithium Ion Diffusion Behavior and Temperature Dependence of Conductivity in Perovskite-Type Solid Electrolyte $Li_{3x}La_{2/3-x}TiO_3$. *Funct. Mater. Lett.* 10:1730002.

Yamada, I., K. Tsuchida, K. Ohgushi, et al. 2011. Giant Negative Thermal Expansion in the Iron Perovskite $SrCu_3Fe_4O_{12}$. *Angew. Chem. Int. Ed.* 50:6579–6582.

Yamada, I., S. Marukawa, N. Hayashi, et al. 2015. Room-Temperature Zero Thermal Expansion in a Cubic Perovskite Oxide $SrCu_3Fe_{4-x}Mn_xO_{12}$. *Appl. Phys. Lett.* 106:151901.

Yamada, Y., H. Yasuda, T. Tayagaki, et al. 2009a. Temperature Dependence of Photoluminescence Spectra of Nondoped and Electron-Doped $SrTiO_3$: Crossover from Auger Recombination to Single-Carrier Trapping. *Phys. Rev. Lett.* 102:247401.

Yamada, Y., H. Yasuda, T. Tayagaki, et al. 2009b. Photocarrier Recombination Dynamics in Highly Excited $SrTiO_3$ Studied by Transient Absorption and Photoluminescence Spectroscopy. *Appl. Phys. Lett.* 95:121112.

Yamada, Y., and Y. Kanemitsu. 2013. Photoluminescence Spectra of Perovskite Oxide Semiconductors. *J. Luminescence* 133:30–34.

Yang, Z., and J. Zu. 2016. Comparison of PZN-PT, PMN-PT Single Crystals and PZT Ceramic for Vibration Energy Harvesting. *Energy Convers. Manage.* 122:321–329.

Yasuda, H., and Y. Kanemitsu. 2008. Dynamics of Nonlinear Blue Photoluminescence and Auger Recombination in $SrTiO_3$. *Phys. Rev. B* 77:193202.

Ye, S., C.-H. Wang, and X.-P. Jing. 2008. Photoluminescence and Raman Spectra of Double-Perovskite $Sr_2Ca(Mo/W)O_6$ with A- and B-Site Substitutions of Eu^{3+}. *J. Electrochem. Soc.* 155:J148.

Zhang, W., J. Tang, and J. Ye. 2007. Structural, Photocatalytic, and Photophysical Properties of Perovskite $MSnO_3$ (M = Ca, Sr, and Ba) Photocatalysts. *J. Mater. Res.* 22:1859–1871.

Zhang, W., K. Fujii, E. Niwa, et al. 2020. Oxide-Ion Conduction in the Dion–Jacobson Phase $CsBi_2Ti_2NbO_{10-\delta}$. *Nat. Commun.* 11: 1–8.

Zuev, A. Y., A. I. Vylkov, A. N. Petrov, et al. 2008. Defect Structure and Defect-Induced Expansion of Undoped Oxygen Deficient Perovskite $LaCoO_{3-\delta}$. *Solid State Ion.* 179:1876–1879.

8 Applications of Perovskite Oxides

Perovskite oxides exhibit variable chemical compositions, due to the availability of a wide range of A- and B-site cations and valences in their relatively simple structures. The partial replacement of the A- and B-site cations with donor-acceptor elements leads to tailored bandgap and improved accompanying properties. The synthesis of perovskites is facile, and the flexibility to allocate various ions in their lattice leads the perovskite structured materials to exhibit diverse properties favorable for a wide range of science and engineering applications. For instance, owing to the good electrical, ionic, and electronic conductivity of some perovskite oxides, they are highly recommended as a replacement of noble metals in several catalytic reactions. The typical properties of perovskite structured materials are ferroelectricity, magnetism, superconductivity, catalytic activity, etc. which are explored well in different applications, as listed in Table 8.1. Some of the important characteristics of perovskites materials suitable for several applications are strange valence states, mixed valences, anion excess or deficiencies, oxygen vacancies, and valence alternations without structural change.

TABLE 8.1
Applications of Perovskite Oxides

Ground of Application	Applications	The Uniqueness of Perovskite-Type Materials
Biomedical	Biocatalysts and enzyme carriers	Ease of production
	Cancer treatment	Low cost compared to noble metals
	Orthopedic implants	Tunable properties by substitution
	Biosensors	Functionalization by organic molecules
Catalysis	Electrocatalysis	
	Photocatalysis	Tunable electron, proton, and ionic conductivity
	Chemical catalysis	
Energy storage/ conversion	Batteries	Thermal and chemical stability
	Supercapacitors	Sensing and actuating capabilities
	Fuel cells	Reliability
	Hydrogen storage	The high surface area due to 3D network-type structure
	Photovoltaics	
Others	Sensors	
	RRAM	

8.1 BIOMEDICAL APPLICATIONS

8.1.1 Biocatalysts or Enzyme Carriers

The multifunctionality of the perovskite oxides is explored in different applications in the biomedical field. Perovskite oxides are reported to mimic the catalytic activities of enzymes effectively. Many perovskite oxides with transition elements are identified as an alternative to the enzyme peroxidase, which catalyzes the oxidation of a substance in the presence of peroxide. The peroxidase-like activity of perovskite materials originated from the *d*-electron population of the e_g (σ^*) antibonding orbitals of the transition metal sites (Wang et al. 2019b), and the highest activity is observed when e_g is ~1.2 and the lowest when e_g is 0 or 2. Perovskite structured $LaNiO_3$ (Wang et al. 2017c), $LaFeO_3$ (Wang et al. 2019b), $LaCoO_3$ (Wang et al. 2017a), etc. exhibit characteristics of peroxidase enzyme. Substitution of La^{3+} ions with Sr^{2+} ions leads to a shift in the overall oxidation state of Fe from 3+ to higher values, at the same time reducing the e_g occupancy from 2 to lower values (Wang et al. 2019b). The peroxidase-like activity of the above perovskites makes them a promising candidate for the non-enzymatic detection of biomolecules such as glucose and dopamine (DA).

8.1.2 Cancer Treatment

Magnetic perovskite materials are identified as a promising candidate for early-stage detection and treatment of cancer, and the treatment involves imaging and destruction of cancerous cells.

8.1.2.1 Imaging

Magnetic Resonance Imaging (MRI) is a widely used imaging technique in biomedical applications due to its high resolution. However, its application is limited in the detection of cancer due to its low specificity. The specificity of MRI towards cancerous cells can be enhanced using paramagnetic and superparamagnetic nanoparticles as the cell markers. Superparamagnetic nanoparticles enable faster water-proton relaxation, thus disturbing the magnetic fields and the detection with MRI (Kačenka et al. 2011). $La_{1-x}Sr_xMnO_3$ is identified as the most promising perovskite for such applications (Kulkarni et al. 2015). In both the above applications, the perovskite materials need to be biocompatible by reducing their cytotoxicity. It is possible to achieve biocompatibility either by modifying their surface with biocompatible molecules such as bovine serum albumin (BSA) or dextran sulfate (Bhayani et al. 2007) or capping with an inactive material such as SiO_2 (Kačenka et al. 2011). Fluorescent imaging of the targeted region is also possible using perovskite structured fluorescent materials.

8.1.2.2 Hyperthermia Therapy

In hyperthermia therapy, the local temperature in the body is elevated to 40°C–43°C to induce cancer cell death and to enhance the effects of radiotherapy and chemotherapy. Nanosized magnetic materials are capable of heating cancer-affected areas with the help of alternating external energy without any surgical intervention (Kulkarni et al. 2015). Perovskite structured magnetic nanomaterials are reported as a promising candidate for magnetic hyperthermia treatment. In magnetic

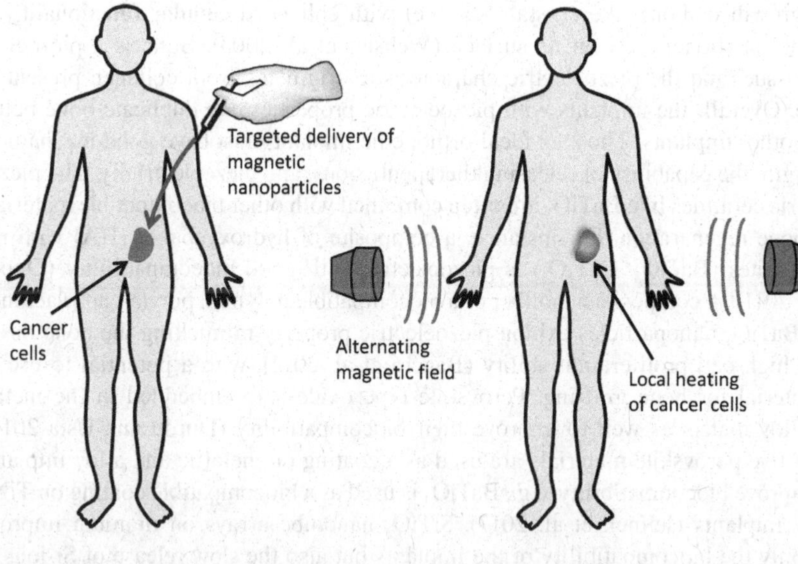

FIGURE 8.1 A schematic representation of hyperthermia treatment.

hyperthermia, the magnetic nanoparticles are targeted to the cancerous cells, and the temperature of the magnetic nanoparticles located in the tumor tissue is then raised by the application of an alternating magnetic field to destroy the cancerous cells (Chang et al. 2018), and a schematic representation of magnetic hyperthermia treatment is shown in Figure 8.1.

Perovskite structured materials with manganese in the B-site are extensively studied as a suitable candidate for the hyperthermia application, e.g., $LaMnO_3$, owing to their low Curie temperature (40°C–50°C), ideal for hypothermia application. Curie temperature (T_c) determines the nature of magnetization in materials, and above T_c the material becomes purely paramagnetic. In a hypothermia application, the heating of the magnetic particles is possible up to Curie temperature, and by selecting the magnetic particles with appropriate Curie temperature, one can achieve self-controlled heating and avoiding the damage of unaffected regions. $LaMnO_3$, an antiferromagnetic insulator, which is transformed into metallic with ferromagnetic ordering by doping the trivalent A-site with aliovalent Sr, Ag, Na, etc. $La_{1-x}Ag_xMnO_3$, $La_{1-y}Na_yMnO_3$ (Shlyakhtin et al. 2007) and $La_{1-x}Sr_xMnO_3$ are attractive candidates for self-controlled hyperthermia applications (Vasseur et al. 2006).

8.1.3 Orthopedic Implants

When considering a material for orthopedic implants, the biocompatibility of the materials is primarily important. The materials must support the proliferation of bone marrow and fibroblast cells on its surface. The above qualities are necessary to avoid any inflammation or toxicity by the implants at the bone-implant interface. Perovskite structured $CaTiO_3$ is demonstrated as a potential biocompatible material for orthopedic implants. $CaTiO_3$ is not only biocompatible but also promotes the growth of bone-like crystals (apatite) with enhanced cellular functionality of osteoblast (bone) cells on its surface (Webster et al. 2003). Bone is a piezoelectric tissue, and the piezoelectric characteristic originates from collagen present in bone. Overall, the implants with piezoelectric properties can duplicate bone better than other implants. Thus, an ideal orthopedic implant must have good mechanical strength, the capability of releasing therapeutic ions, and piezoelectricity. The piezoelectric ceramics like $BaTiO_3$ are often combined with other biocompatible materials for bone regeneration. For instance, a composite of hydroxyapatite (HA) with the perovskites ($BaTiO_3$, $CaTiO_3$) is piezoelectric with good biocompatibility (Dubey et al. 2011). A composite nanofiber of a biocompatible polymer, poly(ε-caprolactone), and $BaTiO_3$ nanoparticles exhibit piezoelectric property mimicking the bone tissue with high cell proliferation ability (Bagchi et al. 2014), with a potential to use as a material for bone grafting. Perovskite-type oxides are embedded in the metallic alloy matrix as well to improve their biocompatibility (Durdu and Usta 2014). Bioactive perovskite materials are used as a coating on metallic and alloy implants to improve biocompatibility, e.g., $BaTiO_3$ is used as a biocompatible coating on TiNb alloy implants (Jelínek et al. 2017). $SrTiO_3$ nanotube arrays on titanium improve not only the biocompatibility of the implants but also the slow release of Sr-ions to enhance osseointegration. The nanoporous structure can also allow the loading of other elements and drugs in the pores (Xin et al. 2009).

8.1.4 Biosensors

8.1.4.1 Hydrogen Peroxide and Glucose Sensing

H_2O_2 is an essential molecule in the human body and performs several physiological activities such as cellular growth, cellular signaling, apoptosis, and immune activation, although at higher H_2O_2 concentrations, it will cause cancer, inflammation, and damage to cells. On the other hand, glucose is the major and basic metabolite present in all living organisms. Hence it is important to monitor H_2O_2 and glucose levels via sensing (He et al. 2017). Non-enzymatic detection of H_2O_2 and glucose is most accurate and reliable compared to enzymatic detection, owing to the susceptibility towards poisonous chemicals, temperature, pH, humidity, etc. Non-enzymatic detection can be done using electrochemical sensors, which are made of nanostructured precious noble metals. Perovskite oxide materials are one of the best alternatives to replace these expensive noble metals in electrochemical sensors.

$La_{0.6}Sr_{0.4}CoO_{3-\delta}$ (LSC) perovskite oxide can show comparable sensing activity to these noble metal-based sensors and schematically represented as Figure 8.2 (He et al. 2017). The oxidation of H_2O_2 and glucose on LSC perovskite proceeds through a Co^{3+}/Co^{4+} redox couple. LSC displays enhanced electrooxidation activities for H_2O_2 and glucose sensing when it is combined with reduced graphene oxide (RGO). The LSC+RGO/glassy carbon electrode (GCE) facilitates sensitivity of 500 and 330 μA mM^{-1} cm^{-2} for H_2O_2 and glucose, respectively, with excellent detection limits and linear response. Further, the $LaCo_{0.4}Fe_{0.6}O_3$ modified with carbon paste (CpE) shows promising sensing ability towards H_2O_2 and glucose sensing in alkaline medium. In this perovskite also Co^{3+}/Co^{4+} redox couple plays a key role in sensing, and iron doping improves this ability further in KOH electrolyte. This sensor exhibits an

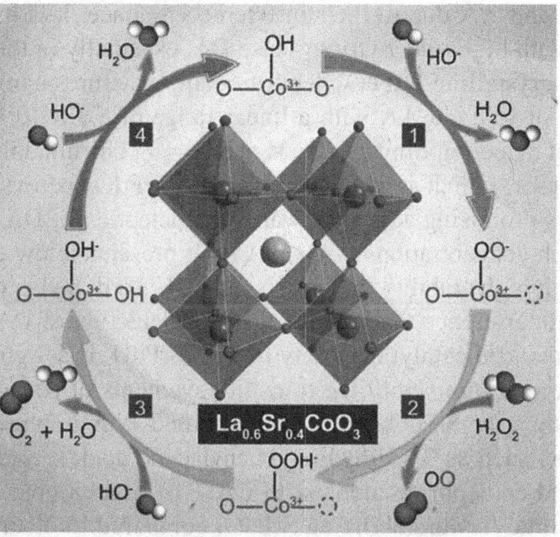

FIGURE 8.2 Schematic representation of H_2O_2 sensing by $La_{0.6}Sr_{0.4}O_3$ perovskite, reproduced with permission of the Elsevier (He et al. 2017).

excellent linear range for H_2O_2 (800–0.01 μM with 2 nM of minimum detection limit) and glucose (500–0.05 μM with 10 nM of minimum detection limit) sensing. This sensor depicts excellent long-term stability along with good anti-interference towards other biomolecules such as DA, uric acid (UA), and ascorbic acid (AA) (Zhang et al. 2013c).

The non-enzymatic biosensors, which are made up of both perovskite oxides and noble metals, exhibit better sensing activity compared to sole perovskite oxide-based sensors. For instance, a glucose biosensor composed of $LaTiO_3$ with the combination of Ag metal, as in $LaTiO_3$-$Ag_{0.2}$ (LTA), showed remarkable sensing ability (Wang et al. 2013b). This sensor displayed superior sensitivity of 784.14 μAmM^{-1} cm^{-2} with 2.1×10^{-7} M and having a broad linear range of 4 mM to 2.5 μM. Furthermore, the sensing application of this sensor in the real world shows promising results with excellent reproducibility, and the results are consistent with the clinical data. The above LTA sensor is highly insensitive to the general interfering molecules like UA, AA, and DA. Similarly, $SrPdO_3$ perovskite modified with a thin film of gold nanoparticles exhibits augmented sensing ability towards glucose owing to the synergetic effect between gold nanoparticles and $SrPdO_3$. The electrocatalytic activity of $SrPdO_3$ perovskite is assigned to the oxygen vacancies present on its surface, and they display intrinsic catalytic activity towards the oxidation of glucose. This $SrPdO_3$ gold nanocomposite shows enhanced reproducibility with good selectivity, minimum detection limit, and broad linearity (El-Ads et al. 2015).

8.1.4.2 Sensing of Neurotransmitters

Sensing of neurotransmitter like DA is essential to screen the mental wellness of humans, so it can help us to detect neural diseases like brain cancers, mental illness, and Parkinson's disorder. But the sensing of DA is very difficult in the presence of UA and AA due to their interference. Hence, it is highly essential to find the sensor with high selectivity to sense DA, especially in the living cells. For instance, single-crystalline $LaFeO_3$ dendritic nanostructures can detect the DA in the interference of UA and AA with a linear range of 8.2×10^{-8} to 1.6×10^{-7} M with an excellent detection limit of 62 nM at $S/N = 3$ (Thirumalairajan et al. 2013). Another DA sensor, which is made of CpE and $SrPdO_3$ perovskite oxide (CpE/$SrPdO_3$), delivers promising activity towards the detection of DA even in the presence of very high concentrations AA and UA. It presents a low detection limit of 9.3 nmol L^{-1} with high stability and precision. CpE/$SrPdO_3$ also displays excellent selectivity, anti-interference, and recovery in the detection of DA in human urine samples. The enhanced catalytic activity of CpE/$SrPdO_3$ is assigned to the interaction between oxygen atoms with the transition elements of perovskite oxide (Atta et al. 2014). Further, the $SrPdO_3$ perovskite modified with graphite can detect other neurotransmitters such as 3,4-dihydroxyphenylacetic acid, serotonin, norepinephrine, L-dopa, and epinephrine along with DA. The graphite electrode is found to be a better substrate for immobilizing $SrPdO_3$ compared to others in detecting all the above neurotransmitters. With the help of this sensor, one can detect AA, UA, L-dopa, and serotonin due to their large peak potential difference (Atta et al. 2013). This composite sensor displays high precision, excellent sensitivity, and selectivity with high endurance (nearly 2 months).

8.2 CATALYSIS

The catalytic activity of the perovskite is dependent on the A- and B-site cations and their activity in various environments. The perovskite oxides are reported as highly active catalysts for oxidation reactions for air and water pollution control, high- and low-temperature electrocatalysts for hydrogen evolution reaction (HER) and oxygen evolution reaction (OER), photocatalysis, etc. Perovskite-based catalysts are stable at high temperatures and various reaction mediums, besides their low cost, making them alternative to noble metal catalysts. The presence of oxygen vacancies, mobility of lattice oxygen, or the B–O bond strength, and the redox properties of B-site elements determine the catalytic activity of the perovskite oxides. In a perovskite structure, the A-site cation stabilizes the structure without any redox catalytic activity, and the B-site transition metal ions act as the activation sites, which can undergo reversible redox reactions without affecting the structure. In general, lanthanum-based perovskite materials with transition metals in the B-site are reported extensively as a catalyst in the majority of applications. The abundance of lanthanum and the transition metals offers a cheap alternative to the noble metal catalysts in several catalytic related applications.

8.2.1 Electrocatalysis

The electrocatalytic properties of perovskite materials are applied in various fields. They are HER, OER, oxygen reduction reaction (ORR), and degradation of organic molecules.

8.2.1.1 Hydrogen Evolution Reaction

Hydrogen is the most promising environment benign source of energy. The electrochemical evolution of hydrogen is the most viable way of storing hydrogen energy. The noble metals are an efficient catalyst for hydrogen evolution through water splitting. The high cost of noble metal catalysts can overcome by using inexpensive earth-abundant materials as catalysts. Perovskite structured oxides with unique multivalent transition metals in the B-site offers a replacement of noble metal catalyst for HER. Perovskite structured materials offer many active sites for catalyzing HER based on their crystal structure. The schematic illustration is shown in Figure 8.3.

It is commonly assumed that the activity of metallic oxides towards HER is weak in acidic media; however, perovskite materials are reported as HER catalysts in acid medium. The H_2 evolution in acidic medium using perovskite structured materials is reported for the first time in the 1990s using A-site deficient $Sr_{1-x}NbO_{3-\delta}$ ($0.05 \leq x \leq 0.3$) (Manoharan and Goodenough 1990). $LaBO_3$ (B = Ni, Co, Fe, and Mn) perovskites are also identified as HER catalysts in acid medium. The B-site cations significantly influence the HER catalysis of these perovskite materials. The catalytic activity towards HER is in the order $LaFeO_3 > LaCoO_3 > LaNiO_3 > LaMnO_3$ (Galal et al. 2011). Perovskite structured materials with noble metals in the B-site are also identified as stable catalysts for HER in acids, e.g., $SrPdO_3$ and $ARuO_3$ (A = Ca, Sr, or Ba) (Galal et al. 2010; Atta et al. 2012).

In general, the HER activities of oxide catalysts in alkaline medium is lower than that in the acid medium by ≈2–3 orders in magnitude. In alkaline medium, the sluggish kinetics of the cathodic HER ($2H_2O + 2e^- \rightarrow H_2 + 2OH^-$) slows the H_2 evolution leading to large overpotential and energy loss (Subbaraman et al. 2011). Recent studies reveal that the perovskite structured materials are potential HER catalysts in basic medium with high activity and overpotentials at par with noble metals. $Pr_{0.5}(Ba_{0.5}Sr_{0.5})_{0.5}Co_{0.8}Fe_{0.2}O_{3-\delta}$ is one such material with an overpotential

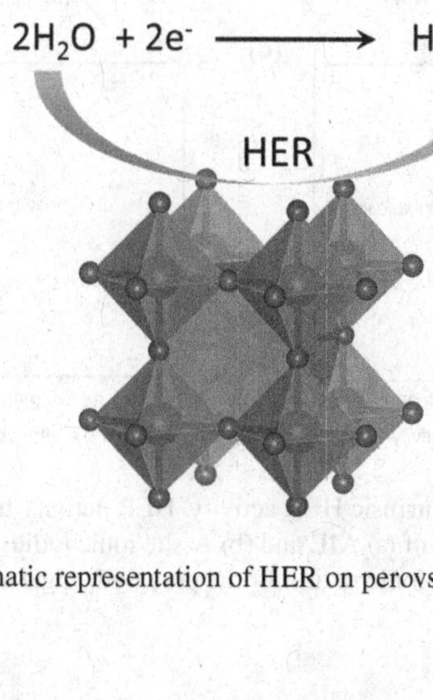

FIGURE 8.3 Schematic representation of HER on perovskite oxide.

of 237 mV at a current density of 10 mA cm^{-2} (Xu et al. 2016c). A recent study on HER activity of double perovskites suggests that optimizing the A-site ionic electronegativity (AIE) will greatly enhance the catalytic activity, even better than state-of-the-art Pt/C catalyst. AIE is considered as a unifying descriptor for predicting the superior HER catalysis. From the volcano plot (a plot shows statistical significance) of $(Gd_{0.5}La_{0.5})BaCo_2O_{5.5+\delta}$, it is predicted that the optimal AIE for HER catalysis is ~2.33 (Figure 8.4). HER activity of perovskite oxides at 10 mA cm^{-2} as a function of an AIE shown in Figure 8.4a and b displays the influence of an A-site ionic radius on determining the HER activity of single and double perovskites. Plots of Tafel slope values as a function of AIE (the rate-determining steps estimated from the Tafel slope are shown in pink and purple dashed lines) are shown in Figure 8.4c and for the HER measurements performed in 1.0 M KOH at 25°C and the gray dashed lines are shown for guidance only. Triangles and squares represent the single and double perovskites, respectively. Sr-doped perovskites (green) and perovskites doped with different lanthanides ($Sm_{0.5}La_{0.5}$ and $Pr_{0.5}Gd_{0.5}$ in black) are shown in distinguished colors. This perovskite shows excellent turnover frequency (TOF) of 22.9 s^{-1} with overpotential and Tafel values of 0.24 V and 27.6 mV dec^{-1}, respectively (Guan et al. 2019).

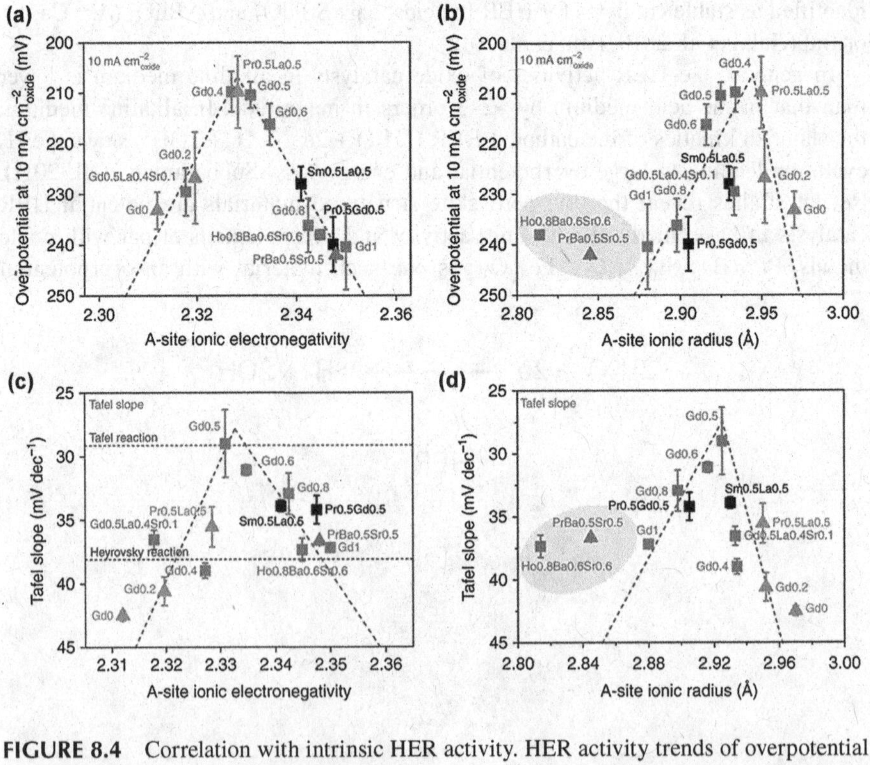

FIGURE 8.4 Correlation with intrinsic HER activity. HER activity trends of overpotential at 10 mA cm^{-2} oxide as a function of (a) AIE and (b) A-site ionic radius for single and double perovskites. (c) AIE and (d) A-site ionic radius for the prepared single and double perovskites (Guan et al. 2019).

8.2.1.2 Oxygen Evolution Reaction (OER) and Oxygen Reduction Reaction (ORR)

Electrochemical ORR is an important aspect of metal-air batteries, and OER is for fuel cell electrodes, and a schematic representation is shown in Figure 8.5. Both OER ($2H_2O \rightarrow O_2 + 4H^+ + 4e^-$) and ORR ($O_2 + 4H^+ + 4e^- \rightarrow 2H_2O$) are four-electron transfer reactions. Noble metal catalysts are well established for ORR (e.g., Pt) and OER (e.g., IrO_2 and RuO_2) applications. The noble metal catalysts are expensive and scarce, and their use as electrocatalysts is not economical at a commercial scale. Moreover, these catalysts tend to aggregate over time and often poisoned by other reactive species during the reaction. Perovskite structured oxides offer bifunctional catalysis towards OER and ORR, which is abundant and available at a low cost. Carbon-based materials decorated with metallic particles exhibit excellent ORR and OER activities (Yang et al. 2017a); however, the long-term stability of carbon-based electrocatalysts are inferior to use them in practical applications.

Transition metal oxides with *d*-orbitals can bind oxygen species on its surface, an important characteristic of the OER/ORR catalyst. Unlike transition metals and other complex oxides, perovskite structured oxides accommodate multiple A-site and B-site cations with different oxidation states without affecting their structure. Substitution of A or B sites allows the fine-tuning of electronic, catalytic properties by generating oxygen vacancies. Perovskite oxides are more stable than the carbon-based catalysts under oxidative environments and offer a competitive catalytic property comparable to noble metals. The OER/ORR catalytic properties of perovskites are dependent on the B-site transition metal.

$LaCoO_3$ is the first perovskite oxide, reported as oxygen electrocatalyst (Meadowcroft 1970; Hardin et al. 2014). The doping of La site with Sr and Co site with Ni was found enhance oxygen catalytic activity by improving its conductivity (Matsumoto et al. 1977). The other transition metal-based catalysts such as $LnMnO_3$, $LaVO_3$, and $SrFeO_3$ displayed OER/ORR activity in basic medium (KOH)

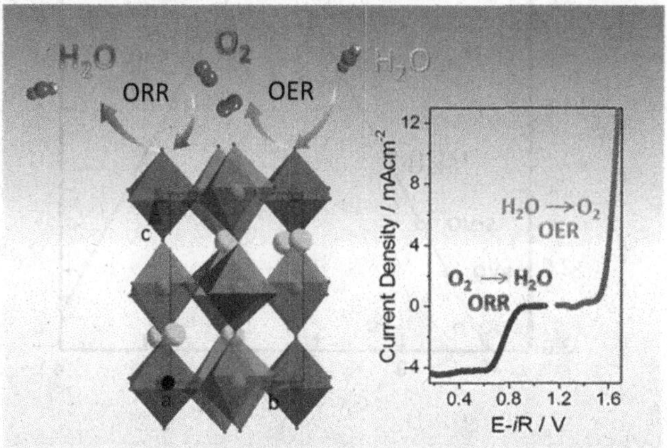

FIGURE 8.5 A schematic representation of OER and ORR on perovskite oxide, reproduced with permission of the American Chemical Society (Retuerto et al. 2019).

(Bockris and Otagawa 1984; Oh et al. 2015). For instance, $LnMnO_3$ shows an ORR activity in KOH, and their activity is directly proportional to the ionic radius of lanthanide cations (La > Pr > Nd > Sm > Gd > Y > Dy > Yb) (Hyodo et al. 1996). The symmetry of crystal and tolerance factor increases with the increase in Ln cation size. The symmetry of crystals and tolerance factors indirectly influences the electrocatalytic activity of these lanthanide manganites. To understand the mechanism of oxygen electrocatalytic activity, several descriptors are proposed, among them oxygen adsorption and e_g value parameters are most important. The oxygen adsorption is related to the Sabatier principle, i.e., the interaction between the oxygen species and catalyst should be neither too strong nor too weak (HOO* and HO*). The interaction between the oxygen species and catalyst is often represented as a volcano plot, the interaction (ΔG_{O*}–ΔG_{HO*}) on the x-axis, and overpotentials of perovskites in the y-axis, as shown in Figure 8.6. The perovskite oxide, which presents at the very top of this volcano, shows superior activity (Man et al. 2011). The two branches represent two rate-determining steps. The oxides on the left-hand side mean too strong binding with oxygen, whereas the right side represents oxides having too weak binding with oxygen species. This volcano plot is valid for the mechanisms which involve a four-electron transfer.

The e_g value is another important descriptor for oxygen electrocatalysts, and which can be tuned by doping of either A or B sites or both in perovskite oxides. Here the integer of overlapping of e_g orbital of B-site with $2p$ orbital of oxygen is considered as a descriptor for oxygen electrocatalytic activity. For example, the partial or complete substitution of A and B sites in $LaCoO_3$ indicates minute changes in the

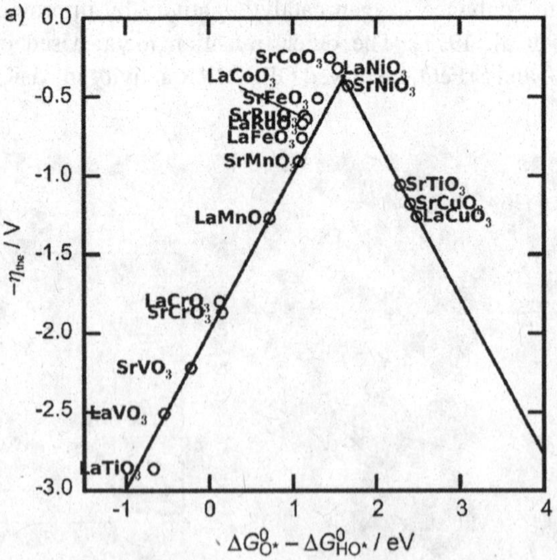

FIGURE 8.6 A representative volcano plot of overpotential was plotted *vs.* the standard free energy for different perovskites, reproduced with permission of the John Wiley & Sons (Man et al. 2011).

crystal structure, influencing the OER activity significantly because of the change in e_g value with the substitution ($La_{0.58}Sr_{0.4}Co_{0.2}Fe_{0.8}O_3$ > $La_{0.76}Sr_{0.2}Co_{0.2}Fe_{0.8}O_3$ > $La_{0.83}Ca_{0.15}Mn_{0.6}Co_{0.4}O_3$ > $La_{0.6}$ $Sr_{0.4}FeO_3$ > $La_{0.74}Sr_{0.2}Co_{0.2}Fe_{0.8}O_3$ > $La_{0.58}Sr_{0.4}Co_{0.2}$ $Cu_{0.1}Fe_{0.7}O_3$ > $La_{0.75}Sr_{0.2}Mn_{0.9}Co_{0.1}O_3$ > $La_{0.97}Mn_{0.4}Co_{0.3}Cu_{0.3}O_3$) (Rincón et al. 2014). Among all perovskites $Ba_{0.5}$ $Sr_{0.5}$ $Co_{0.8}Fe_{0.2}O_{3-\delta}$ (BSCF) has shown superior OER activity and one order magnitude higher activity in alkaline electrolyte than the state-of-the-art catalyst IrO_2. The activity of BSCF is attributed to the presence of large amounts of oxygen deficiencies, high oxygen exchange kinetics, high ionic, and electronic conductivities (Shao and Haile 2004). Recent studies reveal that BSCF is having an e_g filling value near to one, and this value highly influences the interaction of catalyst with reactive intermediates. Hence, the oxygen intermediates form a bond with the catalyst surface neither too strong and nor too weak (Sabatier principle) (Suntivich et al. 2011). Similarly, the ORR activity of $LaNi_{0.5}M_{0.5}O_3$ (M = Co, Fe, Mn, and Cr) perovskite also shows a variation with respect to the dopant. Among them, the oxide doped with Mn showed better catalytic activity (Yuasa et al. 2013). In $La_{0.6}Ca_{0.4}M_{1-x}Fe_xO_3$ perovskite, the oxide with the highest Mn content shows superior ORR performance with a low onset potential.

Like B-site doping, A-site doping does not contribute much to improving electrocatalytic activity. However, they may alter the valence state and Fermi level of B cations. Further, they contribute electrocatalytic activity indirectly by influencing the crystal structures, conductivity, oxygen defects, and electrochemical surface area (Obayashi and Kudo 1978). For example, Sr doping of $La_{1-x}Sr_xCoO_3$ perovskite enhances its electrocatalytic activity, which is assigned to the increase in electrochemical surface area (Tiwari et al. 1995). In another report, the doping of A-site of $CaMnO_{3-\delta}$ with the Lanthanides (Ln = La, Nd, Sm, Gd, Y, and Ho) increases the OER activity by improving the electronic conductivity of catalyst. The enhancement in conductivity might be due to the presence of empty 4*f* orbitals of lanthanides or the creation of oxygen defects (Kozuka et al. 2015).

The double perovskites with both A- and B-site substitutions show superior oxygen catalytic activity. For instance, $Sr_{1.5}La_{0.5}FeMoO_6$ displayed both OER and ORR activity; the results suggested that the enhanced oxygen catalytic activity is due to improved electronic conductivity. Similarly, $Ba_2CoMo_{0.5}Nb_{0.5}O_{6-\delta}$ (Sun et al. 2018), $Ba_2Bi_xSc_{0.2}Co_{1.8-x}O_{6-\delta}$ (Sun et al. 2017), and $PrBa_{0.5}Sr_{0.5}Co_{2-x}Fe_xO_{5+\delta}$ (Bu et al. 2017) exhibit high catalytic activities suitable for commercialization. In general, making electrocatalyst in their nanoforms improves the catalytic activity due to the high surface-to-volume ratio. For example, $LaCoO_3$ perovskite oxide showed improved OER activity by decreasing its particle size (~80 nm). The high OER activity is attributed to e_g filling value and high surface area. The $PrBa_{0.5}Sr_{0.5}Co_{2-x}Fe_xO_{5+\delta}$ (PBSCF) nanofibers prepared via electrospinning method showed promising activity in both OER and ORR due to the presence of a large surface area (Zhao et al. 2017). PBSCF shows 20 times higher mass activity compared with its bulk counterpart in OER (Figure 8.7). In La_2NiMnO_6 double perovskite, the electronic state is optimized by decreasing the particle due to cationic bonding transition from Mn^{4+}–O–Ni^{2+} to Mn^{3+}–O–Ni^{3+}, which results in enhanced OER activity in basic medium. A summary of a few perovskites with OER and ORR activity is presented in Table 8.2.

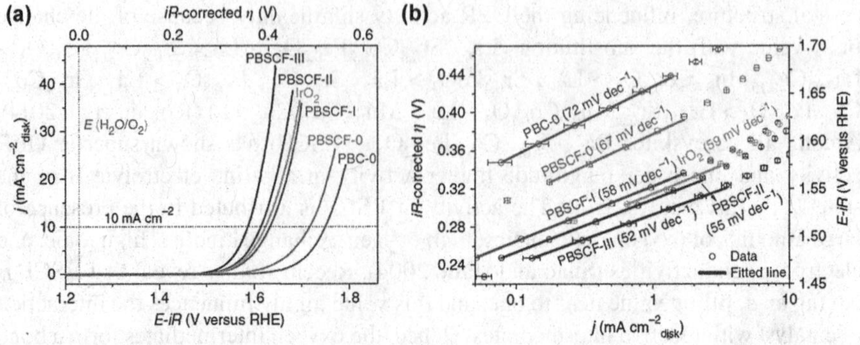

FIGURE 8.7 (a) Capacitance- and ohmic resistance-corrected OER activity curves of IrO_2, $PrBaCo_2O_{5+\delta}$ (PBC), $PrBa_{0.5}Sr_{0.5}Co_{1.5}Fe_{0.5}O_{5+\delta}$ (PBSCF) powders, and PBSCF nanofibers in 0.1 M KOH at 10 mVs^{-1} with a rotation rate of 1,600 r.p.m. These curves were averaged from three independent measurements; (b) Tafel plots are obtained from the steady-state measurements (Zhao et al. 2017).

8.2.2 Photocatalysis

Renewable and eco-friendly energy resources are getting prime importance in order to avoid a global energy crisis and environmental pollution due to the depletion of fossil fuels. Among all renewable energies, solar energy is the cleanest and abundant of all, although intermittent and diffusive nature of sunlight curtails its potential use. Semiconductor photocatalysis is the most promising technology that converts and stores the sunlight in the form of hydrogen and oxygen by splitting water without producing any harmful byproducts. In photocatalytic water splitting, semiconductors will initiate the reaction by producing electrons and holes via the capturing of sunlight. Subsequently, only electrons move from the valence band (VB) to conduction band (CB), and the holes remain at VB. This separation of charge carriers plays a key role in splitting water or degradation of organic dyes by generating redox-active species from photogenerated carriers. If the recombination of electrons and holes happen, the input solar energy is wasted in the form of light or light irradiation. This reaction must be avoided; otherwise, this recombination limits the full efficiency of the semiconductor photocatalysts. The schematic of both mechanisms is shown in Figure 8.8.

8.2.2.1 Photoelectrochemical Catalysis

The perovskite oxides are well explored in photocatalytic water splitting due to their large bandgap. $SrTiO_3$ displayed photocatalytic activity for both overall water splitting and reduction of CO_2 due to the presence of a wide bandgap of 3.2 eV (Nakanishi et al. 2017). Bulk $SrTiO_3$ displayed water splitting behavior with NiO as co-catalyst, which generates hydrogen gas at the rate of 28 μmol g^{-1} h^{-1} (Shoji et al. 2016). The water splitting behavior of $SrTiO_3$ nanoparticles is also evaluated, in which the catalytic ability decreases along with nanoparticle size, which can be attributed to the decrease in light absorption efficiency because of quantum confinement. Under UV irradiation, $SrTiO_3$ displays a reduction of CO_2 to CO with Cu_xO

TABLE 8.2
List of Perovskite Oxides as OER/ORR Catalysts

Perovskite	OER (η@ 10 mA cm^{-2})	ORR (V_{RHE}@ −3 mA cm^{-2})	Reference
$LaNiO_3$	430 mV	0.64 V	Hardin et al. (2013)
$LaCoO_3$	410 mV	0.64 V	Hardin et al. (2014)
$LaMnO_3$		0.52 V @ −1 mA cm^{-2}	Matsumoto and Sato (1986)
$LaNi_{0.8}Fe_{0.2}O_3$	450 mV	0.61 V	Zhang et al. (2015a)
$LaNi_{0.75}Fe_{0.25}O_3$	450 mV	0.67 V	Hardin et al. (2014)
$La_{0.5}Sr_{0.5}CoO_{3-\delta}$	600 mV	0.76 V	Zhao et al. (2012)
$La_{0.6}Sr_{0.4}CoO_{3-\delta}$	590 mV	0.67 V	Oh et al. (2015)
$La_{0.5}Sr_{0.5}Co_{0.8}Fe_{0.2}O_3$	590 mV	0.63 V	Park et al. (2015)
$La_{0.58}\ Sr_{0.4}Co_{0.2}Fe_{0.8}O_3$	450 mV	0.77 V	Rincón et al. (2014)
$La_{0.3}\ (Ba_{0.5}Sr_{0.5})_{0.7}Co_{0.8}\ Fe_{0.2}O_{3-\delta}$	380 mV	0.61 V	Jung et al. (2014)
$LaTi_{0.65}Fe_{0.35}O_{3-\delta}$	-	0.51 V	Prabu et al. (2015)
$LaNi_{0.8}Fe_{0.2}O_3$	510 mV	0.64 V	Zhang et al. (2015a)
$LaNi_{0.75}Fe_{0.25}O_3$	-	0.56 V	Hardin et al. (2014)
$CaMnO_3$	-	0.48 V	Han et al. (2012)
$Ca_2Mn_2O_5$	470 mV	-	Kim et al. (2014a)
$SrFeO_3$	380 mV	-	Matsumoto et al. (1979)
Fe-and Sc-doped $SrCoO_{3-\delta}$	410 mV	-	Wygant et al. (2016)
$SrNb_{0.1}Co_{0.7}Fe_{0.2}O_{3-\delta}$	370 mV	-	Zhu et al. (2017)
$SrCo_{0.4}Fe_{0.2}W_{0.4}O_{3-\delta}$	296 mV	-	Chen et al. (2018)
$SrCo_{0.95}P_{0.05}O_3$	480 mV	-	Zhu et al. (2016)
$Sr_{0.95}\ Nb_{0.1}Co_{0.7}\ Fe_{0.2}O_{3-\delta}$	460 mV	-	Liu et al. (2018b)
$Ba_{0.5}Sr_{0.5}Co_{0.8}Fe_{0.2}O_{3-\delta}$	500 mV	0.61 V	Jung et al. (2015)
$Ba_2CoMo_{0.5}Nb_{0.5}O_{6-\delta}$	445 mV		Sun et al. (2018)
$Ba_2Bi_xSc_{0.2}Co_{1.8-x}O_{6-\delta}$	520 mV@5 mA cm^{-2}		Sun et al. (2017)
$(Pr_{0.5}Ba_{0.5})CoO_{3-\delta}$	340 mV		Grimaud et al. (2013)
$PrBa_{0.5}Sr_{0.5}Co_{1.5}\ Fe_{0.5}O_{5+\delta}$	358 mV		Zhao et al. (2017)
$PrBa_{0.5}\ Sr_{0.5}Co_{2-x}Fe_xO_{5+\delta}$	300 mV	0.68 V	Bu et al. (2017)
$NdBaMn_2O_{5.5}$	430 mV		Wang et al. (2018a)

as co-catalyst, and the presence of this co-catalyst improves the CO selectivity by 20%. The heterojunction of $TiO_2/SrTiO_3$ reduces the CO_2 into gaseous CH_4 and CO gases, and if the noble metals like Pt and Pd loaded, the efficiency of this heterojunction largely improves (Bi et al. 2015). Similarly, $CaTiO_3$ with Ag coating shows a significant CO_2 reduction with a CO generation rate of 2.25 μmol h^{-1} g^{-1} (Yoshida et al. 2015).

Besides alkaline-earth titanates, alkaline tantalates like $NaTaO_3$ shows superior water splitting behavior with hydrogen rate up to 36,750 μmol h^{-1} g^{-1}, due to

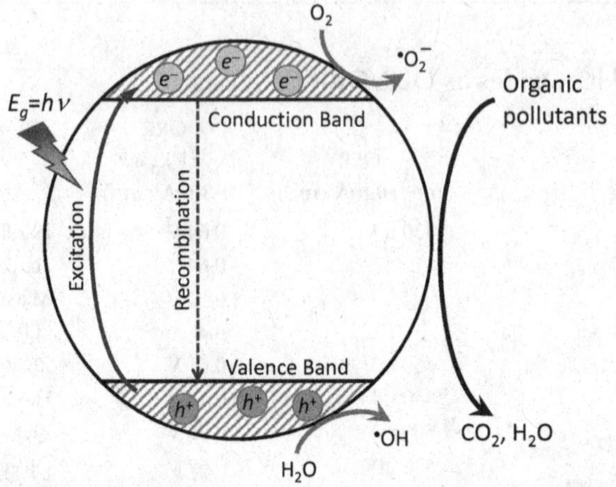

FIGURE 8.8 Mechanism of photocatalytic reaction on perovskite oxide.

the presence of a large bandgap of ~4.0 eV (Liu et al. 2007). In the presence of NiO as co-catalyst, the production rate is decreased drastically, which shows only 19,800 μmol h^{-1} g^{-1} (Kato and Kudo 2003). Along with water splitting, $NaTaO_3$ also displays CO production via reducing CO_2. For instance, under UV illumination Sr-doped $NaTaO_3$ with 2 wt% Ag as co-catalyst shows a CO generation rate of 176 μmol h^{-1} g^{-1} with high selectivity (86%). But with Ca doping, it only shows a generation rate of 148 μmol h^{-1} g^{-1} with 91% selectivity (Nakanishi et al. 2017). $LiTaO_3$ has a larger bandgap (4.7 eV) than $NaTaO_3$, but it displays a lower hydrogen production rate of 430 μmol h^{-1} g^{-1} (Kato and Kudo 2001; Kato et al. 2002). The other perovskite tantalates, such as $KTaO_3$ and $AgTaO_3$, have bandgaps of 3.4 and 3.6 eV, respectively. They display photocatalytic water splitting behavior with the hydrogen generation rate of 138 and 29 μmol h^{-1} g^{-1} (Kato and Kudo 2001; Kato et al. 2002). Perovskite niobates have smaller bandgaps than tantalates and are expected to exhibit more efficient photocatalytic properties. However, niobates failed to deliver the expected photocatalytic behavior; for example, $KNbO_3$ with 3.12 eV is able to generate 5,170 μmol h^{-1} g^{-1} hydrogen gas under UV light irradiation using 0.25% Pt as a catalyst (Ding et al. 2008). Perovskite ferrites also exhibit photocatalytic behavior, and these ferrites usually have a bandgap in the range of 1.8–2.7 eV. The $LaFeO_3$ with bandgap of ~2.34 eV displays superior catalytic activity in photocatalytic water splitting among all ferrites with the hydrogen generation rate of 8,600 μmol h^{-1} g^{-1} by illuminating with visible light (Parida et al. 2010).

Doping is an efficient way to improve the catalytic behavior of perovskites, which narrows the large bandgap of perovskite semiconductors and thus improves the catalytic activity. Several dopants like Zn, Cu, Mn, Cr, Ti, Rh, Ir, Ru, etc. are employed in doping the perovskite oxides (Konta et al. 2004; Zhang et al. 2010a; Yu et al. 2011; Zou et al. 2012; Tan et al. 2014). although no considerable improvement in catalytic activity is observed, especially with Cu and Mn doping in $SrTiO_3$. For example, Ta and Bi doping of $NaTiO_3$ narrows its bandgap to 1.7 and 2.64, respectively

(Li et al. 2009; Wang et al. 2013a). The doped $NaTiO_3$ displays low efficiency compared with the pristine one. Along with cation doping, anion doping is also tried to improve the catalytic efficiency. Nitrogen (N) is used to dope the anion site; for instance, $BaTaO_2N$ is able to harvest the visible light region up to 660 nm. This is because of the reduced bandgap due to the large size of N 2*p* orbitals compared to O 2*p* orbitals (Higashi et al. 2013). Double perovskites offer more flexibility compared to single perovskites, and we can also avoid the undesired defects which are created during doping. Recently, $Ba_2Bi_xNb_{1-x}O_6$ double perovskite displayed improved photocatalytic behavior after the introduction of Bi, which reduces the bandgap. The narrowing of the bandgap is attributed to the introduction of a new CB by the empty 6*s* orbital Bi^{6+} ions (Weng et al. 2017). Perovskite oxides as photocatalysts for hydrogen gas production are listed in Table 8.3.

8.2.2.2 Photocatalytic Dye Degradation

To meet the demand for clean water, recycling of industrial wastewater is mandatory as the carcinogenic organic pollutants limit the use of such water. The textile, pharmaceutical, and agriculture are the dominant source of organic contaminants in water. Semiconducting oxides such as TiO_2, ZnO, and Bi_2O_3 are well-known photocatalysts for organic pollutant degradation. The typical semiconductor photocatalytic activity can be divided into three steps: (1) photogeneration of charge carriers; (2) charge carrier separation and diffusion to photocatalyst surface; and (3) redox reactions on the catalyst surface. A semiconductor photocatalyst possesses a bandgap (E_g) and absorbs energy depending on the bandgap. As the photons with energy equal to E_g are absorbed, electrons (e^-) are excited from the VB to the CB, leaving holes (h^+) behind. These photogenerated charges migrate to the catalyst surface, where they undergo redox reactions with the adsorbed pollutant molecules on the surface. Since the photogenerated e^- and h^+ are powerful reducing and oxidizing agents, h^+ oxidizes the pollutant molecules directly or reacts with hydroxyl anion (OH^-) to generate $^{\bullet}OH$ radicals, which subsequently degrades the pollutant molecules. In a similar way, e^- in the CB of the material reduces the pollutant molecules or the oxygen adsorbed on its surface to generate highly reactive superoxide radicals ($^{\bullet}O_2^-$), which further reacts with the organic pollutants (Kong et al. 2019).

Many perovskite materials are wide bandgap materials with the photon absorption in the UV region, which is only a small part of the visible spectrum; however, they have enough potential to reduce O_2 to $^{\bullet}O_2^-$ and oxidize OH to $^{\bullet}OH$. In order to harvest visible light for photocatalysis, the bandgap of the perovskites needs to be engineered by doping. Incorporation of donor-acceptor ions to the lattice site of the perovskite is a widely appreciated strategy to engineer the E_g value either by elevating the VB energy or by reducing the CB energy. Doping B-site cations with the transition metal of high 4*d* and 5*d* orbital energies can retain the CB minimum energy, but a high electron mobility can be achieved. The recombination of photogenerated charges can be a major problem in such a case. Doping A-site cations can also reduce the bandgap; for instance, Ag doping in $NaSbO_3$ reduces the bandgap by a factor of 0.8 (Wang et al. 2008). Besides A- and B-site doping, simultaneous doping of A- and B-site exhibit a significant impact on E_g. Anion doping is also adopted as a promising strategy for bandgap reduction, and nitrogen is commonly used as a dopant in the

TABLE 8.3
List of Perovskite Oxides as Catalysts for Photocatalytic H_2 Production

Perovskite	Bandgap (eV)	Irradiated Light	Hydrogen Produced (μmolg⁻¹h⁻¹)	Reference
$NaTaO_3$	4.00	UV	2,180	Kato and Kudo (1999, 2003) and Nakanishi et al. (2017)
$NaTaO_3$	3.96	UV	36,750	Liu et al. (2007)
La-doped (A-site) $NaTaO_3$	4.09	UV	19,800	Kato and Kudo (2003)
Ca, Sr, Ba-doped (A-site) $NaTaO_3$	4.00	UV	27,200	Iwase et al. (2009)
$Na_{1-x}K_xTaO_3$	3.75	UV	11,000	Hu et al. (2011)
$LiTaO_3$	4.70	UV	430	Kato and Kudo (2001)
$KTaO_3$	3.60	UV	29	Kato and Kudo (2001)
$AgTaO_3$	3.40	UV	138	Kato et al. (2002)
$KNbO_3$ nanowires	3.20	UV	5,170	Ding et al. (2008)
$CaTiO_3$	3.50	UV	52	Mizoguchi et al. (2002)
$SrTiO_3$	3.20	UV	28	Townsend et al. (2012)
Ti^{3+}-doped $SrTiO_3$	3.20	UV	2,200	Tan et al. (2014)
Zn-doped $SrTiO_3$	3.15	UV	732	Zou et al. (2012)
$PbTiO_3$	2.95	UV	70	Zhen et al. (2014)
$BiFeO_3$ nanowires	2.35	>380 nm	400	Li et al. (2013)
$LaNi_{1-x}Cu_xO_3$ nanoparticles	2.5–2.8	>400 nm	1,180	Li et al. (2010b)
$LaFeO_3$ nanoparticles	2.10	>400 nm	8,600	Parida et al. (2010)
Bi-doped (B-site) $NaTaO_3$	2.88	>400 nm	59.5	Li et al. (2009)
Cu-doped $CaTiO_3$	-	>400 nm	22.7	Kato and Kudo (1999)
$NaTa_{1-x}Cu_xO_3$	2.8–3.4	>415 nm	1,335	Xu et al. (2012b)
$AgNbO_3$	2.80	>420 nm	5.9	Arney et al. (2010)
Ta^{4+} doped (B-site) $NaTaO_3$	1.70	>420 nm	61	Wang et al. (2013a)
Cr-doped $SrTiO_3$ nanoparticles	2.30	>420 nm	330	Yu et al. (2011)
$CaNbO_2N$	2.00	>420 nm	10	Siritanaratkul et al. (2011)
$CaTiO_2N$	2.50	>420 nm	250	Yamasita et al. (2004)
$SrTiO_2N$	2.10	>420 nm	250	Yamasita et al. (2004)
$BaTiO_2N$	2.00	>420 nm	500	Higashi et al. (2009)
$LaTiO_2N$	2.10	>420 nm	3,680	Zhang et al. (2012a)
Mn-doped $SrTiO_3$	2.70	>440 nm	8.9	Konta et al. (2004)

O-site (Kumar et al. 2011b). When the O-site is doped with anions containing 2*p* or 3*p* orbitals with high atomic energies than that of oxygen, it results in the increase of the VB maximum without affecting the CB; as a result, the overall bandgap of the materials is reduced (Wang et al. 2015). Even though doping can significantly

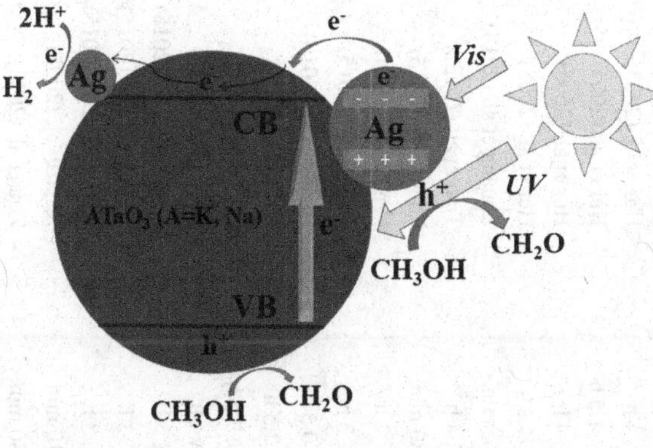

FIGURE 8.9 Promotion effects of Ag particles on the photocatalytic activity of $ATaO_3$ (A = K, Na) under UV–visible light irradiation, reproduced with permission of the American Chemical Society (Xu et al. 2015).

reduce the bandgaps, the fast recombination of electron-hole pairs is a challenge in perovskite materials. The recombination time can be increased by decreasing the diffusion length by reducing the size to nanoscale or by coupling with conducting material like graphene or noble metals like Ag. For instance, the coupling of a small amount of Ag nanoparticles with the $ATaO_3$ (A = Na, K) influences the trapping of photogenerated electrons from the CB of $ATaO_3$. Thus, the addition of Ag nanoparticles promotes the separation of hole pairs and photogenerated electrons, as displayed in Figure 8.9 (Xu et al. 2015). The summary of the photodegradation of organic dyes is tabulated as Table 8.4.

8.2.3 Chemical Catalysis

The catalytic activity of perovskite oxides for the treatment of industrial and automobile originated gaseous effluents, and volatile organic compounds (VOC) are extensively studied. The compounds such as NO, NO_2, CO, hydrocarbons, etc. are considered as the major pollutants released to air from the industry and automobiles. Perovskite structured materials are successfully employed for the decomposition, oxidation, or reduction of the above compounds with activities close to the noble metal catalysts.

Perovskite oxides with oxygen defects and the redox ability of B-site ions determine the performance of the perovskite towards the catalytic decomposition of NO (Zhu et al. 2007). Perovskites like La_2MO_4 (M = Co, Ni, Co, and Cu) and $LaSrMO_4$ (M = Co, Ni, Co, and Cu) are identified as the promising candidate for NO decomposition through a self-redox reaction in which no oxidizing or reducing agent is added to the reaction (Zhu et al. 2007, 2014). The decomposition of NO by the self-redox reaction is the ideal strategy for the removal of NO, but the practical way is by reducing NO in the presence of CO, NH_3, and hydrocarbons. $LaSrMO_4$ (M = Co, Ni, Cu), $LaMnO_3$, $LaFeO_3$ (Giannakas et al. 2004), etc. are suitable catalysts for the reduction of NO in the presence of CO. The catalytic activity of $LaSrMO_4$

TABLE 8.4
List of Perovskite Oxides as Photocatalysts for Dye Degradation

Perovskite	Dyes or Organic Molecules	Bandgap (eV)	Irradiated Light (nm)	Degradation (%) and Time	Reference
$Ba(In_{1/2}Nb_{1/2})O_3$	4-CP	3.3	UV–vis	76%, 4 h	Hur et al. (2005)
$Ba(In_{1/3}\ Pb_{1/3}\ Nb_{1/3})O_3$	4-CP	1.5	UV–vis	~100%, 1.5 h	Hur et al. (2005)
$Ba(In_{1/2}Ta_{1/2})O_3$	4-CP	3.0	UV–vis	90%, 4 h	Hur et al. (2005)
$Ba(In_{1/3}\ Pb_{1/3}\ Ta_{1/3})O_3$	4-CP	1.5	UV–vis	~100%, 2 h	Hur et al. (2005)
$CaTiO_3$	As(III)	3.6	<400	99%,10 min	Zhuang et al. (2014)
Bi_2MoO_6	ARS	2.6	> 420	55%, 4 h	Tian et al. (2013a)
Bi_2WO_6	BPA	2.9	Sunlight	70%, 80 min	Yang et al. (2015)
$Bi_{1-x}Ba_xFeO_3$	CR	2.2	≥400	38%, 2 h	Feng et al. (2014)
Bi_2WO_6	DCP	2.8	>420	32%, 3 h	Chen et al. (2015)
$K_2La_2Ti_3O_{10-x}N_x$	ER	3.4	>400	38%, 0.5 h	Huang et al. (2010)
$BiFeO_3$	MB	2.1	≥420	20%, 2.5 h	Yang et al. (2014)
$BiFeO_3$	MO	2.2	≥400	~100%, 7.5 h	Gao et al. (2014)
$BiFeO_3$	RhB	2.1	>420	42%, 4 h	Chen et al. (2015)
$CaTiO_3$	MB	3.2	>320	30%, 3 h	Alammar et al. (2015)
$SrTiO_3$	MB	3.1	>320	45%, 3 h	Alammar et al. (2015)
$BaTiO_3$	MB	3.0	>320	45%, 3 h	Alammar et al. (2015)
$NaBiO_3$	MB	2.6	≥420	99%, 10 min	Kako et al. (2007)
$SrTiO_{3-x}N_x$	MB	2.9	>420	85%, 80 min	Zou et al. (2012)
$La_{1-x}Ca_xFeO_3$	MB	2.0	≥400	86%, 1.5 h	Li et al. (2010a)
Bi_2MoO_6	MB	2.8	>420	85%, 2 h	Zhang et al. (2010b)
Bi_2MoO_6	MB	2.6	≥410	50%, 0.5 h	Dai et al. (2015)
Bi_2MoO_6	MB	2.6	Sunlight	82%, 4 h	Zhao et al. (2015)

(Continued)

TABLE 8.4 (*Continued*)
List of Perovskite Oxides as Photocatalysts for Dye Degradation

Perovskite	Dyes or Organic Molecules	Bandgap (eV)	Irradiated Light (nm)	Degradation (%) and Time	Reference
Bi_2WO_6	MB	2.7	>420	30%, 2.5 h	Fan et al. (2015)
$AgNb_7O_{18}$	MB	2.8	>420	95%, 4 h	Liu et al. (2017)
$LaNiO_3$	MO	2.2	>400	80%, 5 h	Li et al. (2010c)
Bi_2WO_6	MO	2.8	≥420	60%, 40 min	Tian et al. (2013b)
$La_{1.5}Ln_{0.5}Ti_2O_7$ (Ln = Pr, Gd, Er)	MO	3.4	<400	96%, 1.5 h	Li et al. (2006)
$Bi_4Ti_3O_{12}$	MO	3.1	360	~100%, 2 h	Yao et al. (2003)
$CeCo_xTi_{1-x}O_{3+\delta}$	NB	1.6	UV–vis	91%, 3 h	Fei et al. (2005)
Bi_2WO_6	Phenol	2.8	>400	90%, 40 min	Li et al. (2015b)
Bi_2WO_6	Phenol	2.8	>420	12%, 2 h	Zhang et al. (2012d)
Bi_2MoO_6	Phenol	2.6	>420	99%, 1.5 h	Yin et al. (2010)
$Bi_4Ti_3O_{12}$	Phenol	3.2	UV	99%, 2 h	Tu et al. (2017)
$ZnTiO_3$	RB	3.1	>420	18%, 4 h	Zhang et al. (2012b)
$BiFeO_3$	RhB	2.5	≥400	18%, 7 h	Zhang et al. (2015b)
$BiFeO_3$	RhB	2.2	Sunlight	55.1%, 7 h	Di et al. (2014)
$BaTiO_3$	RhB	3.0	Sunlight	15%, 45 min	Cui et al. (2013)
$NaNb_{0.5}Ta_{0.5}O_3$	RhB	3.3	UV	99%, 1h	Wang et al. (2017b)
Bi_2WO_6	RhB	2.9	>420	80%, 2 h	Zhang et al. (2013a)
Bi_2WO_6	RhB	2.6	>420	58%, 5 h	Xu et al. (2012a)
Bi_2WO_6	RhB	2.9	≥420	65%, 3 h	Li et al. (2012)
Bi_2WO_6	RhB	2.8	>400	96%, 40 min	Li et al. (2015a)
Bi_2WO_6 nanocrystals	RhB	2.9	≥420	13%, 30 min	Zhou et al. (2015)

(*Continued*)

TABLE 8.4 (*Continued*)
List of Perovskite Oxides as Photocatalysts for Dye Degradation

Perovskite	Dyes or Organic Molecules	Bandgap (eV)	Irradiated Light (nm)	Degradation (%) and Time	Reference
Bi_2WO_6 monolayers	RhB	2.7	≥420	98%, 25 min	Zhou et al. (2015)
Bi_2WO_6 microsphere	RhB	2.7	>400	99%, 35 min	Li et al. (2015a)
$Bi_{2-x}Cd_xWO_6$	RhB	2.6	Sunlight	99%, 40 min	Song et al. (2015)
Bi_2MoO_6	RhB	2.7	>420	95%, 2 h	Tian et al. (2010)
Bi_2MoO_6	RhB	2.6	≥420	~100%, 0.5 h	Ma et al. (2015)
Er^{3+}/Bi_2MoO_6	RhB	2.6	≥420	~100, 50 min	Zhou et al. (2011)
$Bi_4Ti_3O_{12}$	RhB	2.9	375~475	90%, 1.5 h	Chen et al. (2016)
$Bi_4Ti_3O_{12}$	RhB	2.9	>420	92%, 1 h	He et al. (2014)
$CaCu_3Ti_4O_{12}$	TC	2.0	≥420	99%, 30 min	Hailili et al. (2018)

Notes: MO: Methyl orange; MB: Methylene blue; RhB: Rhodamine B; NB: Nile blue; CR: Congo red; As(III): Arsenite; ARS: Alizarin red; 4-CP: 4-chlorophenol; ER: Erionyl red; BPA: Bisphenol A; DCP: 2,4-dichlorophenol; TC: Tetracycline.

(M = Co, Ni, Cu, Fe and Mn) materials is in the order $LaSrCoO_4$ > $LaSrNiO_4$ > $LaSrCuO_4$ > $LaSrFeO_4$ > $LaSrMnO_4$ (Zhong and Zeng 2006).

The catalytic reduction of NO or NO_2 in the presence of NH_3 is possible using either perovskite structured ABO_3 or perovskite-like structured A_2BO_4 (A = La, B = Cu, Co, Mn, and Fe) materials. Among the above materials, $LaMnO_3$ exhibits an NO conversion efficiency of 78% at 250°C (Zhang et al. 2013b). In a separate study, it is found that $BiMnO_3$ is more efficient than $LaMnO_3$ at a lower temperature (100°C–180°C), with an NO conversion efficiency of 85% at 100°C (Zhang et al. 2012c). Since NO_2 is the intermediate of NO decomposition and reduction reactions, several attempts are made to oxidize NO to NO_2 using perovskite structured materials. $LaCoO_3$ is identified as a promising material for such application, and the activity is improved remarkably by doping 10% Sr, as in $La_{0.9}Sr_{0.1}CoO_3$, with an oxidation efficiency of conversion of ca. 86% at 300°C (Kim et al. 2010).

N_2O is the major greenhouse exhaust gas with the highest contribution to ozone layer depletion. Perovskite structured $LaBO_3$ (B = Cr, Mn, Fe, and Co) is extensively studied for N_2O decomposition, and the studies reveal that $LaCoO_3$ exhibits the best characteristics for N_2O decomposition, with 50% conversion of N_2O at 455 and 490°C in the absence and presence of 5% oxygen, respectively (Russo et al. 2007). The catalytic properties of $LaCoO_3$ are surface area dependent (Dacquin et al. 2009), and the surface reconstructions on La- and Co-deficient perovskites induce surface La or Co enrichment and a change in oxygen mobility (Wu et al. 2012). The substitution in the A-site cations also improves the performance of the catalyst. When doped with Sr or Ba, the activity of the above materials is tremendously improved (Kumar et al. 2011a). The observed activity is in the following order: $La_{0.8}Sr_{0.2}CoO_{3-\delta}$ > $La_{0.8}Sr_{0.2}FeO_{3-\delta}$ > $La_{0.8}Sr_{0.2}MnO_{3-\delta}$ > $La_{0.8}Sr_{0.2}CrO_{3-\delta}$ (Gunasekaran et al. 1995).

CO and CH_4 are also the major components of exhaust effluents from a high-temperature internal combustion engine. The oxidation of CO to CO_2 by a perovskite material is a combined action of the perovskite material and CO. The oxygen vacancy in the perovskite oxides are active sites for splitting molecular oxygen into atomic oxygen, and CO is a good reducing agent that can react directly with atomic oxygen. Therefore, materials with more oxygen vacancy are more likely to exhibit a higher reaction rate. The most widely accepted strategy for increasing the oxygen vacancy is by substituting the high-valence A-site cation with a low-valence ion such as Sr, Ba, and Ag.

Doping of Sr^{2+} in place of La^{3+} in La-based transition metal perovskites increases the number of oxygen vacancies tremendously, as in the case of $La_{2-x}Sr_xMO_4$ (M = Cu, Ni) (Zhu et al. 2005). One can also create oxygen vacancies by changing the stoichiometric ratio of A- and B-site ions to improve the catalytic activity as in $LaCuO_{3-\delta}$ (Falcón et al. 2000). The redox properties of the transition metals in the B-site play an important role in CO oxidation. For instance, in the absence and presence of Fe in the B-site of $LaAlO_3$, the catalytic activity changes significantly (Ciambelli et al. 2002). Additionally, among the series of $LaCo_{0.5}M_{0.5}O_3$ (M = Mn, Cr, Fe, Ni, Cu) catalysts, $LaCo_{0.5}Mn_{0.5}O_3$ exhibits the best activity as a result of the high redox potential of cobalt (Yan et al. 2013).

In the case of CH_4 oxidation, the catalytic reaction is influenced by the redox properties of the B-site elements than the oxygen vacancies. The lanthanide perovskites with the general formula $LaMO_3$ (M = Mn, Co, Fe, Ni) are excellent catalysts for methane combustion and the M-cation at B-site has a major role in catalytic activity. Like oxidation of NO and CO, $La_{1-x}A_xBO_{3\pm\delta}$, and $La_{1-x}A_xB_{1-y}B_yO_{3\pm\delta}$, perovskites substituted with other ions with a mixed valence state exhibit superior performance. The methane oxidation by perovskite catalysts to CO_2 and H_2O has two underlying reactions: one is suprafacial, employing adsorbed surface oxygen from the gas phase active at low temperatures, and another one is interfacial, employing oxygen from the perovskite lattice apparent at high temperatures (Ladavos and Pomonis 2015).

Catalytic combustion of VOCs using perovskite oxides is another important application of perovskite material. VOCs originated from industrial processes, power generation, automobiles, and solvent assisted processes. The conventional thermal incineration needs a high operating temperature (800°C–1,200°C), which in turn produces other harmful chemical species, and they often involve high operating costs. Like other catalysts La-based perovskites $LaMnO_3$ (Blasin-Aubé et al. 2003), $LaFeO_3$ (Barbero et al. 2006), and $LaCoO_3$ (Spinicci et al. 2003) are studied extensively in the complete oxidation of VOCs at a low temperature <600°C. Perovskite structure with multivalent B-site elements is more favorable for the oxidation VOCs even at a low concentration. For instance, $LaMnO_3$ with multivalence Mn^{4+} and Mn^{3+} in the B-site has more surface oxygen species and better activity than the $LaCoO_3$ with single valence Co^{3+} ions (Spinicci et al. 2003). Doping the trivalent A-site with divalent ions can induce multivalence in B-site, as in the case of $LaFeO_3$. La^{3+} substituted with Ca^{2+} in $LaFeO_3$ resulted in Fe ions with dual oxidation state Fe^{3+} to Fe^{4+} in its lattice (Barbero et al. 2006), resulting in a better activity in the oxidation of propane and ethanol (Figure 8.10).

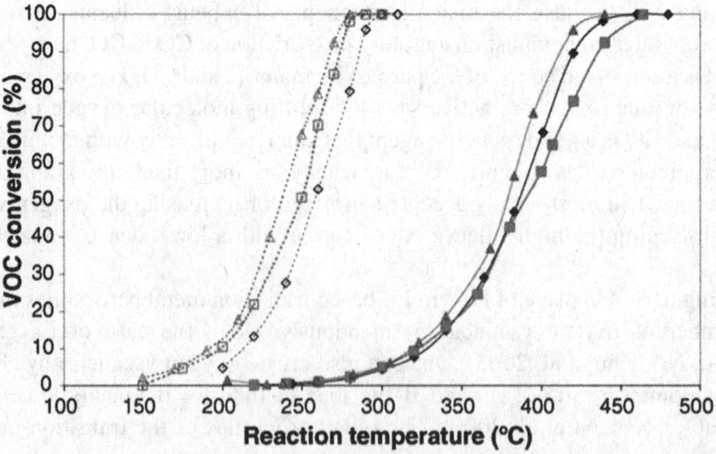

FIGURE 8.10 (See online for color version.) Catalytic activity in total oxidation of propane (solid line, closed symbols) and ethanol (dotted line, open symbols) on (black) $LaFeO_3$, (pink) $La_{0.8}Ca_{0.2}FeO_3$, and (green) $La_{0.6}Ca_{0.4}FeO_3$, reproduced with permission of the Elsevier (Barbero et al. 2006).

8.2.3.1 Reforming of Hydrocarbons

Hydrogen is assumed as the next-generation fuel to replace fossil fuels; however, the energy density of hydrogen with respect to volume is smaller than many fossil fuels; as a result, the storage of hydrogen is a major concern. In addition to the low density, low melting point and the danger of explosion when react with oxygen are the major drawbacks to store hydrogen as a fuel in the stationary and mobile applications. Therefore, partial oxidation of hydrocarbons in the air is an alternate method to generate hydrogen when needed. Methane is a fossil fuel with the highest percentage of hydrogen in it. The complete oxidation of CH_4 to CO_2 and H_2O is possible using perovskite oxides as a catalyst, but the partial oxidation of methane to syngas (with an H_2/CO ratio of about two) in the air is also possible using perovskite oxides. Perovskite oxides can oxidize CH_4 directly into CO and H_2 using its lattice oxygen as oxygen resource (in the feed gas, only CH_4 was present) (Khine et al. 2013). La-based perovskites with the formula $LaMO_3$ (M = Co, Cr, Ni, Rh, Fe) (Slagtern and Olsbye 1994), $AFeO_3$ (A = La, Nd, Eu) (Dai et al. 2006), or with mixed A and B sites (de Santos et al. 2018) elements are extensively studied for such an application.

8.2.3.2 Catalyst Supports

Perovskite oxides are also identified as catalyst supports for precious catalysts. The performance of perovskite structured materials as support is better than other oxides, especially at high temperatures. In such a study, Pt catalysts supported on γ-Al_2O_3 is compared with $LaFeO_3$ for simultaneous reduction of N_2O and NO at a high temperature. The study reveals that $LaFeO_3$ exhibits better resistance to thermal sintering of Pt particles than γ-Al_2O_3. The Pt particles in γ-Al_2O_3 are grown significantly, leading to a negative effect on the catalysis. Thus the stabilizing of Pt catalyst by perovskite can minimize the metal loadings and improve the catalytic performance (Dacquin et al. 2010). There are attempts to embed the noble metals in the perovskite lattice as well. Incorporating precious metal Pd at the B-site ($BaCe_{1-x}Pd_xO_{3-\delta}$) exhibits a high activity comparable to commercial PdO/Al_2O_3 catalyst for CO oxidation (Singh et al. 2007). $LaTi_{0.5}Mg_{0.5-x}Pd_xO_3$ (Petrović et al. 2005), $LaFe_{0.95}Pd_{0.05}O_3$ (Uenishi et al. 2005), $CaTi_{0.95}Pt_{0.05}O_3$, and $LaFe_{0.95}Rh_{0.05}O_3$ (Tanaka et al. 2006) are some notable examples of noble metal incorporated perovskite structured catalyst for automobile exhaust treatment.

8.3 ENERGY STORAGE AND CONVERSION

8.3.1 Batteries

8.3.1.1 Lithium-Ion Batteries

Lithium-ion batteries (LIBs) are portable power sources for large-scale production such as electronic devices, power tools, back-up power units, and automobiles, where high-power and high-energy-density are the primary focus. In LIBs, the charge/discharge process commenced by the movement of lithium ions from the cathode to anode and the subsequent intercalation of ions in the electrodes and *vice versa*. The anode and cathode are separated by an ion conductive membrane called a separator, and an ion conductive electrolyte is used for the smooth progression of ions during

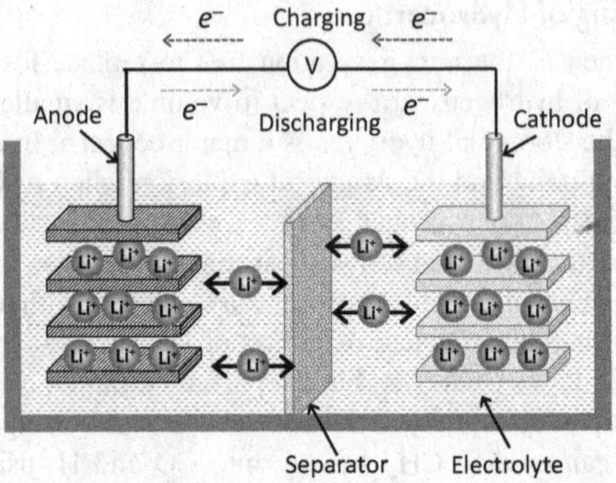

FIGURE 8.11 Schematic of the components in LIB.

the charge-discharge process. The construction of an LIB is shown in Figure 8.11. The performance of LIBs is determined by the diffusivity of Li ions, ionic mobility, conductivity, and specific capacity of electrodes. Recently perovskite-based electrode materials are identified as better electrodes in LIBs, which can deliver better performance through improved cyclic stability, specific capacity, specific energy, and charge/discharge rate. The perovskite structured oxide materials are reported as a potential anode, cathode, and electrolyte in LIBs.

8.3.1.1.1 Anode

The primary characteristics of anodes for LIBs are low molecular weights, low density, the ability to accommodate large amounts of Li ions, cyclic stability, insolubility and stability in electrolytes, and reversible gravimetric and volumetric capacities. The anode materials must also possess potential as close to that of Li metal, good electronic and ionic conductivity, cheap, and environmentally friendly. The graphite meets all the above requirements as a unique anode material and is used in commercial LIBs. However, better Li-ion cycling is possible by redefining the intercalation–deintercalation mechanism. Materials with two-dimensional (2D) layered structure, and 3D network structure can intercalate and deintercalate Li-ions reversibly into their lattice, without altering the crystal structure.

The Perovskite-type oxide materials can enhance Li-ion interaction in different ways. The perovskite structured materials exhibit a 3D-network-type crystal structure with a tailorable cation vacancy, and these cation vacancies can act as potential active sites for the occupancy of incoming Li-ions. For example, $LaNiO_3$ perovskite oxide, which is prepared using a sol–gel method, exhibit superior rate capability with long stability. In a C-rate test, it shows the specific capacity of 77 mAh g^{-1} at 6 C rate and exhibits high retention specific capacity of 92 mAh g^{-1} after 200 cycles at 1 C rate (Zhang et al. 2020). Another way of improving the efficiency of perovskite oxide is doping; for instance, Li doping of $CaMnO_3$ enhances the electrochemical performance. The perovskite with the composition of $Ca_{0.95}Li_{0.05}MnO_{3-\delta}$ is found to be superior over pristine and other compositions. This is because of enhanced lithium-ion diffusion due to the formation of oxygen vacancies (Fang et al. 2018). The efficiency of perovskite electrodes further improved by making composites with conductive carbon frameworks ($CoTiO_3$/graphene) and synthesizing nanoforms ($CeMnO_3$ nanofibers). $CoTiO_3$/graphene composite prepared by ball milling and solid-state method displays enhanced cyclic performance compared to pristine $CoTiO_3$, which is assigned to the improved conductive behavior due to graphene (Li et al. 2017b). The $CeMnO_3$ nanofibers synthesized utilizing the electrospinning technique exhibit enhanced electrochemical performance in terms of specific capacity (2,159 mAh g^{-1}) and Coulombic efficiency (93.79%) with high cyclic stability at 1,000 mAh g^{-1} (Yue et al. 2019).

8.3.1.1.2 Electrolytes

The function of electrolytes in LIBs is to transport the Li-ions between the cathode and anode during the charge/discharge process. Salts of Li in an organic solvent are commonly used as the electrolyte. But in all-solid-state LIBs, perovskite oxide-based solid electrolytes are utilized as electrolytes. The most used perovskite electrolyte is lithium lanthanum titanate with the composition $Li_{3x}La_{(2/3)-x}TiO_3$ (LLTO); the x ranges from 0.07 to 0.13. The high ionic conductivity of LLTO is attributed to the A-site vacancy (Zhao et al. 2019). Recently, researchers reported another perovskite $Li_{0.38}Sr_{0.44}Ta_{0.7}Hf_{0.3}O_{2.95}F_{0.05}$ as an electrolyte for the all-solid-state LIBs. Unlike other solid electrolytes, this perovskite is insensitive to moisture, which makes it a more efficient Li^+ ion conductor compared with other solid electrolytes. This perovskite has a conductivity value of 4.8×10^{-4} S cm^{-1}, and it won't react with water having a pH in between 3 and 14 (Li et al. 2018).

8.3.1.2 Metal-Air Batteries

A metal-air battery uses an anode made of pure metal, the cathode of ambient air, and an electrolyte. The simultaneous reduction of ambient air at the cathode and the oxidation of metal at anode results in the generation of electricity. The metal-air batteries are projected candidates for electric vehicles owing to their high specific capacity and energy density than LIBs. However, the best performance is encountered by the selection of electrode catalysts and electrolytes to improve the capacity, energy efficiency, rechargeability, and cycle life. Due to the tunable composition and properties of perovskite-type materials, they are extensively studied as cathode materials for metal-air batteries to achieve superior battery performance.

Based on the anode material used, the primary metal-air batteries are classified as Li-air, Zn-air, Mg-air, Na-air, and Al-air batteries. The metallic elements such as Zn, Al, Mg, and Na are abundant in the earth, and therefore they are identified as potential anode materials. The schematic of a typical metal-air battery is shown in Figure 8.12. The performance of a metal-air battery majorly depends on the transport and electrochemical properties of cathode materials. The ideal cathode material for a metal-air battery should exhibit good conductivity, high structural stability allowing the deposition of discharge products, excellent permeability to oxygen and electrolyte, and excellent ORR and OER catalytic activity. The air-breathing

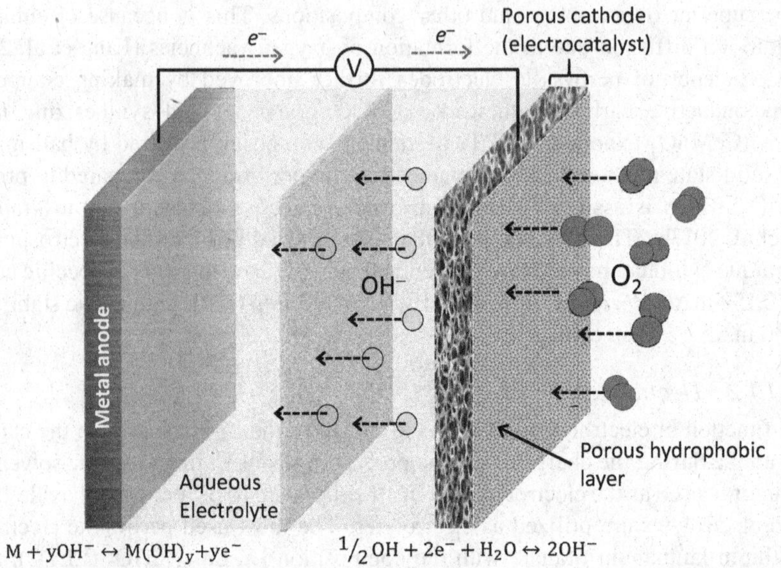

FIGURE 8.12 Construction of a metal-air battery.

hydrophobic membrane must be able to separate oxygen from other components in the air, such as CO_2, H_2O, and other gaseous components. A rechargeable metal-air battery requires an effective OER catalyst to lower the overpotential and increase the energy efficiency (Li et al. 2017a).

Perovskite structured oxides with the transition metal ions in the B-sites are (e.g., Co, Mn, Fe, etc.) reported for excellent ORR catalytic performance in both acid and alkaline media as the transition metal cations can adopt different valence states, which makes them suitable as cathodes in metal-air batteries. The substitution of A-site elements enhances the ability to absorb oxygen, and B-site substitution influences the activity of adsorbed oxygen. Consequently, the properties pertaining to cathode materials can be achieved by partial substitution of A- and B-site elements in the perovskite lattice without disturbing the crystal structure. In general, the transition metal containing perovskite oxides exhibit a well-defined oxygen transport; however, the ORR activity of perovskites is associated to the σ^*-antibonding (e_g) orbital filling of surface transition metal cations and the metal-oxygen covalent bonding. The hypothetical increase in the ORR activity in perovskites is related to the oxygen adsorption (B–O covalency) to form $B–O_2$ competes with the adsorption of hydroxide on the B-site. After three electrons are passed and three hydroxides produced, the B–OH site must be regenerated with water and another electron to continue the cycle. An optimum ORR activity is observed when e_g is near unity, as observed in the case of $LaMnO_3$ and $LaNiO_3$. Moreover, tuning the surface electronic features of the materials can lead to high-performance ORR catalysts, as in the case of nanomaterials. The perovskite nanomaterials, which are used Li-air battery, are summarized in Table 8.5.

TABLE 8.5
List of Perovskite Oxides as Electrodes in Li-Air Battery

Perovskite Oxide	Morphology	Capacity ($mAh\ g^{-1}$)	Condition of Discharge ($mAh\ g^{-1}$)	Number of Cycles	Reference
$LaNiO_3$	Nanoparticles	7,076	500	155	Xu et al. (2016a)
$LaNiO_3$	Nanocube	3,407	1,000–500	23–75	Zhang et al. (2014)
$La_{0.5}\ Sr_{0.5}CoO_{3-\delta}$	Nanotube	5,799	5,799–4,391	5	Liu et al. (2015)
$La_{0.5}\ Sr_{0.5}CoO_{2.91}$	Mixture of nanowire and nanorod	11,059	-	-	Zhao et al. (2012)
$La_{0.5}\ Sr_{0.5}CoO_{2.91}$	Nanotube	7,205	1,000	85	Li et al. (2015c)
$La_{0.75}\ Sr_{0.25}MnO_{3-\delta}$	Nanotube	11,000	11,000–9,000	5	Xu et al. (2013)
$La_{0.8}Sr_{0.2}MnO_3$	Nanoparticles	1,438	-	-	Fu et al. (2012)
$La_{0.8}Sr_{0.2}MnO_3$	Nanorod	8,890	7,360–4,160	5	Lu et al. (2015)
$LaCo_{1-x}Ni_xO_3$ ($x = 0$–1)	Nanoparticles	3,750	1,000	49	Kalubarme et al. (2014)
$LaNi_{1-x}Mg_xO_3$ ($x = 0, 0.08, 0.15$)	Nanoparticles	300	-	-	Du et al. (2014)
$La_{0.6}Sr_{0.4}Co_{1-x}Ni_xO_3$ ($x = 0, 0.05, 0.1$)	Nanoparticles	1,812	500	17	Sun et al. (2016b)
$La_{0.6}Sr_{0.4}Co_{1-x}Mn_xO_3$ ($x = 0, 0.05, 0.1$)	Nanoparticles	2,015	500–370	22	Sun et al. (2016a)
$La_{0.8}Sr_{0.2}Mn_{1-x}Ni_xO_3$ ($x = 0, 0.2, 0.4$)	Nanoparticles	4,408	500	54	Wang et al. (2016c)
$La_{0.6}Sr_{0.4}Co_{0.2}Fe_{0.8}O_3$	Nanoparticles	20,039	500	17	Cheng et al. (2016)
$Ba_{0.9}Co_{0.5}Fe_{0.4}Nb_{0.1}O_{3-\delta}$	Nanoparticles	1,235	1,235–580	3	Jin et al. (2014)
$Ba_{0.9}Co_{0.7}Fe_{0.2}Nb_{0.1}O_{3-\delta}$	Nanoparticles	11,978	1,000	24	Xu et al. (2016b)

8.3.2 Supercapacitors

Supercapacitors provide a reversible energy storage solution for miniaturized devices with high power density, high cyclic stability, and fast storage at low costs. The supercapacitor is anticipated for high volumetric and gravimetric energy density requirements of energy storage systems. Supercapacitors exhibit high power and energy density capabilities and long cycle life with rapid charge-discharge capabilities, which is higher than batteries (George et al. 2018). The power and energy characteristics of supercapacitors fall between electrolytic capacitors and batteries with a high-power density than batteries but lower than capacitors, at the same time lower energy densities than batteries but higher than conventional capacitors. Therefore, the supercapacitors intended to use in the applications that require power surges rather than a high energy storage capacity. Moreover, a significant reduction in the size of energy storage systems is possible by combining supercapacitors with batteries while retaining the power and energy requirements with an increased lifetime. The construction of a supercapacitor is shown in Figure 8.13.

The charge storage mechanism in the supercapacitors is a fast surface dependent process that occurs on the electrode surface, unlike electric double-layer capacitors (EDLCs), where the charge storage mechanism relies on fast redox reactions. The charge storage mechanism in perovskite materials is different from the transition metal oxides, and the performance of the supercapacitors based on perovskites can be potentially improved by controlling the ratio of metallic ions used during the synthesis. Additionally, the perovskite structure provides intrinsic 3D diffusion channels formed by intersecting tetragonal cavity chains for ionic transport.

Several lanthanum-based perovskites are investigated for their performance as anode material for supercapacitors. $LaMnO_3$, $LaFeO_3$, $LaCoO_3$, and $LaNiO_3$ perovskites with different morphology and composition are widely investigated as anode materials for supercapacitor. They are extensively studied as the most potential candidate as the supercapacitor electrode based on charge-storage mechanisms resulting from oxygen anion intercalation and deintercalation. The high specific pseudo-capacitances of these perovskite materials are mainly attributed to anion

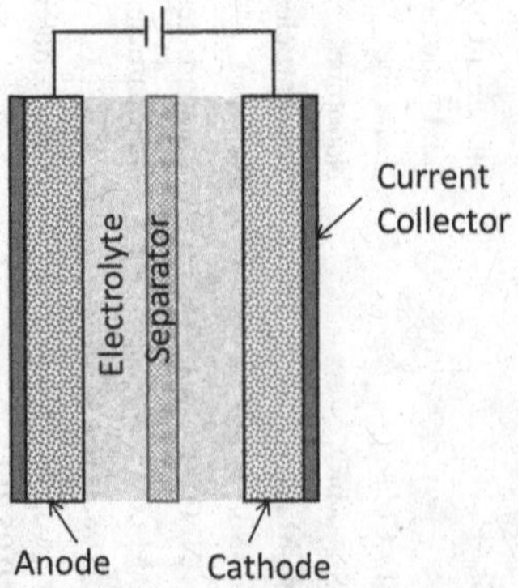

FIGURE 8.13 The structure of a supercapacitor.

intercalation, which can be improved by increasing the number of oxygen vacancies, as the oxygen vacancies act as the diffusion pathways of the ions (Nemudry et al. 2002) and are considered as charge storage sites in perovskites. The electrochemical oxygen intercalation was first identified by Kudo et al. in $Nd_{1-x}Sr_xCoO_3$ perovskites (Kudo et al. 1975) in the alkaline medium. Nanostructured $LaMnO_{3\pm\delta}$ demonstrated anion-based pseudo-capacitance with a specific capacitance about 610 F g^{-1} at 2 mV s^{-1}. The specific capacity of the $LaMnO_3$ perovskite is higher in the basic medium than the acidic counterparts. Therefore, the conventional charge storage in metal oxide pseudo-capacitors through either protonation and deprotonation of the surface oxygen or intercalation of alkali cations is less likely, and the mechanism of charge storage is dominated by anion intercalation. Whenever there is an oxygen vacancy in a perovskite structure (say $LaMnO_{3-\delta}$), some of the Mn^{3+} states change to Mn^{2+} state ($La[Mn^{3+}Mn^{2+}]O_{3-\delta}$). Apparently, the oxygen vacancy site exhibits an affinity towards electrolyte anions as it is readily available. Thus, the vacancy will be occupied by anionic intercalation followed by a change in oxidation states of Mn atom ($Mn^{2+} \rightarrow Mn^{3+}$), which enhances the pseudo-capacitance of the perovskite material (Elsiddig et al. 2017). To intercalate large O^{2-} ions to the dense perovskite lattice, the resistance to insertion should overcome (Mefford et al. 2014). Therefore, the adsorbed hydroxide ion is assumed to transfer its proton to a neighboring lattice oxide to form peroxide-type species, O^- (Mefford et al. 2014; Grenier et al. 1996). Since the oxidation of the adsorbed oxygen ion is reliant on the covalency of the oxygen-transition metal bond in perovskites and the degree of hybridization between the transition metal 3*d* and oxygen 2*p* band (Mefford et al. 2014; Abbate et al. 1992). The mechanism of oxygen intercalation for a given transition metal B can be represented as (Alexander et al. 2019)

$$La\left[B_{2\delta}^{z+};B_{2\delta}^{(Z+1)+}\right]O_{3-\delta}+2\delta e^{-}+\delta H_2O\leftrightarrow La\left[B_{2\delta}^{z+};B_{(1-2\delta)}^{(Z+1)+}\right]O_{3-\delta}+2\delta OH^{-} \quad (8.1)$$

Even though $LaMnO_3$ is attractive due to its rapid anion-intercalation-type charge storage, $LaMnO_3$ exhibits low conductivity and short cycle stability like other oxides (Wang et al. 2016b). Also, aggregation in the nanostructured $LaMnO_3$ can be a major concern as the agglomerations can interfere with the ionic intercalation. In addition, the independence of size on the supercapacitance of anion-intercalation-type perovskite structured materials is a matter of debate. Shafi et al (2018b) compared the supercapacitance of $LaMnO_3$ nanoparticles in the agglomerated state and the same without any agglomeration. The well-dispersed nanoparticles exhibit a threefold increase in the supercapacitance (173–520 F g^{-1}), at the same time energy and power densities (Shafi et al. 2018b) could be attributed to the easier transport of ions through the electrode-electrolyte interface of highly dispersed nanoparticles with improved surface area to volume ratio. Additionally, the performance of the highly porous $LaMnO_3$ perovskite as a supercapacitance electrode is superior to the spherical nanoparticles. Mesopores and micropores structures can act as channels for electrolyte diffusion and, therefore, an enhanced electrochemical capacitance through fast charge transfer. Besides the remarkable capacity retention after 110,000 cycles, the power and energy density of the device are outstanding (Shafi et al. 2018a).

Tailoring the oxygen vacancies allows the superior performance of the oxides in the intended applications. In the case of $LaMnO_3$, oxygen vacancies can be modulated by controlling the Mn^{4+} ions, and the obtained non-stoichiometric $LaMnO_3$ perovskite is an excellent candidate for high-performance supercapacitor electrodes. $LaMn_{1\pm x}O_3$ perovskite structure obtained by the sol–gel process is porous in nature, and the $LaMn_{1.1}O_3$ sample showed a much higher specific capacity higher than stoichiometric $LaMnO_3$ perovskite. This enhancement is obvious by the increased oxygen and cationic vacancies, and a high Mn^{4+}/Mn^{3+} ratio (Elsiddig et al. 2017). It is worthwhile here to mention that room temperature conductivity of La-deficient $La_{0.9}MnO_3$ perovskite increases nearly ten times compared with that of the parent $LaMnO_3$ (Xie et al. 2017). This oxygen content-driven enhancement could also help to improve the supercapacitance of the pure $LaMnO_3$ system.

The partial substitution of the A-site in $LaMnO_3$ perovskite systems with lower valence cation (Sr^{2+}, Ca^{2+}, Ba^{2+}, etc. or simultaneous doping of A and B sites) is another approach to increase the oxygen vacancies and the associated electronic conductivity (Najjar and Batis 2016) and the electrochemical properties. The divalent ions in the A-site can generate holes in the B-site; as a result, Mn^{3+} can be oxidized into Mn^{4+} to maintain a neutral charge, which ultimately results in an increased supercapacitance. Doping 15 mol.% Sr^{2+} ions in the La^{3+} lattice site decreases the internal resistance to 1/3 times of that for $LaMnO_3$ and a higher specific capacitance than pure $LaMnO_3$ (Wang et al. 2016b). The performance of $La_{1-x}Sr_xMnO_3$ electrodes depends on the concentration of Sr ions. With an increase in Sr doping, the conductivity as well as the supercapacitance increases due to the increased oxygen vacancies with Sr-doping until a certain critical level of doping. The pseudo-capacitance exhibited by $La_{1-x}Sr_xMnO_3$ is due to the adsorption of OH^- ions of electrolyte to the active electrode surface and the subsequent generation of O^{2-} and H_2O with the oxidation of local Mn^{2+} ions to Mn^{3+} and then to Mn^{4+} ions. Conversely, after a number of charge and discharge cycles, the perovskite structure is collapsed due to the leaching of strontium and manganese ions into the electrolyte, which deteriorates the supercapacitance of the electrode (Lang et al. 2017). It is important to note that there is a significant discrepancy in the reported values of supercapacitance of the same materials prepared through different routes.

Like Sr doping, Ca-doped perovskite lanthanum manganites ($La_{1-x}Ca_xMnO_3$, LCMs) are also studied as electrode materials of supercapacitors. Doping 50% Ca is the optimum level for achieving the highest specific capacitance and low intrinsic resistance. The charge storage through oxygen intercalation improved by Ca^{2+} doping; unfortunately, severe leaching of elements to the electrolyte leads to poor cycling stability, limiting their applications as electrode materials for supercapacitors (Mo et al. 2018).

The simultaneous co-doping at A- and B-sites of $LaMnO_3$ perovskites are also considered to be a promising way of improving the supercapacitance. Sr- and Co-doped $La_xSr_{1-x}Co_{0.1}Mn_{0.9}O_{3-\delta}$ perovskite prepared by the electrospinning method exhibit a maximum capacitance of 485 F g^{-1} at a current density of 1 A g^{-1}, with a long cycle stability. The redox reactions of Mn^{3+}/Mn^{4+} and Co^{3+}/Co^{4+} ions contribute to the charge storage (Cao et al. 2015a). When doped with Sr and Cu in A and B sites, respectively, it can also create oxygen vacancies, resulting in the improved capacitance from 100 to 464.5 F g^{-1} at 2 A g^{-1} (Cao et al. 2015b). More importantly, this symmetric supercapacitor also shows an excellent cycling life after 2,000 charging and discharging cycles.

$LaNiO_3$ is another lanthanum-based perovskite oxide extensively studied for the supercapacitor electrode. $LaNiO_3$ also displays anion-intercalated pseudo-capacitive behavior. $LaNiO_3$ has a specific capacitance of 478.7 F g^{-1} at 0.1 mV s^{-1} and good cycling stability after 15,000 cycles. The charge storage associated with the intercalation of oxygen ion into oxygen vacancies originated from the variation of valence states of Ni ions in the B-site (Che et al. 2018). The transition metal perovskites with higher covalency and more metallic character reveal the presence of a reduction peak for the peroxide species and the intercalation of oxygen in highly covalent perovskites such as $LaNiO_3$, probably proceeding through a peroxide pathway (Mefford et al. 2014). Two-dimensional $LaNiO_3$ nanosheets with rich pores and a large surface area achieve a specific capacitance of 139.2 mAh g^{-1} at a current density of 1.0 A g^{-1}, a good rate capability, and excellent cycle stability. The $LaNiO_3$// graphene asymmetric supercapacitor device demonstrates a high energy density of 65.8 Wh kg^{-1} at a power density of 1.8 kW kg^{-1} and an outstanding cycling stability performance with 92.4% specific capacitance retention after 10,000 cycles (Wang et al. 2017). $LaNiO_3$, with a hollow spherical structure, shows a high specific capacitance of 422 F g^{-1} at a current density of 1 A g^{-1} with good cyclic stability, due to the high ionic and electronic transportation (Shao et al. 2017). In all the studies, $LaNiO_3$-based supercapacitors display excellent cyclic stability for a prolonged number of charge-discharge cycles. Reducing $LaNiO_3$ by hydrazine can improve the electrical conductivity and electrochemical behavior and ultimately leads to a significant increase in specific capacitance, from ~70 to ~280 F g^{-1} at 1 A g^{-1} (after 12 h reduction). The porous structure after hydrazine reduction provides better accessibility for electrolytes, which is helpful in enhancing the overall electrochemical behavior (Ho and Wang 2017).

Like the $LaMnO_3$, the capacitance of $LaNiO_3$ is boosted by doping and/or by forming a composite with materials with capacitive behavior. $La_xSr_{1-x}NiO_{3-\delta}$ ($0.3 \leq x \leq 1$) nanofibers obtained by the electrospinning method shows a specific capacitance value of 719 F g^{-1} at a current density of 2 A g^{-1} when $x = 0.7$. A capacitance value of 505 F g^{-1} is recorded at a high current density of 20 A g^{-1}. The capacitor device can operate at a cell voltage as high as 2 V, and it exhibits an energy density of 30.5 Wh kg^{-1} at a high power density of 10 kW kg^{-1} and a high energy density of 81.4 Wh kg^{-1} at a low power density of 500 W kg^{-1}. More importantly, this symmetric supercapacitor also shows an excellent cycling performance with 90% specific capacitance retention after 2,000 charging and discharging cycles (Cao et al. 2015c). NiO/$LaNiO_3$ composite film electrodes exhibit an improved supercapacitance than their pure counterparts. The NiO/lanthanum nickel oxide (LNO) electrode with appropriate LNO content possess high specific capacitance as high as 2,030 F g^{-1} at 0.5 A g^{-1} and good cyclability (Liu et al. 2016a). $LaNiO_3$/NiO electrodes with hollow nanofibrous (Hu et al. 2015) and thin film (Liang et al. 2013) morphology also exhibit an enhanced specific capacitance and electrochemical stability. Yet another study, a hybrid stacked electrode of manganese oxide/$LaNiO_3$, exhibited a high specific capacitance of ~160 F g^{-1} at 10 mV s^{-1} (Hwang et al. 2011).

Perovskite structured $LaFeO_3$ is also a potential pseudo-capacitance electrode material as that exhibits a specific capacitance of 313.21 F g^{-1} at a current density of 0.8 A g^{-1} and retain 86.1% of their original capacitance after 5,000 charge-discharge cycles at a scan rate of 100 mV s^{-1} (Li et al. 2017d). The pseudo-capacitance of $LaFeO_3$ can be related to the redox couple Fe^{2+}/Fe^{3+} and Fe^{3+}/Fe^{2+}, respectively, during the charge transfer reaction. Doping divalent Sr^{2+} ions in $LaFeO_3$ manifested more favorable properties suitable for supercapacitor electrodes. The specific capacitance of $La_{0.7}Sr_{0.3}FeO_3$ reaches up to 523.2 F g^{-1} at a current density of 1 A g^{-1} in 1 M Na_2SO_4, besides the excellent rate capability and cycling stability over 5,000 cycles (Wang et al. 2019a). When Al is doped in the $LaFeO_3$ matrix, the specific capacitance is increased to ~260 F g^{-1} as compared to pure $LaFeO_3$ ~200 F g^{-1}. The improved specific capacitance accompanied by Columbic efficiency and cycle life on Al doping can be related to a relative decrease in equivalent series resistance, resulting in smoother ion transport (Rai et al. 2014). Both magnetic and electrochemical properties of $LaFeO_3$ nanoparticles can be improved by Co substitution at Fe^{3+} sites. The $LaFe_{0.9}Co_{0.1}O_3$ nanoparticles show a discharge capacity ten times higher than pure $LaFeO_3$ particles (Phokha et al. 2017), and it is assumed that the anion intercalated charge transfer is endorsed by the existence of both Co^{3+} and Co^{2+} ions in Co-doped $LaFeO_3$ samples. When Na and Fe ions are co-doped in $LaFeO_3$ as in $La_{0.8}Na_{0.2}Fe_{0.8}Mn_{0.2}O_3$, the supercapacitor exhibited an improved capacitance with high rate capability. The presence of two anodic peaks and two cathodic peaks might be related to redox processes taking place with Fe^{2+}/Fe^{3+} and Mn^{3+}/Mn^{4+}, respectively, contributing to the pseudo-capacitance behavior (Rai and Thakur 2017).

$LaCoO_3$ perovskite is also proposed as an active material for the supercapacitor electrode, as the oxygen-vacancy-mediated redox reaction stores the charge in the electrode. Nanospheres of $LaCoO_3$ exhibit a specific capacitance of 203 F g^{-1} at a current density of 1 A g^{-1}, with good cyclic stability (Guo et al. 2017). In Sr-doped $La_xSr_{1-x}CoO_{3-\delta}$ in nanofibrous form improves the energy density of supercapacitor tremendously. $La_xSr_{1-x}CoO_{3-\delta}$, when $x = 0.7$, exhibits a specific capacitance of 747.75 F g^{-1} at a current density of 2 A g^{-1} in 1 M Na_2SO_4 aqueous electrolyte. The capacitor device in the operating voltage range of 0–1.6 V displays excellent electrochemical performance with a high energy density of 34.8 Wh kg^{-1} at a low power density of 400 W kg^{-1} with superior cycling stability ~97% (Cao et al. 2015d). Nickel oxide (NiO) nanoparticles loaded on porous strontium-substituted lanthanum cobaltite ($La_{0.7}Sr_{0.3}CoO_{3-\delta}$) substrate is a binder-free electrode for supercapacitors. The composite electrode exhibits a specific capacitance of 1064.1 F g^{-1} and remarkable cycling stability (80.1% retention after 3,000 cycles at 20 mA cm^{-2}) (Liu et al. 2018a). Therefore, coupling with other materials perovskite structured materials can boost the supercapacitance.

In a comparison made among the supercapacitance of lanthanum-based perovskite powders $LaMnO_3$, $LaFeO_3$, $LaCrO_3$, and $LaNiO_3$ when used as an anode, $LaNiO_3$ exhibits the highest specific capacitance of 106.58 F g^{-1} in 3 M LiOH. Furthermore, 98% of the initial capacitance of $LaNiO_3$ was retained after 500 charge-discharge life cycles at the maximum current density of 1 A g^{-1}. The efficient charge storage of $LaNiO_3$ can be attributed to the better anion intercalated redox reactions owing to low resistance and high conducting nature of $LaNiO_3$ (Arjun et al. 2017). A similar study

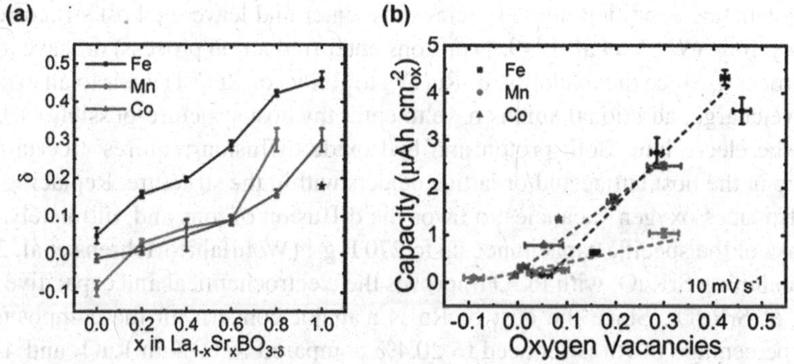

FIGURE 8.14 (a) Oxygen vacancy content (δ) *versus* Sr^{2+} content (x). The ★ is the perovskite $SrFeO_{3-\delta}$ and (b) capacity *versus* their oxygen vacancy content. Using charge capacity rather than pseudo-capacitance accounts for the different voltage window sizes between the B-site elements, reproduced with permission of the American Chemical Society (Alexander et al. 2019).

with Sr^{2+}-doped perovskites, $La_{1-x}Sr_xBO_{3-\delta}$ ($x = 0–1$; B = Fe, Mn, Co), as anion-based pseudo-capacitor electrodes, reveals that $La_{0.2}Sr_{0.8}MnO_{2.7}$ exhibited the highest specific capacitance of 492 F g^{-1} at 5 mV s^{-1} with respect to the Fe and Co perovskites. The incorporation of Sr^{2+} ions leads to an increase in oxygen vacancy (Figure 8.14a) content, which linearly increases the surface area-normalized pseudo-capacitance with a slope controlled by the B-site element as in Figure 8.14b. Upon comparing the three B-sites with different transition metals, the $La_{1-x}Sr_xFeO_{3-\delta}$ samples have the highest peak currents that suggest fast lattice oxygen diffusion rates due to high oxygen vacancy concentrations in them, and if lattice oxygen diffusion is fast, the charge transfer is kinetically limited. However, in oxide anion intercalated pseudo-capacitive mechanism, oxygen vacancies in bulk can be utilized for charge storage, despite differences in charge transfer limitations being kinetic or diffusive in nature (Alexander et al. 2019).

The prospective of strontium-based perovskites, $SrBO_3$ (B = Ti, Mn, Co, Fe, Ru, etc.), with divalent Sr ions in the A-site and a tetravalent transition metal in the B-site are also identified as the promising candidates for supercapacitance electrodes. Ruthenium-based materials are reported to be excellent materials for supercapacitors. The high cost of ruthenium is a major disadvantage. However, the cost can be reduced by incorporating ruthenium in perovskite lattice as in $SrRuO_3$, metallic conducting oxides with the perovskite-type structure. $SrRuO_3$ exhibits pseudo-capacitive behavior and the capacitance changes with the nature and number of dopants. Doping B-site Fe or Co does not influence the capacitance but decreases the stability, but doping Mn increases the capacitance values. A two-fold increase in the capacitance with La doping on the A-site is observed. The charge storage mechanism in $SrRuO_3$ involves protonic conduction via the Grotthus mechanism inside the perovskite structure. The charge storage after the double layer charging is through redox transitions of the electrode material, which involve faradaic charge transfer between the electrode material and the electrolyte. In order to keep the charge balance condition, ionic species must enter and leave the host structure, and upon cycling, (Wilde et al. 1999, 1) protons enter (reduction process) or leave (oxidation process). Since the oxidation of Ru(IV) to Ru(V) or Ru(VI) leads to an excess of positive charge, additional anions have to enter the host structure-possibly OH^- ions from the electrolyte. Both proton and hydroxide diffusion requires a certain free volume in the host lattice and/or lattice defects within the structure. Replacing Sr by La introduces oxygen vacancies to favor the diffusion of ions and, ultimately, to an increase of the specific capacitance up to 270 F g^{-1} (Wohlfahrt-Mehrens et al. 2002).

Combining $SrRuO_3$ with RGO improves the electrochemical and capacitive properties of $SrRuO_3$. Since the cost of Ru is a major concern, in the composite, the mass percentage of Ru is reduced to 20.4% compared to 60% in RuO_2 and 43% in the individual $SrRuO_3$ of commercial Ru-based supercapacitors. The performance of the $SrRuO_3$/RGO composite supercapacitor electrode is significantly affected by the electrolyte. Among the electrolytes, $NaNO_3$, H_3PO_4, and KOH, H_3PO_4 is recommended for high energy density, and KOH is for high power density applications when the aforesaid composite is used as supercapacitor electrode (Galal et al. 2018a, 2018b). *In situ* fabrication of RGO/$SrRuO_3$ composite yielded better capacitance than physically mixing RGO and $SrRuO_3$ (Galal et al. 2019).

$SrCoO_{3-\delta}$ perovskites are capable of oxygen intake/release for the charge-discharge process. Magnetic and electrical transport properties of $SrCoO_{3-\delta}$ are determined by oxygen stoichiometry. The structure of the $SrCoO_3$ changes from $SrCoO_3$ to $SrCoO_{2.5}$, as the valence of Co ions changes from Co^{4+} to Co^{3+}, modifying the electrical transport, magnetic, and thermoelectric properties (Demirel et al. 2017). Therefore, oxygen ions can be exploited as a charge carrier in $SrCoO_3$ electrodes as an anion-intercalation supercapacitor. The specific capacitance of $SrCoO_3$ electrodes gradually increased with cycling, to 572 F g^{-1} after 3,500 cycles, and then a slight decrease in capacitance with an increase in the number of cycles. However, a specific capacitance of 500 F g^{-1} is maintained after 5,000 cycles. The initial progressive increase in the specific capacitance can be related to the increased specific surface area of $SrCoO_3$ from the leaching of Sr^{2+} into the alkaline solution. The increased surface area enhanced the surface reaction rate, which improved the capacitance of $SrCoO_3$ with cycling time during the initial cycles as the degree of cation leaching reaches beyond some point, the perovskite structure damages and result in a decrement in the capacitance (Liu et al. 2016b). Therefore, a meticulous evaluation of cation leaching is necessary during the design and development of perovskite materials for supercapacitor electrodes.

Doping of molybdenum can enrich the oxygen vacancies in $SrCoO_3$, which can be an excellent oxygen anion-intercalated charge-storage electrode. It is demonstrated that doping at B-site with cations having high valence state augments oxygen vacancy concentration and, at the same time, restores the structural integrity (Nagai et al. 2007). The presence of oxygen vacancies by Mo-doping in $SrCo_{0.9}Mo_{0.1}O_{3-\delta}$, improves the ion diffusion rate, which results in an excellent specific capacitance of 1,223.34 Fg^{-1} at a current density of 1 Ag^{-1}. The O^{2-} diffusion is the rate-limiting factor for charge storage in $SrCo_{0.9}Mo_{0.1}O_{3-\delta}$. The hybrid cell with $SrCo_{0.9}Mo_{0.1}O_{3-\delta}$ as the positive and RGO nanoribbon as the negative electrode exhibits a high energy density of 74.8 Wh kg^{-1} at a power density of 734.5 W kg^{-1}; also the device sustains an energy density of 33 Wh kg^{-1} at a high power delivery rate of 6,600 W kg^{-1}. The specific capacitance retention is 97.6% after 10,000 charge-discharge cycles (Tomar et al. 2018).

Altin et al. reported the synthesis of $SrCo_{1-x}Ru_xO_{3-\delta}$ nominal compositions, where $x = 0.0$–1.0, using solid-state reaction technique. X-ray diffraction (XRD) analysis confirms the structure of $x = 0$ sample as hexagonal $Sr_6Co_5O_{15}$. As the Co ions are substituted by Ru, a two-phase structure (hexagonal *R*32 and orthorhombic *Pbnm*) emerges up to $x \leq 0.5$. As the Ru content is increased further, the hexagonal *R*32 phase disappears completely, and an orthorhombic *Pbnm* phase becomes the main phase. SEM images show that the grain size of the samples decreases with increasing Ru content. Temperature-dependent electrical conductivity studies indicate upon Ru substitution in the nominal $SrCo_{1-x}Ru_xO_{3-\delta}$ compounds, and resistivity decreases due to the appearance of the metallic $SrRuO_3$ phase. The cyclic voltammogram (CV) of the samples show capacitive properties upon Ru substitution. The cycle measurements of the capacitors yield promising results for potential supercapacitor applications (Altin et al. 2018).

Materials with the alkaline solution were focused on the structure and specific surface area of the electrode material, and ultimately the electrochemical performance was emphasized. Both BSCF and $SrCoO_3$ (SC) were found to experience cation leaching in alkaline solution, resulting in an increase in the specific surface area of the material, but overleaching caused the damage of the perovskite structure of BSCF. Barium leaching was more serious than strontium, and the cation leaching was composition dependent. Although high initial capacitance was achieved for BSCF, it was not a good candidate as an intercalation-type electrode for supercapacitor because of poor cycling stability from serious Ba^{2+} and Sr^{2+} leaching. Instead, SC was a favorable electrode candidate for practical use in supercapacitors due to its high capacity and proper cation leaching capacity, which brought beneficial effect on cycling stability.

Perovskites with barium at A-site such as $BaMnO_3$, $BaTiO_3$, etc. can act as an electrochemical supercapacitor electrode. When $BaMnO_3$ is used as the supercapacitor electrode in 1 M, sodium nitrate electrolyte exhibited a specific capacitance of 170.8 F g^{-1} at the scan rate of 10 mV s^{-1} (Shanmugavadivel et al. 2016). Though the Coulombic efficiency of the $BaMnO_3$ is 93.9% with high rate performances, the capacity retention is dropped drastically to 141.6 F g^{-1} after 500 cycles, which may be due to the leaching of Ba ions to the electrolyte as observed in the case of $BaSrCoFeO_3$. A composite electrode of polypyrrole/modified multiwalled carbon nanotube (CNT) nanocomposites/$BaTiO_3$ exhibits a specific capacitance of 155.5 F g^{-1} and an energy density of 21.6 W h kg^{-1}, at a scan rate of 10 mV s^{-1} when used as a supercapacitor. The maximum power density is 385.7 W kg^{-1} at a scan rate of 200 mV s^{-1} (Moniruzzaman et al. 2013). Piezoelectric $BaTiO_3$ is also potentially used as a solid-state electrolyte, Hydrogen-doped $BaTiO_3$ films are successfully used as a solid-state electrolyte in supercapacitor. The presence of highly mobile and reactive protons allows the diffusion of positively charged oxygen vacancies easily with a low energetic barrier (Gonon and El Kamel 2007).

Double perovskites are also studied as potential materials for supercapacitor electrodes. An interesting structural aspect of the double perovskites $A_2BB'O_6$ is that some possess the ordering of cations in B sites, characterized by the presence of chains $-B^{II}-OB^{VI}-O$ (Vasala and Karppinen 2015). This ordering allows a significant improvement in ion transport in the presence of mixed valences and distinguishes them as a suitable candidate for electrode applications. Many perovskite oxides exhibit charge storage through an anion-intercalation mechanism, and the characteristic anion exchange is also observed in some double perovskites. The actual replacement of cations can suggestively modify these exchange interactions (Ivanov et al. 2011). Moreover, B-site cation with a low energy barrier for valence change, the stability of the perovskite lattice, and the presence of alkaline earth cations (Sr, Ba) are few important aspects in tailoring the material for supercapacitor electrodes. In addition, the size dependence on energy storage in the electrodes is decided by the mechanism of energy storage in them. If the energy storage mechanism is through anion intercalation, the size of the electrode material does not play a significant role in deciding the supercapacitance as the 3D network structure of perovskite allows the diffusion of anions throughout the electrode. Materials with high oxygen vacancy concentrations exhibit anion-based intercalation charge storage, and they are identified as promising candidates as the supercapacitor electrode in recent years. Reduced $PrBaMn_2O_{6-\delta}$ (r-PBM) double perovskite is one such material and does not require high surface areas to achieve a high energy storage capacity due to the bulk intercalation mechanism. $PrBaMn_2O_{6-\delta}$ possesses a layered double perovskite structure and shows ultrahigh pseudo-capacitance functions and an oxygen anion-intercalation-type electrode material for supercapacitors. The layered double perovskite structure is attained by treating in a hydrogen atmosphere at high temperature, which in turn enhances the capacitance. $PrBaMn_2O_{6-\delta}$ demonstrates a very high specific capacitance of 1,034.8 F g^{-1} and an excellent volumetric capacitance of ≈2,535.3 F cm^{-3} at a current density of 1 A g^{-1} due to the layered structure and the higher oxygen diffusion rate and oxygen vacancy concentration associated with it (Liu et al. 2018c). The initial specific capacitance of r-PBM is 445 F g^{-1}, which is increased steadily to a maximum value of 1,034.8 F g^{-1} during the first 600 cycles, and then the capacitance is reduced gradually but retained a capacitance of 960.8 F g^{-1} after 5,000 cycles. The optimal hydrogen treatment has a significant role in capacitance values. The synergic impact of oxygen vacancy concentration and the double perovskite structure result in the excellent charge storage performance of this electrode. However, the leaching of cations into the electrolyte solution can be a concern for the long-term operation of these supercapacitors when used with an aqueous alkaline electrolyte (Liu et al. 2016b).

In the anion intercalation mechanism, the capacitance is contributed by surface redox reactions and oxygen ion intercalation in the bulk materials (Xu et al. 2018). During anion intercalation in $PrBaMn_2O_{6-\delta}$, the oxygen ions are intercalated into the crystal lattice by the occupation of the oxygen vacancies, and partial oxygen ions diffuse along the pathway at the edges through the lattice, subsequently accompanied by the oxidation of Mn^{2+} to Mn^{3+}. Consequently, additional oxygen ions are adsorbed on the surface of the crystal by Mn ion diffusion and further oxidation of Mn^{3+} to Mn^{4+}. Therefore B-site cation with a low energy barrier for valence change is crucial while designing such material for supercapacitor application (Liu et al. 2018b).

$Ba_2Bi_{0.1}Sc_{0.2}Co_{1.7}O_{6-\delta}$ is yet another promising double perovskite oxide as oxygen-ion-intercalation-type supercapacitors owing to their high oxygen vacancy concentration, oxygen diffusion rate, and tap density. A maximum capacitance of 1,050 F g^{-1} is recorded in this material with a capacity retention of ~74% after 3,000 charge-discharge cycles at a current density of 1 A g^{-1} when 6 M KOH is used as the electrolyte. A specific energy density of 70 Wh kg^{-1} at the power density of 787 W kg^{-1} is observed when a supercapacitor made using this material as the anode (Xu et al. 2018). The specific capacitance of both the above materials is dependent on the concentration KOH. An increase in the OH^- concentration enhances the oxygen anion diffusion rate in the double perovskite oxide lattice and, as a result, an enhanced electrical performance (Liu et al. 2018b; Xu et al. 2018).

The size is immaterial to electrodes with anion intercalation charge storage in many perovskite-type materials. However, when Y_2NiMnO_6 double perovskite is used as an active material for the positive electrode of electrochemical supercapacitors, the specific capacitance is tremendously improved in an electrode composed of nanostructured Y_2NiMnO_6 than its bulk counterpart. Alam et al. used a low-temperature hydrothermal route for the fabrication of Y_2NiMnO_6 nanowires, and a sol–gel method for bulk samples. A comparative study reveals that the Y_2NiMnO_6 nanowire-based electrode is superior to its bulk counterpart, exhibiting a higher specific capacitance of 77.76 F g^{-1} at 30 mA g^{-1}, the energy density of 0.89 W h kg^{-1} at 30 mA g^{-1}, and a power density of 19.27 W kg^{-1} at 150 mA g^{-1}. The specific capacitance is retained as 63.6 F g^{-1} when a current density of 150 mA g^{-1} is used. The percentage of specific capacity retained by the nanowires is higher than the bulk Y_2NiMnO_6 electrode, after 1800 cycles, which confirms the good stability of the nanowire electrodes. Moreover, the symmetric triangular charge/discharge profile reveals the stable electrochemical performance without any significant structural change (Alam et al. 2016).

Sol–gel electrospun La_2CoMnO_6 nanofibers are another example of potential double perovskite electrode material for supercapacitors. The La_2CoMnO_6 nanofiber electrodes are stable, with a specific capacitance of 109.7 F g^{-1} at a current density of 0.5 A g^{-1} with the capacity retention of 90.9% of its initial value after 1,000 cycles. The continuous CoO_6 and MnO_6 octahedral networks in La_2CoMnO_6 material act as channels for the rapid migration of charges in the electrode with improved conductivity and, ultimately, an increased electrode utilization efficiency (Fu et al. 2018). A similar study conducted on electrospun La_2CoNiO_6 nanofiber as a supercapacitor electrode material reveals that this material exhibits a specific capacitance of 335.0 and 129.1 F g^{-1} for three and two electrode systems, respectively, at a current density of 0.25 A g^{-1} (Wu et al. 2015).

Recently, it is identified that the double perovskite La_2NiMnO_6 can be potentially used as a stable electrode material in supercapacitors in both alkaline and acidic electrolytes. This material exhibits a specific capacitance up to 1,681 F g^{-1} at a current density of 3 A g^{-1} when used as KOH (0.5 M) electrolyte. Whereas in the acidic medium of H_2SO_4 (0.5 M), the capacitance dropped to 492 F g^{-1}. The corresponding power and energy densities are 2,903 W kg^{-1} and 378 Wh kg^{-1} in alkaline, and 3,225 W kg^{-1}, 137 Wh kg^{-1} in acidic medium, respectively. The reported values are well above the performance observed in the case of other double perovskites with identical structures. From the galvanostatic charge-discharge curves, it is identified that this material has a low internal resistance and deviates from an ideal supercapacitor electrode due to the faradaic process at the material surface (Bavio et al. 2018).

8.3.3 Fuel Cells

The fuel cells are alternate energy sources that are developed to account for the increasing energy demand and rising energy prices, and the detrimental effects of fossil fuels such as pollution, global warming, and climate changes. Moreover, the sources of fossil fuels are rapidly deteriorating. The fuel cells use fossil fuel derivatives, hydrogen, and alcohols as fuels to produce electrical energy with high efficiency and reliability. Fuel cells can be classified based on the electrolyte, and the classification of fuel cells and their characteristics are presented in Table 8.6.

The construction of a fuel cell is shown in Figure 8.15. The major parts of a fuel cell are the anode, cathode, and an electrolyte membrane. The fuel is passed through the anode side of the fuel cell and air through the cathode side. The anode is permeable to the fuel where it is oxidized to ions by releasing the subsequent number of electrons, for example, H_2 gas is oxidized to H^+ ions and electrons in polymer electrolyte membrane fuel cell. The H^+ ions pass through the anode to the electrolyte, and the electrons flow through an external circuit, generating an electric current. At the cathode, the oxygen from the air is reduced in the presence of H^+ ions and electrons to form water molecules. A series of such cells are stacked to meet the high-power production units. The interconnects separate each cell from a chemical short circuit and transfer the electric current to the current collectors in each cell or the external load.

8.3.3.1 Perovskites Used in Proton Exchange Membrane Fuel Cell

The catalysts based on noble metals, such as Pt or Au, are well known ORR and OER catalysts with high stability. Replacing the noble metal catalysts can significantly reduce the installation cost of proton exchange membrane fuel cells (PEMFC). Many perovskite materials exhibit comparable catalytic activities in alkaline medium towards ORR and OER activities comparable to platinum-based noble metal catalysts. Perovskites with La at A-site and transition metals at B-site demonstrated good ORR activity. Partially substituting the B-site ions with other transition metal elements significantly improves the catalytic behavior as well as the stability of La-based perovskite materials. $LaCoO_3$ and $LaNiO_3$ perovskites with B-site partial substitution of Mn, Co, Fe, or Ni are the most promising perovskites materials for PEMFCs. These perovskite materials are stable up to a temperature range of 60°C–200°C, and they are promising as cathodes in alkaline PEMFCs (Tarragó et al. 2016).

8.3.3.2 Perovskites Used in Solid Oxide Fuel Cells

In general, the cations in the perovskite structured materials exhibit different valences, and therefore they can occupy different sites in the structure; as a result, unique physical and chemical properties, high oxygen vacancies, and ionic conductivity can

TABLE 8.6
Types of Fuel Cells and Their Characteristics

Fuel Cell	Commercial Electrolytes and Electrodes	Temperature of Operation	Fuel Type	Remarks
Polymer electrolyte membrane/ PEMFC)	Solid polymer electrolyte and porous carbon structures containing platinum as a catalyst	60°C–200°C	Hydrogen at anode and oxygen at the cathode	Deliver high power density
Alkaline fuel cells	Modified PEMFC with potassium hydroxide in water as the electrolyte and a variety of non-precious metals as a catalyst at the anode and cathode	60°C–90°C	Pure hydrogen and oxygen gas.	These fuel cells use an alkaline membrane with demonstrated efficiencies above 60% in space applications.
Solid oxide fuel cell	Solid oxide material as the electrolyte and electrodes.	500°C–1,000°C	Gasoline, diesel, biofuels. Or mixtures of hydrogen, carbon monoxide, carbon dioxide, steam, and methane,	High combined heat and power efficiency, long-term stability, fuel flexibility, low emissions, and relatively low cost. No noble metal catalyst and up to 85% efficiency
DMFC	Platinum and/or ruthenium particles embedded on nanostructured carbons, and polymer as electrolytes	50°C–120°C	Methanol	Transportation of methanol is easier than hydrogen, and energy-density is higher than hydrogen. High power density
Molten carbonate fuel cell	Electrolyte composed of a molten carbonate salt mixture suspended in a porous, chemically inert ceramic lithium aluminum oxide matrix and non-precious metals as catalysts.	550°C–750°C	Natural gas and biogas	Methane and other light hydrocarbon fuels are converted to hydrogen within the fuel cell itself by a process called internal reforming.

(Continued)

TABLE 8.6 (*Continued*)
Types of Fuel Cells and Their Characteristics

Fuel Cell	Commercial Electrolytes and Electrodes	Temperature of Operation	Fuel Type	Remarks
Phosphoric acid fuel cells (PAFCs)	100% phosphoric acid electrolyte retained in a silicon carbide matrix and carbon paper coated with a finely dispersed platinum as a catalyst.	150°C–220°C	Hydrogen-rich gases	These fuel cells are used for stationary power generation. PAFCs are more tolerant of impurities in fossil fuels that have been reformed into hydrogen than proton exchange membrane (PEM) cells. PAFCs are expensive due to the high loading of the platinum catalyst.
Microbial fuel cells	Microbes such as bacteria catalyze electrochemical oxidations or reductions at an anode or cathode	~25°C	Organic matter such as wastewater or nutrients	Microbial fuel cells create electricity using electrode reducing microorganisms. Electrode-oxidizing organisms take electrons from the cathode to reduce various substances to acetates.

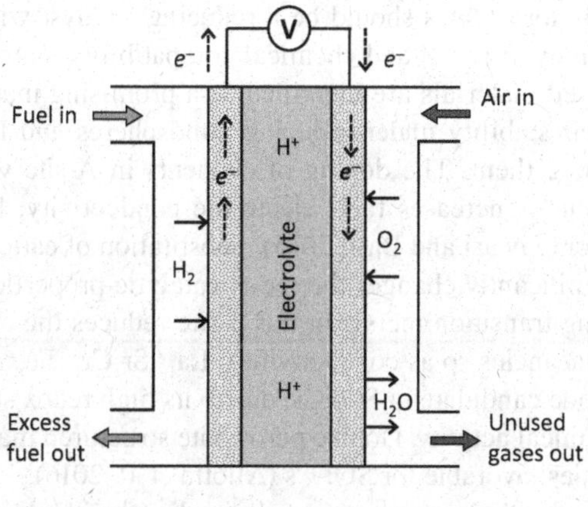

FIGURE 8.15 Schematic of a PEMFC.

be attained. A large number of oxygen vacancies and the associated ionic conductivity make them suitable as electrode and electrolyte materials in SOFCs. In SOFCs, the perovskite materials are identified as suitable candidates for cathode, anode, electrolyte, and interconnectors. The operating and making cost of SOFCs can be considerably reduced by lowering the operating temperature and replacing noble metal alloys. The oxidizing and reducing atmospheres at high temperatures necessarily require stable interconnectors and current collectors. Therefore, by decreasing the operating temperature, less expensive materials can be used in the above applications, thereby reducing the initial cost of SOFCs.

The primary function of a cathode in SOFC is to provide active sites for the electrochemical reduction of oxygen with high catalytic activity. The ideal cathode materials also must have a high electronic and ionic conductivity, a compatible thermal expansion coefficient, thermal and chemical properties of other components of SOFCs such as electrolyte and interconnect materials, porosity, and low cost. Perovskite structured $La_{1-x}Sr_xMnO_{3-\delta}$ (LSM) (Tian et al. 2008) and $La_{1-x}Sr_xCoO_{3-\delta}$ (LSC) (Gwon et al. 2014) are the promising high-temperature cathodes in SOFC due to their high electronic conductivity despite their low ionic conductivity. However, tuning the oxygen vacancies and cation vacancies can significantly affect the transport properties of these materials. The substitution of rare-earth ions at A-site lowers the energy barriers for adsorption and diffusion of oxygen species; at the same time, they exhibit excellent compatibility with yttria-stabilized zirconia (YSZ) electrolyte and good electronic conductivity. The cobalt-ferrite perovskites such as $Ba_{0.5}Sr_{0.5}Co_{0.8}Fe_{0.2}O_3$ (Li et al. 2019) and $La_{1-x}Sr_xFe_{1-y}Co_yO_3$ (Ma et al. 2018) are also promising candidates as cathode materials. The rare-earth orthoferrites exhibit high catalytic activity and reasonable ionic and electronic conductivities at lower temperatures (600°C–800°C). Doping Ca in these materials increases the electrical conductivity and lower cathodic overpotential, with a compatible thermal expansion coefficient. Recent studies show that double perovskite materials with the general formula $ABaCo_2O_{5+x}$ (A = Gb, Pr, Sm, and Nd) exhibit higher catalytic activity than single perovskites (Hong et al. 2017).

The ideal anode for SOFCs should be a reducing catalyst with good electronic and ionic conductivity, thermal and chemical compatibility, porosity, and low cost. Perovskite structured materials are identified as a promising material as anodes in SOFCs due to their stability under reducing atmospheres and high temperatures, $SrTiO_3$ is one among them. The doping of elements in A-site with La and B-site with Nb tremendously increases their electronic conductivity; however, the ionic conductivity is poor (Tiwari and Basu 2015). Substitution of cations like Al, Fe, Ga, Mg, Mn, or Sc significantly changes the redox catalytic properties and the conductivities. Substituting transition metals in the B-site reduces the energy required for the formation of vacancies so as conductivities. $La_{1-x}Sr_xCr_{1-y}Fe_yO_3$ is also identified as a promising anode candidate in SOFCs, due to its high redox stability, conductivity, and electrochemical activity. Double perovskite structured materials also exhibit interesting properties favorable for SOFCs (Aliotta et al. 2016).

The electrolyte of SOFCs is also potentially replaced with perovskite oxide materials. A typical SOFC electrolyte should have high ionic conductivity and low electronic conductivity, in addition to the necessary characteristics of the electrodes such as chemical stability and compatible thermal expansion coefficients. Perovskites with gallium, zirconium, and cerium at the B-site are identified as suitable candidates for electrolytes, e.g., $LaGeO_3$ (Ishihara et al. 1995). Substituting the A-site cations with Sr or Ba can alter the conductivities to a certain level. However, doping Mg at B-site can significantly affect the favorable properties as an electrolyte by increasing the concentration of oxygen vacancies. The proton-conducting oxides such as $BaZrO_3$ (Iguchi et al. 2007) and $BaCeO_3$ (Naeem Khan et al. 2017) are also promising as electrolyte materials. Moreover, these materials allow doping with precious metals to increase oxygen mobility.

The interconnects of SOFCs must be stable to oxidizing and reducing atmospheres, highly electronic conductive, low ionic conductive, gas-tight, and compatible thermal and chemical behavior with other components of SOFCs. Perovskite structured lanthanum and yttrium chromates (Weber et al. 1987; Duran et al. 2004), both *p*-type semiconductors, are identified as suitable candidates for SOFC interconnects. The A-site dopants such as Sr, Mg, and Ca and the B-site dopants Co or Fe can improve the conductivity and thermal stability of chromite perovskites.

8.3.3.3 Perovskite-Based Catalysts for Direct Methanol Fuel Cells

In Direct Methanol Fuel Cells (DMFC), methanol is used as a fuel, which is oxidized electrochemically at the anode and generates electricity, and oxygen is reduced (ORR) at the cathode. The perovskites with transition metals at B-sites can be a promising non-noble metal cathode in DMFCs. Lanthanum-based perovskites with the general formula $LaMO_3$ (M = Co, Mn, Ni, Fe) are promising candidates for ORR in alkaline medium with the characteristics of that of noble metal catalysts. The partial substitution of A-site cations with Ca and Sr enhances the current densities at a lower overpotential; however, the transitions metals at B-site plays a significant role in improving the ORR catalysis any further. The presence of two transition elements in the B-site exhibits better activity than the one with the single element at the B-site. The ORR current density of La-based perovskites is in the following order: $LaCoO_3 > LaMnO_3 > LaNiO_3 > LaFeO_3 > LaCrO_3$. Moreover, the perovskite materials on carbon-based support materials such as graphene and CNTs are more active than the pure perovskites due to the synergic activation mechanism in the composites (Zhu et al. 2015).

Perovskite-type materials can be potentially used as anodes in DMFCs. Pt-containing DMFC anodes exhibit poor reaction kinetics and can be contaminated by carbon species such as CO. Metal oxides are an anticipated alternative as anode due to their chemical stability. The presence of Ru in the B-site of La -or Sr-based perovskites is the most efficient catalyst for methanol oxidation reaction (MOR) (Lan and Mukasyan 2008). The activity towards MOR is further improved by controlling the particle size and doping noble metal, such as Pd or depositing Pt on the surface of the catalysts.

8.3.4 Photovoltaics

The perovskite oxides with ferroelectric properties are suitable for highly stable photovoltaic cells with large open-circuit voltages. In ferroelectric perovskite oxides, the photoexcited carriers are separated by the polarization-induced internal electric field, unlike the electric field developed at the *p-n* junctions in conventional solar cells (Chakrabartty et al. 2018). These photovoltaic devices can generate an open-circuit voltage (V_{oc}) above the bandgap of the perovskite material. Ferroelectric perovskite materials such as $BaTiO_3$ (Dharmadhikari and Grannemann 1982), $LiNbO_3$ (Glass et al. 1995), (Pb, La)TiO_3, Pb(Zr, Ti)O_3, $(Pb_{0.97}La_{0.03})(Zr_{0.52}Ti_{0.48})O_3$ (Qin et al. 2009), and $BaFeO_3$ are studied for their application in such a photovoltaic application. The solar cells based on ferroelectric perovskites exhibit low power conversion efficiencies (PCEs), due to large bandgaps, which absorb UV light, and the UV region contributes only 3.5% of the solar spectrum. The bandgap of $BaFeO_3$ single crystal is 2.2 eV (absorbs 25% of the spectrum), the smallest among the ferroelectric perovskites, which is higher than the ideal bandgap 1.34 eV for photovoltaic applications.

The light-harvesting ability of the ferroelectric oxide perovskite materials can be improved by narrowing the bandgap of these materials, thereby enhancing the range of visible light absorption. The bandgap reduction in perovskite oxides can be achieved by the appropriate selection of the A- and B-site elements as in Bi_2FeCrO_6 double perovskite (Nechache et al. 2015) or by combining perovskite materials together as in $[KNbO_3]_{1-x}[BaNi_{1/2}Nb_{1/2}O_{3-\delta}]_x$ (KBNNO) (Grinberg et al. 2013) to form perovskite solid solutions. $Bi_4Ti_3O_{12}$–$LaCoO_3$ (Choi et al. 2012), $BiMnO_3$-$BiMn_2O_5$, etc. are the solid solutions of a low-bandgap ferroelectric perovskite and a non-ferroelectric material, which exhibit better performance than single perovskites.

The perovskite structured ferroelectric materials exhibit a low conductivity; otherwise, the dielectric properties are affected due to leakage of charge carriers. The poor conductivity of the ferroelectric perovskite materials is detrimental to the photovoltaic charge transport. Improving the electrical conductivity is important to increase the efficiency of bulk photovoltaic perovskites. Unfortunately, the improved conductivity increases the leakage current and affects the ferroelectric behavior of the perovskite. An adjustment between the conductivity and the ferroelectric properties is necessary to enhance the efficiency of ferroelectric photovoltaics. Oxide perovskites containing 3*d* transition metals exhibit low carrier mobilities and high carrier recombination. Non-3*d* transition metals containing oxide double perovskites with $A_2M^{III}M^{V}O_6$ structure can have lower bandgaps and smaller carrier effective masses, and are therefore good absorbers for photovoltaics (Yin et al. 2019).

8.3.5 Hydrogen Storage

The ion-exchange ability of the perovskite materials makes them potential electrolyte candidates for various electrochemical applications. This ability of perovskite oxides such as $SrZrO_3$ and $SrCeO_3$ at elevated temperatures was discovered by Iwahara et al. (1981). The proton dissociated from the gas-phase water molecules was incorporated into the vacancies of ABO_3 oxides, which thus formed covalent bonds with the oxygen atoms, and the proton conduction occurs through the binding proton rotation and migration in the lattice oxygen ions. The vacancies of lattice oxygen play an important role in promoting proton conductivity and storage in ABO_3 perovskite oxides. Additionally, the dopants act as discrete "traps" for the photon absorption (Wang et al. 2012). $ACe_{1-x}M_xO_{3-\delta}$ (A = Sr or Ba, and M = rare-earth element) was found to store hydrogen in bulk, and has the ability to store and discharge hydrogen at room temperature, repeatedly, making them as suitable candidates as anode materials of hydrogen batteries (Esaka et al. 2004). Perovskite structured $LaCrO_3$ is also a potential electrochemical hydrogen storage material (Deng et al. 2010).

Perovskite-type hydrides with the general formula ABH_3 and $ABB'H_6$ are identified as the hydrogen storage materials due to their high hydrogen densities in their lattice. $NaMgH_3$ is one such perovskite-type materials with up to 6 wt.% hydrogen storage capacity with a possibility of reversible hydrogenation/dehydrogenation reaction (Wu et al. 2008). $LiSrH_3$, $LiBaH_3$, $KMgH_3$, $RbMgH_3$, $CsMgH_3$, $RbCaH_3$, $CsCaH_3$, $LiBeH_3$, and $NaBeH_3$ are few other examples for potential hydrogen storage perovskite hydrides. These materials are synthesized by treating the respective hydrides in a hydrogen atmosphere maintained under high pressure.

8.4 OTHER APPLICATIONS

8.4.1 Sensors

The sensor is a device that detects or measures a physical quantity of stimulus and converts it to a measurable form. Sensors are used in a wide range of applications for sensing pressure, temperature, humidity, hazardous gases, drinking water quality monitoring, weather monitoring, detection of explosives, medical diagnostics, food inspection, etc. Perovskite oxides are extensively studied as a sensor material for solid-state and electrochemical sensing of the different physical or chemical stimulus. The high specific surface area of the nanostructured perovskite oxides exhibits superior properties for the aforesaid application. Perovskite materials can be potentially used in electric, magnetic, piezoelectric, thermoelectric, magnetoresistive and electrochemical modes of sensing, as listed in Table 8.7. However, the resistive mode

TABLE 8.7
Classification of Perovskite-Based Sensors

Mode of Operation	Response	Stimuli
Optical (optodes)	Change in optical phenomena, such as absorbance, reflectance, luminescence, and light scattering.	Sensing VOCs, gases, explosives, temperature, etc.
Electrochemical (Bio)	Change in electrochemical behavior, such as voltammetric and amperometric characteristics.	Sensing of glucose, H_2O_2, DA, etc.
Electrical	Change in resistance or conductivity	Sensing of gases, volatile organic compounds, humidity, etc.
Magnetic	Change of paramagnetic properties Magnetoresistive – Change in electrical resistance with the applied magnetic field	Oxygen monitors, magnetic field, etc.
Piezoelectric/ mass	Change in the acoustic wave properties – surface acoustic wave	Bacteria spores, humidity, gases, strain, etc.
	Measuring the change in frequency of a quartz crystal resonator – quartz crystal microbalance	Gases, humidity, pressure, strain, etc.
Thermoelectric	Change in electric voltage	Temperature sensing

of sensing is commonly used for gas and vapor sensing applications and electrochemical mode for analytes with biological origin.

As already discussed, the defect concentration in perovskite oxides determines its conductivity, and this peculiarity allows them to use as catalysts and sensors. Moreover, perovskite oxides have high melting and degradation temperatures and stable to poisoning by different chemical species, which allows them to exhibit long-term reliability and repeatability. The addition of dopants in the perovskite lattice allows the tuning of selectivity, ion transport, conductivity, and catalytic properties in favor of sensor applications.

8.4.1.1 Gas Sensing

The dependence of defect concentration on oxygen partial pressure in perovskite materials, in turn, determines the conductivity of these materials. In addition to the sensitivity toward oxygen, the conductivity of perovskite materials is affected by other gases or vapors such as carbon monoxide, ammonia, ethanol, and hydrocarbon, allowing them to be used as sensors for several gases. However, the selectivity can be controlled by the appropriate selection of the dopants. Perovskites with several transition metals at B-site are potential candidates for gas sensors for the detection of CO, NO_2, methanol, ethanol, and hydrocarbons (Fergus 2007).

8.4.1.2 Optical Sensors

The perovskite oxides, when doped with rare-earth ions, emits light characteristic to the rare-earth ions. $YAlO_3$ is one such perovskite material that acts as an excellent host for rare-earth dopants. The emission from such materials when doped with upconverting Er^{3+} or Nd^{3+} is sensitive to pressure (Hua and Vohra 1997). Similarly, emissions from $LaGdO_3$ co-doped with Er^{3+} and Yb^{3+} (Siaï et al. 2018), $BaTiO_3$ doped with Er^{3+} (Singh and Manam 2018) and $Na_{0.5}Bi_{0.5}TiO_3$ doped with Er^{3+} (Wang et al. 2016a) are sensitive to temperature.

8.4.2 Resistive Random-Access Memory (RRAM)

Resistive Random-Access Memory (RRAM) is a non-volatile memory that is going to play a key role in next-generation digital technology. RRAM comprises of resistance switchable components sandwiched between two terminal electrodes with high storage capability. The storage capacity of these components highly depends upon their intrinsic properties. The information storage of RRAM can be realized by switching high resistive state (HRS) and low resistive state (LRS) alternatively through stimulation of voltage. Intel has announced the production of its first commercial RRAM in 2017, which displays fast performance, high density, and longer stability. Moreover, it is predicted to work as both solid-state storage disk (SSD) and random-access memory (RAM). The mechanism of RRAM encompasses charge transfer, traps, conformal change, filamentary conduction, tunneling effect, conformal charge, and ionic conduction (Wang et al. 2018b). The research on perovskite oxides paves the way for producing efficient and promising RRAM devices.

The first perovskite oxide $Pr_{0.7}Ca_{0.3}MnO_3$ (PCMO), which applied as RRAM device, is reported in the late 1990s (Asamitsu et al. 1997). The further exploration of storage property can be tuned by changing the bandgap using light. Perovskite titanates such as $BaTiO_3$ and $SrTiO_3$ having Ba/SrO_{12} cuboctahedra and TiO_6 octahedra are well known for their dielectric (Yang et al. 2013), ferroelectric, piezoelectric, and resistive switching properties (Spanier et al. 2006; Kim et al. 2007, 2009; Dicken et al. 2008; Ionescu 2012; Son et al. 2013). For instance, the layer-by-layer assembly of cube-like $BaTiO_3$ nanoparticles/polymer composite exhibits both ferroelectric and resistive bipolar switching. This property further improved by optimization of ferroelectric sites density in either vertical or lateral directions of the film, and the schematic representation is shown in Figure 8.16 (Kim et al. 2014b). The theoretical study reveals that the high mobility nature of oxygen vacancies enhances the performance of perovskite device. Recently, researchers fabricated a flexible memory device from the combination of $BaTi_{0.95}Co_{0.05}O_3$ film and a buffer layer of $SrRuO_3$ using mica with 10 μm thickness having <80% transmittance as substrate (Yang et al. 2017b). This flexible nature satisfies the commercial requirements of flexible screens and wearable devices. The combination of $BaTi_{0.95}Co_{0.05}O_3/SrRuO_3$/mica able to cover the region of 500–800 nm due to the ability of buffer layer $SrRuO_3$ to absorb the most of the incident visible light. The flexible nature of this device is examined by bending with a 2.2 mm bending radius, which reveals no considerable change in resistive switching properties. Further, the stability of this device is tested at high temperature by annealing at 500°C, and it is found to be working as normal.

In general, $SrTiO_3$ is the most stable perovskite oxide structure, which comprises the Ti sublattice along with the dislocation sites. In RRAM devices, amorphous $SrTiO_3$ thin film having variable oxygen vacancies with superior uniformity

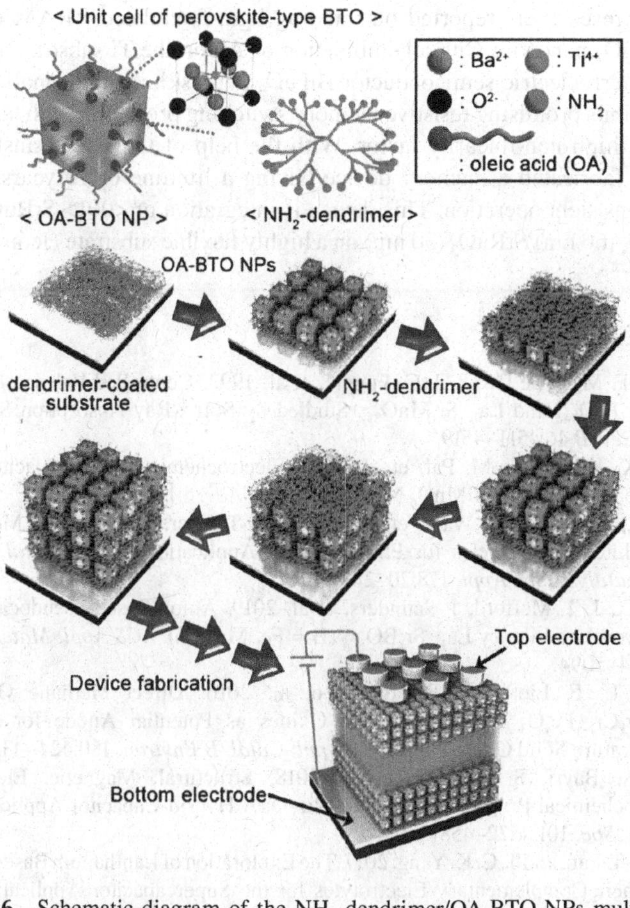

FIGURE 8.16 Schematic diagram of the NH_2-dendrimer/OA-BTO NPs multilayer-based non-volatile memory devices, reproduced with permission of the American Chemical Society (Kim et al. 2014b).

plays an essential role in resistive switching properties (Nili et al. 2014). The conductive nanofilaments which are present in the $SrTiO_3$ layer induce the enhanced resistive switching characteristics by acting as independent nanoswitches. Further, oxygen deficiency in this memory device can be enhanced via electroforming step. Hence the oxygen ions present along the electroforming direction are the keys for improving the bipolar switching behavior in the memory devices. Another way of improving the performance of a memory device is the synergetic combination of nanomaterials with the perovskite oxides. For example, the multifunctional RRAM device fabricated using Nb-doped $SrTiO_3$ and ZnO nanorods demonstrates the multipurpose light-tunable resistive switching properties and photodetector, integrating the functionalities of diode and RRAM (Bera et al. 2013). Selector-free RRAMs are fabricated using Nb-doped $SrTiO_3$ and $BiFeO_3$ nano-island arrays. This memory device has shown a high resistance ratio along with a high ON/OFF ratio of 4,420 and a non-linearity factor of ~1,100 (Bera et al. 2013).

In 2015, researchers reported outstanding light-modulated RRAM devices utilizing $BiMnO_3$ nanowire with a combination of Ag on the Ti substrate (Sun and Li 2015). The ferroelectric semiconductor $BiFeO_3$ perovskite with a small bandgap of ~2.7 eV exhibits promising resistive memory switching properties with large forward current and high non-linearity factor. With the help of the layer transfer method, researchers fabricated a memory device having a lifetime of 10 years with 1,010 cycles of consistent operation. This device is integration of <001> $SrRuO_3$ (20 nm)/$PbZr_xTi_{1-x}O_3$ (60 nm)/$SrRuO_3$ (30 nm) on a highly flexible substrate (Jeon et al. 2016).

REFERENCES

Abbate, M., F. M. F. de Groot, J. C. Fuggle, et al. 1992. Controlled-Valence Properties of $La_{1-x}Sr_xFeO_{3-\delta}$ and $La_{1-x}Sr_xMnO_{3-\delta}$ Studied by Soft-x-Ray Absorption Spectroscopy. *Phys. Rev. B* 46:4511–4519.

Alam, M., K. Karmakar, M. Pal, et al. 2016. Electrochemical Supercapacitor Based on Double Perovskite Y_2NiMnO_6 Nanowires. *RSC Adv.* 6:114722–114726.

Alammar, T., I. Hamm, M. Wark, et al. 2015. Low-Temperature Route to Metal Titanate Perovskite Nanoparticles for Photocatalytic Applications. *Appl. Catal. B Environ. Photocatalysis Sci. Appl.* 178:20–28.

Alexander, C., J. T. Mefford, J. Saunders, et al. 2019. Anion-Based Pseudocapacitance of the Perovskite Library $La_{1-x}Sr_xBO_{3-\delta}$ (B = Fe, Mn, Co). *ACS Appl. Mater. Interfaces.* 11:5084–5094.

Aliotta, C., L. F. Liotta, F. Deganello, et al. 2016. Direct Methane Oxidation on $La_{1-x}Sr_xCr_{1-y}Fe_yO_{3-\delta}$ Perovskite-Type Oxides as Potential Anode for Intermediate Temperature Solid Oxide Fuel Cells. *Appl. Catal. B Environ.* 180:424–433.

Altin, S., A. Bayri, S. Demirel, et al. 2018. Structural, Magnetic, Electrical, and Electrochemical Properties of Sr–Co–Ru–O: A Hybrid-Capacitor Application. *J. Am. Ceram. Soc.* 101:4572–4581.

Arjun, N., G.-T. Pan, and T. C. K. Yang. 2017. The Exploration of Lanthanum Based Perovskites and Their Complementary Electrolytes for the Supercapacitor Applications. *Results Phys.* 7:920–926.

Arney, D., C. Hardy, B. Greve, et al. 2010. Flux Synthesis of $AgNbO_3$: Effect of Particle Surfaces and Sizes on Photocatalytic Activity. *J. Photochem. Photobiol. Chem.* 214:54–60.

Asamitsu, A., Y. Tomioka, H. Kuwahara, et al. 1997. Current Switching of Resistive States in Magnetoresistive Manganites. *Nature* 388:50–52.

Atta, N. F., S. M. Ali, E. H. El-Ads, et al. 2013. The Electrochemistry and Determination of Some Neurotransmitters at $SrPdO_3$ Modified Graphite Electrode. *J. Electrochem. Soc.* 160:G3144.

Atta, N. F., S. M. Ali, E. H. El-Ads, et al. 2014. Nano-Perovskite Carbon Paste Composite Electrode for the Simultaneous Determination of Dopamine, Ascorbic Acid and Uric Acid. *Electrochimica Acta Adv. Electrochem. Matersci. Manfact.* 128:16–24.

Atta, N. F., A. Galal, and S. M. Ali. 2012. The Catalytic Activity of Ruthenates $ARuO_3$ (A = Ca, Sr or Ba) for the Hydrogen Evolution Reaction in Acidic Medium. *Int J Electrochem Sci.* 7:22.

Bagchi, A., S. R. K. Meka, B. N. Rao, et al. 2014. Perovskite Ceramic Nanoparticles in Polymer Composites for Augmenting Bone Tissue Regeneration. *Nanotechnology* 25:485101.

Barbero, B. P., J. A. Gamboa, and L. E. Cadús. 2006. Synthesis and Characterisation of $La_{1-x}Ca_xFeO_3$ Perovskite-Type Oxide Catalysts for Total Oxidation of Volatile Organic Compounds. *Appl. Catal. B Environ.* 65:21–30.

Bavio, M. A., J. E. Tasca, G. G. Acosta, et al. 2018. La_2NiMnO_6 Double Perovskite Nanostructure Prepared by Citrate Route for Supercapacitors. *Matér. Rio Jan.* 23:1–12.

Ben C., Z., F. El Kamel, O. Gallot-Lavallée, et al. 2017. Hydrogen Doped $BaTiO_3$ Films as Solid-State Electrolyte for Micro-Supercapacitor Applications. *J. Alloys Compd.* 721:276–284.

Bera, A., H. Peng, J. Lourembam, et al. 2013. A Versatile Light-Switchable Nanorod Memory: Wurtzite ZnO on Perovskite $SrTiO_3$. *Adv. Funct. Mater.* 23:4977–4984.

Bhayani, K. R., S. N. Kale, S. Arora, et al. 2007. Protein and Polymer Immobilized $La_{0.7}Sr_{0.3}MnO_3$ nanoparticles for Possible Biomedical Applications. *Nanotechnology* 18:345101.

Bi, Y., L. Zong, C. Li, et al. 2015. Photoreduction of CO_2 on $TiO_2/SrTiO_3$ Heterojunction Network Film. *Nanoscale Res. Lett.* 10:345.

Blasin-Aubé, V., J. Belkouch, and L. Monceaux. 2003. General Study of Catalytic Oxidation of Various VOCs over $La_{0.8}Sr_{0.2}MnO_{3+x}$ Perovskite Catalyst–Influence of Mixture. *Appl. Catal. B Environ.* 43:175–186.

Bockris, J. O., and T. Otagawa. 1984. The Electrocatalysis of Oxygen Evolution on Perovskites. *J. Electrochem. Soc.* 131:290.

Bu, Y., O. Gwon, G. Nam, et al. 2017. A Highly Efficient and Robust Cation Ordered Perovskite Oxide as a Bifunctional Catalyst for Rechargeable Zinc-Air Batteries. *ACS Nano* 11:11594–11601.

Cao, Y., B. Lin, Y. Sun, et al. 2015a. Structure, Morphology and Electrochemical Properties of $La_xSr_{1-x}Co_{0.1}Mn_{0.9}O_{3-\delta}$ Perovskite Nanofibers Prepared by Electrospinning Method. *J. Alloys Compd.* 624:31–39.

Cao, Y., B. Lin, Y. Sun, et al. 2015b. Synthesis, Structure and Electrochemical Properties of Lanthanum Manganese Nanofibers Doped with Sr and Cu. *J. Alloys Compd.* 638:204–213.

Cao, Y., B. Lin, Y. Sun, et al. 2015c. Sr-Doped Lanthanum Nickelate Nanofibers for High Energy Density Supercapacitors. *Electrochimica Acta* 174:41–50.

Cao, Y., B. Lin, Y. Sun, et al. 2015d. Symmetric/Asymmetric Supercapacitor Based on the Perovskite-Type Lanthanum Cobaltate Nanofibers with Sr-Substitution. *Electrochimica Acta* 178:398–406.

Chakrabartty, J., C. Harnagea, M. Celikin, et al. 2018. Improved Photovoltaic Performance from Inorganic Perovskite Oxide Thin Films with Mixed Crystal Phases. *Nat. Photonics* 12:271–276.

Chang, D., M. Lim, J. A. C. M. Goos, et al. 2018. Biologically Targeted Magnetic Hyperthermia: Potential and Limitations. *Front. Pharmacol.* 9:831.

Che, W., M. Wei, Z. Sang, et al. 2018. Perovskite $LaNiO_{3-\delta}$ Oxide as an Anion-Intercalated Pseudocapacitor Electrode. *J. Alloys Compd.* 731:381–388.

Chen, G., Z. Hu, Y. Zhu, et al. 2018. Ultrahigh-Performance Tungsten-Doped Perovskites for the Oxygen Evolution Reaction. *J. Mater. Chem. A* 6:9854–9859.

Chen, Y., J. Fang, S. Lu, et al. 2015. One-Step Hydrothermal Synthesis of $BiOI/Bi_2WO_6$ Hierarchical Heterostructure with Highly Photocatalytic Activity. *J. Chem. Technol. Biotechnol.* 90:947–954.

Chen, Z., H. Jiang, W. Jin, et al. 2016. Enhanced Photocatalytic Performance over $Bi_4Ti_3O_{12}$ Nanosheets with Controllable Size and Exposed {001} Facets for Rhodamine B Degradation. *Appl. Catal. B Environ.* 180:698–706.

Chen, Z., W. Jin, Z. Lu, et al. 2015. Ferromagnetic and Photocatalytic Properties of Pure $BiFeO_3$ Powders Synthesized by Ethylene Glycol Assisted Hydrothermal Method. *J. Mater. Sci. Mater. Electron.* 26:1077–1086.

Cheng, J., M. Zhang, Y. Jiang, et al. 2016. Perovskite $La_{0.6}Sr_{0.4}Co_{0.2}Fe_{0.8}O_3$ as an Effective Electrocatalyst for Non-Aqueous Lithium Air Batteries. *Electrochimica Acta* 191:106–115.

Choi, W. S., M. F. Chisholm, D. J. Singh, et al. 2012. Wide Bandgap Tunability in Complex Transition Metal Oxides by Site-Specific Substitution. *Nat. Commun.* 3:1–6.

Ciambelli, P., S. Cimino, G. Lasorella, et al. 2002. CO Oxidation and Methane Combustion on $LaAl_{1-x}Fe_xO_3$ Perovskite Solid Solutions. *Appl. Catal. B Environ.* 37:231–241.

Cui, Y., J. Briscoe, and S. Dunn. 2013. Effect of Ferroelectricity on Solar-Light-Driven Photocatalytic Activity of $BaTiO_{3-\delta}$ Influence on the Carrier Separation and Stern Layer Formation. *Chem. Mater.* 25:4215–4223.

Dacquin, J. P., M. Cabié, C. R. Henry, et al. 2010. Structural Changes of Nano-Pt Particles during Thermal Ageing: Support-Induced Effect and Related Impact on the Catalytic Performances. *J. Catal.* 270:299–309.

Dacquin, J. P., C. Lancelot, C. Dujardin, et al. 2009. Influence of Preparation Methods of $LaCoO_3$ on the Catalytic Performances in the Decomposition of N_2O. *Appl. Catal. B Environ.* 91:596–604.

Dai, K., D. Li, L. Geng, et al. 2015. Facile Preparation of Bi_2MoO_6/Multi-Walled Carbon Nanotube Nanocomposite for Enhancing Photocatalytic Performance. *Mater. Lett.* 160:124–127.

Dai, X. P., R. J. Li, C. C. Yu, et al. 2006. Unsteady-State Direct Partial Oxidation of Methane to Synthesis Gas in a Fixed-Bed Reactor Using $AFeO_3$ (A = La, Nd, Eu) Perovskite-Type Oxides as Oxygen Storage. *J. Phys. Chem. B* 110:22525–22531.

Demirel, S., E. Oz, S. Altin, et al. 2017. Structural, Magnetic, Electrical and Electrochemical Properties of $SrCoO_{2.5}$, $Sr_9Co_2Mn_5O_{21}$ and $SrMnO_3$ Compounds. *Ceram. Int.* 43:14818–14826.

Deng, G., Y. Chen, M. Tao, et al. 2010. Study of the Electrochemical Hydrogen Storage Properties of the Proton-Conductive Perovskite-Type Oxide $LaCrO_3$ as Negative Electrode for Ni/MH Batteries. *Electrochimica Acta* 55:884–886.

Dharmadhikari, V. S., and W. W. Grannemann. 1982. Photovoltaic Properties of Ferroelectric $BaTiO_3$ Thin Films Rf Sputter Deposited on Silicon. *J. Appl. Phys.* 53:8988–8992.

Di, L. J., H. Yang, G. Hu, et al. 2014. Enhanced Photocatalytic Activity of $BiFeO_3$ Particles by Surface Decoration with Ag Nanoparticles. *J. Mater. Sci. Mater. Electron.* 25:2463–2469.

Dicken, M. J., L. A. Sweatlock, D. Pacifici, et al. 2008. Electrooptic Modulation in Thin Film Barium Titanate Plasmonic Interferometers. *Nano Lett.* 8:4048–4052.

Ding, Q.-P., Y.-P. Yuan, X. Xiong, et al. 2008. Enhanced Photocatalytic Water Splitting Properties of $KNbO_3$ Nanowires Synthesized through Hydrothermal Method. *J. Phys. Chem. C* 112:18846–18848.

Du, Z., P. Yang, L. Wang, et al. 2014. Electrocatalytic Performances of $LaNi_{1-x}Mg_xO_3$ Perovskite Oxides as Bi-Functional Catalysts for Lithium Air Batteries. *J. Power Sources* 265:91–96.

Dubey, A. K., B. Basu, K. Balani, et al. 2011. Multifunctionality of Perovskites $BaTiO_3$ and $CaTiO_3$ in a Composite with Hydroxyapatite as Orthopedic Implant Materials. *Integr. Ferroelectr.* 131:119–126.

Duran, P., J. Tartaj, F. Capel, et al. 2004. Formation, Sintering and Thermal Expansion Behaviour of Sr- and Mg-Doped $LaCrO_3$ as SOFC Interconnector Prepared by the Ethylene Glycol Polymerized Complex Solution Synthesis Method. *J. Eur. Ceram. Soc.* 24:2619–2629.

Durdu, S., and M. Usta. 2014. The Tribological Properties of Bioceramic Coatings Produced on Ti_6Al_4V Alloy by Plasma Electrolytic Oxidation. *Ceram. Int.* 40:3627–3635.

El-Ads, E. H., A. Galal, and N. F. Atta. 2015. Electrochemistry of Glucose at Gold Nanoparticles Modified Graphite/$SrPdO_3$ Electrode – Towards a Novel Non-Enzymatic Glucose Sensor. *J. Electroanal. Chem.* 749:42–52.

Elsiddig, Z. A., H. Xu, D. Wang, et al. 2017. Modulating Mn^{4+} Ions and Oxygen Vacancies in Nonstoichiometric $LaMnO_3$ Perovskite by a Facile Sol–Gel Method as High-Performance Supercapacitor Electrodes. *Electrochimica Acta* 253:422–429.

Esaka, T., H. Sakaguchi, and S. Kobayashi. 2004. Hydrogen Storage in Proton-Conductive Perovskite-Type Oxides and Their Application to Nickel–Hydrogen Batteries. *Solid State Ion.* 166:351–357.

Falcón, H., M. J. Martinez-Lope, J. A. Alonso, et al. 2000. Defect $LaCuO_{3-\delta}$ ($\Delta = 0.05–0.45$) Perovskites: Bulk and Surface Structures and Their Relevance in CO Oxidation. *Appl. Catal. B Environ.* 26:131–142.

Fan, H., D. Wang, Z. Liu, et al. 2015. Self-Assembled $BiVO_4/Bi_2WO_6$ Microspheres: Synthesis, Photoinduced Charge Transfer Properties and Photocatalytic Activities. *Dalton Trans.* 44:11725–11731.

Fang, M., X. Yao, W. Li, et al. 2018. The Investigation of Lithium Doping Perovskite Oxide $LiMnO_3$ as Possible LIB Anode Material. *Ceram. Int.* 44:8223–8231.

Fei, D. Q., T. Hudaya, and A. A. Adesina. 2005. Visible-Light Activated Titania Perovskite Photocatalysts: Characterization and Initial Activity Studies. *Catal. Commun.* 6:253–258.

Feng, Y.-N., H.-C. Wang, Y. Shen, et al. 2014. Magnetic and Photocatalytic Behaviors of Ba-Doped $BiFeO_3$ Nanofibers. *Int. J. Appl. Ceram. Technol.* 11:676–680.

Fergus, J. W. 2007. Perovskite Oxides for Semiconductor-Based Gas Sensors. *Sens. Actuators B* 123:1169–1179.

Fu, J., H. Zhao, J. Wang, et al. 2018. Preparation and Electrochemical Performance of Double Perovskite La_2CoMnO_6 Nanofibers. *Int. J. Miner. Metall. Mater.* 25:950–956.

Fu, Z., X. Lin, T. Huang, et al. 2012. Nano-Sized $La_{0.8}Sr_{0.2}MnO_3$ as Oxygen Reduction Catalyst in Nonaqueous Li/O_2 Batteries. *J. Solid State Electrochem.* 16:1447–1452.

Galal, A., N. F. Atta, and S. M. Ali. 2011. Investigation of the Catalytic Activity of $LaBO_3$ (B = Ni, Co, Fe or Mn) Prepared by the Microwave-Assisted Method for Hydrogen Evolution in Acidic Medium. *Electrochimica Acta* 56:5722–5730.

Galal, A., N. F. Atta, S. A. Darwish, et al. 2010. Electrocatalytic Evolution of Hydrogen on a Novel $SrPdO_3$ Perovskite Electrode. *J. Power Sources* 195:3806–3809.

Galal, A., H. K. Hassan, N. F. Atta, et al. 2018a. Effect of Redox Electrolyte on the Specific Capacitance of $SrRuO_{3-\delta}$ Reduced Graphene Oxide Nanocomposites. *J. Phys. Chem. C* 122:11641–11650.

Galal, A., H. K. Hassan, N. F. Atta, et al. 2019. Energy and Cost-Efficient Nano-Ru-Based Perovskites/RGO Composites for Application in High Performance Supercapacitors. *J. Colloid Interface Sci.* 538:578–586.

Galal, A., H. K. Hassan, T. Jacob, et al. 2018b. Enhancing the Specific Capacitance of SrRuO3 and Reduced Graphene Oxide in $NaNO_3$, H3PO_4 and KOH Electrolytes. *Electrochimica Acta* 260:738–747.

Gao, T., Z. Chen, Y. Zhu, et al. 2014. Synthesis of $BiFeO_3$ Nanoparticles for the Visible-Light Induced Photocatalytic Property. *Mater. Res. Bull.* 59:6–12.

George, G., S. L. Jackson, C. Q. Luo, et al. 2018. Effect of Doping on the Performance of High-Crystalline $SrMnO_3$ Perovskite Nanofibers as a Supercapacitor Electrode. *Ceram. Int.* 44:21982–21992.

Giannakas, A. E., A. K. Ladavos, and P. J. Pomonis. 2004. Preparation, Characterization and Investigation of Catalytic Activity for NO^+, CO Reaction of $LaMnO_3$ and $LaFeO_3$ Perovskites Prepared via Microemulsion Method. *Appl. Catal. B Environ.* 49:147–158.

Glass, A. M., D. von der Linde, and T. J. Negran. 1974. High-Voltage Bulk Photovoltaic Effect and the Photorefractive Process in $LiNbO_3$. In *Appl. Phys. Lett.* 25:233–235.

Gonon, P., and F. El Kamel. 2007. High-Density Capacitors Based on Amorphous $BaTiO_3$ Layers Grown under Hydrogen Containing Atmosphere. *Appl. Phys. Lett.* 90:232902.

Grenier, J.-C., M. Pouchard, and A. Wattiaux. 1996. Electrochemical Synthesis: Oxygen Intercalation. *Curr. Opin. Solid State Mater. Sci.* 1:233–240.

Grimaud, A., K. J. May, C. E. Carlton, et al. 2013. Double Perovskites as a Family of Highly Active Catalysts for Oxygen Evolution in Alkaline Solution. *Nat. Commun.* 4:1–7.

Grinberg, I., D. V. West, M. Torres, et al. 2013. Perovskite oxides for Visible-light-absorbing Ferroelectric and Photovoltaic Materials. *Nature* 503:509–512.

Guan, D., J. Zhou, Y.-C. Huang, et al. 2019. Screening Highly Active Perovskites for Hydrogen-Evolving Reaction via Unifying Ionic Electronegativity Descriptor. *Nat. Commun.* 10:1–8.

Gunasekaran, N., S. Rajadurai, and J. J. Carberry. 1995. Catalytic Decomposition of Nitrous Oxide over Perovskite Type Solid Oxide Solutions and Supported Noble Metal Catalysts. *Catal. Lett.* 35:373–382.

Guo, Y., T. Shao, H. You, et al. 2017. Polyvinylpyrrolidone-assisted Solvothermal Synthesis of Porous $LaCoO_3$ Nanospheres as Supercapacitor Electrode. *Int. J. Electrochem. Sci.* 12:7121–7127.

Gwon, O., S. Yoo, J. Shin, et al. 2014. Optimization of $La_{1-x}Sr_xCoO_{3-\delta}$ Perovskite Cathodes for Intermediate Temperature Solid Oxide Fuel Cells through the Analysis of Crystal Structure and Electrical Properties. *Int. J. Hydrog. Energy* 39:20806–20811.

Hailili, R., Z.-Q. Wang, Y. Li, et al. 2018. Oxygen Vacancies Induced Visible-Light Photocatalytic Activities of $CaCu_3Ti_4O_{12}$ with Controllable Morphologies for Antibiotic Degradation. *Appl. Catal. B Environ.* 221:422–432.

Han, X., T. Zhang, J. Du, et al. 2012. Porous Calcium–Manganese Oxide Microspheres for Electrocatalytic Oxygen Reduction with High Activity. *Chem. Sci.* 4:368–376.

Hardin, W. G., J. T. Mefford, D. A. Slanac, et al. 2014. Tuning the Electrocatalytic Activity of Perovskites through Active Site Variation and Support Interactions. *Chem. Mater.* 26:3368–3376.

Hardin, W. G., D. A. Slanac, X. Wang, et al. 2013. Highly Active, Nonprecious Metal Perovskite Electrocatalysts for Bifunctional Metal–Air Battery Electrodes. *J. Phys. Chem. Lett.* 4:1254–1259.

He, H., J. Yin, Y. Li, et al. 2014. Size Controllable Synthesis of Single-Crystal Ferroelectric $Bi_4Ti_3O_{12}$ Nanosheet Dominated with {001} Facets toward Enhanced Visible-Light-Driven Photocatalytic Activities. *Appl. Catal. B Environ.* 156–157:35–43.

He, J., J. Sunarso, Y. Zhu, et al. 2017. High-Performance Non-Enzymatic Perovskite Sensor for Hydrogen Peroxide and Glucose Electrochemical Detection. *Sens. Actuators B Chem.* 244:482–491.

Higashi, M., R. Abe, T. Takata, et al. 2009. Photocatalytic Overall Water Splitting under Visible Light Using $ATaO_2N$ (A = Ca, Sr, Ba) and WO_3 in a IO_3^-/I^- Shuttle Redox Mediated System. *Chem. Mater.* 21:1543–1549.

Higashi, M., K. Domen, and R. Abe. 2013. Fabrication of an Efficient $BaTaO_2N$ Photoanode Harvesting a Wide Range of Visible Light for Water Splitting. *J. Am. Chem. Soc.* 135:10238–10241.

Ho, K.-H., and J. Wang. 2017. Hydrazine Reduction of $LaNiO_3$ for Active Materials in Supercapacitors. *J. Am. Ceram. Soc.* 100:4629–4637.

Hong, W. T., K. A. Stoerzinger, Y.-L. Lee, et al. 2017. Charge-Transfer-Energy-Dependent Oxygen Evolution Reaction Mechanisms for Perovskite Oxides. *Energy Environ. Sci.* 10:2190–2200.

Hu, C.-C., Y.-L. Lee, and H. Teng. 2011. Efficient Water Splitting over $Na_{1-x}K_xTaO_3$ Photocatalysts with Cubic Perovskite Structure. *J. Mater. Chem.* 21:3824–3830.

Hu, L., Y. Deng, K. Liang, et al. 2015. $LaNiO_3$/NiO Hollow Nanofibers with Mesoporous Wall: A Significant Improvement in NiO Electrodes for Supercapacitors. *J. Solid State Electrochem.* 19:629–637.

Hua, H., and Y. K. Vohra. 1997. Pressure-Induced Blueshift of Nd^{3+} Fluorescence Emission in $YAlO_3$: Near Infrared Pressure Sensor. *Appl. Phys. Lett.* 71:2602–2604.

Huang, Y., Y. Wei, S. Cheng, et al. 2010. Photocatalytic Property of Nitrogen-Doped Layered Perovskite $K_2La_2Ti_3O_{10}$. *Sol. Energy Mater. Sol. Cells* 94:761–766.

Hur, S. G., T. W. Kim, S.-J. Hwang, et al. 2005. Synthesis of New Visible Light Active Photocatalysts of $Ba(In_{1/3}Pb_{1/3}M_{1/3})O_3$ (M′ = Nb, Ta): A Band Gap Engineering Strategy Based on Electronegativity of a Metal Component. *J. Phys. Chem. B* 109:15001–15007.

Hwang, D. K., S. Kim, J.-H. Lee, et al. 2011. Phase Evolution of Perovskite $LaNiO_3$ Nanofibers for Supercapacitor Application and P-Type Gas Sensing Properties of LaOCl–NiO Composite Nanofibers. *J. Mater. Chem.* 21:1959–1965.

Hyodo, T., M. Hayashi, N. Miura, et al. 1996. Catalytic Activities of Rare-Earth Manganites for Cathodic Reduction of Oxygen in Alkaline Solution. *J. Electrochem. Soc.* 143:L266.

Iguchi, F., T. Tokikawa, T. Miyoshi, et al. 2007. Performance of $BaZrO_3$ Based Proton Conductors as an Electrolyte for Intermediate Temperature Operating SOFC. *ECS Trans.* 7:2331–2336.

Ionescu, A. M. 2012. Ferroelectric Devices Show Potential. *Nat. Nanotechnol.* 7:83–85.

Ishihara, T., H. Minami, H. Matsuda, et al. 1995. Application of the New Oxide Ionic Conductor, $LaGaO_3$, to the Solid Electrolyte of Fuel Cells. *ECS Proc. Vol.* 1995–1:344–352.

Ivanov, S., P. Nordblad, R. Mathieu, et al. 2011. Short-Range Spin Order and Frustrated Magnetism in Mn_2InSbO_6 and Mn_2ScSbO_6. *Eur. J. Inorg. Chem.* 2011:4691–4699.

Iwahara, H., T. Esaka, H. Uchida, et al. 1981. Proton Conduction in Sintered Oxides and Its Application to Steam Electrolysis for Hydrogen Production. *Solid State Ion.* 3–4:359–363.

Iwase, A., H. Kato, and A. Kudo. 2009. The Effect of Alkaline Earth Metal Ion Dopants on Photocatalytic Water Splitting by $NaTaO_3$ Powder. *ChemSusChem* 2:873–877.

Jelínek, M., P. Vaněk, Z. Tolde, et al. 2017. PLD Prepared Bioactive $BaTiO_3$ Films on TiNb Implants. *Mater. Sci. Eng. C* 70:334–339.

Jeon, J. H., H.-Y. Joo, Y.-M. Kim, et al. 2016. Selector-Free Resistive Switching Memory Cell Based on $BiFeO_3$ Nano-Island Showing High Resistance Ratio and Nonlinearity Factor. *Sci. Rep.* 6:1–10.

Jin, C., Z. Yang, X. Cao, et al. 2014. A Novel Bifunctional Catalyst of $Ba_{0.9}Co_{0.5}Fe_{0.4}Nb_{0.1}O_{3-\delta}$ Perovskite for Lithium–Air Battery. *Int. J. Hydrog. Energy* 39:2526–2530.

Jung, J.-I., H. Y. Jeong, M. G. Kim, et al. 2015. Fabrication of $Ba_{0.5}Sr_{0.5}Co_{0.8}Fe_{0.2}O_{3-\delta}$ Catalysts with Enhanced Electrochemical Performance by Removing an Inherent Heterogeneous Surface Film Layer. *Adv. Mater.* 27:266–271.

Jung, J.-I., H. Y. Jeong, J.-S. Lee, et al. 2014. A Bifunctional Perovskite Catalyst for Oxygen Reduction and Evolution. *Angew. Chem. Int. Ed.* 53:4582–4586.

Kačenka, M., O. Kaman, J. Kotek, et al. 2011. Dual Imaging Probes for Magnetic Resonance Imaging and Fluorescence Microscopy Based on Perovskite Manganite Nanoparticles. *J. Mater. Chem.* 21:157–164.

Kako, T., Z. Zou, M. Katagiri, et al. 2007. Decomposition of Organic Compounds over $NaBiO_3$ under Visible Light Irradiation. *Chem. Mater.* 19:198–202.

Kalubarme, R. S., G.-E. Park, K.-N. Jung, et al. 2014. $LaNi_xCo_{1-x}O_{3-\delta}$ Perovskites as Catalyst Material for Non-Aqueous Lithium-Oxygen Batteries. *J. Electrochem. Soc.* 161:A880.

Kato, H., K. Asakura, and A. Kudo. 2003. Highly Efficient Water Splitting into H_2 and O_2 over Lanthanum-Doped $NaTaO_3$ Photocatalysts with High Crystallinity and Surface Nanostructure. *J. Am. Chem. Soc.* 125:3082–3089.

Kato, H., H. Kobayashi, and A. Kudo. 2002. Role of Ag^+ in the Band Structures and Photocatalytic Properties of $AgMO_3$ (M: Ta and Nb) with the Perovskite Structure. *J. Phys. Chem. B* 106:12441–12447.

Kato, H., and A. Kudo. 1999. Highly Efficient Decomposition of Pure Water into H_2 and O_2 over $NaTaO_3$ Photocatalysts. *Catal. Lett.* 58:153–155.

Kato, H., and A. Kudo. 2001. Water Splitting into H_2 and O_2 on Alkali Tantalate Photocatalysts $ATaO_3$ (A = Li, Na, and K). *J. Phys. Chem. B* 105:4285–4292.

Kato, H., and A. Kudo. 2003. Photocatalytic Water Splitting into H_2 and O_2 over Various Tantalate Photocatalysts. *Catal. Today*, 78:561–569.

Khine, M. S. S., L. Chen, S. Zhang, et al. 2013. Syngas Production by Catalytic Partial Oxidation of Methane over $(La_{0.7}A_{0.3})BO_3$ (A = Ba, Ca, Mg, Sr, and B = Cr or Fe) Perovskite Oxides for Portable Fuel Cell Applications. *Int. J. Hydrog. Energy* 38:13300–13308.

Kim, C. H., G. Qi, K. Dahlberg, et al. 2010. Strontium-Doped Perovskites Rival Platinum Catalysts for Treating NO_x in Simulated Diesel Exhaust. *Science* 327:1624–1627.

Kim, J., X. Yin, K.-C. Tsao, et al. 2014a. $Ca_2Mn_2O_5$ as Oxygen-Deficient Perovskite Electrocatalyst for Oxygen Evolution Reaction. *J. Am. Chem. Soc.* 136:14646–14649.

Kim, P., N. M. Doss, J. P. Tillotson, et al. 2009. High Energy Density Nanocomposites Based on Surface-Modified $BaTiO_3$ and a Ferroelectric Polymer. *ACS Nano* 3:2581–2592.

Kim, P., S. C. Jones, P. J. Hotchkiss, et al. 2007. Phosphonic Acid-Modified Barium Titanate Polymer Nanocomposites with High Permittivity and Dielectric Strength. *Adv. Mater.* 19:1001–1005.

Kim, Y., K. Kook, S. K. Hwang, et al. 2014b. Polymer/Perovskite – Type Nanoparticle Multilayers with Multielectric Properties Prepared from Ligand Addition-Induced Layer – by – Layer Assembly. *ACS Nano* 8:2419–2430.

Kong, J., T. Yang, Z. Rui, et al. 2019. Perovskite-Based Photocatalysts for Organic Contaminants Removal: Current Status and Future Perspectives. *Catal. Today* 327:47–63.

Konta, R., T. Ishii, H. Kato, et al. 2004. Photocatalytic Activities of Noble Metal Ion Doped $SrTiO_3$ under Visible Light Irradiation. *J. Phys. Chem. B* 108:8992–8995.

Kozuka, H., K. Ohbayashi, and K. Koumoto. 2015. Electronic Conduction in La-Based Perovskite-Type Oxides. *Sci. Technol. Adv. Mater.* 16:026001.

Kudo, T., H. Obayashi, and T. Gejo. 1975. Electrochemical Behavior of the Perovskite-Type $Nd_{1-x}Sr_xCoO_3$ in an Aqueous Alkaline Solution. *J. Electrochem. Soc.* 122:159–163.

Kulkarni, V. M., D. Bodas, and K. M. Paknikar. 2015. Lanthanum Strontium Manganese Oxide (LSMO) Nanoparticles: A Versatile Platform for Anticancer Therapy. *RSC Adv.* 5:60254–60263.

Kumar, S., Y. Teraoka, A. G. Joshi, et al. 2011a. Ag Promoted $La_{0.8}Ba_{0.2}MnO_3$ Type Perovskite Catalyst for N_2O Decomposition in the Presence of O_2, NO and H_2O. *J. Mol. Catal. Chem.* 348:42–54.

Kumar, V., Govind, and S. Uma. 2011b. Investigation of Cation (Sn^{2+}) and Anion (N^{3-}) Substitution in Favor of Visible Light Photocatalytic Activity in the Layered Perovskite $K_2La_2Ti_3O_{10}$. *J. Hazard. Mater.* 189: 502–508.

Ladavos, A., and P. Pomonis. 2015. Methane Combustion on Perovskites. In *Perovskites and Related Mixed Oxides*, 367–388. Weinheim, Germany: John Wiley & Sons, Ltd.

Lan, A., and A. S. Mukasyan. 2008. Complex $SrRuO_3$–Pt and $LaRuO_3$–Pt Catalysts for Direct Alcohol Fuel Cells. *Ind. Eng. Chem. Res.* 47:8989–8994.

Lang, X., H. Mo, X. Hu, et al. 2017. Supercapacitor Performance of Perovskite $La_{1-x}Sr_xMnO_3$. *Dalton Trans.* 46:13720–13730.

Li, C., G. Chen, J. Sun, Y. Feng, et al. 2015a. Ultrathin Nanoflakes Constructed Erythrocyte-like Bi_2WO_6 Hierarchical Architecture via Anionic Self-Regulation Strategy for Improving Photocatalytic Activity and Gas-Sensing Property. *Appl. Catal. B Environ.* 163:415–423.

Li, C., G. Chen, J. Sun, J. Rao, et al. 2015b. A Novel Mesoporous Single-Crystal-Like Bi_2WO_6 with Enhanced Photocatalytic Activity for Pollutants Degradation and Oxygen Production. *ACS Appl. Mater. Interfaces* 7:25716–25724.

Li, F., Y. Liu, R. Liu, et al. 2010a. Preparation of Ca-Doped $LaFeO_3$ Nanopowders in a Reverse Microemulsion and Their Visible Light Photocatalytic Activity. *Mater. Lett.* 64:223–225.

Li, J., J. Zeng, L. Jia, et al. 2010b. Investigations on the Effect of Cu^{2+}/Cu^{1+} Redox Couples and Oxygen Vacancies on Photocatalytic Activity of Treated $LaNi_{1-x}Cu_xO_3$ (X = 0.1, 0.4, 0.5). *Int. J. Hydrog. Energy* 35:12733–12740.

Li, L., Z. Chang, and X.-B. Zhang. 2017a. Recent Progress on the Development of Metal-Air Batteries. *Adv. Sustain. Syst.* 1:1700036.

Li, L., H. Yang, Z. Gao, et al. 2019. Nickel-Substituted $Ba_{0.5}Sr_{0.5}Co_{0.8}Fe_{0.2}O_{3-\delta}$: A Highly Active Perovskite Oxygen Electrode for Reduced-Temperature Solid Oxide Fuel Cells. *J. Mater. Chem. A* 7:12343–12349.

Li, M., X. Xiao, Y. Liu, et al. 2017b. Ternary Perovskite Cobalt Titanate/Graphene Composite Material as Long-Term Cyclic Anode for Lithium-Ion Battery. *J. Alloys Compd.* 700:54–60.

Li, P., J. Zhang, Q. Yu, J. Qiao, et al. 2015c. One-Dimensional Porous $La_{0.5}Sr_{0.5}CoO_{2.91}$ Nanotubes as a Highly Efficient Electrocatalyst for Rechargeable Lithium-Oxygen Batteries. *Electrochimica Acta* 165:78–84.

Li, S., J. Zhang, M. G. Kibria, et al. 2013. Remarkably Enhanced Photocatalytic Activity of Laser Ablated Au Nanoparticle Decorated $BiFeO_3$ Nanowires under Visible-Light. *Chem. Commun.* 49:5856–5858.

Li, X., R. Huang, Y. Hu, et al. 2012. A Templated Method to Bi_2WO_6 Hollow Microspheres and Their Conversion to Double-Shell Bi_2O_3/Bi_2WO_6 Hollow Microspheres with Improved Photocatalytic Performance. *Inorg. Chem.* 51:6245–6250.

Li, Y., S. Yao, W. Wen, et al. 2010c. Sol–Gel Combustion Synthesis and Visible-Light-Driven Photocatalytic Property of Perovskite $LaNiO_3$. *J. Alloys Compd.* 491:560–564.

Li, Y., H. Xu, P.-H. Chien, et al. 2018. A Perovskite Electrolyte That Is Stable in Moist Air for Lithium-Ion Batteries. *Angew. Chem. Int. Ed.* 57:8587–8591.

Li, Z., H. Xue, X. Wang, et al. 2006. Characterizations and Photocatalytic Activity of Nanocrystalline $La_{1.5}Ln_{0.5}Ti_2O_7$ (Ln = Pr, Gd, Er) Solid Solutions Prepared via a Polymeric Complex Method. *J. Mol. Catal. Chem.* 260:56–61.

Li, Z., Y. Wang, J. Liu, et al. 2009. Photocatalytic Hydrogen Production from Aqueous Methanol Solutions under Visible Light over $Na(Bi_xTa_{1-x})O_3$ Solid-Solution. *Int. J. Hydrog. Energy* 34:147–152.

Li, Z., W. Zhang, C. Yuan, et al. 2017d. Controlled Synthesis of Perovskite Lanthanum Ferrite Nanotubes with Excellent Electrochemical Properties. *RSC Adv.* 7:12931–12937.

Liang, K., N. Wang, M. Zhou, et al. 2013. Mesoporous $LaNiO_3/NiO$ Nanostructured Thin Films for High-Performance Supercapacitors. *J. Mater. Chem. A* 1:9730–9736.

Liu, G., H. Chen, L. Xia, et al. 2015. Hierarchical Mesoporous/Macroporous Perovskite $La_{0.5}Sr_{0.5}CoO_{3-x}$ Nanotubes: A Bifunctional Catalyst with Enhanced Activity and Cycle Stability for Rechargeable Lithium Oxygen Batteries. *ACS Appl. Mater. Interfaces* 7:22478–22486.

Liu, H., X. Ding, L. Wang, et al. 2018a. Cation Deficiency Design: A Simple and Efficient Strategy for Promoting Oxygen Evolution Reaction Activity of Perovskite Electrocatalyst. *Electrochimica Acta* 259:1004–1010.

Liu, J. W., G. Chen, Z. H. Li, et al. 2007. Hydrothermal Synthesis and Photocatalytic Properties of $ATaO_3$ and $ANbO_3$ (A = Na and K). *Int. J. Hydrog. Energy* 32:2269–2272.

Liu, P., Z. Liu, P. Wu, et al. 2018b. Enhanced Capacitive Performance of Nickel Oxide on Porous $La_{0.7}Sr_{0.3}CoO_{3-\delta}$ Ceramic Substrate for Electrochemical Capacitors. *Int. J. Hydrog. Energy* 43:19589–19599.

Liu, X., G. Du, J. Zhu, et al. 2016a. $NiO/LaNiO_3$ Film Electrode with Binder-Free for High Performance Supercapacitor. *Appl. Surf. Sci.* 384:92–98.

Liu, X., C. Qin, L. Cao, et al. 2017. A Silver Niobate Photocatalyst $AgNb_7O_{18}$ with Perovskite-like Structure. *J. Alloys Compd.* 724:381–388.

Liu, Y., J. Dinh, M. O. Tade, et al. 2016b. Design of Perovskite Oxides as Anion-Intercalation-Type Electrodes for Supercapacitors: Cation Leaching Effect. *ACS Appl. Mater. Interfaces* 8:23774–23783.

Liu, Y., Z. Wang, J.-P. M. Veder, et al. 2018c. Highly Defective Layered Double Perovskite Oxide for Efficient Energy Storage via Reversible Pseudocapacitive Oxygen-Anion Intercalation. *Adv. Energy Mater.* 8:1702604.

Lu, F., Y. Wang, C. Jin, et al. 2015. Microporous $La_{0.8}Sr_{0.2}MnO_3$ Perovskite Nanorods as Efficient Electrocatalysts for Lithium–Air Battery. *J. Power Sources* 293:726–733.

Ma, Q., M. Balaguer, D. Pérez-Coll, et al. 2018. Characterization and Optimization of $La_{0.97}Ni_{0.5}Co_{0.5}O_{3-\delta}$ -Based Air-Electrodes for Solid Oxide Cells. *ACS Appl. Energy Mater.* 1:2784–2792.

Ma, Y., Y. Jia, Z. Jiao, et al. 2015. Hierarchical Bi_2MoO_6 Nanosheet-Built Frameworks with Excellent Photocatalytic Properties. *Chem. Commun.* 51:6655–6658.

Man, I. C., H.-Y. Su, F. Calle-Vallejo, et al. 2011. Universality in Oxygen Evolution Electrocatalysis on Oxide Surfaces. *ChemCatChem* 3:1159–1165.

Manoharan, R., and J. B. Goodenough. 1990. Hydrogen Evolution on $Sr_xNbO_{3-\delta}$ ($0.7 \leq x \leq 0.95$) in Acid. *J. Electrochem. Soc.* 137:910–913.

Matsumoto, Y., J. Kurimoto, and E. Sato. 1979. Oxygen Evolution on $SrFeO_3$ Electrode. *J. Electroanal. Chem. Interfacial Electrochem.* 102:77–83.

Matsumoto, Y., and E. Sato. 1986. Electrocatalytic Properties of Transition Metal Oxides for Oxygen Evolution Reaction. *Mater. Chem. Phys.* 14:397–426.

Matsumoto, Y., H. Yoneyama, and H. Tamura. 1977. Catalytic Activity for Electrochemical Reduction of Oxygen of Lanthanum Nickel Oxide and Related Oxides. *J. Electroanal. Chem. Interfacial Electrochem.* 79:319–326.

Meadowcroft, D. B. 1970. Low-Cost Oxygen Electrode Material. *Nature* 226:847–848.

Mefford, J. T., W. G. Hardin, S. Dai, et al. 2014. Anion Charge Storage through Oxygen Intercalation in $LaMnO_3$ Perovskite Pseudocapacitor Electrodes. *Nat. Mater.* 13:726–732.

Mizoguchi, H., K. Ueda, M. Orita, et al. 2002. Decomposition of Water by a $CaTiO_3$ Photocatalyst under UV Light Irradiation. *Mater. Res. Bull.* 37:2401–2406.

Mo, H., H. Nan, X. Lang, et al. 2018. Influence of Calcium Doping on Performance of $LaMnO_3$ Supercapacitors. *Ceram. Int.* 44:9733–9741.

Moniruzzaman, M., S. Sahoo, D. Ghosh, et al. 2013. Preparation and Characterization of Polypyrrole/Modified Multiwalled Carbon Nanotube Nanocomposites Polymerized in Situ in the Presence of Barium Titanate. *J. Appl. Polym. Sci.* 128:698–705.

Naeem Khan, M., A. K. Azad, C. D. Savaniu, et al. 2017. Robust Doped $BaCeO_{3-\delta}$ Electrolyte for IT-SOFCs. *Ionics* 23:2387–2396.

Nagai, T., W. Ito, and T. Sakon. 2007. Relationship between Cation Substitution and Stability of Perovskite Structure in $SrCoO_{3-\delta}$-Based Mixed Conductors. *Solid State Ion.* 177:3433–3444.

Najjar, H., and H. Batis. 2016. Development of Mn-Based Perovskite Materials: Chemical Structure and Applications. *Catal. Rev.* 58:371–438.

Nakanishi, H., K. Iizuka, T. Takayama, et al. 2017. Highly Active $NaTaO_3$-Based Photocatalysts for CO_2 Reduction to Form CO Using Water as the Electron Donor. *ChemSusChem* 10:112–118.

Nechache, R., C. Harnagea, S. Li, et al. 2015. Bandgap Tuning of Multiferroic Oxide Solar Cells. *Nat. Photonics* 9:61–67.

Nemudry, A., E. L. Goldberg, M. Aguirre, et al. 2002. Electrochemical Topotactic Oxidation of Nonstoichiometric Perovskites at Ambient Temperature. *Solid State Sci.* 4:677–690.

Nili, H., S. Walia, S. Balendhran, et al. 2014. Nanoscale Resistive Switching in Amorphous Perovskite Oxide (a-$SrTiO_3$) Memristors. *Adv. Funct. Mater.* 24:6741–6750.

Obayashi, H., and T. Kudo. 1978. Perovskite-Type Compounds as Electrode Catalysts for Cathodic Reduction of Oxygen. *Mater. Res. Bull.* 13:1409–1413.

Oh, M. Y., J. S. Jeon, J. J. Lee, et al. 2015. The Bifunctional Electrocatalytic Activity of Perovskite $La_{0.6}Sr_{0.4}CoO_{3-\delta}$ for Oxygen Reduction and Evolution Reactions. *RSC Adv.* 5:19190–19198.

Parida, K. M., K. H. Reddy, S. Martha, et al. 2010. Fabrication of Nanocrystalline $LaFeO_3$: An Efficient Sol–Gel Auto-Combustion Assisted Visible Light Responsive Photocatalyst for Water Decomposition. *Int. J. Hydrog. Energy* 35:12161–12168.

Park, H. W., D. U. Lee, M. G. Park, et al. 2015. Perovskite–Nitrogen-Doped Carbon Nanotube Composite as Bifunctional Catalysts for Rechargeable Lithium–Air Batteries. *ChemSusChem* 8:1058–1065.

Petrović, S., L. Karanović, P. K. Stefanov, et al. 2005. Catalytic Combustion of Methane over Pd Containing Perovskite Type Oxides. *Appl. Catal. B Environ.* 58:133–141.

Phokha, S., S. Hunpratub, B. Usher, et al. 2017. Effect of Synthesis Temperature on the Magneto-Electrochemical Properties of $LaFe_{0.9}Co_{0.1}O_3$ Nanoparticles. *J. Alloys Compd.* 708:605–611.

Prabu, M., P. Ramakrishnan, P. Ganesan, et al. 2015. $LaTi_{0.65}Fe_{0.35}O_{3-\delta}$ Nanoparticle-Decorated Nitrogen-Doped Carbon Nanorods as an Advanced Hierarchical Air Electrode for Rechargeable Metal-Air Batteries. *Nano Energy* 15:92–103.

Qin, M., K. Yao, and Y. C. Liang. 2009. Photovoltaic Mechanisms in Ferroelectric Thin Films with the Effects of the Electrodes and Interfaces. *Appl. Phys. Lett.* 95:022912.

Rai, A., A. L. Sharma, and A. K. Thakur. 2014. Evaluation of Aluminium Doped Lanthanum Ferrite Based Electrodes for Supercapacitor Design. *Solid State Ion.* 262:230–233.

Rai, A., and A. K. Thakur. 2017. Effect of Na and Mn Substitution in Perovskite Type $LaFeO_3$ for Storage Device Applications. *Ionics* 23:2863–2869.

Retuerto, M., F. Calle-Vallejo, L. Pascual, et al. 2019. $La_{1.5}Sr_{0.5}NiMn_{0.5}Ru_{0.5}O_6$ Double Perovskite with Enhanced ORR/OER Bifunctional Catalytic Activity. *ACS Appl. Mater. Interfaces* 11:21454–21464.

Rincón, R. A., J. Masa, S. Mehrpour, et al. 2014. Activation of Oxygen Evolving Perovskites for Oxygen Reduction by Functionalization with Fe–N_x/C Groups. *Chem. Commun.* 50:14760–14762.

Rincón, R. A., E. Ventosa, F. Tietz, et al. 2014. Evaluation of Perovskites as Electrocatalysts for the Oxygen Evolution Reaction. *ChemPhysChem* 15:2810–2816.

Russo, N., D. Mescia, D. Fino, et al. 2007. N_2O Decomposition over Perovskite Catalysts. *Ind. Eng. Chem. Res.* 46:4226–4231.

de Santos, M. S., R. C. R. Neto, F. B. Noronha, et al. 2018. Perovskite as Catalyst Precursors in the Partial Oxidation of Methane: The Effect of Cobalt, Nickel and Pretreatment. *Catal. Today* 299:229–241.

Shafi, P. M., A. C. Bose, and A. Vinu. 2018a. Electrochemical Material Processing via Continuous Charge-Discharge Cycling: Enhanced Performance upon Cycling for Porous $LaMnO_3$ Perovskite Supercapacitor Electrodes. *ChemElectroChem* 5:3723–3730.

Shafi, P. M., N. Joseph, A. Thirumurugan, et al. 2018b. Enhanced Electrochemical Performances of Agglomeration-Free $LaMnO_3$ Perovskite Nanoparticles and Achieving High Energy and Power Densities with Symmetric Supercapacitor Design. *Chem. Eng. J.* 338:147–156.

Shanmugavadivel, M., V. V. Dhayabaran, and M. Subramanian. 2016. Nanosized $BaMnO_3$ as High Performance Supercapacitor Electrode Material: Fabrication and Characterization. *Mater. Lett.* 181:335–339.

Shao, T., H. You, Z. Zhai, et al. 2017. Hollow Spherical $LaNiO_3$ Supercapacitor Electrode Synthesized by a Facile Template-Free Method. *Mater. Lett.* 201:122–124.

Shao, Z., and S. M. Haile. 2004. A High-Performance Cathode for the next Generation of Solid-Oxide Fuel Cells. *Nature* 431:170–173.

Shlyakhtin, O. A., V. G. Leontiev, Y.-J. Oh, et al. 2007. New Manganite-Based Mediators for Self-Controlled Magnetic Heating. *Smart Mater. Struct.* 16: N35–N39.

Shoji, S., G. Yin, M. Nishikawa, et al. 2016. Photocatalytic Reduction of CO_2 by Cu_xO Nanocluster Loaded $SrTiO_3$ Nanorod Thin Film. *Chem. Phys. Lett.* 658:309–314.

Siaï, A., P. Haro-González, K. Horchani Naifer, et al. 2018. Optical Temperature Sensing of Er^{3+}/Yb^{3+} Doped $LaGdO_3$ Based on Fluorescence Intensity Ratio and Lifetime Thermometry. *Opt. Mater.* 76:34–41.

Singh, D. K., and J. Manam. 2018. Efficient Dual Emission Mode of Green Emitting Perovskite $BaTiO_3$: Er^{3+} Phosphors for Display and Temperature Sensing Applications. *Ceram. Int.* 44:10912–10920.

Singh, U. G., J. Li, J. W. Bennett, et al. 2007. A Pd-Doped Perovskite Catalyst, $BaCe_{1-x}Pd_xO_{3-\delta}$, for CO Oxidation. *J. Catal.* 249:349–358.

Siritanaratkul, B., K. Maeda, T. Hisatomi, et al. 2011. Synthesis and Photocatalytic Activity of Perovskite Niobium Oxynitrides with Wide Visible-Light Absorption Bands. *ChemSusChem* 4:74–78.

Slagtern, A., and U. Olsbye. 1994. Partial Oxidation of Methane to Synthesis Gas Using La-M-O Catalysts. *Appl. Catal. Gen.* 110:99–108.

Son, J. Y., J.-H. Lee, S. Song, et al. 2013. Four-States Multiferroic Memory Embodied Using Mn-Doped $BaTiO_3$ Nanorods. *ACS Nano* 7:5522–5529.

Song, X. C., W. T. Li, W. Z. Huang, et al. 2015. Enhanced Photocatalytic Activity of Cadmium-Doped Bi_2WO_6 Nanoparticles under Simulated Solar Light. *J. Nanoparticle Res.* 17:134.

Spanier, J. E., A. M. Kolpak, J. J. Urban, et al. 2006. Ferroelectric Phase Transition in Individual Single-Crystalline $BaTiO_3$ Nanowires. *Nano Lett.* 6:735–739.

Spinicci, R., M. Faticanti, P. Marini, et al. 2003. Catalytic Activity of $LaMnO_3$ and $LaCoO_3$ Perovskites towards VOCs Combustion. *J. Mol. Catal. Chem.* 197:147–155.

Subbaraman, R., D. Tripkovic, D. Strmcnik, et al. 2011. Enhancing Hydrogen Evolution Activity in Water Splitting by Tailoring Li^+-$Ni(OH)_2$-Pt Interfaces. *Science* 334:1256–1260.

Sun, B., and C. M. Li. 2015. Retracted Article: Light-Controlled Resistive Switching Memory of Multiferroic $BiMnO_3$ Nanowire Arrays. *Phys. Chem. Chem. Phys.* 17:6718–6721.

Sun, H., G. Chen, J. Sunarso, et al. 2018. Molybdenum and Niobium Codoped B-Site-Ordered Double Perovskite Catalyst for Efficient Oxygen Evolution Reaction. *ACS Appl. Mater. Interfaces* 10:16939–16942.

Sun, H., G. Chen, Y. Zhu, et al. 2017. B-Site Cation Ordered Double Perovskites as Efficient and Stable Electrocatalysts for Oxygen Evolution Reaction. *Chem. – Eur. J.* 23:5722–5728.

Sun, N., H. Liu, Z. Yu, et al. 2016a. Mn-Doped $La_{0.6}Sr_{0.4}CoO_3$ Perovskite Catalysts with Enhanced Performances for Non-Aqueous Electrolyte Li–O_2 Batteries. *RSC Adv.* 6:13522–13530.

Sun, N., H. Liu, Z. Yu, et al. 2016b. The Electrochemical Performance of $La_{0.6}Sr_{0.4}Co_{1-x}Ni_xO_3$ Perovskite Catalysts for Li-O_2 Batteries. *Ionics* 22:869–876.

Suntivich, J., K. J. May, H. A. Gasteiger, et al. 2011. A Perovskite Oxide Optimized for Oxygen Evolution Catalysis from Molecular Orbital Principles. *Science* 334:1383–1385.

Tan, H., Z. Zhao, W. Zhu, et al. 2014. Oxygen Vacancy Enhanced Photocatalytic Activity of Pervoskite $SrTiO_3$. *ACS Appl. Mater. Interfaces* 6:19184–19190.

Tanaka, H., M. Taniguchi, M. Uenishi, et al. 2006. Self-Regenerating Rh- and Pt-Based Perovskite Catalysts for Automotive-Emissions Control. *Angew. Chem. Int. Ed.* 45:5998–6002.

Tarragó, D. P., B. Moreno, E. Chinarro, et al. 2016. Perovskites Used in Fuel Cells. In *Perovskite Synthesis, Properties and Their Related Biochemical and Industrial Application*, 619–637. London, UK: InTechOpen.

Thirumalairajan, S., K. Girija, V. Ganesh, et al. 2013. Novel Synthesis of $LaFeO_3$ Nanostructure Dendrites: A Systematic Investigation of Growth Mechanism, Properties, and Biosensing for Highly Selective Determination of Neurotransmitter Compounds. *Cryst. Growth Des.* 13:291–302.

Tian, G., Y. Chen, R. Zhai, et al. 2013a. Hierarchical Flake-like Bi_2MoO_6/TiO_2 Bilayer Films for Visible-Light-Induced Self-Cleaning Applications. *J. Mater. Chem. A* 1:6961–6968.

Tian, G., Y. Chen, W. Zhou, et al. 2010. Facile Solvothermal Synthesis of Hierarchical Flower-like Bi_2MoO_6 Hollow Spheres as High Performance Visible-Light Driven Photocatalysts. *J. Mater. Chem.* 21:887–892.

Tian, J., Y. Sang, G. Yu, et al. 2013b. A Bi_2WO_6-Based Hybrid Photocatalyst with Broad Spectrum Photocatalytic Properties under UV, Visible, and Near-Infrared Irradiation. *Adv. Mater.* 25:5075–5080.

Tian, R., J. Fan, Y. Liu, et al. 2008. Low-Temperature Solid Oxide Fuel Cells with $La_{1-x}Sr_xMnO_3$ as the Cathodes. *J. Power Sources* 185:1247–1251.

Tiwari, P. K., and S. Basu. 2015. La, Y and Nb Doped $SrTiO_3$ Anodes for Electrolyte Supported SOFC. *ECS Trans.* 68:1435–1446.

Tiwari, S. K., P. Chartier, and R. N. Singh. 1995. Preparation of Perovskite-Type Oxides of Cobalt by the Malic Acid Aided Process and Their Electrocatalytic Surface Properties in Relation to Oxygen Evolution. *J. Electrochem. Soc.* 142:148.

Tomar, A. K., G. Singh, and R. K. Sharma. 2018. Fabrication of a Mo-Doped Strontium Cobaltite Perovskite Hybrid Supercapacitor Cell with High Energy Density and Excellent Cycling Life. *ChemSusChem* 11:4123–4130.

Townsend, T. K., N. D. Browning, and F. E. Osterloh. 2012. Nanoscale Strontium Titanate Photocatalysts for Overall Water Splitting. *ACS Nano* 6:7420–7426.

Tu, S., H. Huang, T. Zhang, et al. 2017. Controllable Synthesis of Multi-Responsive Ferroelectric Layered Perovskite-like $Bi_4Ti_3O_{12}$: Photocatalysis and Piezoelectric-Catalysis and Mechanism Insight. *Appl. Catal. B Environ.* 219:550–562.

Uenishi, M., M. Taniguchi, H. Tanaka, et al. 2005. Redox Behavior of Palladium at Start-up in the Perovskite-Type $LaFePdO_x$ Automotive Catalysts Showing a Self-Regenerative Function. *Appl. Catal. B Environ.* 57:267–273.

Vasala, S., and M. Karppinen. 2015. $A_2B'B''O_6$ Perovskites: A Review. *Prog. Solid State Chem.* 43:1–36.

Vasseur, S., E. Duguet, J. Portier, et al. 2006. Lanthanum Manganese Perovskite Nanoparticles as Possible in Vivo Mediators for Magnetic Hyperthermia. *J. Magn. Magn. Mater.* 302:315–320.

Wang, D., T. Kako, and J. Ye. 2008. Efficient Photocatalytic Decomposition of Acetaldehyde over a Solid-Solution Perovskite $(Ag_{0.75}Sr_{0.25})(Nb_{0.75}Ti_{0.25})O_3$ under Visible-Light Irradiation. *J. Am. Chem. Soc.* 130:2724–2725.

Wang, J., Y. Gao, D. Chen, et al. 2018a. Water Splitting with an Enhanced Bifunctional Double Perovskite. *ACS Catal.* 8:364–371.

Wang, J., S. Su, B. Liu, et al. 2013a. One-Pot, Low-Temperature Synthesis of Self-Doped $NaTaO_3$ Nanoclusters for Visible-Light-Driven Photocatalysis. *Chem. Commun.* 49:7830–7832.

Wang, K., J. Song, X. Duan, et al. 2017a. Perovskite $LaCoO_3$ Nanoparticles as Enzyme Mimetics: Their Catalytic Properties, Mechanism and Application in Dopamine Biosensing. *New J. Chem.* 41:8554–8560.

Wang, M., M. Fang, X. Min, et al. 2017b. Molten Salt Synthesis of $NaNb_xTa_{1-x}O_3$ Perovskites with Enhanced Photocatalytic Activity. *Chem. Phys. Lett.* 686:18–25.

Wang, Q., Z. Chen, Y. Chen, et al. 2012. Hydrogen Storage in Perovskite-Type Oxides ABO_3 for Ni/MH Battery Applications: A Density Functional Investigation. *Ind. Eng. Chem. Res.* 51:11821–11827.

Wang, S., H. Zhou, X. Wang, et al. 2016a. Up-Conversion Luminescence and Optical Temperature-Sensing Properties of Er^{3+}-Doped Perovskite $Na_{0.5}Bi_{0.5}TiO_3$ Nanocrystals. *J. Phys. Chem. Solids* 98:28–31.

Wang, W., B. Lin, H. Zhang, et al. 2019a. Synthesis, Morphology and Electrochemical Performances of Perovskite-Type Oxide $La_xSr_{1-x}FeO_3$ Nanofibers Prepared by Electrospinning. *J. Phys. Chem. Solids* 124:144–150.

Wang, W., M. O. Tadé, and Z. Shao. 2015. Research Progress of Perovskite Materials in Photocatalysis- and Photovoltaics-Related Energy Conversion and Environmental Treatment. *Chem. Soc. Rev.* 44:5371–5408.

Wang, X., W. Cao, L. Qin, et al. 2017c. Boosting the Peroxidase-Like Activity of Nanostructured Nickel by Inducing Its 3+ Oxidation State in $LaNiO_3$ Perovskite and Its Application for Biomedical Assays. *Theranostics* 7:2277–2286.

Wang, X., X. J. Gao, L. Qin, et al. 2019b. e_g Occupancy as an Effective Descriptor for the Catalytic Activity of Perovskite Oxide-Based Peroxidase Mimics. *Nat. Commun.* 10:1–8.

Wang, X. W., Q. Q. Zhu, X. E. Wang, et al. 2016b. Structural and Electrochemical Properties of $La_{0.85}Sr_{0.15}MnO_3$ Powder as an Electrode Material for Supercapacitor. *J. Alloys Compd.* 675:195–200.

Wang, Y., Z. Lv, L. Zhou, et al. 2018b. Emerging Perovskite Materials for High Density Data Storage and Artificial Synapses. *J. Mater. Chem. C* 6:1600–1617.

Wang, Y.-Z, H. Zhong, X. Li, et al. 2013b. Perovskite $LaTiO_3$–$Ag_{0.2}$ Nanomaterials for Nonenzymatic Glucose Sensor with High Performance. *Biosens. Bioelectron.* 48:56–60.

Wang, Z., Y. You, J. Yuan, et al. 2016c. Nickel-Doped $La_{0.8}Sr_{0.2}Mn_{1-x}Ni_xO_3$ Nanoparticles Containing Abundant Oxygen Vacancies as an Optimized Bifunctional Catalyst for Oxygen Cathode in Rechargeable Lithium–Air Batteries. *ACS Appl. Mater. Interfaces* 8:6520–6528.

Weber, W. J., C. W. Griffin, and J. L. Bates. 1987. Effects of Cation Substitution on Electrical and Thermal Transport Properties of $YCrO_3$ and $LaCrO_3$. *J. Am. Ceram. Soc.* 70:265–270.

Webster, T. J., C. Ergun, R. H. Doremus, et al. 2003. Increased Osteoblast Adhesion on Titanium-Coated Hydroxylapatite That Forms $CaTiO_3$. *J. Biomed. Mater. Res. A* 67A:975–980.

Weng, B., Z. Xiao, W. Meng, et al. 2017. Bandgap Engineering of Barium Bismuth Niobate Double Perovskite for Photoelectrochemical Water Oxidation. *Adv. Energy Mater.* 7:1602260.

Wilde, P. M., T. J. Guther, R. Oesten, et al. 1999. Strontium Ruthenate Perovskite as the Active Material for Supercapacitors. *J. Electroanal. Chem.* 461:154–160.

Wohlfahrt-Mehrens, M., J. Schenk, P. M. Wilde, et al. 2002. New Materials for Supercapacitors. *J. Power Sources* 105:182–188.

Wu, H., W. Zhou, T. J. Udovic, et al. 2008. Crystal Chemistry of Perovskite-Type Hydride $NaMgH_3$: Implications for Hydrogen Storage. *Chem. Mater.* 20:2335–2342.

Wu, Y., X. Ni, A. Beaurain, et al. 2012. Stoichiometric and Non-Stoichiometric Perovskite-Based Catalysts: Consequences on Surface Properties and on Catalytic Performances in the Decomposition of N_2O from Nitric Acid Plants. *Appl. Catal. B Environ.* 125:149–157.

Wu, Y.-B., J. Bi, and B.-B. Wei. 2015. Preparation and Supercapacitor Properties of Double-Perovskite La_2CoNiO_6 Inorganic Nanofibers. *Acta Phys. Chim. Sin.* 31:315–321.

Wygant, B. R., K. A. Jarvis, W. D. Chemelewski, et al. 2016. Structural and Catalytic Effects of Iron- and Scandium-Doping on a Strontium Cobalt Oxide Electrocatalyst for Water Oxidation. *ACS Catal.* 6:1122–1133.

Xie, C., L. Shi, J. Zhao, et al. 2017. Insight into the Enhancement of Transport Property for Oriented $La_{0.9}MnO_3$ Films. *J. Phys. Appl. Phys.* 50:205306.

Xin, Y., J. Jiang, K. Huo, et al. 2009. Bioactive $SrTiO_3$ Nanotube Arrays: Strontium Delivery Platform on Ti-Based Osteoporotic Bone Implants. *ACS Nano* 3:3228–3234.

Xu, D., S. Yang, Y. Jin, et al. 2015. Ag-Decorated $ATaO_3$ ($A = K, Na$) Nanocube Plasmonic Photocatalysts with Enhanced Photocatalytic Water-Splitting Properties. *Langmuir* 31:9694–9699.

Xu, J., W. Wang, S. Sun, et al. 2012a. Enhancing Visible-Light-Induced Photocatalytic Activity by Coupling with Wide-Band-Gap Semiconductor: A Case Study on Bi_2WO_6/TiO_2. *Appl. Catal. B Environ.* 111–112:126–132.

Xu, J.-J., D. Xu, Z.-L. Wang, et al. 2013. Synthesis of Perovskite-Based Porous $La_{0.75}Sr_{0.25}MnO_3$ Nanotubes as a Highly Efficient Electrocatalyst for Rechargeable Lithium–Oxygen Batteries. *Angew. Chem. Int. Ed.* 52:3887–3890.

Xu, L., C. Li, W. Shi, et al. 2012b. Visible Light-Response $NaTa_{1-x}Cu_xO_3$ Photocatalysts for Hydrogen Production from Methanol Aqueous Solution. *J. Mol. Catal. Chem.* 360:42–47.

Xu, Q., X. Han, F. Ding, et al. 2016a. A Highly Efficient Electrocatalyst of Perovskite $LaNiO_3$ for Nonaqueous Li–O_2 Batteries with Superior Cycle Stability. *J. Alloys Compd.* 664:750–755.

Xu, Q., S. Song, Y. Zhang, et al. 2016b. $Ba_{0.9}Co_{0.7}Fe_{0.2}Nb_{0.1}O_{3-\delta}$ Perovskite as Oxygen Electrode Catalyst for Rechargeable Li-Oxygen Batteries. *Electrochimica Acta* 191:577–585.

Xu, X., Y. Chen, W. Zhou, et al. 2016c. A Perovskite Electrocatalyst for Efficient Hydrogen Evolution Reaction. *Adv. Mater.* 28:6442–6448.

Xu, Z., Y. Liu, W. Zhou, et al. 2018. B-Site Cation-Ordered Double-Perovskite Oxide as an Outstanding Electrode Material for Supercapacitive Energy Storage Based on the Anion Intercalation Mechanism. *ACS Appl. Mater. Interfaces* 10:9415–9423.

Yamasita, D., T. Takata, M. Hara, et al. 2004. Recent Progress of Visible-Light-Driven Heterogeneous Photocatalysts for Overall Water Splitting. *Solid State Ion.* 172:591–595, Proceedings of the Fifteenth International Symposium on the Reactivity of Solids.

Yan, X., Q. Huang, B. Li, et al. 2013. Catalytic Performance of $LaCo_{0.5}M_{0.5}O_3$ (M = Mn, Cr, Fe, Ni, Cu) Perovskite-Type Oxides and $LaCo_{0.5}Mn_{0.5}O_3$ Supported on Cordierite for CO Oxidation. *J. Ind. Eng. Chem.* 19:561–565.

Yang, J., T. Fujigaya, and N. Nakashima. 2017a. Decorating Unoxidized-Carbon Nanotubes with Homogeneous Ni-Co Spinel Nanocrystals Show Superior Performance for Oxygen Evolution/Reduction Reactions. *Sci. Rep.* 7:1–9.

Yang, J., X. Wang, X. Zhao, et al. 2015. Synthesis of Uniform Bi_2WO_6-Reduced Graphene Oxide Nanocomposites with Significantly Enhanced Photocatalytic Reduction Activity. *J. Phys. Chem. C* 119:3068–3078.

Yang, K., X. Huang, Y. Huang, et al. 2013. Fluoro-Polymer@$BaTiO_3$ Hybrid Nanoparticles Prepared via RAFT Polymerization: Toward Ferroelectric Polymer Nanocomposites with High Dielectric Constant and Low Dielectric Loss for Energy Storage Application. *Chem. Mater.* 25:2327–2338.

Yang, Y., G. Yuan, Z. Yan, et al. 2017b. Flexible, Semitransparent, and Inorganic Resistive Memory Based on $BaTi_{0.95}Co_{0.05}O_3$ Film. *Adv. Mater.* 29:1700425.

Yang, Y. C., Y. Liu, J. H. Wei, et al. 2014. Electrospun Nanofibers of P-Type $BiFeO_3$/n-Type TiO_2 Hetero-Junctions with Enhanced Visible-Light Photocatalytic Activity. *RSC Adv.* 4:31941–31947.

Yao, W. F., H. Wang, X. H. Xu, et al. 2003. Synthesis and Photocatalytic Property of Bismuth Titanate $Bi_4Ti_3O_{12}$. *Mater. Lett.* 57:1899–1902.

Yin, W., W. Wang, and S. Sun. 2010. Photocatalytic Degradation of Phenol over Cage-like Bi_2MoO_6 Hollow Spheres under Visible-Light Irradiation. *Catal. Commun.* 11:647–650.

Yin, W.-J., B. Weng, J. Ge, et al. 2019. Oxide Perovskites, Double Perovskites and Derivatives for Electrocatalysis, Photocatalysis, and Photovoltaics. *Energy Environ. Sci.* 12:442–462.

Yoshida, H., L. Zhang, M. Sato, et al. 2015. Calcium Titanate Photocatalyst Prepared by a Flux Method for Reduction of Carbon Dioxide with Water. *Catal. Today* 251:132–139.

Yu, H., S. Ouyang, S. Yan, et al. 2011. Sol–Gel Hydrothermal Synthesis of Visible-Light-Driven Cr-Doped $SrTiO_3$ for Efficient Hydrogen Production. *J. Mater. Chem.* 21:11347–11351.

Yuasa, M., N. Tachibana, and K. Shimanoe. 2013. Oxygen Reduction Activity of Carbon-Supported $La_{1-x}Ca_xMn_{1-y}Fe_yO_3$ Nanoparticles. *Chem. Mater.* 25:3072–3079.

Yue, B., Q. Hu, L. Ji, et al. 2019. Facile Synthesis of Perovskite $CeMnO_3$ Nanofibers as an Anode Material for High Performance Lithium-Ion Batteries. *RSC Adv.* 9:38271–38279.

Zhang, C., C. Wu, Z. Zhang, et al. 2020. $LaNiO_3$ as a Novel Anode for Lithium-Ion Batteries. *Trans. Tianjin Univ.* 26: 142–147.

Zhang, D., Y. Song, Z. Du, et al. 2015a. Active $LaNi_{1-x}Fe_xO_3$ Bifunctional Catalysts for Air Cathodes in Alkaline Media. *J. Mater. Chem. A* 3:9421–9426.

Zhang, F., A. Yamakata, K. Maeda, et al. 2012a. Cobalt-Modified Porous Single-Crystalline $LaTiO_2N$ for Highly Efficient Water Oxidation under Visible Light. *J. Am. Chem. Soc.* 134:8348–8351.

Zhang, H., G. Chen, Y. Li, et al. 2010a. Electronic Structure and Photocatalytic Properties of Copper-Doped $CaTiO_3$. *Int. J. Hydrog. Energy*, 35:2713–2716.

Zhang, J., Z.-H. Huang, Y. Xu, et al. 2013a. Hydrothermal Synthesis of Graphene/Bi_2WO_6 Composite with High Adsorptivity and Photoactivity for Azo Dyes. *J. Am. Ceram. Soc.* 96:1562–1569.

Zhang, J., Y. Zhao, X. Zhao, et al. 2014. Porous Perovskite $LaNiO_3$ Nanocubes as Cathode Catalysts for Li-O_2 Batteries with Low Charge Potential. *Sci. Rep.* 4:1–6.

Zhang, L., T. Xu, X. Zhao, et al. 2010b. Controllable Synthesis of Bi_2MoO_6 and Effect of Morphology and Variation in Local Structure on Photocatalytic Activities. *Appl. Catal. B Environ.* 98:138–146.

Zhang, P., C. Shao, M. Zhang, et al. 2012b. Bi_2MoO_6 Ultrathin Nanosheets on $ZnTiO_3$ Nanofibers: A 3D Open Hierarchical Heterostructures Synergistic System with Enhanced Visible-Light-Driven Photocatalytic Activity. *J. Hazard. Mater.* 217–218:422–428.

Zhang, R., N. Luo, W. Yang, et al. 2013b. Low-Temperature Selective Catalytic Reduction of NO with NH_3 Using Perovskite-Type Oxides as the Novel Catalysts. *J. Mol. Catal. Chem.* 371:86–93.

Zhang, X., B. Wang, X. Wang, et al. 2015b. Preparation of M@$BiFeO_3$ Nanocomposites (M = Ag, Au) Bowl Arrays with Enhanced Visible Light Photocatalytic Activity. *J. Am. Ceram. Soc.* 98:2255–2263.

Zhang, Y., D. Wang, J. Wang, et al. 2012c. $BiMnO_3$ Perovskite Catalyst for Selective Catalytic Reduction of NO with NH_3 at Low Temperature. *Chin. J. Catal.* 33:1448–1454.

Zhang, Z., S. Gu, Y. Ding, et al. 2013c. Determination of Hydrogen Peroxide and Glucose Using a Novel Sensor Platform Based on $Co_{0.4}Fe_{0.6}LaO_3$ Nanoparticles. *Microchim. Acta* 180:1043–1049.

Zhang, Z., W. Wang, L. Wang, et al. 2012d. Enhancement of Visible-Light Photocatalysis by Coupling with Narrow-Band-Gap Semiconductor: A Case Study on Bi_2S_3/Bi_2WO_6. *ACS Appl. Mater. Interfaces* 4:593–597.

Zhao, B., L. Zhang, D. Zhen, et al. 2017. A Tailored Double Perovskite Nanofiber Catalyst Enables Ultrafast Oxygen Evolution. *Nat. Commun.* 8:14586.

Zhao, J., Q. Lu, C. Wang, et al. 2015. One-Dimensional Bi_2MoO_6 Nanotubes: Controllable Synthesis by Electrospinning and Enhanced Simulated Sunlight Photocatalytic Degradation Performances. *J. Nanoparticle Res.* 17:189.

Zhao, W., J. Yi, P. He, et al. 2019. Solid-State Electrolytes for Lithium-Ion Batteries: Fundamentals, Challenges and Perspectives. *Electrochem. Energy Rev.* 2:574–605.

Zhao, Y., L. Xu, L. Mai, et al. 2012. Hierarchical Mesoporous Perovskite $La_{0.5}Sr_{0.5}CoO_{2.91}$ Nanowires with Ultrahigh Capacity for Li-Air Batteries. *Proc. Natl. Acad. Sci.* 109:19569–19574.

Zhen, C., J. C. Yu, G. Liu, et al. 2014. Selective Deposition of Redox Co-Catalyst(s) to Improve the Photocatalytic Activity of Single-Domain Ferroelectric $PbTiO_3$ Nanoplates. *Chem. Commun.* 50:10416–10419.

Zhong, H., and R. Zeng. 2006. Structure of $LaSrMO_4$(M = Mn, Fe, Co, Ni, Cu) and Their Catalytic Properties in the Total Oxidation of Hexane. *J. Serbian Chem. Soc.* 71:1049–1059.

Zhou, T., J. Hu, and J. Li. 2011. Er_{3+} Doped Bismuth Molybdate Nanosheets with Exposed {010} Facets and Enhanced Photocatalytic Performance. *Appl. Catal. B Environ.* 110:221–230.

Zhou, Y., Y. Zhang, M. Lin, et al. 2015. Monolayered Bi_2WO_6 Nanosheets Mimicking Heterojunction Interface with Open Surfaces for Photocatalysis. *Nat. Commun.* 6:1–8.

Zhu, H., P. Zhang, and S. Dai. 2015. Recent Advances of Lanthanum-Based Perovskite Oxides for Catalysis. *ACS Catal.* 5:6370–6385.

Zhu, J., H. Li, L. Zhong, et al. 2014. Perovskite Oxides: Preparation, Characterizations, and Applications in Heterogeneous Catalysis. *ACS Catal.* 4:2917–2940.

Zhu, J., X. Yang, X. Xu, et al. 2007. Active Site Structure of NO Decomposition on Perovskite(-like) Oxides: An Investigation from Experiment and Density Functional Theory. *J. Phys. Chem. C* 111:1487–1490.

Zhu, J., Z. Zhao, D. Xiao, et al. 2005. CO Oxidation over the Perovskite-Like Oxides $La_{2-x}Sr_xMO_4$ (x = 0.0, 0.5, 1.0; M = Cu, Ni): A Study from Cyclic Voltammetry. *Z. Für Phys. Chem.* 219:807–815.

Zhu, Y., W. Zhou, J. Sunarso, et al. 2016. Phosphorus-Doped Perovskite Oxide as Highly Efficient Water Oxidation Electrocatalyst in Alkaline Solution. *Adv. Funct. Mater.* 26:5862–5872.

Zhu, Y., W. Zhou, Y. Zhong, et al. 2017. A Perovskite Nanorod as Bifunctional Electrocatalyst for Overall Water Splitting. *Adv. Energy Mater.* 7:1602122.

Zhuang, J., Q. Tian, S. Lin, et al. 2014. Precursor Morphology-Controlled Formation of Perovskites $CaTiO_3$ and Their Photo-Activity for As(III) Removal. *Appl. Catal. B Environ.* 156–157:108–115.

Zou, F., Z. Jiang, X. Qin, et al. 2012. Template-Free Synthesis of Mesoporous N-Doped $SrTiO_3$ Perovskite with High Visible-Light-Driven Photocatalytic Activity. *Chem. Commun.* 48:8514–8516.

Zou, J.-P., L.-Z. Zhang, S.-L. Luo, et al. 2012. Preparation and Photocatalytic Activities of Two New Zn-Doped $SrTiO_3$ and $BaTiO_3$ Photocatalysts for Hydrogen Production from Water without Cocatalysts Loading. *Int. J. Hydrog. Energy* 37:17068–17077.

Appendix A1

Examples of Single Perovskites with the Structural Lattice Parameters

Perovskite	Space Group	Crystal System	Length (Å)			Angle		
			a	*b*	*c*	*α*	*β*	*γ*
α-$AgVO_3$	*C*12/*c*1	Monoclinic	10.437	9.890	5.532	90	99.69	90
β-$AgVO_3$	*C*12/*m*1	Monoclinic	18.106	3.5787	8.043	90	104.44	90
$AgAsO_3$	$Pca2_1$	Orthorhombic	19.488	6.600	12.661	90	90	90
$AgBiO_3$	$R\bar{3}:H$	Trigonal	5.641		16.118	90	90	120
$AgBrO_3$	*I*4/*mmm*	Tetragonal	8.590		8.080	90	90	90
$AgIO_3$	$Pbc2_1$	Orthorhombic	7.265	15.17	5.786	90	90	90
$AgNbO_3$	*Pbcm*	Orthorhombic	5.546	5.603	15.636	90	90	90
$AgPO_3$	*P*12/*n*1	Monoclinic	11.860	6.060	7.310	90	93.5	90
$AgSbO_3$	$Fd\bar{3}m$	Cubic	10.230			90	90	90
$AgTaO_3$	*R*3*c*:*H*	Trigonal	5.528		13.716	90	90	120
$AlBO_3$	$P6_3/m$	Hexagonal	8.470		8.090	90	90	120
$BaBiO_3$	*I*12/*m*1	Monoclinic	6.191	6.145	8.678	90	90.16	90
$BaCeO_3$	*Pbnm*	Orthorhombic	6.212	6.235	8.781	90	90	90
$BaCoO_3$	$P6_3/mmc$	Hexagonal	5.590		4.820	90	90	120
$BaFeO_3$	$R\bar{3}c:H$	Trigonal	5.691		28.01	90	90	120
$BaGeO_3$	*C*12/*c*1	Monoclinic	13.1895	7.620	11.717	90	112.28	90
$BaIrO_3$	$Pm\bar{3}m$	Cubic	4.101			90	90	90
$BaMnO_3$	$P6_3/mmc$	Hexagonal	5.694		4.806	90	90	120
$BaNbO_3$	$Pm\bar{3}m$	Cubic	4.039			90	90	90
$BaNiO_3$	$P6_3/mmc$	Hexagonal	5.631		4.808	90	90	120
$BaPbO_3$	*C*12/*m*1	Monoclinic	8.555	8.481	6.012	90	134.81	90
$BaPbO_3$	$Pm\bar{3}m$	Cubic	5.459			90	90	90
$BaPrO_3$	*Pbnm*	Orthorhombic	6.181	6.214	8.722	90	90	90
$BaRuO_3$	$P6_3/mmc$	Hexagonal	5.713		14.050	90	90	120
$BaSnO_3$	$Pm\bar{3}m$	Cubic	4.108			90	90	90

(Continued)

Perovskite	Space Group	Crystal System	Length (Å)			Angle		
			a	*b*	*c*	*α*	*β*	*γ*
$BaTbO_3$	$Pm\bar{3}m$	Cubic	4.297			90	90	90
$BaThO_3$	$Pm\bar{3}m$	Cubic	4.480			90	90	90
$BaTiO_3$	*P4mm*	Tetragonal	3.999		4.017	90	90	90
$BaZrO_3$	$Pm\bar{3}m$	Cubic	4.191			90	90	90
$BiAlO_3$	*R3c:H*	Trigonal	5.375		13.393	90	90	120
$BiBO_3$	*P12/c1*	Monoclinic	6.585	5.027	8.349	90	108.91	90
$BiCoO_3$	*P4mm*	Tetragonal	3.729		4.723	90	90	90
$BiCrO_3$	*C2*	Monoclinic	7.77		8.08	90	90	90
$BiFeO_3$	*R3c:H*	Trigonal	5.588		13.867	90	90	120
$BiGaO_3$	*Pcca*	Orthorhombic	5.416	5.134	9.937	90	90	90
$BiInO_3$	$Pna2_1$	Orthorhombic	5.955	5.602	8.386	90	90	90
$BiMnO_3$	*P12/c1*	Monoclinic	9.623	5.460	9.824	90	110.96	90
$BiNiO_3$	*P*	Triclinic	5.385	5.650	7.708	91.95	89.81	91.54
$BiScO_3$	*C121*	Monoclinic	9.890	5.822	10.047	90	108.30	90
$CaFeO_3$	*Pnma*	Orthorhombic	5.352	7.540	5.326	90	90	90
$CaGeO_3$	*P*	Triclinic	7.269	7.526	8.094	103.44	94.42	90.11
$CaGeO_3$	*Pbnm*	Orthorhombic	5.261	5.269	7.445	90	90	90
$CaIrO_3$	*Cmcm*	Orthorhombic	3.145	9.855	7.293	90	90	90
$CaMnO_3$	*Pbnm*	Orthorhombic	5.451	5.582	7.780	90	90	90
$CaPtO_3$	*Cmcm*	Orthorhombic	3.123	9.912	7.346	90	90	90
$CaSiO_3$	*C12/c1*	Monoclinic	6.839	11.870	19.631	90	90.66	90
$CaSnO_3$	$Fd\bar{3}m$	Cubic	7.890			90	90	90
$CaSnO_3$	*Pbnm*	Orthorhombic	5.454	5.616	7.801	90	90	90
$CaTcO_3$	*Pnma*	Orthorhombic	5.533	7.722	5.411	90	90	90
$CaTeO_3$	$P2_1ca$	Orthorhombic	13.365	6.533	8.190	90	90	90

(Continued)

Perovskite	Space Group	Crystal System	Length (Å) a	b	c	Angle α	β	γ
$CaTeO_3$	*P*4_3	Tetragonal	12.107		11.091	90	90	90
$CaTiO_3$	*Pbnm*	Orthorhombic	5.380	5.440	7.639	90	90	90
$CaVO_3$	*Pnma*	Orthorhombic	5.320	7.549	5.344	90	90	90
$CaZrO_3$	*Pcmn*	Orthorhombic	5.583	8.007	5.759	90	90	90
$CaZrO_3$	$Pm\bar{3}m$	Cubi	3.990			90	90	90
$CdGeO_3$	*B*112	Monoclinic	10.60	5.360	9.730	90	104	90
$CdGeO_3$	*Pbnm*	Orthorhombic	5.211	5.261	7.426	90	90	90
$CdSeO_3$	*Pnma*	Orthorhombic	6.277	8.085	5.294	90	90	90
$CdSiO_3$	*P*12/*c*1	Monoclinic	6.946	7.256	15.070	90	94.79	90
$CdTeO_3$	*Pnma*	Orthorhombic	7.458	14.522	11.0458	90	90	90
$CdTiO_3$	$R\bar{3}:R$	Trigonal	5.820		5.820	90	90	120
$CdVO_3$	*Pnma*	Orthorhombic	14.3010	3.5980	5.204	90	90	90
$CeBO_3$	*Pnma*	Orthorhombic	5.812	5.078	8.195	90	90	90
$CeCrO_3$	$Pm\bar{3}m$	Cubic	3.890			90	90	90
$CeVO_3$	*Pnma*	Orthorhombic	5.538	7.786	5.504	90	90	90
$DyBO_3$	$P6_3/mmc$	Hexagonal	3.793		8.847	90	90	120
$DyCoO_3$	*Pbnm*	Orthorhombic	5.170	5.410	7.397	90	90	90
$DyCrO_3$	*Pbnm*	Orthorhombic	5.263	5.520	7.552	90	90	90
$DyFeO_3$	*Pnma*	Orthorhombic	5.596	7.629	5.301	90	90	90
$DyMnO_3$	*Pnma*	Orthorhombic	5.842	7.378	5.280	90	90	90
$DyScO_3$	*Pbnm*	Orthorhombic	5.449	5.726	7.913	90	90	90
$ErBO_3$	$P6_3/mmc$	Hexagonal	3.767		8.807	90	90	120
$ErCoO_3$	*Pbnm*	Orthorhombic	5.121	5.419	7.352	90	90	90
$ErCrO_3$	*Pbnm*	Orthorhombic	5.223	5.516	7.519	90	90	90
$ErFeO_3$	*Pbnm*	Orthorhombic	5.267	5.581	7.593	90	90	90

(*Continued*)

Perovskite	Space Group	Crystal System	Length (Å)			Angle		
			a	b	c	α	β	γ
$ErGdO_3$	$Ia\bar{3}$	Cubic	10.736			90	90	90
$ErLaO_3$	*Pnma*	Orthorhombic	6.070	5.850	8.450	90	90	90
$ErMnO_3$	$P6_3cm$	Hexagonal	6.112		11.420	90	90	120
$ErNiO_3$	*P*12/*n*1	Monoclinic	5.161	5.511	7.399	90	90.11	90
$GdAlO_3$	*Pnma*	Orthorhombic	5.305	7.448	5.254	90	90	90
$GdBO_3$	$P6_3/mmc$	Hexagonal	3.839		8.906	90	90	120
$GdCoO_3$	*Pbnm*	Orthorhombic	5.217	5.388	7.446	90	90	90
$GdCrO_3$	*Pbnm*	Orthorhombic	5.312	5.515	7.600	90	90	90
$GdFeO_3$	*Pbnm*	Orthorhombic	5.349	5.609	7.669	90	90	90
$GdInO_3$	$P6_3cm$	Hexagonal	6.330		12.334	90	90	120
$GdMnO_3$	*Pnma*	Orthorhombic	5.866	7.431	5.318	90	90	90
$GdScO_3$	*Pnma*	Orthorhombic	5.742	7.926	5.482	90	90	90
$HoAlO_3$	*Pbnm*	Orthorhombic	5.180	5.322	7.374	90	90	90
$HoBO_3$	$P6_3/mmc$	Hexagonal	3.784		8.836	90	90	120
$HoCoO_3$	*Pbnm*	Orthorhombic	5.144	5.416	7.375	90	90	90
$HoFeO_3$	*Pbnm*	Orthorhombic	5.278	5.591	7.602	90	90	90
$HoMnO_3$	$P6_3cm$	Hexagonal	6.141		11.412	90	90	120
$HoNiO_3$	*Pbnm*	Orthorhombic	5.181	5.510	7.425	90	90	90
$HoScO_3$	*Pbnm*	Orthorhombic	5.429	5.715	7.901	90	90	90
$HoVO_3$	*Pnma*	Orthorhombic	5.614	7.626	5.303	90	90	90
$KNbO_3$	*Amm*2	Orthorhombic	5.697	3.971	5.720	90	90	90
$KNbO_3$	*P*1*m*1	Monoclinic	4.049	3.992	4.020	90	90.10	90
$KSbO_3$	$R\bar{3}:H$	Trigonal	6.814		6.814	90	90	120
$KTaO_3$	$Pm\bar{3}m$	Cubic	3.988			90	90	90
KVO_3	*Pbcm*	Orthorhombic	5.176	10.794	5.680	90	90	90

(*Continued*)

Perovskite	Space Group	Crystal System	Length (Å)			Angle		
			a	*b*	*c*	*α*	*β*	*γ*
$LaAlO_3$	$R\bar{3}c$	Trigonal	5.386		13.191	90	90	120
$LaAuO_3$	*Pbcm*	Orthorhombic	4.033	13.073	5.695	90	90	90
$LaBO_3$	*Pnma*	Orthorhombic	5.874	5.109	8.258	90	90	90
$LaCoO_3$	$R\bar{3}c{:}H$	Trigonal	5.342		5.341	90	90	120
$LaCrO_3$	*Pnma*	Orthorhombic	5.486	7.768	5.525	90	90	90
$LaCuO_3$	$R\bar{3}c{:}H$	Trigonal	5.499		13.217	90	90	120
LaErO3	*Pnma*	Orthorhombic	6.043	8.417	5.829	90	90	90
$LaFeO_3$	*Pbnm*	Orthorhombic	5.552	5.563	7.843	90	90	90
$LaGaO_3$	$R\bar{3}c{:}H$	Trigonal	5.543		13.437	90	90	120
$LaHoO_3$	*Pnma*	Orthorhombic	6.094	8.499	5.882	90	90	90
$LaInO_3$	*Pnma*	Orthorhombic	5.940	8.216	5.723	90	90	90
$LaMnO_3$	*Pbnm*	Orthorhombic	5.537	5.747	7.693	90	90	90
$LaNiO_3$	$R\bar{3}c{:}H$	Trigonal	5.453		13.101	90	90	120
$LaPdO_3$	*Pbnm*	Orthorhombic	5.589	5.850	7.866	90	90	90
$LaRhO_3$	*Pnma*	Orthorhombic	5.700	7.897	5.524	90	90	90
$LaRuO_3$	*Pnma*	Orthorhombic	5.746	7.871	5.512	90	90	90
$LaScO_3$	*Pbnm*	Orthorhombic	5.680	5.791	8.094	90	90	90
$LaTiO_3$	*Pbnm*	Orthorhombic	5.634	5.615	7.914	90	90	90
$LaVO_3$	*Pnma*	Orthorhombic	5.555	7.848	5.553	90	90	90
$LaYbO_3$	$Pna2_1$	Orthorhombic	6.010	5.810	8.390	90	90	90
LaYO3	$Pna2_1$	Orthorhombic	6.052	5.936	8.512	90	90	90
$Li_{0.93}WO_3$	$Pm\bar{3}m$	Cubic	3.722			90	90	90
$LiAsO_3$	$R\bar{3}{:}H$	Trigonal	4.808		14.210	90	90	120
$LiNbO_3$	$R3c{:}H$	Trigonal	5.138		13.498	90	90	120
$LiTaO_3$	$R\bar{3}c{:}H$	Trigonal	5.220		13.763	90	90	120

(*Continued*)

Perovskite	Space Group	Crystal System	Length (Å)			Angle		
			a	b	c	α	β	γ
$MgGeO_3$	*Pbca*	Orthorhombic	19.011	9.084	5.415	90	90	90
$MgMnO_3$		Trigonal	4.945		13.730	90	90	120
$MgSiO_3$-low pressure	*Pbnm*	Orthorhombic	4.445	4.665	6.454	90	90	90
$MgSiO_3$-high pressure	$R\bar{3}:H$	Trigonal	4.686		13.291	90	90	120
$MgTiO_3$	$R\bar{3}:R$	Trigonal	5.540		5.540	90	90	120
$MgVO_3$	*Cmmm*	Orthorhombic	5.291	10.018	5.239	90	90	90
$NaBiO_3$	$R\bar{3}:H$	Trigonal	5.567		15.989	90	90	120
$NaNbO_3$	$Pm\bar{3}m$	Cubic	3.906			90	90	90
$NaPO_3$	*P*12/*n*1	Monoclinic	12.120	6.200	6.990	90	90	90
$NaSbO_3$	$Fd\bar{3}m$	Cubic	10.200			90	90	90
$NaTaO_3$	*Pbnm*	Orthorhombic	5.477	5.521	7.789	90	90	90
$NaUO_3$	*Pbnm*	Orthorhombic	5.779	5.907	8.283	90	90	90
$NaVO_3$	*C*1*c*1	Monoclinic	10.494	9.434	5.863	90	108.8	90
$NaWO_3$	$Pm\bar{3}m$	Cubic	3.850			90	90	90
$NdAlO_3$	$R\bar{3}c:H$	Trigonal	5.339		12.999	90	90	120
$NdBO_3$	*P*	Triclinic	6.300	6.548	6.549	93.90	107.68	107.30
$NdCoO_3$	$Pm\bar{3}m$	Cubic	3.770			90	90	90
$NdCrO_3$	*Pbnm*	Orthorhombic	5.425	5.478	7.694	90	90	90
$NdFeO_3$	*Pnma*	Orthorhombic	5.589	7.762	5.449	90	90	90
$NdGaO_3$	*Pbnm*	Orthorhombic	5.417	5.495	7.687	90	90	90
$NdMnO_3$	*Pbnm*	Orthorhombic	5.380	5.854	7.557	90	90	90
$NdNiO_3$	*Pbnm*	Orthorhombic	5.389	5.384	7.613	90	90	90
$NdRhO_3$	*Pnma*	Orthorhombic	5.752	7.770	5.376	90	90	90
$NdScO_3$	*Pbnm*	Orthorhombic	5.581	5.776	8.007	90	90	90
$NdVO_3$	*Pmm*	Cubic	3.890			90	90	90

(*Continued*)

Perovskite	Space Group	Crystal System	Length (Å)			Angle		
			a	b	c	α	β	γ
$PbCdO_3$	$Ia\bar{3}$	Cubic	10.453			90	90	90
$PbGeO_3$	P112/b	Monoclinic	11.469	12.555	7.236	90	113.3	90
$PbGeO_3$	$R\bar{3}:R$	Trigonal	9.328		9.328	90	90	120
$PbHfO_3$	Pbam	Orthorhombic	5.840	11.705	8.175	90	90	90
$PbSeO_3$	P12/m1	Monoclinic	4.552	5.525	6.633	90	112.59	90
$PbSiO_3$	P12/n1	Monoclinic	11.209	7.041	12.220	90	113.12	90
$PbTeO_3$	C12/c1	Monoclinic	26.730	4.600	18.060	90	106.0	90
$PbTiO_3$	Pmmm	Orthorhombic	4.211	4.211	3.875	90	90	90
$PbVO_3$	P4mm	Tetragonal	3.800		4.670	90	90	90
$PbZrO_3$	Cm2m	Orthorhombic	5.890	5.897	4.134	90	90	90
$Pr\ Sr_{0.96}O_3$	Pbnm	Orthorhombic	5.988	6.121	8.548	90	90	90
$PrCoO_3$	$Pm\bar{3}m$	Cubic	3.780			90	90	90
$PrCrO_3$	Pmm	Cubic	3.890			90	90	90
$PrFeO_3$	Pbnm	Orthorhombic	5.482	5.578	7.786	90	90	90
$PrMnO_3$	Pnma	Orthorhombic	5.605	7.665	5.462	90	90	90
$PrNiO_3$	Pbnm	Orthorhombic	5.415	5.376	7.620	90	90	90
$PrRhO_3$	Pnma	Orthorhombic	5.740	7.803	5.417	90	90	90
$PrScO_3$	Pnma	Orthorhombic	5.780	8.025	5.608	90	90	90
$PrVO_3$	$Pm\bar{3}m$	Cubic	3.890			90	90	90
$SmAlO_3$	Pbnm	Orthorhombic	5.291	5.290	7.474	90	90	90
$SmBO_3$	$P6_3/mmc$	Hexagonal	3.862		8.978	90	90	120
$SmCoO_3$	Pbnm	Orthorhombic	5.283	5.350	7.496	90	90	90
$SmCrO_3$	$Pm\bar{3}m$	Cubic	3.860			90	90	90
$SmMnO_3$	Pnma	Orthorhombic	5.862	7.477	5.362	90	90	90
$SmNiO_3$	Pbnm	Orthorhombic	5.328	5.437	7.567	90	90	90

(Continued)

Perovskite	Space Group	Crystal System	Length (Å)			Angle		
			a	*b*	*c*	*α*	*β*	*γ*
$SmRhO_3$	Pnma	Orthorhombic	5.323	5.757	7.708	90	90	90
$SmScO_3$	*Pbnm*	Orthorhombic	5.534	5.762	7.967	90	90	90
$SmVO_3$	$Pm\bar{3}m$	Cubic	3.890			90	90	90
$SrBiO_3$	*P*12/*n*1	Monoclinic	5.948	6.095	8.485	90	90.06	90
$SrGeO_3$	*P*	Triclinic	8.699	9.935	11.148	106.04	89.97	102.11
$SrHfO_3$	$Pm\bar{3}m$	Cubic	4.069			90	90	90
$SrIrO_3$	*C*12/*c*1	Monoclinic	5.604	9.618	14.174	90	93.60	90
$SrMnO_3$	$P6_3/mmc$	Hexagonal	5.454		9.092	90	90	120
$SrNbO_3$	*Pnma*	Orthorhombic	5.689	8.068	5.694	90	90	90
$SrPbO_3$	*Pnma*	Orthorhombic	5.964	8.320	5.860	90	90	90
$SrPbO_3$	*Pnma*	Orthorhombic	5.964	8.320	5.860	90	90	90
$SrRuO_3$	*Pnma*	Orthorhombic	5.568	7.845	5.532	90	90	90
$SrSeO_3$	*P*12/*m*1	Monoclinic	6.570	5.475	4.455	90	106.65	90
$SrSiO_3$	*C*12/*c*1	Monoclinic	12.333	7.146	10.885	90	111.51	90
$SrSnO_3$	*Pbnm*	Orthorhombic	5.698	5.697	8.052	90	90	90
$SrTbO_3$	*Pnma*	Orthorhombic	5.962	8.351	5.873	90	90	90
SrTeO3	*C*12/*c*1	Monoclinic	28.341	5.941	28.658	90	114.33	90
$SrTiO_3$	$Pm\bar{3}m$	Cubic	3.899			90	90	90
$SrVO_3$	$Pm\bar{3}m$	Cubic	3.841			90	90	90
$SrZrO_3$	$Pm\bar{3}m$	Cubic	4.089			90	90	90
$TbAlO_3$	*Pbnm*	Orthorhombic	5.230	5.306	7.416	90	90	90
$TbCoO_3$	*Pbnm*	Orthorhombic	5.202	5.398	7.424	90	90	90
$TbFeO_3$	*Pbnm*	Orthorhombic	5.326	5.602	7.635	90	90	90
$TbMnO_3$	*Pbnm*	Orthorhombic	5.297	5.831	7.403	90	90	90
$TbRhO_3$	*Pnma*	Orthorhombic	5.745	7.625	5.254	90	90	90

(Continued)

Perovskite	Space Group	Crystal System	Length (Å)			Angle		
			a	*b*	*c*	*α*	*β*	*γ*
$TbScO_3$	*Pbnm*	Orthorhombic	5.465	5.729	7.917	90	90	90
$YAlO_3$	$P6_3/mmc$	Hexagonal	3.680		10.520	90	90	120
YBO_3	$P6_3/m$	Hexagonal	3.776		8.806	90	90	120
$YCoO_3$	*Pbnm*	Orthorhombic	5.132	5.417	7.367	90	90	90
$YCrO_3$	*Pnma*	Orthorhombic	5.312	5.618	7.655	90	90	90
$YCuO_3$	*Pnma*	Orthorhombic	5.262	5.666	7.494	90	90	90
$YFeO_3$	*Pnma*	Orthorhombic	5.588	7.595	5.274	90	90	90
$YGaO_3$	$P6_3cm$	Hexagonal	6.065		11.615	90	90	120
YInO3	$P6_3cm$	Hexagonal	6.260		12.249	90	90	120
$YMnO_3$	$P6_3cm$	Hexagonal	6.148		11.443	90	90	120
$YTiO_3$	*Pnma*	Orthorhombic	5.689	7.609	5.335	90	90	90
YVO_3	*Pbnm*	Orthorhombic	5.282	5.589	7.550	90	90	90
$YZrO_3$	$Ia\bar{3}$	Cubic	10.539			90	90	90
$ZnSiO_3$	*Pbca*	Orthorhombic	8.204	9.087	5.278	90	90	90
$ZnMnO_3$	*R*3	Trigonal	4.965		13.800	90	90	120
$ZnTeO_3$	*Pbca*	Orthorhombic	7.360	6.380	12.320	90	90	90

Data is collected from Crystallography Open Database http://www.crystallography.net/cod/index.php

Index

Printed in the United States
by Baker & Taylor Publisher Services